BIOSTIMULANTS IN PLANT
AND PERFORMANCE

PLANT BIOLOGY, SUSTAINABILITY, AND CLIMATE CHANGE

Series Editors

AZAMAL HUSEN
MUHAMMAD IQBAL

BIOSTIMULANTS IN PLANT PROTECTION AND PERFORMANCE

Edited by

AZAMAL HUSEN

Department of Biotechnology, Smt. S. S. Patel Nootan Science & Commerce College, Sankalchand Patel University, Visnagar, Gujarat, India; Department of Biotechnology, Graphic Era (Deemed to be University), Dehradun, Uttarakhand, India; Wolaita Sodo University, Wolaita, Ethiopia

ELSEVIER

Elsevier
Radarweg 29, PO Box 211, 1000 AE Amsterdam, Netherlands
125 London Wall, London EC2Y 5AS, United Kingdom
50 Hampshire Street, 5th Floor, Cambridge, MA 02139, United States

ISBN: 978-0-443-15884-1

For Information on all Elsevier publications
visit our website at https://www.elsevier.com/books-and-journals

Publisher: Nikki Levy
Acquisitions Editor: Nancy Maragioglio
Editorial Project Manager: Kyle Gravel
Production Project Manager: Gomathi Sugumar
Cover Designer: Christian Bilbow

Typeset by MPS Limited, Chennai, India

Dedication

Thank you so much to my lovely brother Engr. Ashraf Husain for his continuous encouragement and overall support.

Contents

11. Use of plant water extracts as biostimulants to improve the plant tolerance against abiotic stresses 165

Muhammad Bilal Hafeez, Asma Hanif, Sobia Shahzad, Noreen Zahra, Bilal Ahmad, Abida Kausar, Aaliya Batool and Muhammad Usman Ibrahim

12. Role of biodegradable mulching films and vegetable-derived biostimulant application for enhancing plants performance and nutritive value 185

Atul Loyal, S.K. Pahuja, D.S. Duhan, Naincy Rani, Divya Kapoor, Rakesh K. Srivastava, Gaurav Chahal and Pankaj Sharma

13. Plant biochemistry and yield in response to biostimulants 205

Bushra Ahmad, Arshad Jamil, Dure Shahwar, Aisha Siddique and Umama Syed

14. Biostimulants mediated imprints on seed physiology in crop plants 221

Riya Johnson, Joy M. Joel, E. Janeeshma and Jos T. Puthur

15. Role of biostimulant in adventitious rooting via stimulation of phytohormones 237

Arshdeep Kaur, Manik Devgan, Radhika Sharma, Antul Kumar, Anuj Choudhary, Ravi Pratap Singh, Dadireddy Madhusudan Reddy, Ajaykumar Venkatapuram, Sahil Mehta and Azamal Husen

16. Arbuscular mycorrhizal fungi inoculation in the modulation of plant yield and bioactive compounds 255

Weria Weisany

List of contributors

Nadeesh M. Adassooriya Department of Chemical and Process Engineering, Faculty of Engineering, University of Peradeniya, Peradeniya, Sri Lanka

Bilal Ahmad Agriculture Genomics Institute at Shenzhen, Chinese Academy of Agricultural Sciences, Shenzhen, P.R. China

Bushra Ahmad Department of Biochemistry, Shaheed Benazir Bhutto Women University, Peshawar, Pakistan

Hafiz Muhammad Ahmad Department of Bioinformatics and Biotechnology, Government College University, Faisalabad, Pakistan

Tarek Alshaal Department of Applied Plant Biology, Faculty of Agriculture, Food Science and Environmental Management, Debrecen University, Debrecen, Hungary; Soil and Water Department, Faculty of Agriculture, Kafrelsheikh University, Kafr El-Sheikh, Egypt

Mohammad Amir Department of Botany, Lucknow University, Lucknow, Uttar Pradesh, India

Mohammad Israil Ansari Department of Botany, Lucknow University, Lucknow, Uttar Pradesh, India

Arvind Department of Zoology, Chaudhary Devi Lal University, Sirsa, Haryana, India

Santiago Atero-Calvo Department of Plant Physiology, Faculty of Sciences, University of Granada, Granada, Spain

Abdelhamid Azeroual Agri-Food and Health Laboratory (AFHL), Faculty of Sciences and Techniques of Settat, Hassan First University, Settat, Morocco

Nóra Bákonyi Department of Applied Plant Biology, Faculty of Agriculture, Food Science and Environmental Management, Debrecen University, Debrecen, Hungary

Döme Barna Department of Applied Plant Biology, Faculty of Agriculture, Food Science and Environmental Management, Debrecen University, Debrecen, Hungary

Aaliya Batool Department of Botany, University of Agriculture, Faisalabad, Pakistan

Bouchaib Bencherki Agri-Food and Health Laboratory (AFHL), Faculty of Sciences and Techniques of Settat, Hassan First University, Settat, Morocco

Begoña Blasco Department of Plant Physiology, Faculty of Sciences, University of Granada, Granada, Spain

Noureddine Chaachouay Agri-Food and Health Laboratory (AFHL), Faculty of Sciences and Techniques of Settat, Hassan First University, Settat, Morocco

Gaurav Chahal College of Agriculture, CCS Haryana Agricultural University, Hisar, Haryana, India

Akansha Chauhan Department of Biotechnology, Graphic Era Deemed to be University, Dehradun, Uttarakhand, India

Shreya Chauhan Department of Forest Biology and Tree Improvement, Dr. Y.S. Parmar University of Horticulture and Forestry, Solan, Himachal Pradesh, India

Ramachandran Chelliah Department of Pharmacology, Saveetha Medical and Dental College and Hospitals, Chennai, Tamil Nadu, India; Department of Food Science and Biotechnology, College of Agriculture and Life Sciences, Kangwon National University, Chuncheon, Gangwon-do, South Korea; Department of Food Science and Biotechnology, Kangwon Institute of Inclusive Technology (KIIT), Kangwon National University, Chuncheon, Gangwon-do, South Korea

Hui Yee Chong School of Applied Sciences, Faculty of Integrated Life Sciences, Quest International University, Ipoh, Perak, Malaysia

Anuj Choudhary Department of Botany, Punjab Agriculture University, Ludhiana, Punjab, India

Manik Devgan Department of Plant Breeding and Genetics, Punjab Agricultural University, Ludhiana, Punjab, India

Éva Domokos-Szabolcsy Department of Applied Plant Biology, Faculty of Agriculture, Food Science and Environmental Management, Debrecen University, Debrecen, Hungary

Allal Douira Plant, Animal Productions and Agro-industry Laboratory, Department of Biology, Faculty of Sciences, Ibn Tofail University, Kenitra, Morocco

D.S. Duhan Department of Vegetable Sciences, CCS Haryana Agricultural University, Hisar, Haryana, India

Mutia Erti Dwiastuti Research Center for Horticultural and Estate Crops, National Research and Innovation Agency (BRIN), Cibinong, Indonesia

Miklós Gábor Fári Department of Applied Plant Biology, Faculty of Agriculture, Food Science and Environmental Management, Debrecen University, Debrecen, Hungary

Aarushi Gautam Department of Biotechnology, Graphic Era Deemed to be University, Dehradun, Uttarakhand, India

Namita Goyat Department of Zoology, Chaudhary Devi Lal University, Sirsa, Haryana, India

Muhammad Bilal Hafeez Department of Agronomy, University of Agriculture, Faisalabad, Pakistan

Asma Hanif Department of Botany, Islamia University of Bahawalpur, Bahawalnagar Campus, Pakistan

Wiwiek Harsonowati Research Center for Horticultural and Estate Crops, National Research and Innovation Agency (BRIN), Cibinong, Indonesia

Ragib Husain Department of Botany, Bareilly College, Bareilly, Uttar Pradesh, India

Azamal Husen Department of Biotechnology, Smt. S. S. Patel Nootan Science & Commerce College, Sankalchand Patel University, Visnagar, Gujarat, India; Department of Biotechnology, Graphic Era (Deemed to be University), Dehradun, Uttarakhand, India; Wolaita Sodo University, Wolaita, Ethiopia

Muhammad Usman Ibrahim Department of Agronomy, University of Agriculture, Faisalabad, Pakistan

Saira Ishaq Department of Food Sciences and Technology, University of Poonch, Rawalakot, Pakistan

Arshad Jamil Department of Plant Breeding and Genetics, University of Agriculture, Dera Ismail Khan, Pakistan

E. Janeeshma Department of Botany, MES KEVEEYAM College, Valanchery, Malappuram, Kerala, India

Joy M. Joel Plant Physiology and Biochemistry Division, Department of Botany, University of Calicut, Malappuram, Kerala, India

Riya Johnson Plant Physiology and Biochemistry Division, Department of Botany, University of Calicut, Malappuram, Kerala, India

Divya Kapoor Department of Microbiology, CCS Haryana Agricultural University, Hisar, Haryana, India

Arshdeep Kaur Department of Plant Breeding and Genetics, Punjab Agricultural University, Ludhiana, Punjab, India

Harmanjot Kaur Department of Botany, Punjab Agriculture University, Ludhiana, Punjab, India

Abida Kausar Department of Botany, Government College Women University Faisalabad, Pakistan

Imran Khan Department of Food Science and Technology, The University of Haripur, Haripur, Khyber Pakhtunkhwa, Pakistan

Rida Oktorida Khastini Department of Biology Education, Faculty of Teacher Training and Education, Sultan Ageng Tirtayasa University, Banten, Indonesia

Antul Kumar Department of Botany, Punjab Agriculture University, Ludhiana, Punjab, India

Manisha Lakhanpal Department of Forest Products, Dr. Y.S. Parmar University of Horticulture and Forestry, Solan, Himachal Pradesh, India

Wendy Ying Ying Liu School of Applied Sciences, Faculty of Integrated Life Sciences, Quest International University, Ipoh, Perak, Malaysia

Atul Loyal Department of Genetics and Plant Breeding, CCS Haryana Agricultural University, Hisar, Haryana, India

Nadun H. Madanayake Department of Botany, Faculty of Science, University of Peradeniya, Peradeniya, Sri Lanka

Inamul Hasan Madar Department of Pharmacology, Saveetha Medical and Dental College and Hospitals, Chennai, Tamil Nadu, India

Péter Makleit Department of Applied Plant Biology, Faculty of Agriculture, Food Science and Environmental Management, Debrecen University, Debrecen, Hungary

Dyah Manohara Research Center for Horticultural and Estate Crops, National Research and Innovation Agency (BRIN), Cibinong, Indonesia

Sahil Mehta Department of Botany, Hansraj College, University of Delhi, New Delhi, India

Sumaira Miskeen Department of Food Science and Technology, The University of Haripur, Haripur, Khyber Pakhtunkhwa, Pakistan

Shreya Mundepi Department of Biotechnology, Graphic Era Deemed to be University, Dehradun, Uttarakhand, India

Sharfa Naaz Department of Botany, Lucknow University, Lucknow, Uttar Pradesh, India

Sara Najeeb Department of Food Science and Technology, The University of Haripur, Haripur, Khyber Pakhtunkhwa, Pakistan

Eloy Navarro-León Department of Plant Physiology, Faculty of Sciences, University of Granada, Granada, Spain

Deog Hwan Oh Department of Food Science and Biotechnology, College of Agriculture and Life Sciences, Kangwon National University, Chuncheon, Gangwon-do, South Korea; Department of Food Science and Biotechnology, Kangwon Institute of Inclusive Technology (KIIT), Kangwon National University, Chuncheon, Gangwon-do, South Korea

S.K. Pahuja Department of Genetics and Plant Breeding, CCS Haryana Agricultural University, Hisar, Haryana, India

Priti Pal Environmental Engineering, Shri Ramswaroop Memorial College of Engineering & Management, Lucknow, Uttar Pradesh, India

Manu Pant Department of Biotechnology, Graphic Era Deemed to be University, Dehradun, Uttarakhand, India

Jati Purwani Research Center for Applied Microbiology, National Research and Innovation Agency (BRIN), Cibinong, Indonesia

Jos T. Puthur Plant Physiology and Biochemistry Division, Department of Botany, University of Calicut, Malappuram, Kerala, India

Abdul Raheem Department of Botany, Lucknow University, Lucknow, Uttar Pradesh, India

Shakeelur Rahman Prakriti Bachao Foundation, Ranchi, Jharkhand, India

Naincy Rani Department of Chemistry, CCS Haryana Agricultural University, Hisar, Haryana, India

Dadireddy Madhusudan Reddy Department of Microbiology, Palamuru University, Mahbubnagar, Telangana, India

Juan Jose Rios Department of Plant Physiology, Faculty of Sciences, University of Granada, Granada, Spain

Juan Manuel Ruiz Department of Plant Physiology, Faculty of Sciences, University of Granada, Granada, Spain

Dure Shahwar Department of Biochemistry, Shaheed Benazir Bhutto Women University, Peshawar, Pakistan

Sobia Shahzad Department of Botany, Islamia University of Bahawalpur, Bahawalnagar Campus, Pakistan

Kuldipika Sharma Department of Forest Products, Dr. Y.S. Parmar University of Horticulture and Forestry, Solan, Himachal Pradesh, India

Mayur Mukut Murlidhar Sharma Department of Agriculture and Life Industry, Kangwon National University, Chuncheon, Gangwon, Republic of Korea

Pankaj Sharma Department of Microbiology, CCS Haryana Agricultural University, Hisar, Haryana, India

Radhika Sharma Department of Soil Science, Punjab Agricultural University, Ludhiana, Punjab, India

Aisha Siddique Department of Biochemistry, Shaheed Benazir Bhutto Women University, Peshawar, Pakistan

Akhilesh Kumar Singh Department of Biotechnology, School of Life Sciences, Mahatma Gandhi Central University, Motihari, Bihar, India

Arundhati Singh School of Agriculture and Environment, Bentley Perth Campus, Curtin University, Perth, Western Australia, Australia

Ravi Pratap Singh Department of Ocean Studies and Marine Biology, Pondicherry University, Port Blair, Andaman and Nicobar Islands, India

Sukhmeet Singh Department of Zoology, Chaudhary Devi Lal University, Sirsa, Haryana, India

Vinay Kumar Singh Department of Botany, Bareilly College, Bareilly, Uttar Pradesh, India

Pravitha Kasu Sivanandan Department of Cancer Biology & Therapeutics MVR Cancer Centre & Research Institute, Calicut, Kerala, India

Abdul Hasyim Sodiq Department of Agrotechnology, Faculty of Agriculture, Sultan Ageng Tirtayasa University, Banten, Indonesia

Deciyanto Soetopo Research Center for Horticultural and Estate Crops, National Research and Innovation Agency (BRIN), Cibinong, Indonesia

Sashi Sonkar Department of Botany, Bankim Sardar College, Tangrakhali, West Bengal, India

Rakesh K. Srivastava International Crops Research Institute for the Semi-Arid Tropics (ICRISAT), Patancheru, Telangana, India

Suliasih Research Center for Applied Microbiology, National Research and Innovation Agency (BRIN), Cibinong, Indonesia

Ghazala Sultan Department of Computer Science, Faculty of Science, Aligarh Muslim University, Aligarh, Uttar Pradesh, India

Umama Syed Department of Biochemistry, Shaheed Benazir Bhutto Women University, Peshawar, Pakistan

Ajaykumar Venkatapuram Department of Microbiology, Palamuru University, Mahbubnagar, Telangana, India; Crop Improvement Group, International Center for Genetic Engineering and Biotechnology, New Delhi, India

Szilvia Veres Department of Applied Plant Biology, Faculty of Agriculture, Food Science and Environmental Management, Debrecen University, Debrecen, Hungary

Tarun Verma Department of Silviculture and Agroforestry, Dr. Y.S. Parmar University of Horticulture and Forestry, Solan, Himachal Pradesh, India

Weria Weisany Department of Agriculture and Food Science, Science and Research Branch, Islamic Azad University, Tehran, Iran

Sri Widawati Research Center for Applied Microbiology, National Research and Innovation Agency (BRIN), Cibinong, Indonesia

Sri Widyaningsih Research Center for Horticultural and Estate Crops, National Research and Innovation Agency (BRIN), Cibinong, Indonesia

Puhpanjali Yadav Department of Botany, Lucknow University, Lucknow, Uttar Pradesh, India

Sheetal Yadav Department of Microbiology, CCS Haryana Agricultural University, Hisar, Haryana, India

Noreen Zahra Department of Botany, University of Agriculture, Faisalabad, Pakistan; Department of Botany, Government College Women University Faisalabad, Pakistan

Lahcen Zidane Plant, Animal Productions and Agro-industry Laboratory, Department of Biology, Faculty of Sciences, Ibn Tofail University, Kenitra, Morocco

About the editor

Azamal Husen is a Professor at Sankalchand Patel University, Visnagar, India, and an Adjunct Professor at Graphic Era (Deemed to be University), Dehradun, Uttarakhand, India. He is also working as a Visiting Professor at the University Putra Malaysia, Selangor, Malaysia, and acting as a Foreign Delegate at Wolaita Sodo University, Wolaita, Ethiopia. Previously, he served as a Professor and Head of the Department of Biology, University of Gondar, Ethiopia. He also worked as a Visiting Faculty of the Forest Research Institute and the Doon College of Agriculture and Forest at Dehradun, India. His research and teaching experience of 25 years encompasses biogenic nanomaterial fabrication and application; plant responses to nanomaterials; plant adaptation to harsh environments at the physiological, biochemical, and molecular levels; herbal medicine; and clonal propagation for improvement of tree species.

He has conducted several research projects sponsored by various funding agencies, including the World Bank (FREEP), the National Agricultural Technology Project (NATP), the Indian Council of Agriculture Research (ICAR), the Indian Council of Forest Research Education (ICFRE), and the Japan Bank for International Cooperation (JBIC). He received four fellowships from India and a recognition award from the University of Gondar, Ethiopia, for excellent teaching, research, and community service. Husen has been on the editorial board and the panel of reviewers of several reputed journals published by Elsevier, Frontiers Media, Taylor & Francis, Springer Nature, RSC, Oxford University Press, Sciendo, The Royal Society, CSIRO, PLOS, MDPI, John Wiley & Sons, and UPM Journals. He is on the advisory board of Cambridge Scholars Publishing, United Kingdom. He is a fellow of the Plantae group of the American Society of Plant Biologists and a member of the International Society of Root Research, Asian Council of Science Editors, and INPST. To his credit are more than 250 publications, and he is the editor-in-chief of the *American Journal of Plant Physiology*.

He is also working as a series editor of "*Exploring Medicinal Plants*," published by Taylor & Francis Group, United States; "*Plant Biology, Sustainability, and Climate Change*," published by Elsevier, United States; and "*Smart Nanomaterials Technology*," published by Springer Nature Singapore Pte Ltd., Singapore.

Preface

Agriculture and the related sectors are the main sources of income in developing and developed countries. Fertilizers and other chemical products have been widely used in agricultural practice to boost production to feed an ever-growing population. The use of these chemicals was very prevalent in the past to get maximum yield. However, the excess and accelerated use of chemical fertilizers over an extended period results in many hazardous consequences, including soil deterioration, soil nutrient mobility, a decline in soil porosity, organic matter, and soil microbes which eventually advance into subpar plant growth and development. To overcome these shortcomings, in recent years, it has been noticed that biostimulants (a diverse class of compounds including substances or microorganisms) are helpful in sustainable plant growth and development. They accelerate plant growth, yield, and chemical composition even under unfavorable conditions.

The main biostimulants are nitrogen-containing compounds, humic materials, some specific compounds released by microbes, plants, and animals, various seaweed extracts, bio-based nanomaterials, phosphite, silicon, and so on. Additionally, new-generation products and bioproducts are being developed for sustainable plant growth and protection. Some research works in the area of biotechnology and nanobiotechnology have shown improved sustainable plant growth and production. The protective roles of biostimulants vary depending on the compound and plant species. Exposure to biostimulants has been shown to accelerate plants' growth and developmental processes, for instance, manage stomatal conductance and rate of transpiration, and increase the rate of photosynthesis. They also increased crop plants' immune systems against the adverse situation. Thus the use of innovations of new-generation biostimulants also enhances plant production systems, through a significant reduction of synthetic chemicals such as pesticides and fertilizers. Moreover, bioinoculants' commercial products obtained from seaweed extract, humic acids, amino acids, fulvic acids, and some microbial inoculants have shown their potential role in adventitious root induction in plants.

Microbial inoculants or microbial-based biostimulants, as a promising and eco-friendly technology, can be widely used to address environmental concerns and fulfill the need for developing sustainable or modern agriculture practices. They have great potential to elicit plant tolerance to various climate change–related stresses and thus enhance plant growth and overall performance-related features. However, for successful implementation of biostimulant-based agriculture in the field under changing climate conditions, an understanding of plant functions and biostimulants' interaction or action mechanisms coping with various abiotic as well as biotic stresses at the physicochemical, metabolic, and molecular levels is required. Mycorrhizae are beneficial fungi that form symbiotic associations with plants and aid in plant development, disease resistance, and soil health. Similarly, phyllospheric and rhizospheric microbiomes can increase plant growth and productivity and, thereby, can act as a driving force for increasing the agricultural production in a sustainable manner. Taken together, this book aims to explore the recent understanding associated with the various roles and applications of biostimulants on different plants for their sustainable growth, production, and overall management.

I am grateful to all contributors for readily accepting my invitation, sharing their knowledge in specialized areas of research, and readily adjusting the suggestions for improving the shape of their contributions. With great pleasure, I also extend my sincere thanks to Nancy, Kyle, and all associates at Elsevier Inc., for their active cooperation. Finally, my special thanks go to Shagufta, Zaara, Mehwish, and Huzaifa for providing their time and overall support to put everything together.

Azamal Husen

CHAPTER

1

Current understanding and application of biostimulants in plants: an overview

Radhika Sharma[1], *Antul Kumar*[2], *Harmanjot Kaur*[2], *Kuldipika Sharma*[3], *Tarun Verma*[4], *Shreya Chauhan*[5], *Manisha Lakhanpal*[3], *Anuj Choudhary*[2], *Ravi Pratap Singh*[6], *Dadireddy Madhusudan Reddy*[7], *Ajaykumar Venkatapuram*[7,8], *Sahil Mehta*[9] *and Azamal Husen*[10,11,12]

[1]Department of Soil Science, Punjab Agricultural University, Ludhiana, Punjab, India [2]Department of Botany, Punjab Agriculture University, Ludhiana, Punjab, India [3]Department of Forest Products, Dr. Y.S. Parmar University of Horticulture and Forestry, Solan, Himachal Pradesh, India [4]Department of Silviculture and Agroforestry, Dr. Y.S. Parmar University of Horticulture and Forestry, Solan, Himachal Pradesh, India [5]Department of Forest Biology and Tree Improvement, Dr. Y.S. Parmar University of Horticulture and Forestry, Solan, Himachal Pradesh, India [6]Department of Ocean Studies and Marine Biology, Pondicherry University, Port Blair, Andaman and Nicobar Islands, India [7]Department of Microbiology, Palamuru University, Mahbubnagar, Telangana, India [8]International Center for Genetic Engineering and Biotechnology, New Delhi, India [9]Department of Botany, Hansraj College, University of Delhi, New Delhi, India [10]Department of Biotechnology, Smt. S. S. Patel Nootan Science & Commerce College, Sankalchand Patel University, Visnagar, Gujarat, India [11]Department of Biotechnology, Graphic Era (Deemed to be University), Dehradun, Uttarakhand, India [12]Wolaita Sodo University, Wolaita, Ethiopia

Biostimulants in Plant Protection and Performance
DOI: https://doi.org/10.1016/B978-0-443-15884-1.00003-8

1.1 Introduction

The agricultural sector currently confronts the dual challenge of boosting productivity to meet the demands of a growing global population and enhancing resource use efficiency, all while minimizing adverse impacts on ecosystem and human health. Over the past three decades, numerous technological advancements have been proposed to promote the sustainability of agricultural production systems by substantially reducing reliance on synthetic agrochemicals such as pesticides and fertilizers. But the hackneyed use of chemical fertilizers over an extended period resulted in soil deterioration, adverse effect on plant growth and human health. A sustainable alternative involves the adoption of natural plant biostimulants (PBs) to address the challenges posed by traditional agrochemicals.

Biostimulants are defined as any formulation containing molecules (natural or synthetically derived) or microbes that actively tend to promote the plant's growth and development (excluding fertilizers, pesticides, and nutrients) (Munaro et al., 2021). They enhance the plant's ability to adapt to abiotic stress (drought, salinity, and diverse temperatures) (du Jardin, 2015) escalate nutrient use efficiency and uptake (Calvo et al., 2014; Sharma et al., 2016), promote vigor and shelf life (Caradonia et al., 2019), and yield and quality attributes (Ricci et al., 2019).

In comparison with chemical fertilizers, biostimulants are regarded as an infallible, effective, sustainable, cost-effective, and eco-friendly alternative (Rouphael et al., 2017). The effects of biostimulants on plants vary depending on whether they interact directly with plant signaling pathways or function through the induction of endo and non-endo symbionts bacteria, fungi, and yeasts to produce beneficial molecules for the plant (Brown and Saa, 2015). They come in a variety of forms, including dust, wettable powders, solutions, and emulsifiable concentrates (Seaman, 1990). The plant's propagules, seeds, and rootstocks can be treated with them and they can be applied foliar or as a soil treatment (Munaro et al., 2021; Gupta et al., 2022).

1.2 Sources and techniques of biostimulant production

Biostimulants have divergent origins and originate from a wide range of sources, including microbial as well as nonmicrobial. Microbial biostimulants generally come from bacteria predominantly, PGPRs (i.e., plant growth-promoting rhizobacteria), beneficial fungi, and mycorrhizae. In contrast, nonmicrobial sources include algal extracts (seaweeds), humic and fulvic acids, chitin and chitosan, protein hydrolysates and amino acids, and plant-based extracts (Nardi et al., 2015; Rouphael et al., 2015; Ruzzi and Aroca, 2015). The exploitation of rhizosphere microorganisms is considered an important way for sustainable and healthy crop production. The different types of biostimulants and its specific uses are depicted in Fig. 1.1.

The use of microbes in agriculture is considered to be a vital method that leads to feasible and flourishing vegetation. They boost plant growth and development by expediting nutrient uptake and solubilization. There were different environments from which these were obscured, including water, soil surfaces (rhizosphere and rhizoplane), composted

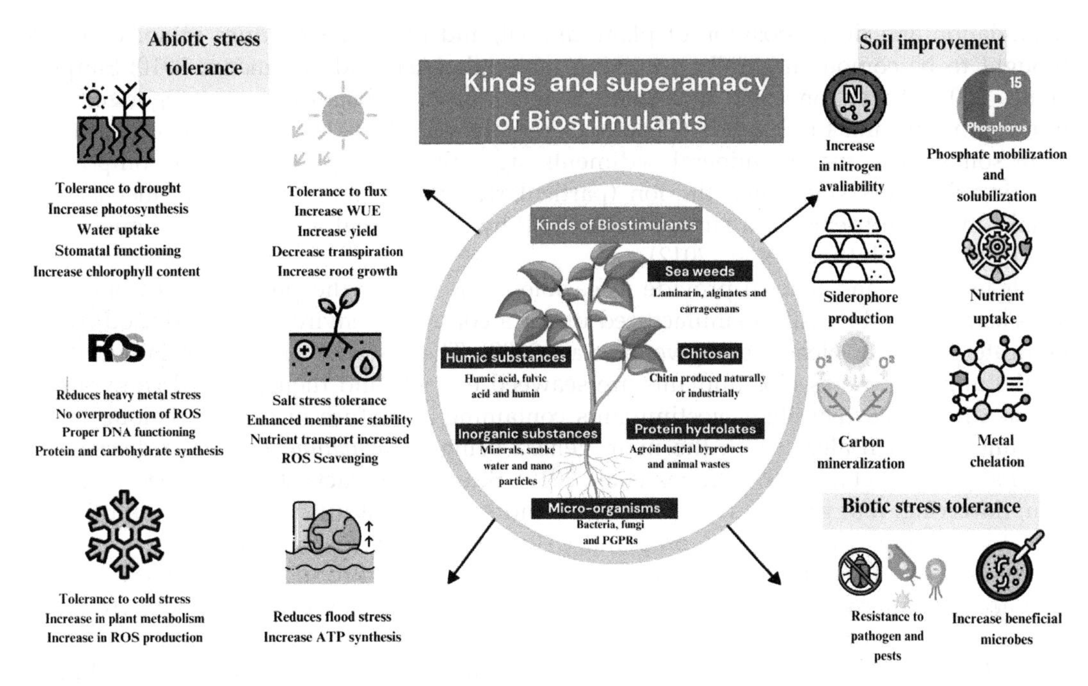

FIGURE 1.1 Diagrammatic depiction of kinds of biostimulants and their effects under unfavorable conditions.

manures, vegetation, and plant residues (Bashan and De–Bashan, 2010). PGPR are either free-living or endophytic bacteria that impact all spectrums of plant life, including subsistence, growth, development, and stress response (Babalola, 2010; Berendsen et al., 2012; Berg et al., 2014). Several species of plant growth-promoting bacteria including Azotobacter, Azospirillum (Bhupenchandra et al., 2020; Garcia de Salamone et al., 1996; Mrkovacki and Milic, 2001), Acinetobacter, Arthrobacter, Bacillus, Enterobacter, Ochrobactrum, Pseudomonas (Gaiero et al., 2013), and Rhizobium (du Jardin, 2015) activate plant growth hormones such as auxins, cytokinins, gibberellins, ethylene, and abscisic acid. These hormones play a crucial role in enhancing plant mineral nutrition by facilitating nitrogen fixation, mobilization of phosphate in the soil and exhibit siderophore production (Dobbelaere et al., 1999), These species also increases the density of lateral roots and root hairs surface that facilitate greater nutrient uptake (Huang et al., 2010; Shabala et al., 2012a,b).

Fungi possess biostimulant activity, especially mycorrhizal fungi that colonize and form a symbiotic association with the plant roots. Fungal consortium formulations containing *Glomus* sp., *Heteroconium chaetospira*, *Trichoderma* sp., *Rhizophagus irregularis*, and *Rhymbocarpus aggregatus* come under the notion of biostimulants (Morales-Payan, 2015; Tejada et al., 2018). They expedite plant proliferation by accelerating the uptake of micro and macronutrients such as nitrogen (N), phosphorus (P), potassium (K), sulfur (S), copper (Cu), calcium (Ca), and zinc (Smith and Read, 2010; Smith et al., 2011).

The nonmicrobial sources of biostimulants are a diverse mixture of molecules that impact crop yield positively. Among them are the humic acids, which are produced by the degradation and decomposition of plant, animal, and microbial residues in soils and are thought to be copious naturally occurring materials (Asli and Neumann, 2010; Simpson et al., 2002). Based on their molecular weight, humic substances are categorized into humins, humic and fulvic acids (Thao and Yamakawa, 2009). They are sourced naturally from soil organic matter, mineral sediments, agricultural by-products, and composting. Humic substances strengthen the ion (particularly cations) mobility in soil, increase the uptake of soil nutrients, and stimulate the ATP hydrolysis that accords cell augmentation and organ growth (Jindo et al., 2012).

Algae (micro and macro) provide a feasible approach for the production of biostimulants as they can be easily manufactured under a controlled environment and modified to fabricate particular bioactive (Ogweno et al., 2008). The majority of commercially utilized algal strains for use as biostimulants are seaweed extracts and their purified compounds. It was discovered that algal biostimulants containing a mixture of extracts belong to the phylum of brown algae, with the main genera being *Ascophyllum, Laminaria, Fucus*, and *Sargassum* (Khan et al., 2009; Drobek et al., 2019). Seaweed extracts stimulate seed germination, flowering, root growth, frost hardiness, and supplement nutrient uptake, limit insect, pest, fungi, and bacteria damage on plants (Alves et al., 2016; Asha et al., 2012).

Chitosan is a deacetylation of biopolymer chitin that is produced both naturally and chemically (du Jardin, 2015). Chitosan is made by removing the N-acetyl groups from chitin using thermo-chemical or enzymatic chitin deacetylation techniques (Kaczmarek et al., 2019). Commercial chitosan preparations are available in a variety of forms including beads, flakes, fibers, solutions, and fine powders. Chitin and chitosan act as distinctive receptors for signaling pathways. They are key players in plant response to stress and cause closure of stomata through an ABA-dependent process that imparts stress response in plants (Iriti et al., 2009; du Jardin, 2015).

Protein hydrolysates (PHs) are a form of plant biostimulant that is defined as a mixture of amino acids, oligopeptides, and polypeptides derived from protein sources via partial hydrolysis (Colla et al., 2016). They are obtained from plant and animal sources through chemical and enzymatic hydrolysis, respectively (Colla and Rouphael, 2015). This manufacturing process induces uncontrolled hydrolysis of the proteins, resulting in the release of numerous free amino acids and the destruction of others (e.g., tryptophan). These released amino acids form complexes with mineral nutrients preventing their insolubilization and thereby increasing nutrient bioavailability. Plant nitrogen assimilation is positively influenced by PHs as they activate certain root enzymes involved in nutrient uptake (Cerdaan et al., 2009)

Plant-derived biostimulants (PDBs) are gaining popularity and are rapidly being incorporated into high-value production systems, where they sustainably enhance yield and quality. Alovera (*Aloe barbadensis*) leaf extract, garlic (*Allium sativum*) leaf extracts (Hayat et al., 2018), licorice (*Glycyrrhiza glabra*) root extract (Rady et al., 2019) legume-derived protein hydrolysate (Rouphael et al., 2018), lemon grass (*Cymbopogon* sp.) (Husen, 2021b) are some of the examples of PDBs. They improve post-harvest fruit quality by increasing nutrient mobilization in the soil, even under stress conditions. The use of PDBs might even result in increased photosynthesis and activate phytohormone signaling pathways that positively impact plant growth and development. Stress resistance endowed by PDBs has been linked to retarded photo inhibition and higher sterol levels, which regulate membrane stability (Lucini et al., 2015; Rouphael and Colla, 2020).

1.3 Active ingredients

Before the sixteenth century, the conventional medical systems were solely stationed on the assistance of nature in strengthening the human body's healing capacity with herbs. The traditional medical practices conjectured a belief in primal energy sustaining life and health. The Chinese called it "qi," and Indians exemplified it as "prana," however, the Western herbalists termed it a "vital force" (Patel and Patel, 2016). Amid the nineteenth century, and the advent of modern medicine, these concepts were unsubstantially signified as remnants of superstition. The era of Western medicine surpassed the traditional practices in China and India. The diversified studies on plant chemistry comprise of viz., photosynthesis, plant respiration, structure, growth, development, and reproduction. The plant needs to be acknowledged as interconnected biological units that have developed beyond the analytical comprehension of science.

Chemical substances, known as phytochemicals, are found naturally in the plant kingdom. Some of them are liable for the organoleptic characteristics of the natural sources where they are present (Patel and Patel, 2016). The active ingredients are envisaged as substances of plant reserves (carbohydrates), executing specific functions within the plant organ function (enzyme) or be the final product secreted by specific organs viz., fruit, flowers (essential oils), or for specific areas such as the epidermis (gums and resins). The physiologically active compounds are distinguished by their chemical structure, predominantly viz., alkaloids, anthocyanins, anthraquinones, cardiac glycosides, coumarins, cyanogenic glycosides, flavonoids, glucosinolate, phenols, saponins, tannins, volatile oils, and terpenoids.

According to the WHO, hereabout 20,000 medicinal plants are reported in 91 countries, inclusive of 12 countries with mega biodiversity. In a way to utilize the biologically active ingredients in the plants, the pioneer steps are extraction, pharmacological screening, isolation, characterization of bioactive component, toxicological assessment, and clinical evaluation (Sasidharan et al., 2011).

1.4 Mechanistic insight

Besides the definition, the classification of biostimulants is also contentious. According to some experts, the mechanism of action is most important, while according to others, the source of the biostimulant may give us more instruments to compare different products and their impacts on different plant species. du Jardin's (2015) classification of biostimulants into seven groups is universally recognized. Carletti et al. (2021), however, pointed out that novel compounds with biostimulant action are required for a deeper comprehension of their biochemical impact and, subsequently, a better classification. The dose administered, the method of application, the timing, and the composition of one or multiple biomolecules, etc., play a role in the outcomes of biostimulant applications (Baltazar et al., 2021). The economic value of biostimulants is rising in importance. According to the European Biostimulant Industry Council (EBIC), the market size is expected to be between 1.5 and 2 billion USD in 2022, with a 10%–12% compound annual growth rate (EBIC, 2020).

A crucial prerequisite for efficient marketing and regulations of manufactured products in agribusiness has been an understanding of their acting mechanisms. In this context, the term mode of action refers to "a distinct influence on a defined biochemical or regulating process." All except a few biostimulants' methods of action are still unclear (Omoarelojie et al., 2021). It is nearly impossible to pinpoint precisely the constituent(s) accountable for bioactivity and to ascertain the involved mode(s) of action because of the heterogeneous nature of the raw materials used in manufacturing and the mixtures of components present in biostimulants. Yakhin et al. (2017) formalized the stages of biostimulant action on plants after their application, including (1) absorption into tissues, translocation, and progression in plants; (2) gene expression, plant signaling, hormone regulation; and (3) coherent whole plant implications.

1.5 Hormonal interaction

Extracellular amino acids have the potential to influence biological functions by serving as signal molecules directly or by altering hormone action through amino acid crosslinking (Tegeder, 2012). According to some studies, amino acid-based biostimulants are rapidly absorbed and translocated by plant tissues. Once internalized, they can act as biocompatible osmolytes, transport regulators, transcription factors, stimulators of stomatal closure, and might even eliminate heavy metals (Kauffman et al., 2007). Some biostimulant phytochemicals may also exhibit signaling activity or stimulate signaling pathways in plants. Several amino acids (Arbona et al., 2013) and peptides serve as signaling molecules in the purview of the development and growth of plants (Mochida and Shinozaki, 2011). Recent research has demonstrated that proteins may also include secret peptide sequences called "cryptides" that can activate the plant stress response.

The emergence of signaling molecules (ligands), their movement, binding to specific receptors, and the subsequent cellular reactions contribute to the transmission of signals (Wang and Irving, 2011). Following the activation of secondary messengers (such as lipids, ions, sugars, nitric oxide, nucleotides, gases, phosphoinositides, etc.) or intracellular signaling mediators by the engagement of the signaling molecule to its receptor, additional cellular reactions are induced. Signaling molecules are believed to possess a catalytic affinity to their receptors, whereas enzymes and their substrates interact geometrically (as in the "lock and key" interaction). Furthermore, in a sophisticated system of interrelated hormones and cofactors, whose actions are intricately dependent upon one another, hormones play a crucial role in the control of metabolic pathways and plant development (Wang and Irving, 2011). It has frequently been demonstrated that biostimulants made from humic compounds, complex organic materials, seaweeds, antiperspirants, phytochemical constituents extracts, etc. (Yakhin et al., 2017) affect the hormonal condition of plants (Kurepin et al., 2014). Although hormone-like molecules may be found in biostimulants, it also seems plausible that such substances may cause the treated plants to produce hormones by themselves (Jannin et al., 2012) and many biostimulants contain amino acids, glycosides, polysaccharides, and organic acids that may function as precursors to or stimulators of intrinsic phytohormones (Parađiković et al., 2011). The activity of natural biostimulants produced from microbes, plants, animals, and humate-based raw material may therefore be mediated by hormones or responses that resemble hormones (Yakhin et al., 2017).

1.6 Role of biostimulant in plant

To meet the food demand for the growing global population, the agricultural sector must increase productivity and resource efficiency while minimizing environmental impacts on ecosystems and human health. Fertilizers and pesticides play an important role in agriculture, providing growers with a powerful tool for increasing yield and ensuring continuous productivity throughout the seasons under both optimal and suboptimal conditions. Several technological innovations have been proposed in the last three decades to improve the sustainability of agricultural production systems by significantly reducing synthetic agrochemicals (Colla and Rouphael, 2015). Biostimulants could be a novel approach to the modification of plant physiological processes to stimulate growth, ameliorate stress-induced constraints, and increase yield. They have been used at different phases of crop production, including seed treatments, foliar sprays during growth, and harvesting. The mechanism of biostimulants in plants is better explained in Fig. 1.2. The mode/mechanisms of action of biostimulants are equally diverse and may include the initiation of nutrient uptake or the release of phosphorus from soils, formulaic activation of soil microbial activity, or enhancement of root growth and improved plant establishment. Various biostimulants have been found to boost plant growth by increasing plant metabolism, germination,

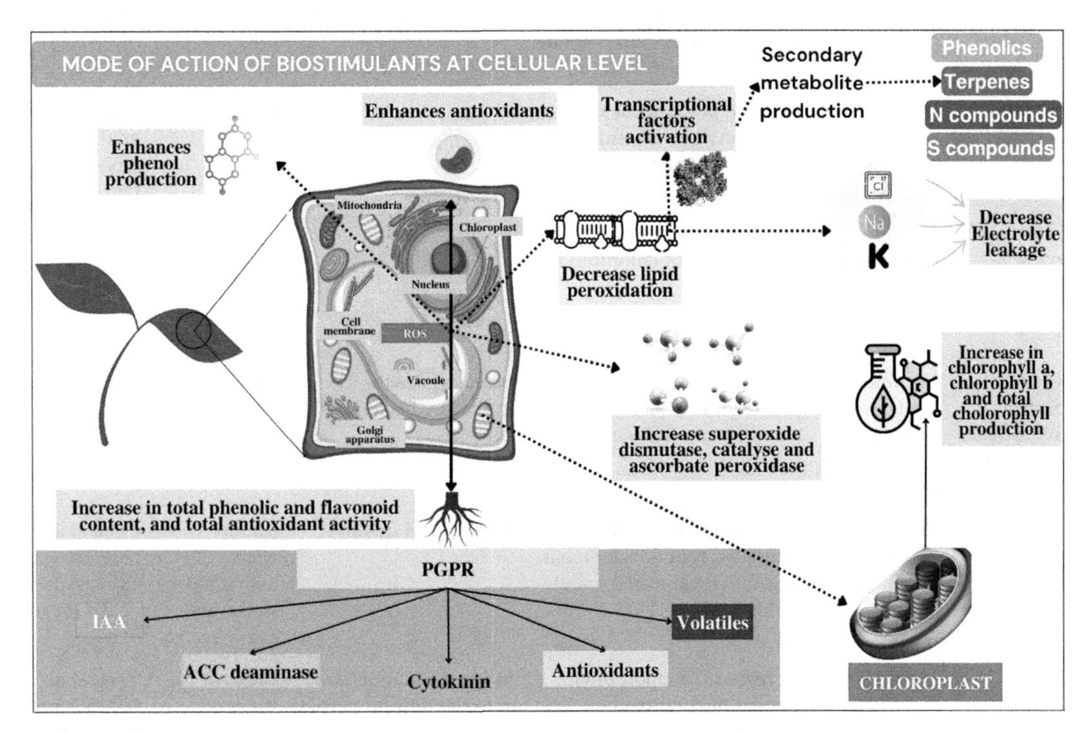

FIGURE 1.2 Mode of action of biostimulants at the cellular level.

photosynthesis, and nutrient absorption from the soil, thereby increasing plant productivity (Yakhin et al., 2017). Categories of biostimulant and their role in agriculture are depicted in Table 1.1.

TABLE 1.1 Categories of biostimulant and their role in agriculture.

Category	Derived from	Crop improvement	Soil improvement	Stress tolerance	Key references
Microbial-derived PBs	Fungal	Increase macronutrient and micronutrients content, phenolic compounds, and plant yield	Facilitate the formation of water aggregate and retain soil moisture	Flooding, drought stress, and salt stress	Saia et al. (2019), Kumar et al. (2020)
	Bacterial	Boost leaf development, chlorophyll content, and promote root development	Bacterial exopolysaccharides enhance soil structure by forming micro and macro aggregates	Cold and heat stress	Theocharis et al. (2012), Sangiorgio et al. (2020)
	Arbuscular mycorrhizal fungi	Modulate root architecture, improve nutrient uptake, and assimilation of nutrients, enhance photosynthesis, and regulate plant hormones	Improved structure of the soil, increased water retention, and reduced leaching of the minerals	Flooding, drought and salt stress	López-Bucio et al. (2015), Kumar et al. (2020)
Seaweed-derived PBs	Brown, red, and green algal species	Increase polyphenol and chlorophyll levels, enhance crop productivity	–	Drought resistant	Elansary et al. (2017), Goñi et al. (2018)
Plant extracts-derived PBs	*Moringa oleifera*, *Cucurbita pepo*	Increasing plant growth and biosynthesis of chlorophylls	–	Drought stress	Elzaawely et al. (2017), Abd El-Mageed et al. (2017)
Vegetal protein hydrolysate-derived PBs	Legume seeds, drench substrate, lettuce	Enhancing biomass, photosynthetic activity and stomatal conductance	Increase growth-promoting bacteria in soil	Drought stress	Luziatelli et al. (2019)
Animal protein hydrolysate-derived PBs	Protein hydrolysates of animal origin	Promote chlorophylls synthesis, root development, vegetative parameters, and yield	Improve soil quality	Salinity, water, temperature stress, and heavy metals	Colla et al. (2017), Casadesús et al. (2019)
Humic substances-derived PBs	Humic acids	Increased the chlorophyll content and antioxidant enzyme activity	Improve soil moisture and retention properties	Drought stress	Yang et al. (2017), Kıran et al. (2019)

1.6.1 Trait improvement

With society's standard of living rising, agricultural traits related to plant quality have become more and more important. In general, agricultural products having better sizes, shapes, firmness levels, colors, and sensory qualities are preferred. Consumers are also concerned about the health benefits, bioactive compounds, and nutritional qualities of agricultural products. As a result, the demand for high-quality products is increasing and quality approaches have emerged (Yuniati et al., 2022). Phytohormones play an important role in the regulation of plant growth and development. Additionally, they serve as chemical signals in response to external variables that otherwise restrict plant growth (Fahad et al., 2015). Multiple factors, such as plant type, climatic conditions, and microbes linked to plant roots, influence a plant's ability to absorb nutrients. A fundamental part of nutrient intake is played by root growth and function, especially in cultivation without chemical fertilizers where nutrients are frequently present in soil solution at relatively low concentrations. Recent studies have reported the ability of plant biostimulants including HS, PHs, and seaweed extracts (SWE) to enhance root growth and development, enabling greater soil exploration and resource acquisition (nutrients and water) by plant roots. It has been speculated that auxins, which are present in many PBs, are most likely to stimulates the roots (De Pascale et al., 2017). When two greenhouse pepper cultivars' seedlings were exposed to *Cladosporium sphaerospermum*, a considerable stimulation of plant growth parameters, yield, and yield components was seen. Another study found that tobacco plants treated with *C. Sphaerospermum* preserved faster rates of growth, which was linked to a variety of possible physiological and molecular processes such as cell expansion and cycle, photosynthesis, phytohormone balance, and defensive responses (Rouphael and Colla, 2020).

1.6.2 Nitrogen use efficiency

The influence of nitrogen on growth and development determines its role in plants. It is a component of nucleic acids, participates in photosynthesis, and produces amino acids which are part of plant proteins (Jämtgård et al., 2010). Biostimulants are frequently used in addition to standard fertilization treatments to improve nutrient use efficiency and crop quality. Due to their ability to affect plant metabolism and low nutrient concentrations, they are markedly different from fertilizers. Increased soil nutrient availability, plant nutrient absorption, and assimilation may all be achieved through the application of bioactive natural substances and microbial inoculants (De Pascale et al., 2017). Rhizobial N-fixation is estimated to contribute between 20 and 22 Tg N yr^{-1} and up to 40 Tg N yr^{-1} to worldwide agricultural systems. Leguminous crop rhizobial inoculants are the most common and historically the first commercially available microbial products in agriculture. Genetic advancements in the N-fixing symbiosis between rhizobia and crop plants have proven challenging to achieve. Since the process of converting atmospheric nitrogen to ammonia requires a lot of energy, increasing nitrogen fixation requires the bacterial cell to prioritize oxidative phosphorylation of carbon sources over glycogen formation. Even though inoculated bean plants produced more dry matter and nodules, studies with *Rhizobium tropici* glycogen synthase deletion mutants failed to survive in soil conditions (Backer et al., 2018). Arginine is important in nitrogen storage and transport in plants and has been shown to accumulate under abiotic and biotic stress (Lea et al., 2006).

1.6.3 Quality control

Quality may be characterized as a combination of agronomic and organoleptic qualities, as well as mineral and vitamin content (Di Vittori et al., 2018). Biostimulants can affect the qualities of fruits and vegetables. When applied as a foliar treatment, growth elicitors such as chitosan, putrescine, and salicylic acid dramatically increased fruit physical attributes such as fruit weight and diameter. Salicylic acid and putrescine enhanced the physical qualities of the stone fruits, according to data from earlier investigations (Ali et al., 2014). Color is a significant visual trait that demonstrates the quality of fruit. The content of anthocyanins has a significant impact on the color of the fruit. Fruits with improved chemical qualities may have higher pro-health values as well as higher sensory values. Guaiacol, for example, was used as a foliar application to improve the wine quality. Guaiacol has been proven to boost the quantity of glycosylated aromatic compounds in grapes from the "Microvine" variety (Sánchez–Gómez et al., 2019). Humic substance application can enhance plant development parameters, and increase the number of photosynthetic pigments, carotenoids, total phenols, flavonoids, and NPK concentration (Bayat et al., 2021).

1.6.4 Biotic and abiotic stress tolerance

Abiotic factors include soil composition, salinity, acidity, hot and cold temperatures, drought, pollution, dampness wind, or UV radiation. Stress from adverse stimuli can drastically diminish harvest yields because plants respond by employing their energy reserves to combat stress rather than focusing on yield (Husen, 2021a,b, 2022a,b). Biotic factors are bacteria, fungi, or viruses that cause a variety of plant diseases. Infections caused by fungi and bacteria can lower productivity and potentially result in the loss of the entire harvest (Drobek et al., 2019). The processes that govern stress tolerance in plants are intricate and complex. While genetic engineering is linked to ethical and societal acceptance difficulties, conventional breeding is a time- and money-consuming method of increasing stress tolerance in crop plants. Secondary metabolites function as antioxidants, inhibiting oxidative processes triggered by stress and increasing the antioxidant property of vegetables, flowers, and fruits. Some research has been done on the effects of biostimulants on secondary metabolites in crops (Bulgari et al., 2015). Salicylic acid (SA) is a signaling molecule produced naturally in plants that are engaged in plant defense-related responses against pathogen invasion. SA's most essential job in plants, among its many effects, is as an endogenous inducer of plant defense mechanisms against pathogen attacks (Mahmood et al., 2017). Ethylene is a gaseous hormone that is active at extremely low concentrations and is a stress hormone, as can be seen by its concentration surges during various abiotic and biotic stresses. Stress-related ethylene buildup may increase plant tolerance or aggravate senescence and other indications of the stress response (Backer et al., 2018).

Some PGPR plants benefit from ethylene's major contribution to enhanced crop stress tolerance (Nadeem et al., 2014). Both stressed and unstressed situations have been researched for PGPR function, and the latter frequently stimulate growth more when the former is the case, such as during drought stress (Rubin et al., 2017). PGPR not only acts as a biocontrol agent but also shields plants from diseases by triggering biochemical and molecular defense reactions inside the plant (Lugtenberg and Kamilova, 2009; Luziatelli et al., 2019). Additionally, PGPR creates antibiotic metabolites—such as lipopeptides, polyketides, and antifungal metabolites—that inhibit pathogens (Prashar et al., 2013). Protein hydrolysates and specific amino acids such as proline and betaine, as well as their derivatives and precursors, have been shown to induce plant defense responses and increase plant tolerance to a variety of abiotic stresses such as salinity, drought, temperature, and oxidative conditions (Calvo et al., 2014).

Recent advances in genomics and transcriptomics have informed our current understanding of how plants respond to environmental stresses. The response is mediated by an intricate network of signals that detect the stress and initiate molecular, biochemical, and physiological processes that may be specific to each stress (Hirayama and Shinozaki, 2010). Stress-inducible genes are expressed at the molecular level, which codes for proteins that directly protect against stress, such as osmoprotectants, detoxifying enzymes, and transporters, as well as genes that code for proteins that are regulatory in nature, such as transcription factors, protein kinases, and phosphatases. Metabolism can be altered by the synthesis of endogenous regulatory molecules such as salicylic and abscisic acids, as well as compatible solutes such as proline and glycine-betaine, which stabilize proteins and cell structures, maintain cell turgor, and scavenge reactive oxygen species. Plant membrane, cell wall architecture, cell cycle, and cell division may all change at the cellular level (Krasensky and Jonak, 2012). Glycine betaine and proline accumulation are generally associated with increased stress tolerance, and foliar spray of these compounds has been shown to improve tolerance to abiotic stresses in a variety of higher plants including maize, barley, soybean, alfalfa, and rice (Reis et al., 2012; Ahmad et al., 2013). Other amino acids influence abiotic stress tolerance. Topical application of glutamate and ornithine, proline precursors, can also strengthen resistance to salt stress (Da Rocha et al., 2012). Proline may function in osmoregulation by compensating for the water deficit caused by heavy metal exposure (Sharma and Dietz, 2006).

1.7 Regulatory framework for biostimulants in plants

Products classified as biostimulants (BS) are highly diverse and also thought of as plant invigorators. Plant biostimulants are anticipated to see a Compound Annual Growth Rate (CAGR) of 13.58% from 2017 to 2022, growing to a predicted value of USD 3.68 billion. A large number of national regulations governing fertilizers require beneficial substances to be registered, granted a license, or both, before being commercialized (Arcadia International, 2014).

In Europe, any product that performs one or more of the functions covered by Regulation (EC) No. 1107/2009 shall continue to be governed by plant protection product laws and is not subject to the fertilizer regulation (European Commission, 2016). In other words, products that do not directly address biotic stress brought on by dangerous organisms fall under the category of plant protection products (PBs), and these goods must be registered as such. According to (La Torre et al., 2016), the PB regulatory framework is highly inconsistent and dispersed throughout Europe. Recently, the regulations that govern these products on a national level have been revised. The new EU Fertilizing Products Regulation, which specifies rules for the commercialization of fertilizer merchandise, including PBs, has been upgraded and is also in line with the circular economy concept. It acknowledges PBs as a useful and cutting-edge tool for farmers and can act as a significant impetus.

The regulatory frameworks in other nations also necessitate that PB products are registered before being put on the market. The legal system around PBs is complex in both the USA and Europe. In actuality, state laws govern PBs, and there is no unified statute or definition of PBs (Arcadia International, 2014). Additionally, each registration is only valid for a specific amount of time (3 years in Brazil, South Africa, Australia, and 5 years in Canada), as the product may well be updated on account of advancements in science and technology. In South Africa, PBs are classified as "Group 3 Fertilisers," which are defined as substances or organisms, that enhance or sustain the physical, chemical, or physiological status (fertility) of the soil. These substances are regulated by Regulations Regarding and by Fertilisers, Farm Feeds, Agricultural Remedies, and Stock Remedies Act, 1947. Similarly, in Brazil, PBs are classified as inoculants and biofertilizers, and their products are subject to regulations passed in 2004 and 1980. PBs are generally regarded as biofertilizers in India and are governed by the Fertilizer (Control) Order, 1985, recently amended in 2009. Since March 2006, biofertilizers have also been governed under the same order. They include mycorrhizal biofertilizers, phosphate-solubilizing bacteria, *Rhizobium*, *Azotobacter*, and *Azospirillum*, etc. The standards of Part A in Schedule III of the Fertilizer (Control) Order of 1985 must be fulfilled by biofertilizers. Any firm aspiring to advertise, sell, or engage in their marketing as a Commercial Distributor must apply to the controller to receive a registration certificate. The validity of each registration certificate is three years from the day of disbursement.

There are several different ways that biostimulants are regulated. The development of biostimulants faces technical, scientific, and regulatory difficulties. Technical problems include the formulation and blending of biostimulants with other plant protection chemicals, while the intricacy of the physiological effects of biostimulants presents the main scientific challenge (du Jardin, 2015). Intellectual property, biostimulant classification, and pre-market evaluation all present regulatory issues. At present, it is uncertain if, and also where, biostimulants will be designated a separate regulatory category under national or international legislation. The majority of countries' current rules on pesticides and fertilizers permit the safe marketing of biostimulants, but there should be clear grounds for characterizing them in a separate regulatory domain.

1.8 Advances in biostimulant production in plants

To have desired growth and yield of crops, address climate change, and feed the growing population from limited use of resources, sustainable agriculture through biostimulants is gaining importance. The nanomaterials-based biostimulants developed in agriculture minimize post-harvest loss, provide protection against biotic and abiotic stress, regulate biogeochemical cycling, and greater benefit-cost ratio (Kumar et al., 2022; Sharma et al., 2022). Target-specific delivery and regulated release of nutrients are made possible by the use of nanomaterials in nutrition delivery.

Humic substances (HS) comprising Humic, fulvic acids, and Humin are a fraction of organic matter produced in response to prolonged humification and enhance the nutritional status of soil by forming complexes. Sprayed HS together with $FeSO_4 \bullet 7H_2O$ with nano ferrihydrite on the leaves increased the yield and Fe content by improving chlorophyll a, chlorophyll b, total chlorophyll, and leaf area contents in spinach (Turan et al., 2022). In apples, HS was reported to be potential biostimulant acid at the rate of a 0.15% increase in yield, and the highest B:C ratio of 5.51 was recorded over the previous year (Khan et al., 2009). In addition, humic acid enables nutrients to chelate to treat nutrient shortages. Under drought stress, a plant's ability to take additional nutrients from the root is aided by enhanced root development, increased cell membrane permeability, increased respiration, and increased photosynthesis. Hayati et al. (2022) revealed that the best grain production was produced in areas with 1–2 g/L of nanochelate iron and 250–500 mg/L of humic acid increasing grain yield, oil yield, and an essential oil absorption of nutrients and micronutrients, and it significantly affects nutrients during drought stress in *Nigella sativa*.

Besides providing plant nutrition, these smart NPs used as biostimulants are extensively used to regulate the physiological process. Other noteworthy characteristics can be observed with the use of salicylic acid-chitosan nanoparticles (SA-CS NPs) for promoting plant defense and growth in maize due to antifungal, antioxidant-defense enzyme activities, balancing reactive oxygen species and seed growth-promoting activities that can even address forthcoming multi-dynamic stresses (Kumaraswamy et al., 2019; Kumar et al., 2020). Trichoderma spp. can also biosynthesize selenium nanoparticles (SeNPs) that play a role in protecting selenoproteins against reactive oxygen species (ROS) and is known to have beneficial benefits on plant and animal metabolism at low doses. Additionally, Se can serve as a defense mechanism against the negative effects of heavy metals (Bărbieru et al., 2019).

Insects are exposed to nanoparticles by direct touch, ingestion, and inhalation. Nanoparticles that come into physical contact with a cell's exoskeleton can attach to sulfur in protein or phosphorus in DNA or inside of the cell, deactivating organelles and enzymes, and ultimately causing cell death. Insect cuticles are covered in nanoclay, nanoalumina, and nanosilica, which are connected to them and draw water from the insect's body (Bărbieru et al., 2019; Sarkar et al., 2022). For instance, the use of SiNPs for the transfer of hormones like abscisic acid, salicylic acid, fungicides, and insecticides and its ability to enhance plant growth, development, and stress tolerance has been documented functionalized. These groups frequently function as a linker for the conjugation of biologically active molecules such as organic acids, enzymes, sugar moieties, and hormones which speed up their uptake in plant cells (Sarkar et al., 2022). Acetylcholine esterase and carboxyl esterase are bound by Zn NP_S and TiO_2 NPS, which have an impact on their activity. These nanoparticles work similarly to insecticides in the carbamate and organophosphorus groups in terms of their method of action. The enzyme glutathione S transferase is impacted by Ag nanoparticles (Anandhi et al., 2020). Nanomaterial in biostimulant production is mentioned in Table 1.2.

The feasibility and fexibilty of biostimulant herbicides as atrazine carriers, chitosan, and alginate nanocapsules were the most efficient and quick weed eradication methods. It is vital to keep working to achieve an adequate concentration of atrazine's active ingredient within the nanocapsules for both the target and nontarget plants (Cordon et al., 2022). Nanotechnology offers sustainable advantages in agriculture such as the detection and killing of pathogens. The number of colonies for several plant pathogenic fungi was significantly reduced as a result of the antifungal application of AgNPs, with an efficiency of up to 90% against *Pyricularia grisea, Fusarium verticillioides, Helminthosporium oryzae, Penicillium brevicompactum*, and *Fusarium moniliforme*.

Microbial CuONPs produced using *Pseudomonas fluorescens* and *T. viride*, two different biocontrol agents, demonstrated the highest percentage growth of antifungal efficacies. On the other hand, CuONPs made from *T. viride* showed significantly more antifungal activity in vivo than the standard Bordeaux mixture (Elamawi et al., 2018; Sawake et al., 2022). The employment of nanotechnology-enabled approaches has significantly reduced the cultural eutrophication of water bodies. It is also essential for the preparation and packaging of food. The practicality and adaptability of nanoparticles contribute to the preservation of food's nutritious value following consumer demand (Sawake et al., 2022). Overcoming the

TABLE 1.2 Nanomaterial in biostimulant production.

Nanomaterial	Dimension (nm)	Plant	Mode of application	Effect on plants	Key references
Silica	12	*Solanum lycopersicum* Mill.	Seed treatment	Increase seed germination index, seed vigor index, seedling fresh and dry weight	Siddiqui and Al-Whaibi (2014)
Platinum	5	*Triticum aestivum* L.	Irrigation	Leaf area, dry weight of shoot, and root	Astafurova et al. (2015)
Silica	20–30	*Oryza sativa* L.	Seed treatment	Pest resistance	Debnath et al. (2011)
Gold	24	*Arabidopsis thaliana*	Seed treatment	Enhance seed germination, growth, and antioxidant potential	Kumar et al. (2020)
Silver	10–100	*Borago officinalis* L.	Foliar spray	Enhance plant shoot, leaf number, plant height, and seed yield	Sah et al. (2011)
Copper	30	*Vicia sativa* L.	Irrigation	Increased plant growth	Ivanycheva et al. (2012)
Manganese	50	*Vigna radiata* L.	Seed treatment	Boost carbon and nitrogen assimilatory processes	Pradhan et al. (2014)
Ferric oxide	10–50	*Arachis hypogaea* L.	Soil treatment	Increased root length, plant height and plant biomass	Rui et al. (2016)
Calcium	15–20	*Glycine max* L.	Growth medium	Promote plant growth and yield	Liu and Lal (2014)

detrimental impacts of abiotic stressors on crop yield has been a top priority for the research community in recent years. The ability of NPs to shield crop plants from various biotic and abiotic challenges through increased cellular antioxidants, nutrient uptake, photosynthetic efficiency, and modulation of biochemical/molecular pathways, has increased stress tolerance in plants. Therefore, nanotechnology is a novel strategy that should be used to address present problems and prepare for future agricultural concerns.

1.9 Conclusion and future outlook

Biostimulants improve a plant's ability to withstand stress and fight off free radicals, as well as its ability to absorb and distribute nutrients and grow and develop its vegetative and generative organs. The pieces of evidence reported here generally support the benefits of using biostimulants in agriculture production, particularly in growth situations that are stressful due to transplantation, reduced fertilization, or other abiotic stress instances. Several studies were able to show that microbial and nonmicrobial PBs can trigger a variety of morpho-anatomical, biochemical, physiological, and molecular plant responses, such as enhancing crop productivity and NUE, and raising resistance to abiotic stressors. It is clear that biostimulants can increase seed vigor and plant productivity, but to use them most effectively, we need to study diverse biostimulant usage in different crops and in different regions, accounting for climate change scenarios. Biostimulant mechanisms, especially under stress, are unclear due to their complex composition. So, researching a single active compound from specific biostimulants will enhance our understanding about its effects.

References

Abd El-Mageed, T.A., Semida, W.M., Rady, M.M., 2017. Moringa leaf extract as biostimulant improves water use efficiency, physio-biochemical attributes of squash plants under deficit irrigation. Agric. Water Manag. 193, 46–54.

Ahmad, R., Lim, C.J., Kwon, S.Y., 2013. Glycine betaine: a versatile compound with great potential for gene pyramiding to improve crop plant performance against environmental stresses. Plant. Biotechnol. Rep. 7, 49–57.

Ali, L., Abbasi, N.A., Hafiz, L.A., 2014. Physiological response and quality attributes of peach fruit cv. Flordaking as affected by different treatments of calcium chloride, putrescine and salicylic acid. Pak. J. Agri. Sci. 51, 33–39.

Alves, R.C., Merces, P.F.F., Souza, I.R.A., Alves, C.M.A., Silva, A.P.S.A., Lima, V.L.M., et al., 2016. Antimicrobial activity of seaweeds of Pernambuco, Northeastern Coast of Brazil. Afr. J. Microbiol. Res. 10, 312–318.

Anandhi, S., Saminathan, V.R., Yasotha, P., Saravanan, P.T., Rajanbabu, V., 2020. Nanopesticides in pest management. J. Entomol. Zool. Stud. 8, 685–690.

Arbona, V., Manzi, M., Ollas, C., Gómez-Cadenas, A., 2013. Metabolomics as a tool to investigate abiotic stress tolerance in plants. Int. J. Mol. Sci. 14, 4885–4911.

Arcadia International, 2014. *A legal framework for plant biostimulants and agronomic fertiliser additives.* <http://ec.europa.eu/DocsRoom/documents/5403/attachments/1/translations/en/renditions/nativen> (accessed 20.06.17).

Asha, A., Rathi, J.M., Raja, P.D., Sahayaraj, K., 2012. Biocidal activity of two marine algal extracts against third instar nymph of *Dysdrcus cingulatus* (Fab.) (*Hemiptera, Pyrrhocoridae*). Biocididal activity of marine algae. J. Biopest 5, 129–134.

Asli, S., Neumann, P.M., 2010. Rhizosphere humic acid interacts with root cell walls to reduce hydraulic conductivity and plant development. Plant. Soil. 336, 313–322.

Astafurova, T., Zotikova, A., Morgalev, Y., Verkhoturova, G., Postovalova, V., Kulizhskiy, S., et al., 2015. Effect of platinum nanoparticles on morphological parameters of spring wheat seedlings in a substrate–plant system. In: IOP Conference Series: Materials Science and Engineering, vol. 98, no. 1. IOP Publishing, pp. 012004.

Babalola, O.O., 2010. Beneficial bacteria of agricultural importance. Biotechnol. Lett. 32, 1559–1570.

Backer, R., Rokem, J.S., Ilangumaran, G., Lamont, J., Praslickova, D., Ricci, E., et al., 2018. Plant growth-promoting rhizobacteria: context, mechanisms of action, and roadmap to commercialization of biostimulants for sustainable agriculture. Front. Plant. Sci. 1473.

Baltazar, M., Correia, S., Guinan, K.J., Sujeeth, N., Bragança, R., Gonçalves, B., 2021. Recent advances in the molecular effects of biostimulants in plants: an overview. Biomolecules 11, 1096.

Bărbieru, O.G., Dimitriu, L., Călin, M., Răut, I., Constantinescu-Aruxandei, D., Oancea, F., 2019. Plant biostimulants based on selenium nanoparticles biosynthesized by *Trichoderma* strains. Proceedings 29, 95.

Bashan, Y., De-Bashan, L.E., 2010. How the plant growth-promoting bacterium *Azospirillum* promotes plant growth – a critical assessment. Adv. Agron. 108, 77–136.

Bayat, H., Shafie, F., Aminifard, M.H., Daghighi, S., 2021. Comparative effects of humic and fulvic acids as biostimulants on growth, antioxidant activity and nutrient content of yarrow (*Achillea millefolium* L.). Sci. Hortic. 279, 109912.

Berendsen, R.L., Pieterse, C.M., Bakker, P.A., 2012. The rhizosphere microbiome and plant health. Trends Plant. Sci. 17, 1360–1385.

Berg, G., Grube, M., Schloter, M., Smalla, K., 2014. Unraveling the plant microbiome: looking back and future perspectives. Front. Microbiol. 5, 1–7.

Bhupenchandra, I., Devi, S.H., Basumatary, A., Dutta, S., Singh, L.K., Kalita, P., et al., 2020. Biostimulants: potential and prospects in agriculture. Int. Res. J. Pure Appl. Chem. 21, 20–35.

Brown, P., Saa, S., 2015. Biostimulants in agriculture. Front. Plant. Sci. 6, 1–3.

Bulgari, R., Cocetta, G., Trivellini, A., Vernieri, P., Ferrante, A., 2015. Biostimulants and crop responses: a review. Biol. Agric. Hortic. 31, 1–17.

Calvo, P., Nelson, L., Kloepper, J.W., 2014. Agricultural uses of plant biostimulants. Plant. Soil. 383, 3–41.

Caradonia, F., Battaglia, V., Righi, L., Pascali, G., Torre, A.L., 2019. Plant biostimulant regulatory framework: prospects in Europe plant biostimulant and current situation at international level. J. Plant. Growth Regul. 38, 438–448.

Carletti, P., García, A.C., Silva, C.A., Merchant, A., 2021. Editorial: towards a functional characterization of plant biostimulants. Front. Plant. Sci. 12, 677772.

Casadesús, A., Polo, J., Munné-Bosch, S., 2019. Hormonal effects of an enzymatically hydrolyzed animal protein-based biostimulant (Pepton) in water-stressed tomato plants. Front. Plant. Sci. 10, 758.

Cerdaan, M., Sanchez-Sanchez, A., Oliver, M., Juarez, M., Sanchez-Andreu, J.J., 2009. Effect of foliar and root applications of amino acids on iron uptake by tomato plants. Acta Hortic. 830, 481–488.

Colla, G., Rouphael, Y., 2015. Biostimulants in horticulture. Sci. Hortic. 196, 1–134.

Colla, G., Rouphael, Y., Lucini, L., Canaguier, R., Stefanoni, W., Fiorillo, A., et al., 2016. Protein hydrolysate-based biostimulants: origin, biological activity and application methods. Acta Hortic. 1148, 27–38.

Colla, G., Hoagland, L., Ruzzi, M., Cardarelli, M., Bonini, P., Canaguier, R., et al., 2017. Biostimulant action of protein hydrolysates: unraveling their effects on plant physiology and microbiome. Front. Plant. Sci. 8, 2202.

Cordon, G., Valiño, I.L., Prieto, A., Costa, C., Marchi, M.C., Diz, V., 2022. Effects of the nanoherbicide made up of atrazine-chitosan on the primary events of photosynthesis. J. Photochem. Photobiol. 12, 100144.

Da Rocha, I.M.A., Vitorello, V.A., Silva, J.S., Ferreira-Silva, S.L., Viégas, R.A., Silva, E.N., et al., 2012. Exogenous ornithine is an effective precursor and the δ-ornithine amino transferase pathway contributes to proline accumulation under high N recycling in salt-stressed cashew leaves. J. Plant. Physiol. 169, 41–49.

De Pascale, S., Rouphael, Y., Colla, G., 2017. Plant biostimulants: innovative tool for enhancing plant nutrition in organic farming. Eur. J. Hortic. Sci. 82, 277–285.

Debnath, N., Das, S., Seth, D., Chandra, R., Bhattacharya, S.C., Goswami, A., 2011. Entomotoxic effect of silica nanoparticles against *Sitophilus oryzae* (L.). J. Pest. Sci. 84, 99–105.

Di Vittori, L., Mazzoni, L., Battino, M., Mezzetti, B., 2018. Pre-harvest factors influencing the quality of berries. Sci. Hortic. 233, 310–322.

Dobbelaere, S., Croonenborghs, A., Thys, A., Broek, A.V., Vanderleyden, J., 1999. Phytostimulatory effect of *Azospirillum brasilense* wild type and mutant strains altered in IAA production on wheat. Plant. Soil. 212, 155–164.

Drobek, M., Frac, M., Cybulska, J., 2019. Plant Biostimulants: Importance of the quality and yield of horticultural crops and the improvement of plant tolerance to abiotic stress – a review. Agronomy 9, 2–18.

du Jardin, P., 2015. Plant biostimulants: definition, concept, main categories and regulation. Sci. Hortic. 196, 3–14.

Elamawi, R.M., Al-Harbi, R.E., Hendi, A.A., 2018. Biosynthesis and characterization of silver nanoparticles using *Trichoderma longibrachiatum* and their effect on phytopathogenic fungi. Egypt. J. Biol. Pest. Control. 28, 28.

Elansary, H.O., Yessoufou, K., Abdel-Hamid, A.M.E., El-Esawi, M.A., Ali, H.M., Elshikh, M.S., 2017. Seaweed extracts enhance salam turfgrass performance during prolonged irrigation intervals and saline shock. Front. Plant. Sci. 8, 830.

Elzaawely, A.A., Ahmed, M.E., Maswada, H.F., Xuan, T.D., 2017. Enhancing growth, yield, biochemical, and hormonal contents of snap bean (*Phaseolus vulgaris* L.) sprayed with moringa leaf extract. Arch. Agron. Soil. Sci. 63, 687–699.

European Biostimulant Industry Council [EBIC], 2020. Economic Overview of the European Biostimulants Market. < https://biostimulants.eu/highlights/economic-overview-of-the-european-biostimulants-market/ > (accessed November 2021).

European Commission, 2016. Proposal for a Regulation Laying Down Rules on the Making Available on the Market of CE Marked Fertilising Products and Amending Regulations (EC)1069/2009 and (EC)1107/2009. COM(2016). pp. 157.

Fahad, S., Hussain, S., Matloob, A., Khan, F.A., Khaliq, A., Saud, S., et al., 2015. Phytohormones and plant responses to salinity stress: a review. Plant. Growth Regul. 75, 391–404.

Gaiero, J.R., McCall, C.A., Thompson, K.A., Day, N.J., Best, A.S., Dunfield, K.E., 2013. Inside the root microbiome: bacterial root endophytes and plant growth promotion. Am. J. Bot. 100, 1738–1750.

Garcia de Salamone, I.E.G., Dobereiner, J., Urquiaga, S., Boddey, R.M., 1996. Biological nitrogen fixation in *Azospirillum strainmaize* genotype associations as evaluated by the ^{15}N isotope dilution technique. Biol. Fertil. Soils 23, 249–256.

Goñi, O., Quille, P., O'Connell, S., 2018. *Ascophyllum nodosum* extract biostimulants and their role in enhancing tolerance to drought stress in tomato plants. Plant. Physiol. Biochem. 126, 63–73.

Gupta, S., Dolezal, K., Kulkarni, M.G., Balazs, E., Staden, J.V., 2022. Role of non-microbial biostimulants in regulation of seed germination and seedling establishment. Plant Growth Regul. 97, 271–313.

Hayat, S., Ahmad, H., Ali, M., Hayat, K., Khan, M., Cheng, Z., 2018. Aqueous garlic extract as a plant biostimulant enhances physiology, improves crop quality and metabolite abundance, and primes the defense responses of receiver plants. Appl. Sci. 8, 1505.

Hayati, A., Hosseini, S.M., Rahimi, M.M., Kelidari, A., 2022. The effect of humic acid and nano-chelate iron on the absorption of some nutrients, grain yield, oil, and essential oil of (Nigella sativa) under drought stress. Commun. Soil Sci. Plant Anal. 27, 1–2.

Hirayama, T., Shinozaki, K., 2010. Research on plant abiotic stress responses in the post-genome era: past, present and future. Plant J. 61 (6), 1041–1052.

Huang, J., Gu, M., Lai, Z., Fan, B., Shi, K., Zhou, H., Yu, J.Q., Chen, Z., 2010. Functional analysis of the *Arabidopsis* PAL gene family in plant growth, development, and response to environmental stress. Plant Physiol. 153, 1526–1538.

Husen, A., 2021a. Plant Performance Under Environmental Stress: Hormones, Biostimulants and Sustainable Plant Growth Management. Springer Nature Switzerland AG, Cham.

Husen, A., 2021b. Harsh Environment and Plant Resilience: Molecular and Functional Aspects. Springer Nature Switzerland AG, Cham.

Husen, A., 2022a. Plants and Their Interaction to Environmental Pollution: Damage Detection, Adaptation, Tolerance, Physiological and Molecular Responses. Elsevier Inc, Cambridge, MA.

Husen, A., 2022b. Environmental Pollution and Medicinal Plants. Taylor & Francis Group, LLC, Boka Raton, FL.

Iriti, M., Picchi, V., Rossoni, M., Gomarasca, S., Ludwig, N., Gargano, M., Faoro, F., 2009. Chitosan antitranspirant activity is due to abscisic acid-dependent stomatal closure. Environ. Exp. Bot. 66, 493–500.

Ivanycheva, Y.N., Zheglova, T.V., Polishchuk, S.D., 2012. Copper nanopowders and copper oxide effect onto phytohormones activity at Vicia and summer wheat sprouts. Vestn. Ryazansk Gos. Agrotekhnol Univ. 1, 12–14.

Jämtgård, S., Näsholm, T., Huss-Danell, K., 2010. Nitrogen compounds in soil solutions of agricultural land. Soil Biol. Biochem. 42 (12), 2325–2330.

Jannin, L., Arkoun, M., Ourry, A., Laîné, P., Goux, D., Garnica, M., 2012. Microarray analysis of humic acid effects on *Brassica napus* growth: involvement of N, C and S metabolisms. Plant Soil 359, 297–319. Available from: https://doi.org/10.1007/s11104-012-1191-x.

Jindo, K., Martim, S.A., Navarro, E.C., Aguiar, N.O., Canellas, L.P., 2012. Root growth promotion by humic acids from composted and non-composted urban organic wastes. Plant Soil 353, 209–220.

Kaczmarek, M.B., Struszczyk-Swita, K., Li, X., Szczęsna-Antczak, M. Daroch. M., 2019. Enzymatic modifications of chitin, chitosan, and chitooligosaccharides. Front Bioeng. Biotechnol. 7, 243.

Kauffman, G.L., Kneivel, D.P., Watschke, T.L., 2007. Effects of a biostimulant on the heat tolerance associated with photosynthetic capacity, membrane thermostability, and polyphenol production of perennial ryegrass. Crop Sci. 47, 261–267. Available from: https://doi.org/10.2135/cropsci2006.03.0171.

Khan, W., Rayirath, U.P., Subramanian, S., Jithesh, M.N., Rayorath, P., Hodges, D.M., et al., 2009. Seaweed extracts as biostimulants of plant growth and development. J. Plant Growth Regul. 28, 386–399.

Kıran, S., Furtana, G.B., Talhouni, M., Ellialtıoğlu, Ş.Ş., 2019. Drought stress mitigation with humic acid in two Cucumis melo L. genotypes differ in their drought tolerance. Bragantia 78, 490–497.

Krasensky, J., Jonak, C., 2012. Drought, salt, and temperature stress-induced metabolic rearrangements and regulatory networks. J. Exp. Bot. 63 (4), 1593–1608.

Kumar, A., Choudhary, A., Kaur, H., Guha, S., Mehta, S., Husen, A., 2022. Potential applications of engineered nanoparticles in plant disease management: a critical update. Chemosphere 295, 133798.

Kumar, R., Trivedi, K., Anand, K.G.V., Ghosh, A., 2020. Science behind biostimulant action of seaweed extract on growth and crop yield: insights into transcriptional changes in roots of maize treated with *Kappaphycus alvarezii* seaweed extract under soil moisture stressed conditions. J. Appl. Phycol. 32, 599–613.

Kumaraswamy, R.V., Kumari, S., Choudhary, R.C., Sharma, S.S., Pal, A., Raliya, R., et al., 2019. Salicylic acid functionalized chitosan nanoparticle: a sustainable biostimulant for plant. Int. J. Biol. Macromol. 123, 59–69.

Kurepin, L.V., Zaman, M., Pharis, R.P., 2014. Phytohormonal basis for the plant growth promoting action of naturally occurring biostimulators. J. Sci. Food Agric. 94, 1715–1722. Available from: https://doi.org/10.1002/jsfa.6545.

La Torre, A., Battaglia, V., Caradonia, F., 2016. An overview of the current plant biostimulant legislations in different European Member States. J. Sci. Food Agric. 96, 727–734.

Lea, P.J., Sodek, L., Parry, M.A.J., Shewry, P.R., Halford, N.G., 2006. Asparagine in plants. Ann. Appl. Biol. 150, 1–26.

Liu, R., Lal, R., 2014. Synthetic apatite nanoparticles as a phosphorus fertilizer for soybean (*Glycine max*). Sci. Rep. 4, 5686.

López-Bucio, J., Pelagio-Flores, R., Herrera-Estrella, A., 2015. Trichoderma as biostimulant: exploiting the multilevel properties of a plant beneficial fungus. Sci. Hortic. 196, 109–123.

Lucini, L., Rouphael, Y., Cardarelli, M., Canaguier, R., Kumar, P., Colla, G., 2015. The effect of a plant-derived biostimulant on metabolic profiling and crop performance of lettuce grown under saline conditions. Sci. Hortic. 182. Available from: https://doi.org/10.1016/j.scienta.2014.11.022.

Lugtenberg, B., Kamilova, F. (2009). Plant-growth-promoting rhizobacteria. Annu. Rev. Microbiol. 63, 541–556. Available from: https://doi.org/10.1146/annurev.micro.62.081307.162918.

Luziatelli, F., Ficca, A.G., Colla, G., Baldassarre Švecová, E., Ruzzi, M., 2019. Foliar application of vegetal-derived bioactive compounds stimulates the growth of beneficial bacteria and enhances microbiome biodiversity in lettuce. Front. Plant. Sci. 10, 60.

Mahmood, N., Abbasi, N.A., Hafiz, I.A., Ali, I., Zakia, S., 2017. Effect of biostimulants on growth, yield and quality of bell pepper cv. Yolo Wonder. Pak. J. Agric. Sci. 54 (2).

Mochida, K., Shinozaki, K., 2011. Advances in omics and bioinformatics tools for systems analyses of plant functions. Plant Cell Physiol. 52, 2017–2038. Available from: https://doi.org/10.1093/pcp/pcr153.

Morales-Payan, J., 2015. Influence of foliar sprays of an amino acid formulation on fruit yield of 'edward' mango. Acta. Hortic. 1075, 157–159.

Mrkovacki, N., Milic, V., 2001. Use of *Azotobacter chroococcum* as useful in agricultural application. Ann. Microbiol. 51, 145–158.

Munaro, D., Nunes, A., Schmitz, C., Bauer, C., Coelho, D.S., Oliveria, E.R., et al., 2021. In: Studies in Natural Products Chemistry, 71. Elsevier, pp. 87–120.

Nadeem, S.M., Ahmad, M., Zahir, Z.A., Javaid, A., Ashraf, M., 2014. The role of mycorrhizae and plant growth promoting rhizobacteria (PGPR) in improving crop productivity under stressful environments. Biotechnol. Adv. 32 (2), 429–448.

Nardi, S., Cardarelli, M., Ertani, A., Lucini, L., Canaguier, R., 2015. Protein hydrolysates as biostimulants in horticulture. Sci. Hortic (Amsterdam) 196, 28–38.

Ogweno, J.O., Song, X.S., Shi, K., Hu, W.H., Mao, W.H., Zhou, Y.H., et al., 2008. Brassinosteroids alleviate heat-induced inhibition of photosynthesis by increasing carboxylation efficiency and enhancing antioxidant systems in *Lycopersicon esculentum*. J. Plant Growth Regul. 27, 49–57.

Omoarelojie, L.O., Kulkarni, M.G., Finnie, J.F., Van Staden, J., 2021. Modes of action of biostimulants in plants. In: *Biostimulants for Crops from Seed Germination to Plant Development*, Academic Press, pp. 445–459.

Parađiković, N., Vinković, T., Vinković Vrček, I., Žuntar, I., Bojić, M., Medić-Šarić, M., 2011. Effect of natural biostimulants on yield and nutritional quality: an example of sweet yellow pepper (*Capsicum annuum* L.) plants. J. Sci. Food Agric. 91, 2146–2152. Available from: https://doi.org/10.1002/jsfa.4431.

Patel, V., Patel, R., 2016. The active constituents of herbs and their plant chemistry, extraction and identification methods. J. Chem. Pharm. Res. 8 (4), 1423–1443.

Pradhan, S., Patra, P., Mitra, S., Dey, K.K., Jain, S., Sarkar, S., et al., 2014. Manganese nanoparticles: impact on nonnodulated plant as a potent enhancer in nitrogen metabolism and toxicity study both in vivo and in vitro. J. Agric. Food Chem. 62, 8777–8785.

Prashar, P., Kapoor, N., Sachdeva, S., 2013. Biocontrol of plant pathogens using plant growth promoting bacteria. Sustain. Agric. Rev. 319–360.

Rady, M.M., Desoky, E.S., Elrys, A.S., Boghdady, M.S., 2019. Can licorice root extract be used as an effective natural biostimulant for salt-stressed common bean plants. S. Afr. J. Bot. 121, 294–305.

Reis, S.P.D., Lima, A.M., De Souza, 2012. Recent molecular advances on downstream plant responses to abiotic stress. Int. J. Mol. Sci. 13 (7), 8628–8647.

Ricci, M., Tilbury, L., Daridon, B., Sukalac, K., 2019. General principles to justify plant biostimulant claims. Front Plant Sci. 10, 1–8.

Rouphael, Y., Colla, G., 2020. Biostimulants in agriculture. Front. Plant. Sci. 11, 40.

Rouphael, Y., Franken, P., Schneider, C., Schwarz, D., Giovannetti, M., Agnolucci, M., et al., 2015. Arbuscular mycorrhizal fungi act as biostimulants in horticultural crops. Sci. Hortic. 196, 91–108.

Rouphael, Y., De Micco, V., Arena, C., Raimondi, G., Colla, G., De Pascale, S., 2017. Effect of Ecklonia maxima seaweed extract on yield, mineral composition, gas exchange, and leaf anatomy of zucchini squash grown under saline conditions. J. Appl. Phycol. 29, 459–470.

Rouphael, Y., Giordano, M., Cardarelli, M., Cozzolino, E., Mori, M., Kyriacou, M.C., et al., 2018. Plant- and seaweed-based extracts increase yield but differentially modulate nutritional quality of greenhouse spinach through biostimulant action. Agronomy 8, 126.

Rubin, R.L., van Groenigen, K.J., Hungate, B.A., 2017. Plant growth promoting rhizobacteria are more effective under drought: a meta-analysis. Plant. Soil. 416, 309–323.

Rui, M., Ma, C., Hao, Y., Guo, J., Rui, Y., Tang, X., et al., 2016. Iron oxide nanoparticles as a potential iron fertilizer for peanut (*Arachis hypogaea*). Front. Plant. Sci. 7, 815.

Ruzzi, M., Aroca, R., 2015. Plant growth-promoting rhizobacteria act as biostimulants in horticulture. Sci. Hortic. 196, 124–134.

Sah, S., Sorooshzadeh, A., Rezazadehs, H., Naghdibadi, H.A., 2011. Effect of nano silver and silver nitrate on seed yield of borage. J. Med. Plants Res. 5, 171–175.

Saia, S., Colla, G., Raimondi, G., Di Stasio, E., Cardarelli, M., Bonini, P., et al., 2019. An endophytic fungi-based biostimulant modulated lettuce yield, physiological and functional quality responses to both moderate and severe water limitation. Sci. Hortic. 256, 108595.

Sánchez-Gómez, R., Torregrosa, L., Zalacain, A., Ojeda, H., Bouckenooghe, V., Schneider, R., et al., 2019. Behavior of glycosylated aroma precursors in Microvine fruits after guaiacol foliar application. Sci. Hortic. 246, e1–e8.

Sangiorgio, D., Cellini, A., Donati, I., Pastore, C., Onofrietti, C., Spinelli, F., 2020. Facing climate change: application of microbial biostimulants to mitigate stress in horticultural crops. Agronomy 10, 794.

Sarkar, M.M., Mathur, P., Mitsui, T., Roy, S., 2022. A review on functionalized silica nanoparticle amendment on plant growth and development under stress. Plant. Growth. Regul. 1–17.

Sasidharan, S., Chen, Y., Saravanan, D., Sundram, K.M., Latha, L.Y., 2011. Extraction, isolation and characterization of bioactive compounds from plants' extracts. Afr. J. Tradit. Complement. Altern. Med. 8, 1.

Sawake, M.M., Moharil, M.P., Ingle, Y.V., Jadhav, P.V., Ingle, A.P., Khelurkar, V.C., et al., 2022. Management of phytophthora parasitica causing gummosis in citrus using biogenic copper oxide nanoparticles. J. Appl. Microbiol. 132, 3142–3154.

Seaman, D., 1990. Trends in the formulation of pesticides—an overview. Pestic. Sci. 29, 437–449.

Shabala, L., Mackay, A., Jacobsen Serik, Z., Shabala, S., 2012a. Oxidative stress protection and stomatal patterning as components of salinity tolerance mechanism in quinoa (*Chenopodium quinoa*). Physiol. Plant. 146, 26–38.

Shabala, L., Mackay, A., Tian, Y., Jacobsen, S.E., Zhou, D., Shabala, S., 2012b. Oxidative stress protection and stomatal patterning as components of salinity tolerance mechanism in quinoa (*Chenopodium quinoa*). Physiol. Plant. 146, 26–38.

Sharma, S.S., Dietz, K.J., 2006. The significance of amino acids and amino acid-derived molecules in plant responses and adaptation to heavy metal stress. J. Exp. Bot. 57, 711–726.

Sharma, H.S.S., Selby, C., Carmichael, E., McRoberts, C., Rao, J.R., Ambrosino, P., et al., 2016. Physicochemical analyses of plant biostimulant formulations and characterisation of commercial products by instrumental techniques. Chem. Biol. Technol. Agric. 3, 1–17.

Sharma, R., Kumar, A., Devgan, M., Kaur, A., Kaur, H., Choudhary, A., et al., 2022. Applications of nanostructured materials in agriculture: a review. Mat. Today Proceed. 69, 549–555.

Siddiqui, M.H., Al-Whaibi, M.H., 2014. Role of nano-SiO_2 in germination of tomato (*Lycopersicum esculentum* seeds Mill.). Saudi J. Biol. Sci. 21, 13–17.

Simpson, A.J., Kingery, W.L., Hayes, M.H., Spraul, M., Humpfer, E., Dvortsak, P., et al., 2002. Molecular structures and associations of humic substances in the terrestrial environment. Sci. Nat. 89, 84–88.

Smith, S.E., Read, D.J., 2010. Mycorrhizal Symbiosis, third ed. Academic Press, London.

Smith, S.E., Jakobsen, I., Grønlund, M., Smith, F.A., 2011. Roles of arbuscular mycorrhizas in plant phosphorus nutrition: interactions between pathways of phosphorus uptake in arbuscular mycorrhizal roots have important implications for understanding and manipulating plant phosphorus acquisition. Plant. Physiol. 156, 1050–1057.

Tegeder, M., 2012. Transporters for amino acids in plant cells: some functions and many unknowns. Curr. Opin. Plant. Biol. 15, 315–321.

Tejada, M., Rodríguez-Morgado, B., Paneque, P., Parrado, J., 2018. Effects of foliar fertilization of a biostimulant obtained from chicken feathers on maize yield. Eur. J. Agron. 96, 54–59.

Thao, H.T.B., Yamakawa, T., 2009. Phosphite (phosphorous acid): fungicide, fertilizer or bio-stimulator? Soil. Sci. Plant. Nutr. 55, 228–234.

Theocharis, A., Bordiec, S., Fernandez, O., Paquis, S., Dhondt-Cordelier, S., Baillieul, F., 2012. *Burkholderia phytofirmans* PsJN primes *Vitis vinifera* L. and confers a better tolerance to low nonfreezing temperatures. Mol. Plant. Microbe Interact. 25, 241–249.

Turan, M., Ekinci, M., Kul, R., Kocaman, A., Argin, S., Zhirkova, A.M., et al., 2022. Foliar applications of humic substances together with Fe/Nano Fe to increase the iron content and growth parameters of spinach (*Spinacia oleracea* L.). Agronomy 12, 2044.

Wang, Y.H., Irving, H.R., 2011. Developing a model of plant hormone interactions. Plant. Signal. Behav. 6, 494–500.

Yakhin, O.I., Lubyanov, A.A., Yakhin, I.A., Brown, P.H., 2017. Biostimulants in plant science: a global perspective. Front. Plant. Sci. 7, 2049.

Yang, W., Li, P., Guo, S., Fan, B., Song, R., Zhang, J., et al., 2017. Compensating effect of fulvic acid and superabsorbent polymer on leaf gas exchange and water use efficiency of maize under moderate water deficit conditions. Plant. Growth Regul. 83, 351–360.

Yuniati, N., Kusumiyati, K., Mubarok, S., Nurhadi, B., 2022. The role of moringa leaf extract as a plant biostimulant in improving the quality of agricultural products. Plants 11, 2186.

CHAPTER

2

Biobased nanomaterials and their role as biostimulants in plants growth, metabolic profile performance, and productions

Nadun H. Madanayake[1] *and Nadeesh M. Adassooriya*[2]

[1]**Department of Botany, Faculty of Science, University of Peradeniya, Peradeniya, Sri Lanka**
[2]**Department of Chemical and Process Engineering, Faculty of Engineering, University of Peradeniya, Peradeniya, Sri Lanka**

2.1 Introduction

Global agriculture has enhanced its productivity remarkably to compete against the ever-increasing food demand. However, rising demand has caused a serious effect on over-exploiting natural resources, influencing sustainable agriculture. Also, the repercussions of anthropogenic activities exert an influence on global climatic change (Puglia et al., 2021). This leads to changes in the weather patterns around the globe in a long-standing manner. Global climate changes have consequences for the environment, such as extreme droughts, water shortages, severe fires, rises in sea levels, flooding, landslides, melting polar glaciers, and storms. Hence, this shift has a detrimental effect on all living organisms, including plants. Also, this leads to the development of better conditions to reduce plant vigor. Also, infections from pests and pathogens, including bacteria, fungi, and viruses, as well as temperature, pollution, salinity, soil pH, soil salinity, water scarcity, and extreme weather conditions due to biotic and abiotic stress, can change the plant physiology immensely (Husen, 2021a,b, 2022a,b; Iriti and Vitalini, 2020; Jiménez-Arias et al., 2019; Karkanis et al., 2018; Läuchli and Grattan, 2012; Zinn et al., 2010). The new paradigm in the environment leads to a new set of biotic and abiotic stresses on plants, expecting a decline in crop productivity (Coleman-Derr and Tringe, 2014; Puglia et al.,

Biostimulants in Plant Protection and Performance
DOI: https://doi.org/10.1016/B978-0-443-15884-1.00024-5

2021). Therefore it is important to have better strategies to sustainably protect crops from adverse environmental conditions. Conventional approaches such as the application of agrochemicals as well as genetic engineering may have certain limits to tolerate these challenges. For instance, salt, drought, and thermal-resistant cultivars have been developed using biotechnological and genetic engineering strategies (Ashraf and Akram, 2009; Bita and Gerats, 2013; Rani et al., 2021; Raza, 2021; Singh et al., 2010; Turan et al., 2012). A symbiont-based approach is another strategy to mitigate these environmental stresses (Coleman-Derr and Tringe, 2014). However, these approaches have been utilized for decades and are time-consuming, labor-intensive, and, most of the time, costly. Hence, it requires novel strategies to minimize the pressure incurred on plants biotically and abiotically in an economical manner. Recently, biostimulant application to enhance the strength of plants against any biotic and abiotic stress has been identified as a hot topic among scientists (Jiménez-Arias et al., 2020).

Nanotechnology has also been utilized to enhance plant vigor against these stresses. Nanotechnology deals with nanoparticles (NPs), which are materials in the range of 1−100 nm, at least from one of their dimensions. The inherent properties of NPs have enabled us to utilize them as delivery systems as well, and certain nanomaterials (NMs) themselves have the potential to enhance plant resistance. NMs have been reported to enhance seed germination, seedling growth, and plant metabolism, thus upregulating the expression of genes related to plant growth improvement. In addition, NMs have the potential to enhance resistance against environmental stresses (Husen, 2020a,b, 2023; Husen and Iqbal, 2019; Jalil and Ansari, 2019; Saxena et al., 2016; Siddiqi and Husen, 2022; Singh and Husen 2019, 2020). Therefore this proves the potential of NMs to act as biostimulants to enhance crop productivity. At present, the focus on implying biobased NPs as biostimulants has been blossoming as a potential mechanism to enhance plant vigor. NMs derived from agricultural wastes, plants, and animal wastes have extensively been utilized to synthesize biobased NMs. Hence, this chapter will elaborate on the use of biobased NMs as biostimulants in plants.

2.2 Biostimulants

Plant biostimulants are used to promote plant vigor against biotic and abiotic stresses. They are promising candidates for promoting plant growth by inducing root and shoot growth, plant nutrient uptake, and the biogenic synthesis of plant growth regulators. Also, biostimulants are reported to lessen the application of inorganic fertilizers (Ali et al., 2020; Hernández-Herrera et al., 2018; Hurtado et al., 2021; Shahrajabian et al., 2021; Zohara et al., 2019). Biostimulants are of diverse origins ranging from biological to non-biological as well as microbial to nonmicrobial. In addition, based on their action biostimulants can also be either plant growth regulators or none (Shahrajabian et al., 2021). Several classes of compounds have been recorded as biostimulants capable of regulating plant physiology against biotic and abiotic stresses while enhancing their growth. Humic acids, fulvic acid, protein hydrolysates, seaweed extracts, chitosan, chitin, polyamines, amino acids, alginate, inorganic compounds, beneficial bacteria, microalgae, fungi, and inorganic compounds are some examples of biostimulants (Amirkhani et al., 2016; Arslan et al., 2021; Bayat et al.,

2021; Casadesús et al., 2019; Colla et al., 2015; Dima et al., 2019; Do Rosário Rosa et al., 2021; Gemin et al., 2019; Kopta et al., 2018; López-Bucio et al., 2015; Novello et al., 2021; Saia et al., 2019). Therefore biostimulants can be recommended as natural and environmentally safe substitutes to synthetic plant growth regulators and agrochemicals such as pesticides for sustainable crop productivity.

Humins and fulvic acid-like molecules have the biostimulant potential to be applied in different agro-formulations to assist in plant development. Canellas et al. (2015) stated that biostimulants based on humic substances are employed to enhance crop productivity in horticulture applications. Humic substances are compounds generated *via* chemical, biological, and microbial degradation of organic matter of plant and animal origin. These transformed materials represent a major proportion of organic carbon in the earth's crust. Humus significantly contributes to the regulation of vital ecophysiological processes in the environment.

Unique properties of bioactive components in seaweed extracts have phytostimulatory effects on plant growth and protection. Bioactive components of seaweed extracts have the potential to induce plant defense responses against biotic and abiotic stress. Sulfated polysaccharides such as fucoidan, sulfated galactan, laminarin in brown algae, mannitol and alginates, proteins, polyunsaturated fatty acids, pigments, polyphenols, minerals, and plant growth regulators such as zetaine, indole-3-propionic acid, indole-3-acetic acid, gibberellins, abscisic acid like lunularic acid, and betaines are found as phytostimulatory bioactives in seaweed extracts (Chojnacka et al., 2012; Fleurence, 2022). Mainly, seaweed extracts have vital effects on shoot and root growth, and the presence of auxin and auxin-like bioactive agents has stimulatory effects on plants (Fleurence, 2022). Hence, this has become a popular and upcoming crop management model in agriculture. For instance, foliar application of *Kappaphycus alvarezii* extracts on tomatoes significantly increased the yield while enhancing the fruit quality and macro- and micronutrient contents (Zodape et al., 2011). Ashour et al. (2021) tested the effectiveness of commercial seaweed liquid extract (True-Algae-Max (TAM)) on *Capsicum annuum*. It was reported that TAM is rich in phytochemical compounds, such as ascorbic acid, phenolics, and flavonoids, showing a significant effect on improving plant growth. Enhancing plant strength or crop productivity *via* biostimulant application generally requires a regular application of these materials to a field. This is ineffective and costly because of the possible microbial degradation of active components. In agriculture, we use nanoencapsulation strategies to deliver plant nutrients as well as pesticides. In the same way, nanoencapsulation approaches can be used to deliver active biostimulants within an inert cargo to protect and slowly release them, thus preventing their environmental degradation. However, it is yet to be explored systematically (Jiménez-Arias et al., 2020).

2.3 Biobased nanomaterials

Biobased NMs, or biological NPs, have emerged as a promising approach in many industrial applications. Biobased NMs are nanometric particles derived from renewable natural resources. These materials mainly include monomer unit materials, fibers, and resins, which may augment the fabrication of new materials with superior characteristics.

Also, their inherent properties—eco-friendliness, renewability, biocompatibility, and lower carbon footprint—develop them as revolutionary biodegradable raw materials (Bagal-Kestwal et al., 2016). Currently, biobased NMs are utilized in cancer therapy, medicine, energy storage, electronics, textiles, food, agriculture, etc. (Al-Tayyar et al., 2020; Barik et al., 2022; Khandelwal et al., 2019; Li et al., 2021; You et al., 2017; Yu et al., 2014; Zare et al., 2019). At present, studies on the application of biobased NMs themselves or as modified platforms have been utilized as biostimulants in agriculture. Biobased NMs derived from chitosan, chitin, cellulose, silica, and lignin have been identified to enhance plant strength against different modes of stress. Cellulose is a linear glucose polymer linked *via* β-1–4-glycosidic bonds that has been used to develop nanocellulose particles. Nanocellulose can be synthesized mainly using plant-based materials and via bacteria (Bagal-Kestwal et al., 2016; De Amorim et al., 2020; Gatenholm and Klemm, 2010). Chitin is a derivative of glucose with N-acetylglucosamine as the structural unit. Chitin is a principal polyose found in fungi and in the exoskeletons of crustaceans and insects. Chitosan is the *N*-deacetylated product of chitin, which has been widely employed in many industrial applications (Bagal-Kestwal et al., 2016).

Lignin is the second-most biopolymer in nature and consists of approximately 30% of wood biomass. It is a complex biopolymer consisting of a web of aromatic structures derived from phenylpropanoid monomers, which mainly consist of p-hydroxyphenyl, guaiacyl, or syringyl. The proportion of these units varies with different plant species (Del Buono et al., 2021; Schneider et al., 2021; Yang et al., 2018). Also, lignin-based bio-NPs are emerging as functional NMs for numerous applications, especially due to their numerous functional moieties. In addition, lignin can offer numerous functionalities, such as ultraviolet protection, antioxidant, and antimicrobial properties. Hence, this has been exponentially studied for developing nanocarriers (Santo Pereira et al., 2022).

2.4 Nanochitosan as biostimulant

Kumaraswamy et al. (2019) developed salicylic acid-functionalized chitosan NPs as a biostimulant system to promote plant defense and enhance the plant growth of maize plants. It was reported that salicylic acid chitosan NPs expressed significant levels of elevated antioxidant-defense enzyme activities, balancing reactive oxygen species, cell wall reinforcement by lignin deposition, disease control, and plant growth in maize. Field experiments have shown that the new nanocomposite substantially increased disease management against post-flowering stalk rot disease while simultaneously enhancing the maize yield. Sharma et al. (2020) tested chitosan functionalized with copper and salicylic acid in maize. It was found that the chitosan nanocomposite enhanced the seedling vigor index and reserve food mobilizing enzymes in seedlings, whereas foliar application remarkably amplified the antioxidant enzyme activity and chlorophyll content while reducing the malondialdehyde content in test plants. Also, this composite induced sucrose translocation in internodes, proving its potential as a biostimulation agent in agricultural applications. Pereira et al. (2019) developed alginate/chitosan and chitosan/tripolyphosphate nanocarriers to deliver gibberellic acid. Field trials with the nanocarrier reported a significant enhancement in *Solanum lycopersicum* yield upon seed priming. Oliveira et al.

(2016) fabricated S-nitroso-mercaptosuccinic acid encapsulated in chitosan NPs to deliver nitric oxide (NO) as a signaling molecule involved in plant response against abiotic stresses. An investigation of the performance of the nanocarrier observed the alleviation of salt stress on maize plants.

2.5 Nanocellulose as biostimulant

Sherif and Hedia (2022) synthesized urea-loaded nanocellulose using rice straw to assess the effect on *Triticum aestivum* seeds. The biobased nanocomposite was developed using bleaching, acid hydrolysis, and flocculation processes, and it was found that primed *Triticum aestivum* seeds with nanocellulose had significantly increased root length, surface area, and vigor index with respect to bulk cellulose treatments. Wang et al. (2021) developed a pH-sensitive nanocellulose/sodium alginate/metal nanocomposite for the slow release of nitrogen. It was reported that a pH-dependent slow release of urea was observed. Also, the newly developed composite had the slowest release rate of urea at pH 11, indicating the potential to applied as a nanoformulation in arid regions. In addition, the composite thus developed shows a promising soil water holding capacity and a significant impact on the growth of wheat with respect to germination rate, number of tillers, photosynthetic rate, and chlorophyll content. Bauli et al. (2021) fabricated nanocellulose-carboxymethyl cellulose hydrogels for the soil conditioning and slow release of NPK fertilizers. It was reported that the incorporation of nanocellulose densified the hydrogel nanocomposite to reduce the water absorption capacity. Also, the successful encapsulation of NPK fertilizers showed a slow and continuous release with the Fickian diffusion mechanism. Moreover, pot trials using cucumber as a test plant have shown an effective impact on plant growth while simultaneously mitigating nutrient loss while enhancing water use in agriculture. Desliu-Avram et al. (2019) fabricated bacterial nanocellulose-chitosan nanoemulsions comprising humic and fulvic acids obtained *via* lignocellulosic biomass as biostimulants.

2.6 Lignin nanoparticles as biostimulant

Recently, few studies have focused on exploring lignin NPs as biostimulants to enhance plant vigor against biotic and abiotic stresses. Del Buono et al. (2021) recently investigated the effect of lignin NPs on maize. It was reported that lignin NPs positively affected maize germination, primary growth, chlorophyll, carotenoid, and anthocyanin contents, and soluble protein contents. Salinas et al. (2021) evaluated the effect of zein and lignin-based polymeric lignin-based NPs on soybean health. It was reported that lignin NPs did not show a positive effect on root and shoot elongation, chlorophyll, and stem biomass.

2.7 Biosilicon nanoparticles as biostimulant

Although silicon is not an essential plant nutrient, it has been widely applied as a plant protectant to deliver numerous molecular cargos. Also, these are regarded as a group of

TABLE 2.1 Effects of some of the reported biobased nanomaterials as biostimulants on plants.

Biobased nanomaterial	Test plant/	Key effects	References
Chitin nanofiber	Tomato	Shoot biomass and chlorophyll contents were significantly increased when hydroponically cultivated under ultralow contents of nutrients. Nitrogen and carbon contents were remarkably amplified in tomatoes. Transcriptomics showed the upregulation of genes responsible for nitrogen acquisition and assimilation, nutrient allocation, and photosynthesis.	Egusa et al. (2020)
Chitin nanofiber	Cabbage and strawberry	Chitin nanofibers were capable of inducing elicitor activity in cabbage and strawberry against *Alternaria brassicicola* and *Colletotrichum fructicola*. Defense-related genes were significantly upregulated in cabbage plants before and after pathogen infection.	Parada et al. (2018)
Protein/$CaCO_3$/chitin nanofiber	Tomato	Nanofiber composites improved the growth of tomatoes in a hydroponics system with respect to the leaf number and stem diameter.	Aklog et al. (2016)
Chitosan NPs	Onion	Chitosan NPs at 50 ppm had significantly increased plant growth characteristics such as plant height, number of leaves, biomass, and specific leaf area. Also, these treatments had positively influenced the bulb quality parameters.	Geries et al. (2020)
Chitosan/tripolyphosphate (TPP) NPs	*Capsicum annuum*	Growth and biomass accumulations of *Capsicum annuum* were improved with the application of biobased nanocomposites. In addition, peroxidase, and catalase activities in *Capsicum annuum* were significantly increased in a concentration and organ-dependent manner. Furthermore, soluble phenols, proline, and alkaloid contents were noticeably enhanced.	Asgari-Targhi, et al. (2018)
Chitosan NPs loaded with indole-3-acetic acid	Lettuce	Application of indole-3-acetic acid containing chitosan NPs on lettuce increased the number of leaves by 30.9%.	Valderrama et al. (2020)
Chitosan-titania nanobiocomposites	*Vicia faba L.*	Disease severity was strikingly reduced against *Bean yellow mosaic virus* while improving the plant growth indices, photosynthetic pigments, membrane stability index, and relative water content of *Vicia faba L.* In addition, antioxidant content and soluble protein contents were **enhanced, whereas electrolyte leakage, hydrogen peroxide, and lipid** peroxidation activities of the test plants were reduced.	Sofy et al. (2020)

Chitosan- graphene oxide nanocomposite	Eggplant	The nematocidal activity of nanocomposites against *Meloidogyne incognita* showed that chitosan-graphene oxide had reduced 85.42%, 75.3%, 55.5%, 87.81%, and 81.32% in the numbers of L2 larvae in galls, females, egg masses, and the developmental stage, respectively. Also, chitosan-graphene oxide enhanced the chlorophyll a, chlorophyll b, total phenols, and free proline contents in eggplants. Moreover, chitosan-graphene oxide increased the catalase (98.3%), peroxidase (97.52%), polyphenol oxidase (113.8%), and superoxide dismutase (42.43%) activities of test plants.	Attia et al. (2021)
Chitosan, chitosan copper oxide and chitosan-zinc oxides, chitosan-silver NPs	Chickpea.	Nanocomposites depicted a better antifungal activity against *Fusarium oxysporum* while promoting the growth of chickpea plants.	Kaur et al. (2018)
Alginate/chitosan and chitosan/tripolyphosphate NPs loaded with gibberellic acid	*Phaseolus vulgaris*	The nanocomposites significantly improved the leaf area and chlorophyll and carotenoid contents.	Santo Pereira et al. (2017)
Lignin NPs	Maize	Lignin NP positively induces seedling growth. In addition, chlorophyll (a and b), carotenoid, and anthocyanin. Soluble protein contents showed a dose-dependent enhancement.	Del Buono et al. (2021)

plant biostimulants and are used for crop protection activities. The content of silicon in plants differs according to the plant species, and this variation is attributed to the specificity of silicon uptake by plants. Studies have proven the biostimulant potential of silicon-based NPs. For instance, Hidalgo-Santiago et al. (2021) evaluated the effect of silicon-based biostimulants against water stress in lettuce plants. It was observed that the silicon composite significantly enhanced the lettuce growth while reducing the lipid peroxidation and H_2O_2 levels and conserving the photosynthesis operation. Hassan et al. (2022) describe nanosilicon as capable of enhancing the growth, improvement, and efficiency of drought-stressed Kalamata' olive trees. Particularly certain biomass wastes are highly rich in silicon and can be used as better sources for synthesis. Biological sources such as rice husks, sugarcane bagasse, groundnut shells, bamboo leaves, sorghum husks, spelt husks, and horsetail have been extensively studied in synthesizing silicon bio-NPs. Despite numerous studies on the biostimulant properties of silicon-based NPs, biostimulant utilization of soluble silica and silica bio-NPs is underutilized. This is mainly due to a lack of data on the availability of soluble silicon, analyzing the silicon availability in soil, and not accepting silicon as an essential element for crops (Guerriero et al., 2021). Table 2.1 summarizes the effects of some of the reported biobased NMs as biostimulants on plants.

2.8 Conclusion and future perspectives

Biobased NMs have emerged as promising, environmentally safe, and sustainable materials derived from renewable natural resources. Biobased NMs consist of monomer unit materials, fibers, and resins combined to form new materials with superior characteristics. Therefore their unique physicochemical properties have enabled them to be employed in numerous industrial applications. Anthropogenic activities have a huge impact on the global climate. Hence, this becomes the major cause of global climatic change and has adverse effects on plant physiology. This leads to diminished the plant's strength to cope with strikes due to both abiotic and biotic stress. Several conventional approaches have been used to address these issues; however, each strategy has its own merits and demerits. Recently, people have identified certain metabolites, compounds, and extracts from the environment that have the potential to induce phytostimulatory effects to counteract biotic and abiotic stress experienced by plants. These molecules, or compounds, are simply defined as biostimulants. Although biostimulants show a promising effect on plant protection, they generally require a regular application to a field. Practically, this is ineffective and costly because of possible microbial degradation of active components. At present scientists have a huge attention on the application of biobased NMs such as chitin, chitosan, cellulose, lignin, and silica to be used as plant stimulants. Still, this is a budding area that requires proper and systemic studies to identify their biostimulant potential. Moreover, it is important to understand the mode of phytostimulatory action on plant systems against biotic and abiotic stress. Even though biobased NMs can be a promising material to upregulate plant vigor there are concerns regarding the potential toxicity towards plants and the environment.

References

Aklog, Y.F., Egusa, M., Kaminaka, H., Izawa, H., Morimoto, M., Saimoto, H., et al., 2016. Protein/$CaCO_3$/chitin nanofiber complex prepared from crab shells by simple mechanical treatment and its effect on plant growth. Int. J. Mol. Sci. 17 (10), 1600.

Ali, M.K.M., Critchley, A.T., Hurtado, A.Q., 2020. The impacts of AMPEP K^+ (*Ascophyllum* marine plant extract, enhanced with potassium) on the growth rate, carrageenan quality, and percentage incidence of the damaging epiphyte Neosiphonia apiculata on four strains of the commercially important carrageenophyte Kappaphycu s, as developed by micropropagation techniques. J. Appl. Phycology 32, 1907–1916.

Al-Tayyar, N.A., Youssef, A.M., Al-Hindi, R., 2020. Antimicrobial food packaging based on sustainable bio-based materials for reducing foodborne pathogens: a review. Food Chem. 310, 125915.

Amirkhani, M., Netravali, A.N., Huang, W., Taylor, A.G., 2016. Investigation of soy protein-based biostimulant seed coating for broccoli seedling and plant growth enhancement. HortScience 51 (9), 1121–1126.

Arslan, E., Agar, G., Aydin, M., 2021. Humic acid as a biostimulant in improving drought tolerance in wheat: the expression patterns of drought-related genes. Plant. Mol. Biol. Report. 39 (3), 508–519.

Asgari-Targhi, G., Iranbakhsh, A., Ardebili, Z.O., 2018. Potential benefits and phytotoxicity of bulk and nano-chitosan on the growth, morphogenesis, physiology, and micropropagation of *Capsicum annuum*. Plant. Physiol. Biochem. 127, 393–402.

Ashour, M., Hassan, S.M., Elshobary, M.E., Ammar, G.A., Gaber, A., Alsanie, W.F., et al., 2021. Impact of commercial seaweed liquid extract (TAM®) biostimulant and its bioactive molecules on growth and antioxidant activities of hot pepper (*Capsicum annuum*). Plants 10 (6), 1045.

Ashraf, M., Akram, N.A., 2009. Improving salinity tolerance of plants through conventional breeding and genetic engineering: an analytical comparison. Biotechnol. Adv. 27 (6), 744–752.

Attia, M.S., El-Sayyad, G.S., Abd Elkodous, M., Khalil, W.F., Nofel, M.M., Abdelaziz, A.M., et al., 2021. Chitosan and EDTA conjugated graphene oxide antinematodes in eggplant: toward improving plant immune response. Int. J. Biol. Macromolecules 179, 333–344.

Bagal-Kestwal, D.R., Kestwal, R.M., Chiang, B.H., 2016. Bio-based nanomaterials and their bionanocomposites. Nanomaterials Nanocomposites: Zero- to Three-Dimensional Materials and Their Composites. pp. 255–330.

Barik, B., Maji, B., Sarkar, D., Mishra, A.K., Dash, P., 2022. Cellulose-based nanomaterials for textile applications. Bio-Based Nanomaterials. Elsevier, pp. 1–19.

Bauli, C.R., Lima, G.F., de Souza, A.G., Ferreira, R.R., Rosa, D.S., 2021. Eco-friendly carboxymethyl cellulose hydrogels filled with nanocellulose or nanoclays for agriculture applications as soil conditioning and nutrient carrier and their impact on cucumber growing. Colloids Surf. A 623, 126771.

Bayat, H., Shafie, F., Aminifard, M.H., Daghighi, S., 2021. Comparative effects of humic and fulvic acids as biostimulants on growth, antioxidant activity and nutrient content of yarrow (*Achillea millefolium* L.). Sci. Horticulturae 279, 109912.

Bita, C.E., Gerats, T., 2013. Plant tolerance to high temperature in a changing environment: scientific fundamentals and production of heat stress-tolerant crops. Front. plant. Sci. 4, 273.

Canellas, L.P., Olivares, F.L., Aguiar, N.O., Jones, D.L., Nebbioso, A., Mazzei, P., et al., 2015. Humic and fulvic acids as biostimulants in horticulture. Sci. horticulturae 196, 15–27.

Casadesús, A., Polo, J., Munné-Bosch, S., 2019. Hormonal effects of an enzymatically hydrolyzed animal protein-based biostimulant (Pepton) in water-stressed tomato plants. Front. Plant. Sci. 10, 758.

Chojnacka, K., Saeid, A., Witkowska, Z., Tuhy, L., 2012. Biologically active compounds in seaweed extracts-the prospects for the application. Open Conf. Proc. J. 3 (1), 20–28.

Coleman-Derr, D., Tringe, S.G., 2014. Building the crops of tomorrow: advantages of symbiont-based approaches to improving abiotic stress tolerance. Front. Microbiol. 5, 283.

Colla, G., Nardi, S., Cardarelli, M., Ertani, A., Lucini, L., Canaguier, R., et al., 2015. Protein hydrolysates as biostimulants in horticulture. Sci. Horticulturae 196, 28–38.

De Amorim, J.D.P., de Souza, K.C., Duarte, C.R., da Silva Duarte, I., de Assis Sales Ribeiro, F., Silva, G.S., et al., 2020. Plant and bacterial nanocellulose: production, properties and applications in medicine, food, cosmetics, electronics and engineering. A review. Environ. Chem. Lett. 18, 851–869.

Del Buono, D., Luzi, F., Puglia, D., 2021. Lignin nanoparticles: a promising tool to improve maize physiological, biochemical, and chemical traits. Nanomaterials 11 (4), 846.

Desliu-Avram, M., Dima, S.O., Turcanu, A.A., Radu, E., Stanciuc, A.M., Doncea, S.M., et al., 2019. Nanoemulsions based on biopolymers loaded with humic and fulvic acids derived from hydrothermally treated biomass. Multidiscip. Digital Publ. Inst. Proc. 29 (1), 82.

Dima, S.O., Turcanu, A.A., Doncea, S.M., Faraon, V., Radu, E., Moraru, A., et al., 2019. Chitosan nanoparticles stabilized with gallic acid, never-dried bacterial nanocellulose, and alginate have biostimulant potential for plants. Multidiscip. Digital Publ. Inst. Proc. 29 (1), 59.

Do Rosário Rosa, V., Dos Santos, A.L.F., da Silva, A.A., Sab, M.P.V., Germino, G.H., Cardoso, F.B., et al., 2021. Increased soybean tolerance to water deficiency through biostimulant based on fulvic acids and *Ascophyllum nodosum* (L.) seaweed extract. Plant. Physiol. Biochem. 158, 228–243.

Egusa, M., Matsukawa, S., Miura, C., Nakatani, S., Yamada, J., Endo, T., et al., 2020. Improving nitrogen uptake efficiency by chitin nanofiber promotes growth in tomato. Int. J. Biol. Macromolecules 151, 1322–1331.

Fleurence, J., 2022. Biostimulant potential of seaweed extracts derived from laminaria and ascophyllum nodosum. Biostimulants: Exploring Sources and Applications. Springer Nature Singapore, Singapore, pp. 31–49.

Gatenholm, P., Klemm, D., 2010. Bacterial nanocellulose as a renewable material for biomedical applications. MRS Bull. 35 (3), 208–213.

Gemin, L.G., Mógor, Á.F., Amatussi, J.D.O., Mógor, G., 2019. Microalgae associated to humic acid as a novel biostimulant improving onion growth and yield. Sci. Horticulturae 256, 108560.

Geries, L., Omnia, H.S., Marey, R.A., 2020. Soaking and foliar application with chitosan and nano chitosan to enhancing growth, productivity and quality of onion crop. Plant. Arch. 20 (2), 3584–3591.

Guerriero, G., Sutera, F.M., Torabi-Pour, N., Renaut, J., Hausman, J.F., Berni, R., et al., 2021. Phyto-courier, a silicon particle-based nano-biostimulant: evidence from Cannabis sativa exposed to salinity. ACS Nano 15 (2), 3061–3069.

Hassan, I.F., Ajaj, R., Gaballah, M.S., Ogbaga, C.C., Kalaji, H.M., Hatterman-Valenti, H.M., et al., 2022. Foliar application of nano-silicon improves the physiological and biochemical characteristics of 'Kalamata'olive subjected to deficit irrigation in a semi-arid climate. Plants 11 (12), 1561.

Hernández-Herrera, R.M., Santacruz-Ruvalcaba, F., Briceño-Domínguez, D.R., Filippo-Herrera, D., Andrea, D., Hernández-Carmona, G., 2018. Seaweed as potential plant growth stimulants for agriculture in Mexico. Hidrobiológica 28 (1), 129–140.

Hidalgo-Santiago, L., Navarro-León, E., López-Moreno, F.J., Arjó, G., González, L.M., Ruiz, J.M., et al., 2021. The application of the silicon-based biostimulant Codasil® offset water deficit of lettuce plants. Sci. Horticulturae 285, 110177.

Hurtado, A.Q., Neish, I.C., Ali, M.K.M., Norrie, J., Pereira, L., Michalak, I., et al., 2021. Extracts of seaweeds used as biostimulants on land and sea crops—an efficacious, phyconomic, circular blue economy: with special reference to Ascophyllum (brown) and Kappaphycus (red) seaweeds. Biostimulants for Crops from Seed Germination to Plant Development. Academic Press, pp. 263–288.

Husen, A., 2020a. Interactions of metal and metal-oxide nanomaterials with agricultural crops: an overview. In: Husen, A., Jawaid, M. (Eds.), Nanomaterials for Agriculture and Forestry Applications. Elsevier Inc, Cambridge, MA, pp. 167–197. Available from: https://doi.org/10.1016/B978-0-12-817852-2.00007-X.

Husen, A., 2020b. Carbon-based nanomaterials and their interactions with agricultural cropsIn: Husen, A., Jawaid, M. (Eds.), Elsevier Inc, Cambridge, MA, pp. 199–218. Available from: https://doi.org/10.1016/B978-0-12-817852-2.00008-1.

Husen, A., 2021a. Plant performance under environmental stress. In: *Hormones, Biostimulants and Sustainable Plant Growth Management*. Springer Nature Switzerland AG, Cham, pp. 1–606. Available from: https://doi.org/10.1007/978-3-030-78521-5.

Husen, A., 2021b. Harsh Environment and Plant Resilience: Molecular and Functional Aspects. Springer Nature Switzerland AG, Cham.

Husen, A., 2022a. Plants and Their Interaction to Environmental Pollution: Damage Detection, Adaptation, Tolerance, Physiological and Molecular Responses. Elsevier Inc, Cambridge, MA.

Husen, A., 2022b. Environmental Pollution and Medicinal Plants. Taylor & Francis Group, LLC.

Husen, H., 2023. Nanomaterials and Nanocomposites Exposures to Plants: Response, Interaction, Phytotoxicity and Defense Mechanisms. Springer Nature Singapore Pte Ltd., Singapore.

Husen, A., Iqbal, M., 2019. Nanomaterials Plant. Potential. Springer International Publishing, Cham. Available from: https://doi.org/10.1007/978-3-030-05569-1.

Iriti, M., Vitalini, S., 2020. Sustainable crop protection, global climate change, food security and safety—plant immunity at the crossroads. Vaccines 8 (1), 42.

Jalil, S.U., Ansari, M.I., 2019. Nanoparticles and Abiotic Stress Tolerance in Plants: synthesis, Action, and Signaling Mechanisms. In: Khan, I.R., Reddy, P.S., Khan, N. A. (Eds.), Plant Signaling Molecules: Role and Regulation under Stressful Environments, pp. 549–561. Available from: https://doi.org/10.1016/B978-0-12-816451-8.00034-4.

Jiménez-Arias, D., Garcia-Machado, F.J., Morales-Sierra, S., Luis, J.C., Suarez, E., Hernández, M., et al., 2019. Lettuce plants treated with L-pyroglutamic acid increase yield under water deficit stress. Environ. Exp. botany 158, 215–222. Available from: https://doi.org/10.1016/J.ENVEXPBOT.2018.10.034.

Jiménez-Arias, D., Morales-Sierra, S., Borges, A.A., Díaz Díaz, D., 2020. Biostimulant nanoencapsulation: the new keystone to fight hunger.

Karkanis, A., Ntatsi, G., Alemardan, A., Petropoulos, S., Bilalis, D., 2018. Interference of weeds in vegetable crop cultivation, in the changing climate of Southern Europe with emphasis on drought and elevated temperatures: a review. J. Agric. Sci. 156 (10), 1175–1185.

Kaur, P., Duhan, J.S., Thakur, R., 2018. Comparative pot studies of chitosan and chitosan-metal nanocomposites as nano-agrochemicals against fusarium wilt of chickpea (*Cicer arietinum* L.). Biocatalysis Agric. Biotechnol. 14, 466–471.

Khandelwal, A., Joshi, R., Mukherjee, P., Singh, S. D., Shrivastava, M., 2019. Use of bio-based nanoparticles in agriculture. Panpatte, D.G., Jhala Y.K. (Eds.), Nanotechnology for Agriculture: Advances for Sustainable Agriculture, pp. 89–100. Available from: https://doi.org/10.1007/978-981-32-9370-0_6/COVER

Kopta, T., Pavlikova, M., Sękara, A., Pokluda, R., Maršálek, B., 2018. Effect of bacterial-algal biostimulant on the yield and internal quality of lettuce (*Lactuca sativa* L.) produced for spring and summer crop. Not. Botanicae Horti Agrobotanici Cluj-Napoca 46 (2), 615–621.

Kumaraswamy, R.V., Kumari, S., Choudhary, R.C., Sharma, S.S., Pal, A., Raliya, R., et al., 2019. Salicylic acid functionalized chitosan nanoparticle: a sustainable biostimulant for plant. Int. J. Biol. Macromolecules 123, 59–69.

Läuchli, A., Grattan, S.R., 2012. Soil pH extremes. Plant Stress Physiology. CABI, Wallingford, pp. 194–209. Available from: https://doi.org/10.1079/9781845939953.0194.

Li, Y., Zheng, X., Chu, Q., 2021. Bio-based nanomaterials for cancer therapy. Nano Today 38, 101134.

López-Bucio, J., Pelagio-Flores, R., Herrera-Estrella, A., 2015. *Trichoderma* as biostimulant: exploiting the multilevel properties of a plant beneficial fungus. Sci. Horticulturae 196, 109–123.

Novello, G., Cesaro, P., Bona, E., Massa, N., Gosetti, F., Scarafoni, A., et al., 2021. The effects of plant growth-promoting bacteria with biostimulant features on the growth of a local onion cultivar and a commercial zucchini variety. Agronomy 11 (5), 888.

Oliveira, H.C., Gomes, B.C., Pelegrino, M.T., Seabra, A.B., 2016. Nitric oxide-releasing chitosan nanoparticles alleviate the effects of salt stress in maize plants. Nitric Oxide 61, 10–19.

Parada, R.Y., Egusa, M., Aklog, Y.F., Miura, C., Ifuku, S., Kaminaka, H., 2018. Optimization of nanofibrillation degree of chitin for induction of plant disease resistance: elicitor activity and systemic resistance induced by chitin nanofiber in cabbage and strawberry. Int. J. Biol. Macromolecules 118, 2185–2192.

Pereira, A.D.E.S., Oliveira, H.C., Fraceto, L.F., 2019. Polymeric nanoparticles as an alternative for application of gibberellic acid in sustainable agriculture: a field study. Sci. Rep. 9 (1), 1–10.

Puglia, D., Pezzolla, D., Gigliotti, G., Torre, L., Bartucca, M.L., Del Buono, D., 2021. The opportunity of valorizing agricultural waste, through its conversion into biostimulants, biofertilizers, and biopolymers. Sustainability 13 (5), 2710.

Rani, S., Kumar, P., Suneja, P., 2021. Biotechnological interventions for inducing abiotic stress tolerance in crops. Plant. Gene 27, 100315.

Raza, A., 2021. Eco-physiological and biochemical responses of rapeseed (*Brassica napus* L.) to abiotic stresses: consequences and mitigation strategies. J. Plant. Growth Regul. 40 (4), 1368–1388.

Saia, S., Colla, G., Raimondi, G., Di Stasio, E., Cardarelli, M., Bonini, P., et al., 2019. An endophytic fungi-based biostimulant modulated lettuce yield, physiological and functional quality responses to both moderate and severe water limitation. Sci. Horticulturae 256, 108595.

Salinas, F., Astete, C.E., Waldvogel, J.H., Navarro, S., White, J.C., Elmer, W., et al., 2021. Effects of engineered lignin-graft-PLGA and zein-based nanoparticles on soybean health. NanoImpact 23, 100329.

Santo Pereira, A.E., Silva, P.M., Oliveira, J.L., Oliveira, H.C., Fraceto, L.F., 2017. Chitosan nanoparticles as carrier systems for the plant growth hormone gibberellic acid. Colloids Surf. B 150, 141–152.

Santo Pereira, A.D.E., de Oliveira, J.L., Savassa, S.M., Rogério, C.B., de Medeiros, G.A., Fraceto, L.F., 2022. Lignin nanoparticles: new insights for a sustainable agriculture. J. Clean. Prod. 131145.

Saxena, R., Tomar, R.S., Kumar, M., 2016. Exploring nanobiotechnology to mitigate abiotic stress in crop plants. J. Pharm. Sci. Res. 8 (9), 974.

Schneider, W.D.H., Dillon, A.J.P., Camassola, M., 2021. Lignin nanoparticles enter the scene: a promising versatile green tool for multiple applications. Biotechnol. Adv. 47, 107685.

Shahrajabian, M.H., Chaski, C., Polyzos, N., Tzortzakis, N., Petropoulos, S.A., 2021. Sustainable agriculture systems in vegetable production using chitin and chitosan as plant biostimulants. Biomolecules 11 (6), 819.

Sharma, G., Kumar, A., Devi, K.A., Prajapati, D., Bhagat, D., Pal, A., et al., 2020. Chitosan nanofertilizer to foster source activity in maize. Int. J. Biol. macromolecules 145, 226–234.

Sherif, F., Hedia, R., M.R., 2022. Priming seeds with urea-loaded nanocellulose to enhance wheat (*Triticum aestivum*) germination. Alex. Sci. Exch. J. 43 (1), 151–160.

Siddiqi, K.S., Husen, A., 2022. Plant response to silver nanoparticles: a critical review. Crit. Rev. Biotechnol. 42 (7), 773–990. Available from: https://doi.org/10.1080/07388551.2021.1975091.

Singh, R.K., Redoña, E., Refuerzo, L., 2010. Varietal improvement for abiotic stress tolerance in crop plants: special reference to salinity in rice. Pareek, A., Sopory, S.K., Bohnert, H.J., Govindjee (Eds.), Abiotic Stress Adaptation in Plants: Physiological, Molecular and Genomic Foundation, pp. 387–415. Available from: https://doi.org/10.1007/978-90-481-3112-9_18.

Singh, S., Husen, A., 2019. Role of nanomaterials in the mitigation of abiotic stress in plants. In: Husen, A., Iqbal, M. (Eds.), Nanomaterials and Plant Potential. Springer International Publishing AG, Cham, pp. 441–471. Available from: https://doi.org/10.1007/978-3-030–05569-1_18.

Singh, S., Husen, A., 2020. Behavior of agricultural crops in relation to nanomaterials under adverse environmental conditions. In: Husen, A., Jawaid, M. (Eds.), Nanomaterials for Agriculture and Forestry Applications. Elsevier Inc, Cambridge, MA, pp. 219–256. Available from: https://doi.org/10.1016/B978-0-12-817852-2.00009-3.

Sofy, A.R., Hmed, A.A., Abd EL-Aleem, M.A., Dawoud, R.A., Elshaarawy, R.F., Sofy, M.R., 2020. Mitigating effects of Bean yellow mosaic virus infection in faba bean using new carboxymethyl chitosan-titania nanobiocomposites. Int. J. Biol. Macromolecules 163, 1261–1275.

Turan, S., Cornish, K., Kumar, S., 2012. Salinity tolerance in plants: breeding and genetic engineering. Australian J. Crop. Sci. 6 (9), 1337–1348.

Valderrama, A., Lay, J., Flores, Y., Zavaleta, D., Delfín, A.R., 2020. Factorial design for preparing chitosan nanoparticles and its use for loading and controlled release of indole-3-acetic acid with effect on hydroponic lettuce crops. Biocatalysis Agric. Biotechnol. 26, 101640.

Wang, Y., Shaghaleh, H., Hamoud, Y.A., Zhang, S., Li, P., Xu, X., et al., 2021. Synthesis of a pH-responsive nanocellulose/sodium alginate/MOFs hydrogel and its application in the regulation of water and N-fertilizer. Int. J. Biol. Macromolecules 187, 262–271.

Yang, W., Fortunati, E., Gao, D., Balestra, G.M., Giovanale, G., He, X., et al., 2018. Valorization of acid isolated high yield lignin nanoparticles as innovative antioxidant/antimicrobial organic materials. ACS Sustain. Chem. Eng. 6 (3), 3502–3514.

You, J., Li, M., Ding, B., Wu, X., Li, C., 2017. Crab chitin-based 2D soft nanomaterials for fully biobased electric devices. Adv. Mater. 29 (19), 1606895.

Yu, S., Jeong, S.G., Chung, O., Kim, S., 2014. Bio-based PCM/carbon nanomaterials composites with enhanced thermal conductivity. Sol. Energy Mater. Sol. Cell 120, 549–554.

Zare, E.N., Makvandi, P., Borzacchiello, A., Tay, F.R., Ashtari, B., Padil, V.V., 2019. Antimicrobial gum bio-based nanocomposites and their industrial and biomedical applications. Chem. Commun. 55 (99), 14871–14885.

Zinn, K.E., Tunc-Ozdemir, M., Harper, J.F., 2010. Temperature stress and plant sexual reproduction: uncovering the weakest links. J. Exp. Botany 61 (7), 1959–1968.

Zodape, S.T., Gupta, A., Bhandari, S.C., Rawat, U.S., Chaudhary, D.R., Eswaran, K., Chikara, J., 2011. Foliar application of seaweed sap as biostimulant for enhancement of yield and quality of tomato (*Lycopersicon esculentum* Mill.). JSIR 70 (03), 215–219. Available from: http://nopr.niscpr.res.in/handle/123456789/11089.

Zohara, F., Surovy, M.Z., Khatun, A., Prince, F.R.K., Ankada, A.M., Rahman, M., et al., 2019. Chitosan biostimulant controls infection of cucumber by Phytophthora capsici through suppression of asexual reproduction of the pathogen. Acta Agrobotanica 72 (1).

CHAPTER

3

Biobased nanomaterials and their interaction with plant growth-promoting rhizobacteria/blue-green algae/Rhizobium for sustainable plant growth and development

Imran Khan[1], *Ghazala Sultan*[2], *Sumaira Miskeen*[1], *Inamul Hasan Madar*[3], *Sara Najeeb*[1], *Pravitha Kasu Sivanandan*[4], *Ramachandran Chelliah*[3,5,6] and *Deog Hwan Oh*[5,6]

[1]Department of Food Science and Technology, The University of Haripur, Haripur, Khyber Pakhtunkhwa, Pakistan [2]Department of Computer Science, Faculty of Science, Aligarh Muslim University, Aligarh, Uttar Pradesh, India [3]Department of Pharmacology, Saveetha Medical and Dental College and Hospitals, Chennai, Tamil Nadu, India [4]Department of Cancer Biology & Therapeutics MVR Cancer Centre & Research Institute, Calicut, Kerala, India [5]Department of Food Science and Biotechnology, College of Agriculture and Life Sciences, Kangwon National University, Chuncheon, Gangwon-do, South Korea [6]Department of Food Science and Biotechnology, Kangwon Institute of Inclusive Technology (KIIT), Kangwon National University, Chuncheon, Gangwon-do, South Korea

3.1 Introduction—nanotechnology for sustainable agriculture

The utilization of advanced technologies, such as nanotechnology, appears to have the potential to revolutionize the agriculture industry (Husen, 2022, 2023; Sharma et al., 2021; Kumar et al., 2021, 2022). Nanotechnology applied to agriculture is referred to as

Biostimulants in Plant Protection and Performance
DOI: https://doi.org/10.1016/B978-0-443-15884-1.00021-X

nanoagriculture, and it is currently focused on targeted farming using nanosized particles like nanofertilizers. This approach provides unique tools for enhancing the productivity of agricultural plants by improving their nutrient uptake (Tarafdar et al., 2013; Egamberdieva et al., 2017). Compared to particles of larger scale, nanosized particles possess unique physiochemical properties that can protect plant life, diagnose plant illnesses, observe plant promotion, improve food quality, increase food's nutritional content, and reduce waste. Recent research has shown that nanofertilizers are much more efficient than traditional fertilizers, as they minimize nitrogen loss through leaching, minimize pollution, and promote long-term absorption by soil microbes. Furthermore, a study by Prasad et al. (2010) has demonstrated the benefits of using nanofertilizers. The study showed how sustained-release fertilizers could improve the topsoil by mitigating the negative effects of excessive use of conventional chemical fertilizers.

The conventional methods of applying plant growth promoting rhizobacteria (PGPR) as a fertilizer are ineffective because approximately 90% of them are lost to the atmosphere after application. Additionally, PGPR cannot withstand environmental factors such as temperature and ultraviolet radiation, increasing farmers' implementation costs due to runoff.

Using nanoencapsulation technology, PGPR's stability, composition, lifespan, and dispersion can be improved while protecting the bacteria. This allows for the controlled release of PGPR, thereby increasing its efficacy (Vejan et al., 2016).

3.2 Role of plant growth promoting rhizobacteria for plant growth enhancement

3.2.1 Plant growth promoting rhizobacteria

PGPR, a genus of bacteria, can be found in the region surrounding plant roots, known as the rhizosphere (Ahmad et al., 2008). PGPR has positive effects in reducing phytopathogenic bacteria and is connected to the root system to strengthen plant development (Kloepper et al., 1980; Son et al., 2014). The bacteria that inhabit the rhizosphere, the area surrounding plant roots, are referred to as "PGPR." These microorganisms can enhance plant growth by creating a favorable soil environment for root development. The high concentration of soil microorganisms in the rhizosphere leads to a depleted nutrient pool with low levels of most macro- and micronutrients (Burdman et al., 2000). The rhizosphere is populated by a rich diversity of microorganisms, including algae, *actinomycetes*, bacteria, fungi, and protozoa, with bacteria being the dominant group (Kaymak, 2011). It is widely recognized and acknowledged that utilizing these microbial populations can enhance plant growth (Saharan and Nehra, 2011; Bhattacharyya and Jha, 2012).

PGPRs have a significant role in the production of biofertilizers, and they can interact with plants. They can broadly be divided into two categories: unrestricted rhizobacteria, which live outside plant cells, and symbiotic bacteria, which exist within plants and directly transfer metabolites (Gray and Smith, 2005) (Table 3.1).

3.2.1.1 Mechanism of action of plant growth-promoting rhizobacteria

PGPR can directly and indirectly enhance plant development in a variety of ways (Ahmed et al., 2017). Direct methods include processes such as phosphate solubilization,

TABLE 3.1 List of plant growth promoting bacteria.

PGPR	PGPR mechanism	Crops	Mode of application	Findings	Key references
Azorhizobium	Nitrogen fixation	Wheat	Rhizobial cultures were added to the wheat plants four times at the time of seed planting.	After five weeks incubation with *A.caulinodonas* IRBG314 there will short lateral roots approx. five times having length of 3mm	Sabry et al. (1997)
Bacillus	Cytokinin synthesis	Cucumber and pepper	1mL Inoculum containing about 108 cells were used in seedling incubation	Findings shows that soil planted with cucumber and pepper have higher amount of K and P.	Han, Lee (2005, 2006)
Herbaspirillum	Nitrogen fixation	Rice	Seed was inoculated	At 7 days of age, O. officinalis W0012 seedlings exhibited GFP-tagged cells of the Herbaspirillum sp. strain B501gfp1 that appeared to be localized in the intercellular spaces between the shoot tissues.	Elbeltagy et al. (2001)
Phyllo bacterium	Siderophore production	Strawberries	A strawberry seedling was inoculated into 1 mL of 108 CFU/mL suspensions	The strain developed on the CAS indicator medium where the colonies had a 3.5 mm yellow-orange halo surrounding them, indicating the formation of siderophores.	Flores-Félix et al. (2015)
Gluconacetobacter	Nitrogen fixation	Sugar cane	Soak seedling roots for an hour	SEM was used to confirm that *G. diazotrophicus* had established itself endophytically into sugarcane stems.	Munoz-Rojas and Caballero-Mellado (2003)
Bacillus	Auxin synthesis	potato	Seed-dipping (1 CFU in 108 mL)	In comparison to non-inoculated plants, both strains increased the auxin concentration of inoculated plants by 71.4% and 433%, respectively.	Ahmed and Hasnain (2010)

nitrogen fixation, siderophore production, and the generation of plant hormones such as auxin, cytokinin, and gibberellin. Indirect methods, on the other hand, do not promote growth directly but rather support the growth process by producing volatile organic compounds and nonvolatile biocides and helping plants withstand bactericidal and abiotic stresses (Aloo et al., 2019; Parewa et al., 2018).

3.2.2 Direct on plant growth promotion factors

PGPR can improve plant growth by aiding the absorption of soil nutrients such as nitrogen, phosphate, and iron. PGPR also generates phytohormones such as cytokinins and ethylene, which help stimulate plant growth. These microorganisms can contribute to the overall health and productivity of crops by creating a favorable soil environment and promoting the uptake of essential nutrients (Glick, 2012; Kumar, 2016).

3.2.2.1 *Biological nitrogen fixation*

More than 80% of atmospheric nitrogen is an inert gas and cannot be taken up by plants (Patel and Minocheherhomji, 2018; Aloo et al., 2019). Nitrogen-fixing microorganisms use an intricate enzyme system known as nitrogenase to transform nitrogen into ammonia (nitrogen in such a form that plants can use) (Shaikh and Sayyed, 2014).

Biological nitrogen fixation is the process by which nearly two-thirds of the global nitrogen supply is fixed through symbiotic or free-living relationships between plants and bacteria to improve plant growth and increase yields (Gouda et al., 2018). Legumes rely on symbiotic nitrogen-fixing bacteria, such as *Bradyrhizobium*, *Mesorhizobium*, *Azorhizobium*, *Rhizobium*, *Allorhizobium*, and *Sinorhizobium*, to improve plant growth and increase yields. These bacteria are capable of fixing nitrogen and promoting plant development and are considered to be highly effective nitrogen-fixing PGPR (Vaikuntapu et al., 2014; Hayat et al., 2010). Free-living PGPR, such as *Burkholderia*, *Azotobacter*, *Bacillus*, *Paenibacillus*, and *Herbaspirillum*, have been shown to colonize plant root surfaces and promote plant growth. These bacteria can effectively adhere to plant roots and play a role in improving plant health (Ahmed et al., 2017).

3.2.2.2 *Phosphate solubilization*

Phosphorus, being a crucial macronutrient, plays a significant role in the growth and development of various plant species, including tomatoes. This element has a profound impact on various metabolic processes that are essential for the overall functioning of the plant, such as photosynthesis, cellular respiration, signal transduction pathways, and energy transfer mechanisms. These processes heavily depend on the presence and availability of phosphorus, and thus, its level can significantly impact the efficacy of these processes and, in turn, the overall health and productivity of the plant (Ahmed et al., 2017).

However, soil primarily contains phosphorus in the form of insoluble organic and inorganic phosphate. The solubilization of insoluble inorganic phosphate or the release of phosphates from organic molecules are important functions of phosphate-solubilizing bacteria (PSB). Additionally, plants can only absorb phosphate in the form of monobasic (HPO_4) and dibasic ($H_2PO_4^{2-}$) ions (Gouda et al., 2018).

The solubilization mechanism of PSB involves the release of low-molecular-weight organic acids, such as tartaric acid, acetate acid, gluconic acid, glycolic acid, succinic acid, ketogluconate, and lactic acid (Patel and Minocheherhomji, 2018). Phosphates are produced from organic compounds in several ways. Phosphatases break down phosphoester bonds; phytases function by generating phytic acid; phosphonates use Mg(II) as a cofactor to catalyze the breakdown of phosphonoacetaldehyde into acetaldehyde and phosphate; and C-P lyases break down phosphonates at the C-P position (Selvakumar et al., 2018; Morais et al., 2000).

There have been reports of the use of phosphate solubilizers such as *Azotobacter chroococcum, Bacillus anthina, Acidovorans delafieldii, P. putida, Bacillus megaterium, Enterobacter cloacae, P. fluorescens, Pseudomonas agglomerans, Microbacterium laevaniformans*, and *Pseudomonas cepacia* (Ahmed et al., 2017; Rai and Nabti, 2017).

3.2.2.3 Siderophores production

Therefore, it is crucial for the survival and growth of these microorganisms in the rhizosphere. To overcome this limitation, microorganisms have evolved mechanisms to acquire iron from the surrounding environment, such as producing siderophores (iron-binding compounds) and reducing iron from ferric to ferrous form, which is more easily solubilized and available for uptake. Iron is also a crucial element in various metabolic processes that are vital for the survival and growth of microorganisms (Gupta et al., 2017; Patel and Minocheherhomji, 2018). These siderophores are iron-binding molecules that are secreted by microorganisms such as PGPR and have a high affinity for iron. By chelating the iron and solubilizing it, siderophores make it easier for plants to absorb the iron, which is essential for their growth and development. The release of siderophores by PGPR helps to increase the availability of iron to plants and maintains the balance between plant and microbial growth in the rhizosphere ecosystem, thereby promoting its overall health and productivity (Singh et al., 2017). The production of siderophores by PGPR, including *Alcaligenes, Pseudomonas, Bradyrhizobium, Bacillus, Enterobacter*, and *Rhizobium*, helps them acquire iron, which is necessary for their growth and survival. The siderophores chelate and solubilize the complex iron, making it available to the plants and giving PGPR a competitive advantage over other microorganisms. The ability to produce siderophores is especially crucial in competitive environments where multiple carbon sources are available, as it may determine the outcome of competition for these resources. This process of iron acquisition by PGPR helps maintain the balance between plant and microbial growth and promotes the overall health and productivity of the rhizosphere ecosystem (Shaikh et al., 2016; Shaikh and Sayyed, 2014; Tsegaye et al., 2017).

3.2.2.4 Phytohormone production

Phytohormones are naturally occurring organic substances that, at very low concentrations, affect a range of physiological or structural processes in plants, including cell expansion and cell proliferation (Parewa et al., 2018). The production of phytohormones by PGPR, such as cytokinins and ethylene, has a significant impact on the metabolic activity of plants. This indirect stimulation can help enhance the plant's defense mechanisms and improve its ability to manage abiotic stress factors. By modifying the plant's metabolism and growth, PGPR-produced phytohormones can promote overall plant health and resilience.

PGPR secretes various phytohormones, including gibberellins (GA), abscisic acid, auxin, cytokinin, ethylene, and others, in response to major abiotic stress factors such as drought, salinity, heat, cold, flooding, and ultraviolet radiation. These stress factors pose a significant threat to crop production globally, leading to reduced yields. By secreting phytohormones, PGPR can help mitigate the impact of abiotic stress and promote plant growth and health under adverse conditions (Egamberdieva et al., 2017; Singh et al., 2019a,b).

GA, a type of phytohormone secreted by PGPR, has a significant impact on various physiological processes in plants. These processes include seed germination and emergence, floral induction, flower and fruit development, stem and leaf growth, and most notably, shoot elongation. GA plays a crucial role in plant growth and development and has been shown to enhance plant resilience under adverse conditions (Egamberdieva et al., 2017). Some PGPRs that are capable of producing GA include *Azospirillum lipoferum*, *Bacillus pumilus*, *Acetobacter diazotrophicus*, *Bacillus macrolides*, *Bacillus cereus*, *Acinetobacter calcoaceticus*, and *Herbaspirillum seropedicae*. By secreting GA, these PGPR can enhance plant growth and resilience, helping to mitigate the impact of abiotic stress factors such as drought, salinity, and extreme temperatures (Patel and Minocheherhomji, 2018; Vaikuntapu et al., 2014).

Cytokinins are a class of phytohormones that play a crucial role in regulating plant growth and development. They promote cell division, cell enlargement, and tissue growth in various plant tissues and play a critical role in balancing the growth-inhibiting effect of auxin. This delicate balance between auxin and cytokinin helps ensure the proper development of shoots and roots and enables plants to respond to environmental cues and stimuli. The effects of cytokinins are particularly noticeable in roots, stems, and leaves, where they stimulate cell division, elongation, and differentiation, promoting overall plant growth (Nasir, 2016). Cytokinins play a critical role in various physiological processes within plants, including seed germination, release of buds from apical dominance, leaf expansion, reproductive development, delay of senescence, increased cell division, improved root development, inhibited root elongation, shoot initiation, and other responses to exogenous application (Patel and Minocheherhomji, 2018).

3.2.2.5 Nutrient availability for plant uptake

PGPR can improve the availability of nutrients in the soil surrounding plant roots (rhizosphere) by reducing nutrient leaching. One important nutrient for plant growth is nitrogen, which is often a limiting factor. By stabilizing nitrogen, PGPR can promote plant growth. Only prokaryotes (single-celled organisms without a nucleus) can convert atmospheric nitrogen into forms that plants can absorb (Lloret and Martínez-Romero, 2005; Raymond et al., 2004).

One genus of PGPR is *Azospirillum*, a free-living nitrogen-fixing bacteria that is commonly found in association with cereal crops like rice in temperate regions. This bacterium can convert atmospheric nitrogen into forms that plants can use, serving as an alternative source of essential nitrogen compounds. The presence of *Azospirillum* in the rhizosphere has been shown to improve rice crop yields through increased growth and improved nutrient uptake. Although *Azospirillum* is widely distributed in temperate regions, it remains a relatively uncommon form of nitrogen-fixing bacteria, making its role in improving crop production even more significant (Tejera et al., 2005).

Because some PGPR can dissolve phosphate (Wani et al., 2007), the presence of *Kocuria turfanensis* strain 2M4, isolated from rhizospheric soil, has been found to increase the concentration of phosphate ions in the soil, thereby facilitating plant uptake of this essential nutrient (Goswami et al., 2014). Yadav et al. (2014) investigated how PGPR affected rice's ability to absorb nutrients. They employed *Pseudomonas fluorescens, Pseudomonas putida, and Pseudomonas fluorescens* PGPR strains.

3.2.2.6 *Plant growth regulators*

Artificial compounds that imitate the natural hormones produced by plants are referred to as "Plant Growth Regulators" or "Plant External Hormones." These synthetic substances have a similar effect on plants as the naturally occurring hormones, stimulating or inhibiting various aspects of plant growth and development. These are crucial tools for increasing agricultural output since they are employed to control plant development. Artificial compounds can mimic the effects of naturally occurring hormones and regulate various aspects of plant growth and development (Lugtenberg et al., 2002; Somers et al., 2004). Plant growth regulators are phytohormones that are synthesized externally by natural and synthetic means rather than by the plants themselves. Another important plant hormone is ethylene, which regulates various physiological processes such as seed germination, leaf abscission, and fruit ripening. Ethylene plays a crucial role in coordinating plant growth and development, and its levels can be artificially manipulated to induce specific responses in plants (Reid, 1988). High amounts of ethylene also cause defoliation, cell functions that limit stem and root development, premature senescence, and other effects that negatively impact crop production (Li et al., 2005; Glick, 2012). When under stress, legumes can produce large amounts of ethylene, which can stop nitrogen fixation and root elongation (Jackson, 1991), which leads to premature senescence (Ahmad et al., 2013).

3.2.2.7 *Production of enzymes*

The activity of growth regulators in generating protective enzymes might be categorized as a biopesticide. PGPR enhances plant growth by suppressing plant pathogenic agents, primarily by producing compounds with antibacterial and antifungal properties that serve as protective measures. These enzymes exhibit broad-spectrum antifungal activity and are crucial for mediating plant growth promotion by inhibiting the proliferation of phytopathogenic fungi. (Kumar et al., 2010). In addition to their chitinase and beta-glucanase production, *Pseudomonas* spp. also effectively inhibits two of the most destructive crop diseases globally, *Rhizoctonia solani* and *Phytophthora capsici*. These PGPR strains can inhibit these phytopathogens' growth and spread, promoting healthy plant growth and mitigating crop losses (Arora et al., 2008).

3.2.3 Indirect mechanisms on plant growth promotion factors

The development of chitinase and other enzymes, lipase, and protease that allow the dissolution of harmful fungi and bacteria, along with the initiation of induced resistance, are three ways that PGPR subsequently increases plant growth (Gouda et al., 2018; Butterbach-Bahl et al., 2013; Ryu et al., 2003).

Several PGPR strains produce volatile organic compounds (VOCs) that can directly or indirectly increase plant biomass production, disease resistance, and tolerance to abiotic stress. Despite species-specific differences in the type and quantity of volatile compounds emitted, VOC emission is a common property among a wide range of soil microbes (Effmert et al., 2012; Kanchiswamy et al., 2015).

3.2.3.1 *Nonvolatile biocidals (antibiotics and fungicidals)*

Even at low concentrations, the low-molecular-weight compounds generated by PGPR are antimicrobial (Fravel, 1988). PGPR is the preferred biological control agent for sustainable agriculture because of its chemicals. To control plant infections, PGPR creates nonvolatile chemicals such as phenazines, cyclic lipopeptides (CPLs), pyoluteorin, and phloroglucinols (Vacheron et al., 2013). According to reports, PGPR from the *Bacillus*, *Pseudomonas*, and *Streptomyces* genera has been utilized to combat plant diseases in various economically important plant yields (Meena et al., 2019; Almoneafy et al., 2021; Ngalimat et al., 2021). Bacterial strains generate phloroglucinols, a broad-spectrum antibiotic that causes systemic resistance among plants by acting as an odd phytoalexin elicitor (Dwivedi and Johri, 2003). Fluorescent pseudomonas strains that generate 2,4-diacetylphloroglucinol are connected to plant root protection against soil-borne phytopathogens (Weller et al., 2007). It is said to stop fungi like *Fusarium*, *Pythium*, *Rhizoctonia*, and *Alternaria* from causing diseases, including dampening off, rot disease, and wilting diseases (De Souza et al., 2003; Mcspadden Gardener, 2007; Muller et al., 2018). *Pseudomonas*, *Burkholderia*, *Brevibacterium*, and *Streptomyces* are the bacteria that produce phenazines, which are heterocyclic nitrogenous compounds (Chen et al., 2014; Dasgupta et al., 2015). They are renowned because of their ability to manage nematodes by way of plant fungal strains (Zhou et al., 2019; Muller et al., 2018). According to reports, fluorescent pseudomonads like *P. aeruginosa*, *P. chlororaphis*, and *P. fluorescens* generate phenazine-speculative phenazine-1-carboxylic acid, which is efficient against a variety of pathogenic bacteria and fungi like *Gaeumannomyces graminis*, *Pythium* sp., and *Polyporus* spp. (Saraf et al., 2014). It was discovered that three different CPLs, iturin, surfactin, and plipastatin-fengycin, were synthesized by the Bacillus species based on their amino acid composition and fatty acid branching (Farace et al., 2015). Numerous phytopathogens, including *Sclerotinia sclerotium*, *Botrytis cinerea*, *Xanthomonas axonopodis*, *Pseudomonas syringae*, *Colletotrichum gloeosporioides*, *species of Aspergillus Alternaria alternata, and Fusarium moniliforme*, are susceptible to these CPLs (Malviya et al., 2020). As a result, PGPR's generation of nonvolatile chemicals offers antagonistic agents for a variety of phytopathogens (Glick et al., 2007).

3.2.3.2 *Abiotic stress tolerance in plant*

Agriculture yield decline is thought to be mostly caused by abiotic stress. However, the degree of abiotic stress varies according to the parameters of the plant (physical abnormalities such as being vulnerable to illnesses and abscission) and soil types (absence of hormonal and dietary abnormalities) (Nadeem et al., 2010). Previous research on the plant's growth-promoting rhizobacteria processes in response to abiotic factors was substantial. In accordance with Pishchik et al. (2002), the negative impact of cadmium exposure on growing plants can be mitigated by the bacteria's ability to bind cadmium ions from the soil, thereby reducing its availability in the soil (Pishchik et al., 2002; Ahmad et al., 2013; Naveed et al., 2014).

P. aeruginosa strains have reportedly increased Vigna radiata (mung bean) plant development under dry circumstances, according to Sarma and Saikia (2014). The size of a plant's stomatal openings determines how well it can use water for growth.

The amount of water vapor that PGPR-inoculated plants were able to release through their stomata leaves in drought-like conditions was greater than that of non-PGPR-inoculated plants (Ahmad et al., 2013; Naveed et al., 2014). Both studies' results demonstrate that plants infected with PGPR tend to use less water. The environment may benefit from this finding's reduction in water consumption. According to Marulanda et al. (2010), the ability of maize stems to absorb water in salty environments was increased by *Bacillus megatertum* variant inoculation. A similar trend was seen by Gond et al. (2015), whereas corn roots were given *Pantoea agglomerans*. The capacity of the maize root to absorb water in saline environments has improved, they found. In this situation, microorganisms that can thrive in hypersaline environments will have an advantage in colonizing the root rhizospheres and exterior spaces of roots that are already subject to high salinity conditions. The initial step was to determine whether the bacterial isolates could survive in hypersaline settings by looking at them.

Azospirillum was also utilized by Fasciglione et al. (2015) to evaluate lettuce growth under salt stress. They found that inoculating lettuce with *Azospirillum* sp. during salinity stress not only improves the quality of the lettuce but also extends its shelf life, resulting in increased yield.

3.2.3.3 *Beneficial and harmful aspects of plant growth promoting rhizobacteria*

Rhizobacteria play a crucial role in maintaining soil fertility and promoting the growth and development of plants. Despite some conflicting findings from other studies, growth enhancement is brought about through the various processes discussed in previous chapters (Saharan and Nehra, 2011). For instance, it is reported that some *Pseudomonas* species have been shown to produce cyanide. Therefore, the bacteria's cyanide generation is regarded as a trait that promotes and inhibits growth. Cyanide also works as a biocontrol agent for several plant diseases (Martínez-Viveros et al., 2010); it can furthermore negatively impact plant development (Bakker and Schippers, 1987). According to Vacheron et al. (2013), PGPR auxin production can have both positive and negative effects on developing plants. Perhaps the amount of auxin determines how efficient it is. For example, it promotes plant development at low doses but limits root development at high quantities (Bakker and Schippers, 1987). Additionally, *Bradyrhizobium elkanii's rhizobitoxine* has a double-end product. As it inhibits the synthesis of ethylene, the stress-induced production of ethylene in nodulation can be decreased (Vijayan et al., 2013). On the other hand, rhizobitoxine is also regarded as a plant toxin because it causes foliar chlorosis in soybeans (Xiong and Fuhrmann, 1996).

The explanation above has demonstrated that while plant growth-promoting rhizobacteria are quite active in encouraging the development of plants, a small number of microorganisms may restrict growth. Furthermore, the above-discussed adverse effects only happen under specified circumstances and due to certain personality traits.

Hence, selecting the appropriate strain is vital for maximizing the benefits related to improved plant growth and development (Vejan et al., 2016).

3.3 Nanomaterials in biofertilizer formulations

Biofertilizers are a viable substitute for chemical fertilizers and offer considerable advantages in terms of sustainability and environmental friendliness. They supply essential nutrients and have a positive impact on the soil, consequently resulting in better plant growth. However, the efficacy of biofertilizers can be limited due to the slow release of nutrients, instability in the field due to a lack of prescribed soil and environmental conditions (temperature, radiation, and pH sensitive), the need for specific storage conditions, a low bacterial population, and their low shelf life (Seeda et al., 2021). To overcome these limitations, nanomaterial-based formulations of biofertilizers have been developed to enhance the efficiency and productivity of crops (Akhtar et al., 2022). Nanomaterials have unique properties such as a large surface area, high reactivity, and high mobility, and their size ranges from 1 to 100 nanometers. Nanomaterials have emerged as a promising tool in agriculture, providing new solutions for crop production, improving the efficiency of fertilizers, and increasing plant growth (Shang et al., 2019).

In recent years, the use of nanomaterials in biofertilizer formulations have attracted significant attention due to their potential to enhance plant growth and increase crop yield. Therefore, the integration of nanomaterials in biofertilizer formulations can overcome the existing limitations and provide additional benefits to plants. This is achieved by the small size of nanomaterials, which allows them to penetrate plant cells and tissues, increasing the efficiency of nutrient uptake (Akhtar et al., 2022). Additionally, nanomaterials are able to enhance the solubility and stability of nutrients, improving their availability to plants and reducing losses due to leaching and volatilization (Elemike et al., 2019). As a result, biofertilizers that incorporate nanomaterials can provide higher nutrient levels to crops, resulting in better growth and higher yields.

Another advantage of using nanomaterials in biofertilizer formulations is their ability to improve the release of nutrients (El-Saadony et al., 2021). This is accomplished by using nanoemulsions, nanocapsules, nanospheres, and nanosuspensions, which allow for the slow release of nutrients, reducing the need for frequent applications of fertilizers (Chhipa, 2016). Furthermore, nanomaterials can also improve the efficiency of fertilizer applications by reducing the amount of fertilizer required to achieve the desired results (Mejias et al., 2021). This reduces the environmental impact of fertilizers and saves farmers money on fertilizer costs. Nanomaterials have also been shown to positively impact the environment by reducing the use of chemical fertilizers. Chemical fertilizers can negatively impact the environment, as they can contribute to soil and water pollution and release greenhouse gases (Hazarika et al., 2022). By reducing the use of chemical fertilizers, biofertilizers that incorporate nanomaterials can help mitigate agriculture's negative environmental impact (Babu et al., 2022). Additionally, using nanomaterials in biofertilizer formulations improves soil health by promoting the growth of beneficial microorganisms, leading to a healthier soil ecosystem. Therefore, the use of nanomaterials in biofertilizer formulations has the potential to revolutionize agriculture by improving nutrient uptake, reducing the use of chemical fertilizers, and increasing crop yields.

3.3.1 Role of plant growth promoting rhizobacteria as a biofertilizer

PGPR are a group of soil-borne bacteria that are capable of promoting plant growth by providing essential nutrients and suppressing soil-borne pathogens (Beneduzi et al., 2012; dos Santos et al., 2020). The use of PGPR as a biofertilizer is gaining recognition as a sustainable and environmentally friendly alternative to traditional chemical fertilizers. According to (Vejan et al., 2016), PGPR can provide plants with essential nutrients such as nitrogen, phosphorus, and iron, as well as other minerals and vitamins that are required for proper growth and development. PGPR can also help plants absorb these nutrients more efficiently, reducing the need for chemical fertilizers and increasing plant growth and productivity.

Another benefit of PGPR is its ability to suppress soil-borne pathogens. Researchers have found that PGPR can suppress the growth of root-knot nematodes, a major soil-borne pathogen that causes significant damage to crops (Gamalero and Glick, 2020; Shah et al., 2021; Backer et al., 2018). This increased microbial diversity can improve soil fertility, reducing the need for chemical fertilizers and promoting healthy plant growth. The use of PGPR as a biofertilizer has also been found to have positive impacts on crop yields. The addition of PGPR to the soil resulted in significant increases in crop yields, including higher yields of wheat, corn, and soybeans. These findings suggest that the use of PGPR as a biofertilizer has the potential to improve the productivity of agricultural systems, reduce the need for chemical fertilizers, and promote more sustainable and environmentally friendly approaches to crop production. Fig. 3.1 shows the impact of PGPR on growth promotion in plants and inhibition of external stress (Fig. 3.2).

3.4 Nanoparticles as biofertilizers

A novel strategy gaining popularity in developing nations' agriculture sectors is nanotechnology. When plants are subjected to unfavorable conditions, nanoparticles (NPs) act as cues to activate a variety of defense mechanisms. Due to their large surface area, high solubility, and lightweight nature, NPs are a better source of fertilization than regular salts (Akhtar et al., 2022). Nanobiofertilizer is composed of an amalgam of NPs and biofertilizers. By gradually releasing nutrients, it can improve the efficiency with which plants utilize nutrients. Biofertilizer cells are incorporated into the nanomaterial capsule by a process called encapsulation. It involves the use of starch with a nontoxic, biodegradable substance like calcium alginate. These bacterial strains' growth is accelerated by starch (Du et al., 2018; Vafa et al., 2021).

The application of nanobiofertilizer dramatically enhances plant growth. The continuous and delayed release of NPs into the plant's rhizosphere is caused by the coating of NPs with biofertilizers, which increases the effectiveness of biofertilizers. They keep fertilizer from decomposing and lessen the possibility of leaching (Ali et al., 2021; Paschalidis et al., 2021;

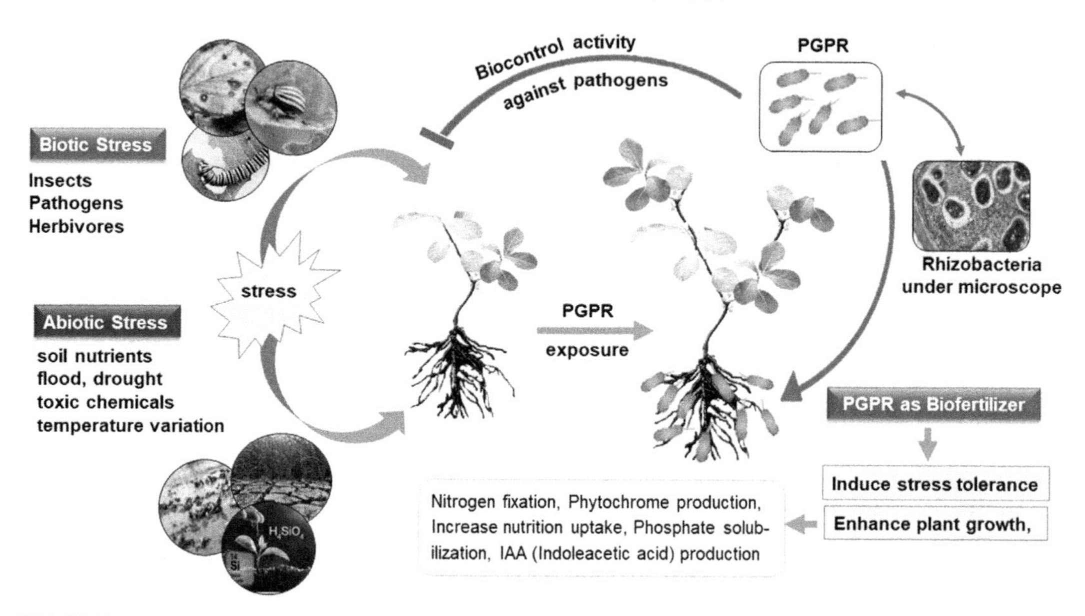

FIGURE 3.1 Schematic representation of external stress and plant growth promoting rhizobacteria (PGPR)-mediated growth promotion in plants. The model shows stress induced in the form of biotic and biotic stress by various sources; insects, pathogens, herbivores; nutrients proportion, water availability, exposure to toxic chemicals, and temperature variations. Plants protected with PGPR show significant growth attributed to nitrogen fixation, phytochrome production, increased nutrition uptake, phosphate solubilization, plant growth hormone IAA production, and also improved stress tolerance. PGPR also hinders the pathogen attacks through its biocontrol properties.

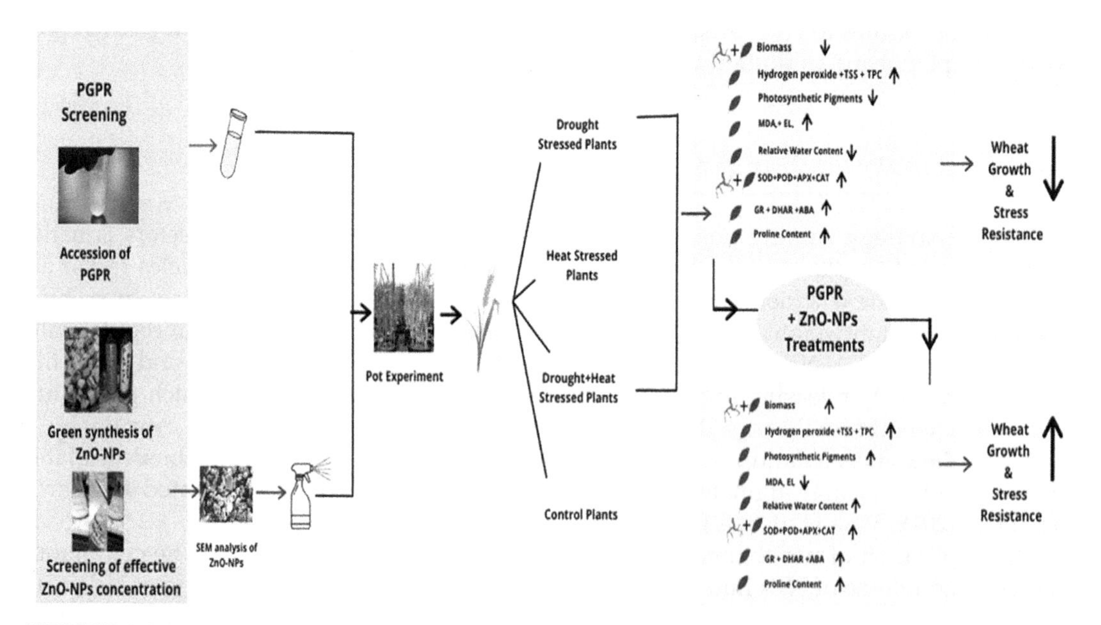

FIGURE 3.2 Coactive role of zinc oxide nanoparticles and plant growth promoting rhizobacteria for mitigation of synchronized effects of heat and drought stress in wheat plants (Azmat et al., 2022).

Vedamurthy et al., 2021; Shcherbakova et al., 2017). This chapter has provided an overview of the widely utilized nanobiofertilizers, including zinc, titania, silicon, copper, iron, and silver, that contribute to sustainable plant growth and development. We have briefly explained their significant role in plants acting as nanobiofertilizers and outlined their relevance to promoting plant growth.

3.5 Titanium oxide

Titanium oxide (TiO_2) NPs are an enigmatic semiconductive substance with a high refractive index, UV absorption, photocatalytic, and antibacterial capabilities. These engineered NPs (ENPs) exhibit a crystal structure that has certain qualities that can be advantageous in a given application; most frequently, the suitable mineral form is chosen based on whether it has a lower or higher capacity for photocatalysis (Šebesta et al., 2021; Macwan et al., 2011). The geometry of TiO_2 NPs can be modified during their synthesis to produce both nanorods and spherical NPs. TiO_2 NPs can be synthesized in sizes ranging from a few nanometers to 100 nm in any dimension. TiO_2 NPs have a wide range of uses in a variety of human endeavors, including agriculture, as a result of their characteristics (Rouhani et al., 2013). Likewise to ZnO NPs, TiO_2 NPs' surface characteristics are frequently altered to improve their stability, heighten their beneficial effects, or lessen their toxicity (Silva et al., 2013; Chen and Mao, 2007). Water purification, pollutant degradation, antimicrobial coating, biosensing, and medication delivery are a few of their environmental uses (Mahlambi et al., 2015; Han et al., 2016). TiO_2 NPs have been used to safeguard seeds, improve plant development and germination, combat crop illnesses (Aragay et al., 2012), break down pesticides, and find pesticide residues (Raliya et al., 2015a,b). These NPs have also been linked to improvements in plant health, seed or produce yield, and root and shoot growth. There were also documented increases in the production of chlorophyll, the amount of soluble leaf protein (Raliya et al., 2015a,b) and carotenoids (Tan et al., 2017), and the uptake of several crucial components (Mustafa et al., 2021). TiO_2 NPs were also used to reduce environmental challenges like dryness in wheat significantly (Lian et al., 2020) and excessive Cd levels in maize (Moaveni et al., 2011a,b).

According to the El-Sagan and Shokry (2019) study, when TiO_2 was applied to a turnip crop (*Brassica rapa*) at a rate of 2 mg/L, followed by a rate of 1 mg/L, the maximum value of growth, yield parameters, and chemical contents were obtained (Nayana et al., 2020; Palmqvist et al., 2015; Mahmoodzadeh et al., 2013; Palmqvist et al., 2015).

3.6 Silicon oxide

Nanotechnology has provided effective alternatives when using silica NPs (silicon oxide (SiO_2)-NPs) as eustressors (Magaña-López et al., 2022). SiO_2-NPs are efficiently distributed throughout tissues due to their small size (5–20 nm) and other features like shape, surface charge, and texture. It was found that plants absorb SiO_2-NPs through diffusion through their roots and leaves, where they are subsequently polymerized or transported to other tissues, where they produce a variety of physiological and cellular functions

(Aguirre-Becerra et al., 2017; Mittal et al., 2020; Goswami et al., 2022). In the case of barley and maize, SiO_2-NP treatment has improved germination and shoot and root growth (Disfani et al., 2017). Additionally, tomato plant germination rates increased when SiO_2-NPs were applied at a concentration of 8 g/L (Siddiqui and Al-Whaibi, 2014). Similarly, research indicates that SiO_2-NPs can successfully control the stress brought on by UV-B radiation in wheat (*Triticum aestivum L.*) (Tripathi et al., 2017). Corn (*Zea mays L.)* growth was enhanced by silica NPs, which encouraged an increase in the total biomass of the soil microbial communities (Tian et al., 2020; Alharbi et al., 2022). Additionally, there was a rise in CAT and PAL activity in chili pepper plants, providing further evidence that bacteria-NP therapy causes plant immunity (Ferrusquía-Jiménez et al., 2022). Recent studies have shown that SiO_2-NPs may directly interact with flora, changing their shape and physiology and helping to increase plant growth and yield (Tables 3.2 and 3.3).

TABLE 3.2 Application of biosynthesized SiO_2 in promoting plant growth.

Application	Plant species	Biological activity
Plant growth and crop productivity	Tomato	Improved seed germination, fresh and dry weight
	Oat	Increase plant growth
	Astragalus fridae	Enhance expression of PAL and lignification in leaf & roots
		Reinforced in the vascular system and leaf thickness
	Potato	Increase the mass of potato
	Wheat and Lupin	Increase plant growth and biomass, chlorophyll content and seed germination.
	Fenugreek *Foenum graecum*	Increase cell wall lignification, the activity of PAL and protein content
		Increase root and chlorophyll content
	Larix olegensis	Increase seed germination and root elongation
		Increase flowering
	Maize	Increase PAL activity and Phenolic compound
Environmental stress	Cucumber	Increase growth and yield N and P nutrient content
Salt stress	Soybean	Increase K concentration, antioxidant activity, and decrease in
Water deficient and salt stress	Tomato	Na concentration, lipid peroxidation, and ROS production
UV-B stress	Cucumber	Improves several physiological attributes
Metal toxicity Lead (Pb) and Cadmium (Cd)	Bean	Increase productivity and chlorophyll
	Phaseolous vulgaris	Highly increased germination percentage
Cromium (Cr)		Provide resistance
	Wheat	Reduce metal concentration, increase plant growth and biomass
	Pea	Reduce Cr accumulation and increase nutrient content
(Fungicidal effects against some fungal pathogen)	Rice	*Fusarium oxysporum*
	Maize	*Aspergillus niger*
	Tomato	*Candida albicans, Ilyonctrial mosspanacis*
	Panax ginseng	*Botrytis cinea*
	Grapes	
Entomotoxic effect against		*Sitophilus oryzae*
		S. Tutaabsoluta
		C. maculates

TABLE 3.3 Impact of nanoparticles on plant gene expression.

Nanoparticles	Plants	Stress	Effect on Genes Expression
Silicon nanoparticles	Solanum lycopersicum	Salinity	Upregulation of NCED3, TAS14, CRK1 and AREB genes Downregulation of MAPK3, DDF2, MPAK2, ERF3, APX2 and RBOH1 genes (Almutairi, 2016)
	Oryza sativa	Salinity	Upregulation of LSi1 and LSi2 genes (Abdel-Haliem et al., 2017)
	Oryza sativa	Biotic stress	Upregulation of A2WZ30, B8AP99, A2YNH4, B8AZZ8, Fo2, B8B. B8A9F5, B8BF84. B8BHM9 and 2XRR2 genes (Cheng et al., 2021)
Copper nanoparticles	Piper nigrum	–	Upregulation of miR159 Downregulation of MVK gene (Tabatabaee et al., 2021)
Iron nanoparticles	Oryza sativa	Cd and drought stress	Downregulation of OsLCT1, OsHMA3 and OsHMA2 genes (Ahmed et al., 2021)
Silver nanoparticles	Nigella sativa	–	Upregulation of PAL and CHS genes (Baniebrahim et al., 2021)
	Arabidopsis thaliana	–	Upregulation of 286 genes Downregulation of 81 genes (Kaveh et al., 2013)
	Arabidopsis thaliana	–	Upregulation of PCS, GS, GR and GSTU12 genes (Nair and Chung, 2014)
	Solanum lycopersicum	–	Upregulation of PAL and EIX genes (Noori et al., 2019)
	Arabidopsis thaliana	–	Upregulation of 438 genes (Kohan-Baghkheirati and Geisler-Lee, 2015)

3.7 Zinc oxide

An amphoteric semiconductive material, zinc oxide (ZnO) NPs, are one of the ENPs (Klingshirn et al., 2010; Kołodziejczak-Radzimska and Jesionowski, 2014). Additionally, the aforementioned qualities are infinitely adjustable. A wide variety of structures, including flower-like structures, nanorods, nanotubes, and spherical or oblong NPs, can be easily synthesized (Kołodziejczak-Radzimska and Jesionowski, 2014; Sabir et al., 2014). Zinc oxide NPs (ZnO-NPs) frequently have their surfaces altered to increase their stability in colloidal suspension, improve their beneficial effects on plants, and lower their potential toxicity. It is possible to modify their surface with inorganic substances such as SiO_2, and Al_2O_3, simple organic substances such as silanes or organic acids, or more sophisticated polymeric matrices. Since it is anticipated that the biosynthesis of ZnO NPs will provide an eco-friendly nanomaterial, this material is frequently chosen for agricultural applications (Saravanan et al., 2021; Kolenčík et al., 2020; Tarafdar et al., 2012). Unlike other ENPs, including TiO_2 NPs, ZnO-NPs are easier to dissolve (Bian et al., 2011), which has an impact on plant health in part due to their nano-specific features but also significantly more so due to the release of Zn, which is crucial for numerous cellular activities (Singh et al., 2021). The impact of environmental pressures on plants, such as salt

(Torabian et al., 2016), metals, metalloids (Rizwan et al., 2019), drought (Dimkpa et al., 2020), and temperature (García-López et al., 2018), is also said to be lessened by ZnO-NPs. ZnO-NPs boost plant seed germination, growth (Singh et al., 2019a,b), antioxidant activity and protein synthesis (Venkatachalam et al., 2017), chlorophyll content (Pullagurala et al., 2018), photosynthesis (Faizan et al., 2018), production of oils and seeds (Kolenčík et al., 2020), and element uptake when used at the right doses (Peralta-Videa et al., 2014).

Azmat et al. studied the potential of PGPR-Pseudomonas sp. and green synthesized ZnO-NPs. Wheat plants were negatively affected by drought and heat stress synergistically, rather than individually. In addition to this observation, hydrogen peroxide (H_2O_2) and MDA content were significantly elevated. The treated plants exposed to all stress groups showed improved biomass, photosynthetic pigments, soluble sugars, protein, and indole acetic acid (IAA) contents, which improved wheat growth and stress resistance. The results of this study proved that plants can be protected from both isolated and combined heat and drought stress by the synchronized activities of PGPR and ZnO-NPs (Azmat et al., 2022).

In a study published in 2020, Hosseinpour et al. examined the protective effects of different concentrations of ZnO-NPs (0, 20, and 40 mg L1) and PGPBs (no bacteria, *Bacillus subtilis, Lactobacillus casei, and Bacillus pumilus*) on DNA damage and cytosine methylation changes in tomato (*Solanum lycopersicum L. "Linda"* (250 mM NaCl)). Salinity stress accelerated the rate of polymorphism measured by RAPD, while PGPB and ZnO-NPs mitigated its negative effects. It was observed that salinity stress-related molecular abnormalities were dramatically decreased in plants treated with various PGPBs at various concentrations of ZnO-NPs in the rhizosphere. The use of PGPBs and ZnO-NPs improved the stability of the genomic template; this improvement was more pronounced when *Lactobacillus casei* and 40 mg L1 of ZnO-NPs were utilized. Together, the usage of PGPB and ZnO-NPs under salt stress exhibited a general beneficial effect, lowering genetic impairment in tomato seedlings (Hosseinpour et al., 2020). The detoxifying potential of PGPR (*B. pumilus*), ZnO, and TiO_2 NPs in maize growing in Cd-rich soil was observed when they were used as single treatments and in combination, according to Shafiq et al. They found that the effects of Cd stress on GP and plant biomass in terms of shoot length, root length, leaf area, fresh weight, and dry weight lowered both (Shafiq et al., 2022; Kolya et al., 2015; Mushtaq et al., 2020). When both NPs were applied, which provided Zn and Ti to plants and controlled the synthesis of auxin, which is crucial for maintaining cell membrane integrity under stressful circumstances, an increase in Cd resistance and better plant growth and biomass were seen (Landa, 2021). Important nutrients like K, Ca, P, Mg, and Fe are restored as a result, helping organelles like mitochondria and chloroplasts to maintain their structural and functional integrity. As a result, plant defense mechanisms against ROS species improve (Bhat et al., 2022).

The synthesis of IAA by Pseudomonas sp. DN18 was unaffected by ZnO-NP NPs, which lacked any bactericidal effect. This nanoformulation, when applied to the seeds of Oryza sativa, was found to have growth-promoting and biocontrolling properties toward the fungus Sclerotium rolfsii. In a different study, Burkholderia caribensis and Pantoea agglomerans were immobilized in a coculture. The nanofibers retained the viability and plant growth-promoting attributes of both encapsulated strains, which had positive effects on the length and dry weight of soybean roots as well as seed germination (Balla et al., 2022).

3.8 Other nanoparticles

The sustainable production of crops enters a new chapter with the use of nanobiofertilizers in agriculture. The use of NPs enhances plant growth and stress resistance. Another approach being investigated in agriculture is the injection of biofertilizers. Biofertilizers enclosed in NPs make up nanobiofertilizers. While NPs are minuscule (1–100 nm) particles with many benefits, biofertilizers are plant-based carriers containing beneficial microbial organisms. The NPs' properties and performance have been improved by their green synthesis. Nanobiofertilizers are more efficient than other conventional methods. They also fulfill their function more effectively than the conventional salts that were previously utilized in agriculture to increase crop productivity. In comparison to conventional chemical fertilizers, nanobiofertilizers produce superior and longer-lasting outcomes (Akhtar et al., 2022).

3.8.1 Copper nanoparticles

Copper (Cu) is a crucial nutrient needed for typical plant development. In plants, it serves both a structural and practical purpose. It is a micronutrient, and the amount in a plant depends on how easily and readily it is absorbed by the soil (Lawre et al., 2014; Dimkpa et al., 2019). Plants keep the concentration of copper in their various tissues at the ideal level. It participates in several internal mitochondrial and chloroplast processes (Elsheery et al., 2020; Lopez-Lima et al., 2021; Priyanka et al., 2019; Lafmejani et al., 2018; Shabbir et al., 2020). By applying exogenous Cu to plants, the copper deficiency can be cured. Crop growth and yield are enhanced by the use of CuNPs. Using *A. indica* leaf extract, copper nanoparticles (CuNPs) can be produced. Drops of the leaf extract were added to a cupric chloride solution while being continuously stirred, and the mixture's color was checked to see if it changed from pale yellow to dark brown. The change in color denotes the emergence of CuNPs. After washing and centrifuging the solution, the CuNPs were dried. To verify the synthesis and purity level of CuNPs, various analytical-based characterization methods were employed (Nagar and Devra, 2018). The method used to shrink copper salt to a nanoscale was Citrus limon (L.) *Burm.f.* fruit extract (Akl and Amer, 2021). Additionally, it was noted in the literature that *Plantago asiatica L.* (Nasrollahzadeh et al., 2017), *Eclipta prostrate (L.)* (Chung et al., 2017), and *Tinospora cordifolia (Willd.) Miers* (Sharma et al., 2019) leaf extracts were utilized to create CuNPs. CuNPs are a great choice for utilization due to their quick absorption and gradual release (Abbasifar et al., 2020). To make crops more tolerant to drought, it promotes antioxidant activity, maintains photosynthetic rate, and improves water status. The fruit quality of *S. lycopersicum* is improved by the foliar application of CuNPs. Additionally, it increases the activity of catalase, superoxide dismutase, and the creation of bioactive molecules (López-Vargas et al., 2018).

3.8.2 Iron nanoparticles

One of the crucial micronutrients needed for plant growth is iron. It is necessary to regulate respiration and photosynthesis as well as to produce certain enzymes. Iron deficiency results in decreased chlorophyll synthesis, which causes the leaves to turn yellow. The leaves develop chlorosis and necrosis as a result. Iron fertilizer application enhances plant growth and lessens the negative impacts of environmental conditions (Pourjamshid, 2021; Vaghar et al., 2020). Nanofertilizer treatment minimizes the negative impacts of chemical fertilizer, and administration through both the soil and foliar techniques is beneficial for enhancing plant growth (Mohamadipoor and Mahboub Khomami, 2013). Utilizing plant extracts of *C. sinensis*, *A. indica*, and *Eucalyptus tereticornis*, green synthesis of iron nanoparticles (FeNPs) is carried out (Herlekar et al., 2014; Mohamadipoor and Mahboub Khomami, 2013). Plants under stress respond favorably to the application of FeNPs (Ebrahiminezhad et al., 2017). Plants readily absorb them, and they lessen the ions' damaging effects. In *S. lycopersicum*, FeNP fertilization results in higher levels of phenol, vitamin C, and glutathione (El-Desouky et al., 2021). In comparison to untreated plants, FeNPs promote T. aestivum germination and lengthen the roots and shoot length (Yasmeen et al., 2015; Adrees et al., 2020).

3.8.3 Silver nanoparticles

Silver ions are important for the development of shoots, roots, somatic embryos, and genetic changes. By using silver, plants produce less ethylene, which lowers their risk of chlorosis and chlorophyll degradation. Additionally, it increases the synthesis of bioactive compounds, slows the aging process, and boosts crop growth and grain yield (Yasmeen et al., 2015; Adrees et al., 2020).

The effectiveness of nanoparticles (NPs) in plants is determined by several factors, including their chemical composition, surface properties, and application rate (Öktem and Keleş, 2018; Kumari et al., 2017). High concentrations of NPs can have detrimental effects. Conversely, an optimal dosage can elicit beneficial responses in plants, and the required amount varies depending on the plant species. Research indicates that NPs can be highly efficient when used appropriately (Mahendran et al., 2019; Ahmed and Mustafa, 2020). The synthesis of NPs was indicated by the color change from yellow to dark brown. Atomic force microscopy, scanning electron microscopy, ultraviolet-visible spectroscopy, and X-ray diffraction were used to carry out the characterization (Soliman et al., 2020). The formation of hydrogen peroxide and hyperhydricity in the cultivation of *Dianthus chinensis* L. were decreased to 13% and 50%, respectively, by the use of silver nanoparticles (AgNPs) (Sreelekshmi et al., 2021). Additionally, they increased the length of G. max's roots and decreased the generation of reactive oxygen species (Mustafa et al., 2020). In *S. lycopersicum* infected with Alternaria alternative, they boosted the carotenoid, chlorophyll, and polyphenol oxidase concentrations to 46%, 45%, and 33%, respectively. Compared to untreated plants, the disease's severity was reduced by 58% (Mahawar et al., 2020). In comparison to soil application, it is a less expensive and more practical approach (Ahmadi-Majd et al., 2021). In S. lycopersicum infected with *A. solani*, the foliar application of AgNPs enhanced the fresh weight and chlorophyll content to 34% and 26%, respectively. The fungal spore count was reduced by 48%, and the stress enzymes that help plants tolerate stress were more active (Kumari et al., 2017; Nejatzadeh 2021; Bastami et al., 2021).

3.9 Conclusion

In this chapter, we discussed insights into various NPs that significantly contribute to plant growth and development. We have described the common NPs like zinc, titanium, silicon, copper, iron, and silver used in the agricultural field to improve plant growth and development and to act as nanobiofertilizer. Nanobiofertilizers promote plant growth while enhancing plant nutrition, productivity, and shelf life and defying biotic and abiotic stressors. By activating multiple mechanisms, the use of nanobiofertilizers maintains the nutrient content in the soil and improves the growth and yield of crops. The interaction between NPs and biofertilizer should be boosted to regulate the distribution and release of nano-based biofertilizer at the intended target site. It is essential to comprehend the manner in which nanobiofertilizers foster agricultural progress to formulate a more competent nanobiofertilizer and put it into practical use. The design of novel approaches, combined with rigorous testing, will be paramount to ensuring their effectiveness.

References

Abbasifar, A., Shahrabadi, F., ValizadehkKaji, B., 2020. Effects of green synthesized zinc and copper nano-fertilizers on the morphological and biochemical attributes of basil plant. J. Plant Nutr. 43, 1104–1118.

Abdel-Haliem, M., Hegazy, H.S., Hassan, N.S., Naguib, D., 2017. Effect of silica ions and nano silica on rice plants under salinity stress. Ecol. Eng. 99, 282–289.

Adrees, M., Khan, Z.S., Ali, S., Hafeez, M., Khalid, S., Rehman, M.Z.U., et al., 2020. Simultaneous mitigation of cadmium and drought stress in wheat by soil application of iron nanoparticles. Chemosphere. 238, 124681.

Aguirre-Becerra, H., Feregrino-Pérez, A.A., Esquivel, K., Perez-Garcia, C.E., Vazquez-Hernandez, M.C.

Ahmad, F., Ahmad, I., Khan, M., 2008. Screening of free-living rhizospheric bacteria for their multiple plant growth promoting activities. Microbiol. Res. 163, 173–181.

Ahmad, M., Zahir, Z.A., Khalid, M., Nazli, F., Arshad, M., 2013. Efficacy of Rhizobium and Pseudomonas strains to improve physiology, ionic balance and quality of mung bean under salt-affected conditions on farmer's fields. Plant Physiol. Biochem. 63, 170–176.

Ahmadi-Majd, M., Nejad, A.R., Mousavi-Fard, S., Fanourakis, D., 2021. Postharvest application of single, multi-walled carbon nanotubes and nanographene oxide improves rose keeping quality. J. Hortic. Sci. Biotechnol. 1–15.

Ahmed, A., Hasnain, S., 2010. Auxin-producing Bacillus sp.: auxin quantification and effect on the growth of Solanum tuberosum. Pure Appl. Chem. 82, 313–319.

Ahmed, R.H., Mustafa, D.E., 2020. Green synthesis of silver nanoparticles mediated by traditionally used medicinal plants in Sudan. Int. Nano Lett. 10, 1–14.

Ahmed, B., Zaidi, A., Khan, M.S., Rizvi, A., Saif, S., Shahid, M., 2017. Perspectives of plant growth promoting rhizobacteria in growth enhancement and sustainable production of tomato. Microb. Strateg. Veg. Prod. 125–149.

Ahmed, T., Noman, M., Manzoor, N., Shahid, M., Abdullah, M., Ali, L., et al., 2021. Nanoparticle-based amelioration of drought stress and cadmium toxicity in rice via triggering the stress responsive genetic mechanisms and nutrient acquisition. Ecotoxicol. Environ. Saf. 209, 111829.

Akhtar, N., Ilyas, N., Meraj, T.A., Pour-Aboughadareh, A., Sayyed, R.Z., Mashwani, Z.U., et al., 2022. Improvement of plant responses by nanobiofertilizer: a step towards sustainable agriculture. Nanomaterials (Basel, Switzerland) 12 (6).

Akl, M.A., Amer, M.W., 2021. Green synthesis of copper nanoparticles by Citrus limon fruits extract, characterization and antibacterial activity. Chem. Int. 7, 1–8.

Alharbi, K., Hafez, E., Omara, A.E.-D., Awadalla, A., Nehela, Y., 2022. Plant growth promoting rhizobacteria and silica nanoparticles stimulate sugar beet resilience to irrigation with saline water in salt-affected soils. Plants 11, 3117. 2022.

Ali, S.S., Darwesh, O.M., Kornaros, M., Al-Tohamy, R., Manni, A., El-Shanshoury, A.E.-R.R., et al., 2021. Nano-biofertilizers: synthesis, advantages, and applications. Biofertilizers. Woodhead Publishing, Cambridge, MA, USA, pp. 359–370.

Almoneafy, A.A., Moustafa-Farag, M., Mohamed, H.I., 2021. The auspicious role of plant growth-promoting rhizobacteria in the sustainable management of plant diseases. Plant. Growth-Promoting Microbes Sustain. Biotic Abiotic Stress. Manag. Springer, Cham, Switzerland AG, pp. 251–283.

Almutairi, Z.M., 2016. Effect of nano-silicon application on the expression of salt tolerance genes in germinating tomato (Solanum lycopersicum L.) seedlings under salt stress. Plant. Omics 9, 106–114.

Aloo, B.N., Makumba, B., Mbega, E.R., 2019. The potential of Bacilli rhizobacteria for sustainable crop production and environmental sustainability. Microbiol. Res. 219, 26–39.

Aragay, G., Pino, F., Merkoçi, A., 2012. Nanomaterials for sensing and destroying pesticides. Chem. Rev. 112, 5317–5338.

Arora, N.K., Khare, E., Oh, J.H., Kang, S.C., Maheshwari, D.K., 2008. Diverse mechanisms adopted by fluorescent Pseudomonas PGC2 during the inhibition of Rhizoctonia solani and Phytophthora capsici. World J. Microbiology Biotechnol. 24, 581–585.

Azmat, A., Tanveer, Y., Yasmin, H., Hassan, M.N., Shahzad, A., Reddy, M., et al., 2022. Coactive role of zinc oxide nanoparticles and plant growth promoting rhizobacteria for mitigation of synchronized effects of heat and drought stress in wheat plants. Chemosphere. 297, 133982. Jun.

Babu, S., Singh, R., Yadav, D., Rathore, S.S., Raj, R., Avasthe, R., et al., 2022. Nanofertilizers for agricultural and environmental sustainability. Chemosphere 292, 133451.

Backer, R., Rokem, J.S., Ilangumaran, G., Lamont, J., Praslickova, D., Ricci, E., et al., 2018. Plant growth-promoting rhizobacteria: context, mechanisms of action, and roadmap to commercialization of biostimulants for sustainable agriculture. Front. Plant Sci. 9, 1473.

Bakker, A.W., Schippers, B., 1987. Microbial cyanide production in the rhizosphere in relation to potato yield reduction and *Pseudomonas* spp-mediated plant growth-stimulation. Soil. Biol. Biochem. 19, 451–457.

Balla, A., Silini, A., Cherif-Silini, H., Chenari Bouket, A., Alenezi, F.N., Belbahri, L., 2022. Recent advances in encapsulation techniques of plant growth-promoting microorganisms and their prospects in the sustainable agriculture. Appl. Sci. 12, 9020.

Baniebrahim, S., Pishkar, L., Iranbakhsh, A., Talei, D., Barzin, G., 2021. Physiological and molecular responses of black cumin (*Nigella sativa* L.) seedlings to silver nanoparticles. J. Plant Nutr. 44, 2885–2896.

Bastami, A., Amirnia, R., Sayyed, R., Enshasy, H., 2021. The effect of mycorrhizal fungi and organic fertilizers on quantitative and qualitative traits of two important Satureja species. Agronomy. 11, 1285.

Beneduzi, A., Ambrosini, A., Passagliaet, L.M.P., 2012. Plant growth-promoting rhizobacteria (PGPR): their potential as antagonists and biocontrol agents. Genet. Mol. Biol. 35 (4), 1044–1051.

Bhat, J.A., Faizan, M., Bhat, M.A., Huang, F., Yu, D., Ahmad, A., et al., 2022. Defense interplay of the zinc-oxide nanoparticles and melatonin in alleviating the arsenic stress in soybean (*Glycine max* L.). Chemosphere 288, 132471.

Bhattacharyya, P.N., Jha, D.K., 2012. Plant growth-promoting rhizobacteria (PGPR): emergence in agriculture. World J. Microbiol. Biotechnol. 28, 1327–1350.

Bian, S.-W.W., Mudunkotuwa, I.A., Rupasinghe, T., Grassian, V.H., 2011. Aggregation and dissolution of 4 nm ZnO nanoparticles in aqueous environments: influence of pH, ionic strength, size, and adsorption of humic acid. Langmuir 27, 6059–6068.

Burdman, S., Jurkevitch, E., Okon, Y., 2000. Recent advances in the use of plant growth promoting rhizobacteria (PGPR) in agriculture, Microb. Interact. Agriculture Forestry, II. Science Publishers, Inc., Enfield, USA, pp. 229–250.

Butterbach-Bahl, K., Baggs, E.M., Dannenmann, M., Kiese, R., Zechmeister-Boltenstern, S., 2013. Nitrous oxide emissions from soils: how well do we understand the processes and their controls? Philos. Trans. R. Soc. B: Biol. Sci. 368, 20130122.

Chen, S., Zou, J., Hu, Z., Chen, H., Lu, Y., 2014. Global annual soil respiration in relation to climate, soil properties and vegetation characteristics: summary of available data. Agric. For. Meteorol. 198, 335–346.

Chen, X., Mao, S.S., 2007. Titanium dioxide nanomaterials: synthesis, properties, modifications, and applications. Chem. Rev. 107, 2891–2959.

Cheng, B., Chen, F., Wang, C., Liu, X., Yue, L., Cao, X., et al., 2021. The molecular mechanisms of silica nanomaterials enhancing the rice (*Oryza sativa* L.) resistance to planthoppers (Nilaparvata lugens Stal). Sci. Total. Environ. 767, 144967.

Chhipa, H., 2016. Nanofertilizers and nanopesticides for agriculture. Environ. Chem. Lett. 15 (1), 15–22.

Chung, I., Rahuman, A.A., Marimuthu, S., Kirthi, A.V., Anbarasan, K., Padmini, P., et al., 2017. Green synthesis of copper nanoparticles using Eclipta prostrata leaves extract and their antioxidant and cytotoxic activities. Exp. Ther. Med. 14, 18–24.

Dasgupta, D., Kumar, A., Mukhopadhyay, B., Sengupta, T.K., 2015. Isolation of phenazine 1, 6-di-carboxylic acid from *Pseudomonas aeruginosa* strain HRW. 1-S3 and its role in biofilm-mediated crude oil degradation and cytotoxicity against bacterial and cancer cells. Appl. Microbiol. Biotechnol. 99, 8653–8665.

De Souza, J.T., Arnould, C., Deulvot, C., Lemanceau, P., Gianinazzi-Pearson, V., Raaijmakers, J.M., 2003. Effect of 2, 4-diacetylphloroglucinol on Pythium: cellular responses and variation in sensitivity among propagules and species. Phytopathology 93, 966–975.

Dimkpa, C.O., Singh, U., Bindraban, P.S., Elmer, W.H., Gardea-Torresdey, J.L., White, J.C., 2019. Zinc oxide nanoparticles alleviate drought-induced alterations in sorghum performance, nutrient acquisition, and grain fortification. Sci. Total. Environ. 688, 926–934. 2019.

Dimkpa, C.O., Andrews, J., Sanabria, J., Bindraban, P.S., Singh, U., Elmer, W.H., et al., 2020. Interactive effects of drought, organic fertilizer, and zinc oxide nanoscale and bulk particles on wheat performance and grain nutrient accumulation. Sci. Total. Environ. 722, 137808.

Disfani, M.N., Mikhak, A., Kassaee, M.Z., Maghari, A., 2017. Effects of nano Fe/SiO_2 fertilizers on germination and growth of barley and maize. Arch. Agron. Soil. Sci. 63, 817–826.

dos Santos, R.M., Diaz, P.A.E., Lobo, L.L.B., Rigobelo, E.C., 2020. Use of plant growth-promoting rhizobacteria in maize and sugarcane: characteristics and applications. Front. Sustain. Food Syst. 4.

Du, C., Abdullah, J.J., Greetham, D., Fu, D., Yu, M., Ren, L., et al., 2018. Valorization of food waste into biofertiliser and its field application. J. Clean. Prod. 187, 273–284.

Dwivedi, D., Johri, B., 2003. Antifungals from fluorescent pseudomonads: biosynthesis and regulation. Curr. Sci. 1693–1703.

Ebrahiminezhad, A., Zare-Hoseinabadi, A., Sarmah, A.K., Taghizadeh, S., Ghasemi, Y., Berenjian, A., 2017. Plant-mediated synthesis and applications of iron nanoparticles. Mol. Biotechnol. 60, 154–168.

Effmert, U., Kalderás, J., Warnke, R., Piechulla, B., 2012. Volatile mediated interactions between bacteria and fungi in the soil. J. Chem. Ecol. 38, 665–703.

Egamberdieva, D., Wirth, S.J., Alqarawi, A.A., Abd_Allah, E.F., Hashem, A., 2017. Phytohormones and beneficial microbes: essential components for plants to balance stress and fitness. Front. Microbiol. 8, 2104.

Elbeltagy, A., Nishioka, K., Sato, T., Suzuki, H., Ye, B., Hamada, T., et al., 2001. Endophytic colonization and in planta nitrogen fixation by a Herbaspirillum sp. isolated from wild rice species. Appl. Environ. Microbiol. 67, 5285–5293.

Elemike, E., Uzoh, I., Onwudiwe, D., Babalola, O., 2019. The role of nanotechnology in the fortification of plant nutrients and improvement of crop production. Appl. Sci. 9 (3), 499.

El-Desouky, H.S., Islam, K.R., Bergefurd, B., Gao, G., Harker, T., Abd-El-Dayem, H., et al., 2021. Nano iron fertilization significantly increases tomato yield by increasing plants' vegetable growth and photosynthetic efficiency. J. Plant Nutr. 44, 1649–1663.

El-Saadony, M.T., ALmoshadak, A.S., Shafi, M.E., Albaqami, N.M., Saad, A.M., El-Tahan, A.M., et al., 2021. Vital roles of sustainable nano-fertilizers in improving plant quality and quantity-an updated review. Saudi J. Biol. Sci. 28 (12), 7349–7359.

El-Sagan, M.A.M., Shokry, A.M., 2019. Impact of bio-fertilizer and TiO_2 nanoparticles spray on growth, productivity and pickle quality of turnip crop (Brassica Rapa). Egypt. J. Desert Res. 69 (1), 101–121.

Elsheery, N.I., Helaly, M.N., El-Hoseiny, H.M., Alam-Eldein, S.M., 2020. Zinc oxide and silicone nanoparticles to improve the resistance mechanism and annual productivity of salt-stressed mango trees. Agronomy 10, 558. 2020.

Faizan, M., Faraz, A., Yusuf, M., Khan, S.T., Hayat, S., 2018. Zinc oxide nanoparticle-mediated changes in photosynthetic efficiency and antioxidant system of tomato plants. Photosynthetica 56, 678–686.

Farace, G., Fernandez, O., Jacquens, L., Coutte, F., Krier, F., Jacques, P., et al., 2015. Cyclic lipopeptides from *Bacillus subtilis* activate distinct patterns of defence responses in grapevine. Mol. Plant Pathol. 16, 177–187.

Fasciglione, G., Casanovas, E.M., Quillehauquy, V., Yommi, A.K., Goñi, M.G., Roura, S.I., et al., 2015. Azospirillum inoculation effects on growth, product quality and storage life of lettuce plants grown under salt stress. Sci. Hortic. 195, 154–162.

Ferrusquía-Jiménez, N.I., González-Arias, B., Rosales, A., Esquivel, K., Escamilla-Silva, E.M., Ortega-Torres, A.E., et al., 2022. Elicitation of Bacillus cereus-Amazcala (B.c-A) with SiO_2 nanoparticles improves its role as a plant growth-promoting bacteria (PGPB) in chili pepper plants. Plants 11, 3445.

Flores-Félix, J.D., Silva, L.R., Rivera, L.P., Marcos-García, M., García-Fraile, P., Martínez-Molina, E., et al., 2015. Plants probiotics as a tool to produce highly functional fruits: the case of Phyllobacterium and vitamin C in strawberries. PLoS One 10, e0122281.

Fravel, D.R., 1988. Role of antibiosis in the biocontrol of plant diseases. Annu. Rev. Phytopathol. 26, 75–91.

Gamalero, E., Glick, B.R., 2020. The use of plant growth-promoting bacteria to prevent nematode damage to plants. Biology 9 (11), 381.

García-López, J.I., Zavala-García, F., Olivares-Sáenz, E., Lira-Saldívar, R.H., Díaz Barriga-Castro, E., Ruiz-Torres, N.A., et al., 2018. Zinc oxide nanoparticles boosts phenolic compounds and antioxidant activity of *Capsicum annuum* L. during germination. Agronomy 8, 215.

Glick, B.R., 2012. Plant growth-promoting bacteria: mechanisms and applications. Scientifica 2012.

Glick, B.R., Cheng, Z., Czarny, J., Duan, J., 2007. Promotion of plant growth by ACC deaminase-producing soil bacteria. N. Perspect. Approaches Plant. Growth-Promoting Rhizobacteria Res. Springer Dordrecht, Media B. V; UK, pp. 329–339.

Gond, S.K., Bergen, M.S., Torres, M.S., White, J.F., Kharwar, R.N., 2015. Effect of bacterial endophyte on expression of defense genes in Indian popcorn against Fusarium moniliforme. Symbiosis 66, 133–140.

Goswami, D., Pithwa, S., Dhandhukia, P., Thakker, J.N., 2014. Delineating Kocuria turfanensis 2M4 as a credible PGPR: a novel IAA-producing bacteria isolated from saline desert. J. Plant. Interact. 9, 566–576.

Goswami, P., Mathur, J., Srivastava, N., 2022. Silica nanoparticles as novel sustainable approach for plant growth and crop protection. Heliyon 8, e09908. 2022.

Gouda, S., Kerry, R.G., Das, G., Paramithiotis, S., Shin, H.-S., Patra, J.K., 2018. Revitalization of plant growth promoting rhizobacteria for sustainable development in agriculture. Microbiol. Res. 206, 131–140.

Gray, E., Smith, D., 2005. Intracellular and extracellular PGPR: commonalities and distinctions in the plant–bacterium signaling processes. Soil. Biol. Biochem. 37, 395–412.

Gupta, S., Kaushal, R., Gupta, S., 2017. Plant growth promoting Rhizobacteria: bioresouce for enhanced productivity of Solanaceous vegetable crops. Acta Sci. Agric. 1, 10–15.

Han, H., Lee, K., 2005. Phosphate and potassium solubilizing bacteria effect on mineral uptake, soil availability and growth of eggplant. Res. J. Agric. Biol. Sci. 1, 176–180.

Han, H.-S., Lee, K., 2006. Effect of co-inoculation with phosphate and potassium solubilizing bacteria on mineral uptake and growth of pepper and cucumber. Plant Soil Environ. 52, 130.

Han, C., Lalley, J., Namboodiri, D., Cromer, K., Nadagouda, M.N., 2016. Titanium dioxide-based antibacterial surfaces for water treatment. Curr. Opin. Chem. Eng. 11, 46–51.

Hayat, R., Ali, S., Amara, U., Khalid, R., Ahmed, I., 2010. Soil beneficial bacteria and their role in plant growth promotion: a review. Ann. Microbiol. 60, 579–598.

Hazarika, A., Yadav, M., Yadav, D.K., Yadav, H.S., 2022. An overview of the role of nanoparticles in sustainable agriculture. Biocatal. Agric. Biotechnol. 43, 1878–8181.

Herlekar, M., Barve, S., Kumar, R., 2014. Plant-mediated green synthesis of iron nanoparticles. J. Nanopart. 2014, 140614.

Hosseinpour, A., Haliloglu, K., Tolga Cinisli, K., Ozkan, G., Ozturk, H.I., Pour-Aboughadareh, A., et al., 2020. Application of zinc oxide nanoparticles and plant growth promoting bacteria reduces genetic impairment under salt stress in tomato (Solanum lycopersicum L. 'Linda'). Agriculture. 10 (11), 521.

Husen, A., 2022. Engineered Nanomaterials for Sustainable Agricultural Production, Soil Improvement and Stress Management. Elsevier Inc, Cambridge, MA, USA.

Husen, A., 2023. Nanomaterials and Nanocomposites Exposures to Plants (*Response, Interaction, Phytotoxicity and Defense Mechanisms*). Springer Nature Singapore Pte Ltd., Singapore.

Jackson, M., 1991. Ethylene in root growth and development. In: Matoo, A.K., Suttle, J.C. (Eds.), The Plant Hormone Ethylene. CRC Press, Boca Raton, FL.

Kanchiswamy, C.N., Malnoy, M., Maffei, M.E., 2015. Chemical diversity of microbial volatiles and their potential for plant growth and productivity. Front. Plant Sci. 6, 151.

Kaveh, R., Li, Y.-S., Ranjbar, S., Tehrani, R., Brueck, C.L., Van Aken, B., 2013. Changes in *Arabidopsis thaliana* gene expression in response to silver nanoparticles and silver ions. Environ. Sci. Technol. 47, 10637–10644.

Kaymak, H.C., 2011. Potential of PGPR in agricultural innovations. Plant Growth Health Promot. Bact. 45–79.

Klingshirn, C., Fallert, J., Zhou, H., Sartor, J., Thiele, C., Maier-Flaig, F., et al., 2010. 65 years of ZnO research—old and very recent results. Phys. Status Solidi. 247, 1424–1447.

Kloepper, J.W., Leong, J., Teintze, M., Schroth, M.N., 1980. Enhanced plant growth by siderophores produced by plant growth-promoting rhizobacteria. Nature 286, 885–886.

Kohan-Baghkheirati, E., Geisler-Lee, J., 2015. Gene expression, protein function and pathways of arabidopsis thaliana responding to silver nanoparticles in comparison to silver ions, cold, salt, drought, and heat. Nanomaterials. 5, 436–467.

Kolenčík, M., Ernst, D., Urík, M., Ďurišová, Ľ., Bujdoš, M., Šebesta, M., et al., 2020. Foliar application of low concentrations of titanium dioxide and zinc oxide nanoparticles to the common sunflower under field conditions. Nanomaterials 10, 1619. 2020.

Kołodziejczak-Radzimska, A., Jesionowski, T., 2014. Zinc oxide—from synthesis to application: a review. Materials. 7, 2833.

Kolya, H., Maiti, P., Pandey, A., Tripathy, T., 2015. Green synthesis of silver nanoparticles with antimicrobial and azo dye (Congo red) degradation properties using Amaranthus gangeticus Linn leaf extract. J. Anal. Sci. Technol. 6, 1–7.

Kumar, A., 2016. Phosphate solubilizing bacteria in agriculture biotechnology: diversity, mechanism and their role in plant growth and crop yield. Int. J. Adv. Res. 4, 116–124.

Kumar, A., Choudhary, A., Kaur, H., Mehta, S., Husen, A., 2021. Smart nanomaterial and nanocomposite with advanced agrochemical activities. Nano Res. Lett. 16 (156), 1–26.

Kumar, A., Choudhary, A., Kaur, H., Guha, S., Mehta, S., Husen, A., 2022. Potential applications of engineered nanoparticles in plant disease management: a critical update. Chemosphere 295, 133798.

Kumar, H., Bajpai, V.K., Dubey, R., Maheshwari, D., Kang, S.C., 2010. Wilt disease management and enhancement of growth and yield of Cajanus cajan (L) var. Manak by bacterial combinations amended with chemical fertilizer. Crop Prot. 29, 591–598.

Kumari, M., Pandey, S., Bhattacharya, A., Mishra, A., Nautiyal, C., 2017. Protective role of biosynthesized silver nanoparticles against early blight disease in Solanum lycopersicum. Plant Physiol. Biochem. 121, 216–225.

Lafmejani, Z.N., Jafari, A.A., Moradi, P., Moghadam, A.L., 2018. Impact of foliar application of copper sulphate and copper nanoparticles on some morpho-physiological traits and essential oil composition of peppermint (Mentha piperita L.). Herba Pol. 64, 13–24.

Landa, P., 2021. Positive effects of metallic nanoparticles on plants: overview of involved mechanisms. Plant Physiol. Biochem. 161, 12–24.

Lawre, S., Laware, S.L., Raskar, S., 2014. Influence of zinc oxide nanoparticles on growth, flowering and seed productivity in onion metal oxide nanomaterials view project influence of zinc and iron nano-prtilces on chickpea: nodulation and growth view project flowering and seed productivity in onion. Int. J. Curr. Microbiol. Sci. 3, 874–881.

Li, Q., Saleh-Lakha, S., Glick, B.R., 2005. The effect of native and ACC deaminase-containing Azospirillum brasilense Cd1843 on the rooting of carnation cuttings. Can. J. Microbiol. 51, 511–514.

Lian, J., Zhao, L., Wu, J., Xiong, H., Bao, Y., Zeb, A., et al., 2020. Foliar spray of TiO_2 nanoparticles prevails over root application in reducing Cd accumulation and mitigating Cd-induced phytotoxicity in maize (Zea mays L.). Chemosphere 239, 124794.

Lloret, L., Martínez-Romero, E., 2005. Evolution and phylogeny of rhizobia. Rev. Latinoam. Microbiol. 47, 43–60.

Lopez-Lima, D., Mtz-Enriquez, A., Carrión, G., Basurto-Cereceda, S., Pariona, N., 2021. The bifunctional role of copper nano-particles in tomato: effective treatment for Fusarium wilt and plant growth promoter. Sci. Hortic. 277, 109810. 2021.

López-Vargas, E.R., Ortega-Ortíz, H., Cadenas-Pliego, G., de Alba Romenus, K., Cabrera de la Fuente, M., Benavides-Mendoza, A., et al., 2018. Foliar application of copper nanoparticles increases the fruit quality and the content of bioactive compounds in tomatoes. Appl. Sci. 8, 1020.

Lugtenberg, B.J., Chin-A-Woeng, T.F., Bloemberg, G.V., 2002. Microbe–plant interactions: principles and mechanisms. Antonie Van Leeuwenhoek 81, 373–383.

Macwan, D.P., Dave, P.N., Chaturvedi, S., 2011. A review on nano-TiO_2 sol–gel type syntheses and its applications. J. Mater. Sci. 46, 3669–3686.

Magaña-López, E., Palos-Barba, V., Zuverza-Mena, N., Vázquez-Hernández, M.C., White, J.C., Nava-Mendoza, R., et al., 2022. Nanostructured mesoporous silica materials induce hormesis on chili pepper (Capsicum annuum L.) under greenhouse conditions. Heliyon 8, e09049.

Mahawar, H., Prasanna, R., Gogoi, R., Singh, S.B., Chawla, G., Kumar, A., 2020. Synergistic effects of silver nanoparticles augmented Calothrix elenkinii for enhanced biocontrol efficacy against Alternaria blight challenged tomato plants. 3 Biotech 10, 102.

Mahendran, D., Geetha, N., Venkatachalam, P., 2019. Role of silver nitrate and silver nano-particles on tissue culture medium and enhanced the plant growth and development. In Vitro Plant Breeding towards Novel Agronomic Traits: Biotic and Abiotic Stress Tolerance. Springer, Singapore, pp. 59–74.

Mahlambi, M.M., Ngila, C.J., Mamba, B.B., 2015. Recent developments in environmental photocatalytic degradation of organic pollutants: the case of titanium dioxide nanoparticles—a review. J. Nanomater. 2015, 790173.

Mahmood zadeh, H., Nabavi, M., Kashefi, H., 2013. Effect of nanoscale titanium dioxide particles on the germination and growth of canola (Brassica napus). J. Ornam. Hortic. Plants 3, 25–32.

Malviya, D., Sahu, P.K., Singh, U.B., Paul, S., Gupta, A., Gupta, A.R., et al., 2020. Lesson from ecotoxicity: revisiting the microbial lipopeptides for the management of emerging diseases for crop protection. Int. J. Environ. Res. Public Health 17, 1434.

Martínez-Viveros, O., Jorquera, M.A., Crowley, D., Gajardo, G., Mora, M., 2010. Mechanisms and practical considerations involved in plant growth promotion by rhizobacteria. J. Soil. Sci. Plant Nutr. 10, 293–319.

Marulanda, A., Azcón, R., Chaumont, F., Ruiz-Lozano, J.M., Aroca, R., 2010. Regulation of plasma membrane aquaporins by inoculation with a Bacillus megaterium strain in maize (Zea mays L.) plants under unstressed and salt-stressed conditions. Planta 232, 533–543.

Mcspadden Gardener, B.B., 2007. Diversity and ecology of biocontrol Pseudomonas spp. in agricultural systems. Phytopathology 97, 221–226.

Meena, M., Swapnil, P., Zehra, A., Aamir, M., Dubey, M.K., Patel, C.B., et al., 2019. Virulence factors and their associated genes in microbes. New and Future Developments in Microbial Biotechnology and Bioengineering. Elsevier, India.

Mejias, J.H., Salazar, F., Pérez Amaro, L., Hube, S., Rodriguez, M., Alfaro, M., 2021. Nanofertilizers: a cutting-edge approach to increase nitrogen use efficiency in grasslands. Front. Environ. Sci. 9 (635114), 1–8. Available from: https://doi.org/10.3389/fenvs.2021.635114.

Mittal, D., Kaur, G., Singh, P., Yadav, K., Ali, S.A., 2020. Nanoparticle-based sustainable agriculture and food science: recent advances and future outlook. Front. Nanotechnol. 2, 10.

Moaveni, P., Farahani, H.A., Maroufi, K., 2011a. Effect of TiO_2 nanoparticles spraying on barley (Hordeum vulgare L.) under field condition. Adv. Environ. Biol. 5, 2220–2223.

Moaveni, P., A. Talebi, A. Farahani and K. Maroufi, 2011b. Study of nanoparticles TiO_2 spraying on some yield components in barley (Hordem vulgare L.). Proceedings of the International Conference on Environmental and Agriculture Engineering IPCBEE, Volume 15, July 29–31, 2011, IACSIT Press, Jurong West, pp: 115–119.

Mohamadipoor, R., Mahboub Khomami, A., 2013. Effect of application of iron fertilizers in two methods "foliar and soil application" on growth characteristics of Spathyphyllum illusion. Eur. J. Exp. Biol. 3, 241–245.

Morais, M.C., Zhang, W., Baker, A.S., Zhang, G., Dunaway-Mariano, D., Allen, K.N., 2000. The crystal structure of Bacillus cereus phosphonoacetaldehyde hydrolase: insight into catalysis of phosphorus bond cleavage and catalytic diversification within the HAD enzyme superfamily. Biochemistry 39, 10385–10396.

Muller, T., Ruppel, S., Behrendt, U., Lentzsch, P., Müller, M.E., 2018. Antagonistic potential of fluorescent pseudomonads colonizing wheat heads against mycotoxin producing alternaria and fusaria. Front. Microbiol. 9, 2124.

Munoz-Rojas, J., Caballero-Mellado, J., 2003. Population dynamics of *Gluconacetobacter diazotrophicus* in sugarcane cultivars and its effect on plant growth. Microb. Ecol. 46, 454–464.

Mushtaq, T., Shah, A.A., Akram, W., Yasin, N.A., 2020. Synergistic ameliorative effect of iron oxide nanoparticles and *Bacillus subtilis* S4 against arsenic toxicity in *Cucurbita moschata*: Polyamines, antioxidants, and physiochemical studies. Int. J. Phytoremediat. 22, 1408–1419.

Mustafa, G., Hasan, M., Yamaguchi, H., Hitachi, K., Tsuchida, K., Komatsu, S., 2020. A comparative proteomic analysis of engineered and bio synthesized silver nanoparticles on soybean seedlings. J. Proteom. 224, 103833.

Mustafa, H., Ilyas, N., Akhtar, N., Raja, N.I., Zainab, T., Shah, T., et al., 2021. Biosynthesis and characterization of titanium dioxide nanoparticles and its effects along with calcium phosphate on physicochemical attributes of wheat under drought stress. Ecotoxicol. Environ. Saf. 223, 112519.

Nadeem, S.M., Zahir, Z.A., Naveed, M., Ashraf, M., 2010. Microbial ACC-deaminase: prospects and applications for inducing salt tolerance in plants. Crit. Rev. Plant Sci. 29, 360–393.

Nagar, N., Devra, V., 2018. Green synthesis and characterization of copper nano-particles using Azadirachta indica leaves. Mater. Chem. Phys. 213, 44–51.

Nair, P.M.G., Chung, I.M., 2014. Assessment of silver nanoparticle-induced physiological and molecular changes in Arabidopsis thaliana. Environ. Sci. Pollut. Res. 21, 8858–8869.

Nasir, S., 2016. Review on major potato disease and their management in Ethiopia. Int. J. Hortic. Floric. 4, 239–246.

Nejatzadeh, F., 2021. Effect of silver nanoparticles on salt tolerance of Satureja hortensis l. during in vitro and in vivo germination tests. Heliyon. 7, e05981.

Nasrollahzadeh, M., Momeni, S.S., Sajadi, S.M., 2017. Green synthesis of copper nanoparticles using Plantago asiatica leaf extract and their application for the cyanation of aldehydes using K4Fe(CN)6. J. Colloid Interface Sci. 506, 471–477.

Naveed, M., Hussain, M.B., Zahir, Z.A., Mitter, B., Sessitsch, A., 2014. Drought stress amelioration in wheat through inoculation with Burkholderia phytofirmans strain PsJN. Plant Growth Regul. 73, 121–131.

Nayana, A.R., Joseph, B.J., Jose, A., Radhakrishnan, E.K., 2020. Nanotechnological advances with PGPR applications. In: Hayat, S., Pichtel, J., Faizan, M., Fariduddin, Q. (Eds.), Sustainable Agriculture Reviews 41. Sustainable Agriculture Reviews, vol 41. Springer, Cham.

Ngalimat, M.S., Mohd Hata, E., Zulperi, D., Ismail, S.I., Ismail, M.R., Mohd Zainudin, N.A.I., et al., 2021. Plant growth-promoting bacteria as an emerging tool to manage bacterial rice pathogens. Microorganisms 9, 682.

Noori, A., Donnelly, T., Colbert, J., Cai, W., Newman, L.A., White, J.C., 2019. Exposure of tomato (Lycopersicon esculentum) to silver nanoparticles and silver nitrate: physiological and molecular response. Int. J. Phytoremediat. 22, 40–51.

Öktem, M., Keleş, Y., 2018. The role of silver ions in the regulation of the senescence process in Triticum aestivum. Turk. J. Biol. 42, 517–526.

Palmqvist, N.G.M., et al., 2015. Nano titania aided clustering and adhesion of beneficial bacteria toplant roots to enhance crop growth and stress management. Nature Publishing Group, pp. 1–12.

Parewa, H.P., Meena, V.S., Jain, L.K., Choudhary, A., 2018. Sustainable crop production and soil health management through plant growth-promoting rhizobacteria. Role Rhizospheric Microbes Soil: Volume 1: Stress. Manag. Agric. Sustainability. Springer, Singapore, pp. 299–329.

Paschalidis, K., Fanourakis, D., Tsaniklidis, G., Tzanakakis, V.A., Bilias, F., Samara, E., et al., 2021. Pilot cultivation of the vulnerable cretan endemic Verbascum arcturus L. (scrophulariaceae): effect of fertilization on growth and quality features. Sustainability 13, 14030.

Patel, T.S., Minocheherhomji, F.P., 2018. Plant growth promoting Rhizobacteria: blessing to agriculture. Int. J. Pure Appl. Biosci. 6, 481–492.

Peralta-Videa, J.R., Hernandez-Viezcas, J.A., Zhao, L., Diaz, B.C., Ge, Y., Priester, J.H., et al., 2014. Cerium dioxide and zinc oxide nanoparticles alter the nutritional value of soil cultivated soybean plants. Plant Physiol. Biochem. 80, 128–135.

Pishchik, V., Vorobyev, N., Chernyaeva, I., Timofeeva, S., Kozhemyakov, A., Alexeev, Y., et al., 2002. Experimental and mathematical simulation of plant growth promoting rhizobacteria and plant interaction under cadmium stress. Plant Soil 243, 173–186.

Pourjamshid, S.A., 2021. Study the effect of iron, zinc and manganese foliar application on morphological and agronomic traits of bread wheat (Chamran cultivar) under different irrigation regimes. Environ. Stress. Crop Sci. 14, 109–118.

Prasad, R., Jain, V., Varma, A., 2010. Role of nanomaterials in symbiotic fungus growth enhancement. Curr. Sci. 99, 1189–1191.

Priyanka N., Geetha N., Ghorbanpour M., Venkatachalam P. Advances in Phytonanotechnology. Academic Press, Cambridge, MA, USA: 2019. Role of engineered zinc and copper oxide nanoparticles in promoting plant growth and yield: Present status and future prospects, pp. 183–201.

Pullagurala, V.L.R., Adisa, I.O., Rawat, S., Kalagara, S., Hernandez-Viezcas, J.A., Peralta-Videa, J.R., et al., 2018. ZnO nanoparticles increase photosynthetic pigments and decrease lipid peroxidation in soil grown cilantro (Coriandrum sativum). Plant Physiol. Biochem. 132, 120–127.

Rai, A., Nabti, E., 2017. Plant growth-promoting bacteria: importance in vegetable production. Microb. Strateg. Veg. Prod. 23–48.

Raliya, R., Biswas, P., Tarafdar, J.C., 2015a. TiO_2 nanoparticle biosynthesis and its physiological effect on mung bean (Vigna radiata L.). Biotechnol. Rep. 5, 22–26.

Raliya, R., Nair, R., Chavalmane, S., Wang, W.-N., Biswas, P., 2015b. Mechanistic evaluation of translocation and physiological impact of titanium dioxide and zinc oxide nanoparticles on the tomato (Solanum lycopersicum L.) plant. Metallomics 7, 1584–1594.

Raymond, J., Siefert, J.L., Staples, C.R., Blankenship, R.E., 2004. The natural history of nitrogen fixation. Mol. Biol. Evol. 21, 541–554.

Reid, M.S. The role of ethylene in flower senescence. IV International Symposium on Postharvest Physiology of Ornamental Plants 261, 1988. 157–170.

Rizwan, M., Ali, S., Zia ur Rehman, M., Adrees, M., Arshad, M., Qayyum, M.F., et al., 2019. Alleviation of cadmium accumulation in maize (Zea mays L.) by foliar spray of zinc oxide nanoparticles and biochar to contaminated soil. Environ. Pollut. 248, 358–367.

Rouhani, M., Samih, M., Kalantari, S., 2013. Insecticidal Effect of Silica and Silver Nanoparticles on the Cowpea Seed Beetle, Callosobruchus maculates F. (Col.:Bruchidae).

Ryu, C.-M., Farag, M.A., Hu, C.-H., Reddy, M.S., Wei, H.-X., Paré, P.W., et al., 2003. Bacterial volatiles promote growth in Arabidopsis. Proc. Natl. Acad. Sci. 100, 4927–4932.

Sabir, S., Arshad, M., Chaudhari, S.K., 2014. Zinc oxide nanoparticles for revolutionizing agriculture: synthesis and applications. Sci. World J. 8. 2014.

Sabry, S.R., Saleh, S.A., Batchelor, C.A., Jones, J., Jotham, J., Webster, G., et al., 1997. Endophytic establishment of Azorhizobium caulinodans in wheat. Proc. R. Soc. Lond. Ser. B: Biol. Sci. 264, 341–346.

Saharan, B., Nehra, V., 2011. Plant growth promoting rhizobacteria: a critical review. Life Sci. Med. Res. 21, 30.

Saraf, M., Pandya, U., Thakkar, A., 2014. Role of allelochemicals in plant growth promoting rhizobacteria for biocontrol of phytopathogens. Microbiol. Res. 169, 18–29.

Saravanan, A., Kumar, P.S., Karishma, S., Vo, D.V.N., Jeevanantham, S., Yaashikaa, P.R., et al., 2021. A review on biosynthesis of metal nanoparticles and its environmental applications. Chemosphere. 264, 128580.

Sarma, R.K., Saikia, R., 2014. Alleviation of drought stress in mung bean by strain Pseudomonas aeruginosa GGRJ21. Plant Soil 377, 111–126.

Šebesta, M., Kolenčík, M., Sunil, B.R., Illa, R., Mosnáček, J., Ingle, A.P., et al., 2021. Field application of ZnO and TiO_2 nanoparticles on agricultural plants. Agronomy. 11 (11), 2281.

Seeda, A., Yassen, M.A., El-Nour, E.Z.A.E.M.A.A., Sahar, M., 2021. Effectiveness of biofertilizers for enhancing nutrients availability in rhizosphere zone, stimulate aggregate formation and their effects by climatic changes. Middle East. J. Appl. Sci. 11-03.

Selvakumar, G., Bindu, G.H., Bhatt, R.M., Upreti, K.K., Paul, A.M., Asha, A., et al., 2018. Osmotolerant cytokinin producing microbes enhance tomato growth in deficit irrigation conditions. Proc. Natl. Acad. Sci. India Sect. B: Biol. Sci. 88, 459–465.

Shabbir, Z., Sardar, A., Shabbir, A., Abbas, G., Shamshad, S., Khalid, S., et al., 2020. Copper uptake, essentiality, toxicity, detoxification and risk assessment in soil-plant environment. Chemosphere. 259, 127436.

Shah, A., Nazari, M., Antar, M., Msimbira, L.A., Naamala, J., Lyu, D., et al., 2021. PGPR in agriculture: a sustainable approach to increasing climate change resilience. Front. Sustain. Food Syst. 5, 667546.

Shafiq, T., Yasmin, H., Shah, Z.A., Nosheen, A., Ahmad, P., Kaushik, P., et al., 2022. Titanium oxide and zinc oxide nanoparticles in combination with cadmium tolerant Bacillus pumilus ameliorates the cadmium toxicity in maize. Antioxidants. 11 (11), 2156.

Shaikh, S., Sayyed, R., 2014. Role of plant growth-promoting rhizobacteria and their formulation in biocontrol of plant diseases. Plant microbes symbiosis: applied facets. Springer, New Delhi, India.

Shaikh, S.S., Sayyed, R.Z., REDDY, M., 2016. Plant growth-promoting rhizobacteria: an eco-friendly approach for sustainable agroecosystem, Plant, Soil Microbes: Implic. Crop Sci., 1. Springer, Springer International Publishing AG in Cham, pp. 181–201.

Shang, Y., Hasan, M.K., Ahammed, G.J., Li, M., Yin, H., Zhou, J., 2019. Applications of nanotechnology in plant growth and crop protection: a review. Molecules 24 (14), 2558.

Sharma, P., Pant, S., Dave, V., Tak, K., Sadhu, V., Reddy, K.R., 2019. Green synthesis and characterization of copper nano-particles by Tinospora cardifolia to produce nature-friendly copper nano-coated fabric and their antimicrobial evaluation. J. Microbiol. Methods 160, 107–116.

Sharma, P., Pandey, V., Sharma, M.M.M., Patra, A., Singh, B., Mehta, S., et al., 2021. A review on biosensors and nanosensors application in agroecosystems. Nano Res. Lett. 16, 136.

Shcherbakova, E.N., Shcherbakov, A.V., Andronov, E., Gonchar, L.N., Kalenska, S., Chebotar, V.K., 2017. Combined pre-seed treatment with microbial inoculants and Mo nanoparticles changes composition of root exudates and rhizosphere microbiome structure of chickpea (Cicer arietinum L.) plants. Symbiosis. 73, 57–69.

Siddiqui, M.H., Al-Whaibi, M.H., 2014. Role of nano-SiO 2 in germination of tomato (Lycopersicum Esculentum seeds mill.). Saudi J. Biol.Sci. 21, 13–17.

Silva, R.M., TeeSy, C., Franzi, L., Weir, A., Westerhoff, P., Evans, J.E., et al., 2013. Biological response to nano-scale titanium dioxide (TiO_2): role of particle dose, shape, and retention. J. Toxicol. Environ. Heal. Part. A 76, 953–972.

Singh, R., Pandey, D., Kumar, A., Singh, M., 2017. PGPR isolates from the rhizosphere of vegetable crop Momordica charantia: characterization and application as biofertilizer. Int. J. Curr. Microbiol. Appl. Sci. 6, 1789–1802.

Singh, J., Kumar, S., Alok, A., Upadhyay, S.K., Rawat, M., Tsang, D.C.W., et al., 2019a. The potential of green synthesized zinc oxide nanoparticles as nutrient source for plant growth. J. Clean. Prod. 214, 1061–1070.

Singh, M., Singh, D., Gupta, A., Pandey, K.D., Singh, P., Kumar, A., 2019b. Plant growth promoting rhizobacteria: application in biofertilizers and biocontrol of phytopathogens. PGPR amelioration in sustainable agriculture. Elsevier.

Singh, P., Arif, Y., Siddiqui, H., Sami, F., Zaidi, R., Azam, A., et al., 2021. Nanoparticles enhances the salinity toxicity tolerance in Linum usitatissimum L. by modulating the antioxidative enzymes, photosynthetic efficiency, redox status and cellular damage. Ecotoxicol. Environ. Saf. 213, 112020.

Soliman, M., Qari, S.H., Abu-Elsaoud, A., El-Esawi, M., Alhaithloul, H., Elkelish, A., 2020. Rapid green synthesis of silver nanoparticles from blue gum augment growth and performance of maize, fenugreek, and onion by modulating plants cellular antioxidant machinery and genes expression. Acta Physiol. Plant. 42, 148.

Somers, E., Vanderleyden, J., Srinivasan, M., 2004. Rhizosphere bacterial signalling: a love parade beneath our feet. Crit. Rev. Microbiol. 30, 205–240.

Son, J.-S., Sumayo, M., Hwang, Y.-J., Kim, B.-S., Ghim, S.-Y., 2014. Screening of plant growth-promoting rhizobacteria as elicitor of systemic resistance against gray leaf spot disease in pepper. Appl. Soil Ecol. 73, 1–8.

Sreelekshmi, R., Siril, E.A., Muthukrishnan, S., 2021. Role of biogenic silver nanoparticles on hyperhydricity reversion in Dianthus chinensis L. an in vitro model culture. J. Plant. Growth Regul. 41, 23–39.

Tabatabaee, S., Iranbakhsh, A., Shamili, M., Ardebili, Z.O., 2021. Copper nanoparticles mediated physiological changes and transcriptional variations in microRNA159 (miR159) and mevalonate kinase (MVK) in pepper, potential benefits and phytotoxicity assessment. J. Environ. Chem. Eng. 9, 106151.

Tan, W., Du, W., Barrios, A.C., Armendariz, R., Zuverza-Mena, N., Ji, Z., et al., 2017. Surface coating changes the physiological and biochemical impacts of nano-TiO_2 in basil (Ocimum basilicum) plants. Environ. Pollut. 222, 64–72.

Tarafdar, J.C., Agrawal, A., Raliya, R., Kumar, P., Burman, U., Kaul, R.K., 2012. ZnO nanoparticles induced synthesis of polysaccharides and phosphatases by Aspergillus Fungi. Adv. Sci. Eng. Med. 4, 324–328.

Tarafdar, A., Raliya, R., Wang, W.-N., Biswas, P., Tarafdar, J., 2013. Green synthesis of TiO_2 nanoparticle using Aspergillus tubingensis. Adv. Sci. Eng. Med. 5, 943–949.

Tejera, N., Lluch, C., Martinez-Toledo, M., Gonzalez-Lopez, J., 2005. Isolation and characterization of Azotobacter and Azospirillum strains from the sugarcane rhizosphere. Plant. Soil. 270, 223–232.

Tian, L., Shen, J., Sun, G., Wang, B., Ji, R., Zhao, L., 2020. Foliar application of SiO_2 nanoparticles alters soil metabolite profiles and microbial community composition in the pakchoi (Brassica chinensis L.) rhizosphere grown in contaminated mine soil. Environ. Sci. Technol. 54, 13137–13146.

Torabian, S., Zahedi, M., Khoshgoftarmanesh, A., 2016. Effect of foliar spray of zinc oxide on some antioxidant enzymes activity of sunflower under salt stress. J. Agric. Sci. Technol. 18, 1013–1025.

Tripathi, D.K., Singh, S., Singh, V.P., Prasad, S.M., Dubey, N.K., Chauhan, D.K., 2017. Silicon nanoparticles more effectively alleviated UV-B stress than silicon in wheat (Triticum aestivum) seedlings. Plant Physiol. Biochem. 110, 70–81.

Tsegaye, Z., Assefa, F., Beyene, D., 2017. Properties and application of plant growth promoting rhizobacteria. Int. J. Curr. Trends Pharmacobiol. Med. Sci. 2, 30–43.

Vacheron, J., Desbrosses, G., Bouffaud, M.-L., Touraine, B., Moënne-Loccoz, Y., Muller, D., et al., 2013. Plant growth-promoting rhizobacteria and root system functioning. Front. Plant Sci. 4, 356.

Vafa, Z., Sohrabi, Y., Sayyed, R., Suriani, N.L., Datta, R., 2021. Effects of the combinations of rhizobacteria, mycorrhizae, and seaweed, and supplementary irrigation on growth and yield in wheat cultivars. Plants 10, 811.

Vaghar, M.S., Sayfzadeh, S., Zakerin, H.R., Kobraee, S., Valadabadi, S.A., 2020. Foliar application of iron, zinc, and manganese nano-chelates improves physiological indicators and soybean yield under water deficit stress. J. Plant Nutr. 43, 2740–2756.

Vaikuntapu, P.R., Dutta, S., Samudrala, R.B., Rao, V.R., Kalam, S., Podile, A.R., 2014. Preferential promotion of Lycopersicon esculentum (Tomato) growth by plant growth promoting bacteria associated with tomato. Indian J. Microbiol. 54, 403–412.

Vedamurthy, A.B., Bhattacharya, S., Das, A., Shruthi, S.D., 2021. The Netherlands: Exploring nanomaterials with rhizobacteria in current agricultural scenario. Adv. Nano-Fertilizers Nano-Pesticides Agriculture. Elsevier, Amsterdam, pp. 487–503.

Vejan, P., Abdullah, R., Khadiran, T., Ismail, S., Nasrulhaq Boyce, A., 2016. Role of plant growth promoting rhizobacteria in agricultural sustainability—a review. Molecules 21, 573.

Venkatachalam, P., Priyanka, N., Manikandan, K., Ganeshbabu, I., Indiraarulselvi, P., Geetha, N., et al., 2017. Enhanced plant growth promoting role of phycomolecules coated zinc oxide nanoparticles with P supplementation in cotton (Gossypium hirsutum L.). Plant. Physiol. Biochem. 110, 118–127.

Vijayan, R., Palaniappan, P., Tongmin, S., Elavarasi, P., Manoharan, N., 2013. Rhizobitoxine enhances nodulation by inhibiting ethylene synthesis of Bradyrhizobium elkanii from Lespedeza species: validation by homology modelingand molecular docking study. World J. Pharm. Pharm. Sci. 2, 4079–4094.

Wani, P.A., Khan, M.S., Zaidi, A., 2007. Synergistic effects of the inoculation with nitrogen-fixing and phosphate-solubilizing rhizobacteria on the performance of field-grown chickpea. J. Plant Nutr. Soil Sci. 170, 283–287.

Weller, D.M., Landa, B., Mavrodi, O., Schroeder, K., De La Fuente, L., Bankhead, S.B., et al., 2007. Role of 2, 4-diacetylphloroglucinol-producing fluorescent Pseudomonas spp. in the defense of plant roots. Plant Biol. 9, 4–20.

Xiong, K., Fuhrmann, J., 1996. Comparison of rhizobitoxine-induced inhibition of β-cystathionase from different bradyrhizobia and soybean genotypes. Plant Soil 186, 53–61.

Yadav, J., Verma, J.P., Jaiswal, D.K., Kumar, A., 2014. Evaluation of PGPR and different concentration of phosphorus level on plant growth, yield and nutrient content of rice (Oryza sativa). Ecol. Eng. 62, 123–128.

Yasmeen, F., Razzaq, A., Iqbal, M.N., Jhanzab, H.M., 2015. Effect of silver, copper and iron nanoparticles on wheat germination. Int. J. Biosci. 6, 112–117.

Zhou, D., Feng, H., Schuelke, T., De Santiago, A., Zhang, Q., Zhang, J., et al., 2019. Rhizosphere microbiomes from root knot nematode non-infested plants suppress nematode infection. Microb. Ecol. 78, 470–481.

CHAPTER

4

Role of protein hydrolysates in plants growth and development

Sashi Sonkar[1], *Priti Pal*[2] *and Akhilesh Kumar Singh*[3]

[1]Department of Botany, Bankim Sardar College, Tangrakhali, West Bengal, India [2]Environmental Engineering, Shri Ramswaroop Memorial College of Engineering & Management, Lucknow, Uttar Pradesh, India [3]Department of Biotechnology, School of Life Sciences, Mahatma Gandhi Central University, Motihari, Bihar, India

4.1 Introduction

As the world's population rises, agriculture will need to find ways to increase food production while reducing its negative impact on the environment and public health. Numerous solutions have been offered to fulfill global food demand, with a primary emphasis on breeding varieties with higher productivity. However, this one-size-fits-all method leads to limited advantages, particularly considering that the genetic potential of staple crops is almost at its maximum. On the other hand, it was assumed that optimizing crop management and enhancing resource utilization under varied environmental zones holds the key to sustainably increasing agricultural production and gives rise to the concept of "more with less" (Colla et al., 2017).

The utilization of protein hydrolysates (PHs) is a cutting-edge technique that shows great promise for addressing these demanding issues. The use of PHs as biostimulants has been lauded for a variety of agronomic and horticultural crops. The PHs were produced from protein sources through the process of partial hydrolysis and were composed of amino acids, oligopeptides, and polypeptides (Schaafsma, 2009). They may be used as a foliar dressing or sprayed on the leaves, and they come in a variety of forms, such as soluble granules and powers as well as liquid extracts (Colla et al., 2015). The PHs are prepared via enzymatic hydrolysis, thermal (heat), and chemical processes (acid and alkaline hydrolysis), which include a diverse array of plant biomass (vegetable by-products, corn wet-milling, alfalfa hay, and legume seeds) and animal wastes (connective or epithelial tissues such as casein, chicken feathers, fish by-products, blood meal, and leather by-

Biostimulants in Plant Protection and Performance
DOI: https://doi.org/10.1016/B978-0-443-15884-1.00009-9

products) (Colla et al., 2015, 2017). Businesses and the scientific community are increasingly interested in PHs derived from the maize wet-milling sector and vegetable byproducts because they may provide a long-term, cost-effective, and environmentally responsible solution to the waste disposal issue (Pecha et al., 2012; Baglieri et al., 2014). At present, the acid hydrolysis of animal proteins accounts for the majority of the market, followed by the enzymatic hydrolysis of plant proteins for PH biostimulants in the markets (Colla et al., 2015). However, countries like India, China, the United States, Spain, and Italy are the primary producers of PHs for agricultural applications worldwide. The meat and leather industries in Europe and East Asia have spawned a number of these businesses to valorize their byproducts by turning them into fertilizer and biostimulants. However, in recent years, several businesses have brought plant-derived PHs to the Asian, European, and U.S. markets, and these PHs are finding increased acceptability by farmers due to their richness in bioactive components and their remarkable effectiveness in boosting crop performances (Colla et al., 2017).

Through the regulation of physiological and molecular processes in plants, PHs have been shown to play important roles as biostimulants, mitigating the effects of abiotic stress as well as causing growth and increasing yields in plants (Yakhin et al., 2017). The term "abiotic stress" may refer to a variety of environmental factors, such as water, nutrients, temperature, heavy metals, and salinity stresses (Colla et al., 2017; Rouphael et al., 2017). The responses of plants to PHs include regulation of the activities of three enzymes in the Krebs cycle (such as malate dehydrogenase, isocitrate dehydrogenase, and citrate synthase), regulation of nitrogen uptake facilitated by significant enzymes involved in the process of nitrogen assimilation, and increased nitrogen and carbon metabolisms (Nardi et al., 2016). Due to the inclusion of bioactive peptides, PHs may potentially affect hormonal activities in plants (Colla et al., 2015). Many commercial products derived from PHs have been shown in many studies to stimulate plant hormone-like actions (gibberellins and auxin), which in turn boost shoot and root growth and crop yields (Lucini et al., 2015). It has been shown that the application of PHs to plants and soils has indirect impacts on nutrient uptake and the development of plants (du Jardin, 2015). Applying nutrients at the root and foliar levels has been demonstrated to increase the abilities of plant species to absorb and use both micronutrients and macronutrients (Ertani et al., 2009; Halpern et al., 2015). The architecture of the root systems was changed (number of lateral roots, length, and density), increased nutrient availability in the soil solution due to the complexation of nutrients by amino acids and peptides, and increased microbial activity, all of which have been linked to improved nutrient uptake performance of PH-treated plants (Colla et al., 2015; du Jardin, 2015). Some researchers have shown that using PHs produced by animals may have negative consequences for fruiting plants, including growth suppression and phytotoxicity (Colla et al., 2017). Overconsumption of free amino acids by leaves results in an intracellular amino acid imbalance, an energy drain from the active transport of amino acids, a reduction in nitrate uptake, and an increase in cellular susceptibility to apoptosis; this phenomenon is known as "general amino acid inhibition" (Bonner and Jensen, 1997).

Researchers have made great strides in deciphering the biostimulant qualities of PHs, but much remains unknown regarding the target metabolic pathways and mechanisms of action evoked by their administration. Furthermore, PHs may differ in their biostimulant activities based on factors such as leaf solubility and permeability, mode (root versus leaf) and time of application, concentration, growing conditions, phenological stages, cultivars, species, characteristics, and origin (Colla et al., 2015). Since PH-based biostimulants are often sprayed on the leaves, it is essential that the active components (amino acids and peptides) make it deep within the treated plants (Colla et al., 2015; Yakhin et al., 2017). The observed biostimulant action after PH treatment could be functioning indirectly via a microbial-mediated improvement of plant health (Colla et al., 2014). In recent years, it has been generally recognized that microorganisms may increase plant fitness by modifying physiological and developmental processes, leading to increased nutrient and water intake and higher resistance against environmental stresses (Philippot et al., 2013). In recent times, several PHs have been documented as plant biostimulants by improving precise traits in plant species, and others are under continuous assessment by omic as well as phenomic-based approaches (Colla et al., 2017).

4.2 Chemical characteristics of protein hydrolysates

Most of the components of PHs are free amino acids and peptides (Calvo et al., 2014). Carbohydrates as well as trace amounts of phytohormones, phenols, mineral elements, and other organic compounds are all possible components of PHs (Ertani et al., 2014; Colla et al., 2015). The essential roles of amino acids in plant biochemistry include acting as signaling in hormone metabolism (alanine and proline), promoting seed germination (aspartic acid, glutamic acid, lysine, methionine, phenylalanine, and threonine), promoting chlorophyll biogenesis (alanine, lysine, and serine), chelating agents (cysteine, glutamic acid, glycine, histidine, and lysine), and anti-stress (hydroxyproline and proline). However, considerable care must be taken during the processing of protein sources to ensure the reliability and functioning of amino acids (Paleckiene et al., 2007; Popko et al., 2018). The PHs have different chemical profiles based on the method of preparation (enzymatic or chemical hydrolysis) and protein source (e.g., collagen from alfaalfa biomass, legume seeds, fish by-products, and leather by-products) (Colla et al., 2017).

Proline and glycine are two of the most abundant amino acids in collagen-derived PHs, whereas glutamic and aspartic acids are the most abundant in PHs generated from fish and legumes (Colla et al., 2015, 2017). In addition, hydroxylysine and hydroxyproline, both of which may be employed as markers for this kind of PH, are often found in high concentrations in collagen-derived PHs (Colla et al., 2015). Chemical hydrolysis using acids (H_2SO_4 and HCl) at high temperatures ($>121°C$) and pressures (>220.6 kPa) is often used to create PHs originating from animals. Due to the dominant nature of acid hydrolysis, the final products are mostly made up of free amino acids and a lesser amount of soluble peptides. Some amino acids, such as threonine, serine, cysteine, and tryptophan, are completely destroyed during acid hydrolysis, while many others have their biological activity reduced because they are changed from the L-form to the D-form (racemization) (Colla et al., 2015, 2017). When compared to PHs created by chemical hydrolysis, those derived from plants have a greater ratio of peptides: free amino acids as well as a higher percentage of L-amino acids since they are synthesized using a mild approach such as enzymatic hydrolysis utilizing temperatures below 60°C and proteolytic enzymes. Peptides range in size from several hundred to a thousand Daltons, with the smaller ones having more biological activity

(Quartieri et al., 2002). Peptides of biological interest, particularly those found in plants, have been extracted and chemically described. The soybean-derived PHs consist of a short peptide of 12 amino acids known as "root hair promoting peptide" has been identified (Matsumiya and Kubo, 2011). Plant tissues have been shown to contain a wide variety of bioactive peptides, such as clavata3, phytosulfokines, and systemins, that operate as signaling molecules in plant development, growth, and defense (Ryan et al., 2002). The soluble peptides and free amino acids account for 75% of the total protein content in the legume-derived PH biostimulant Trainer (Italpollina S.P.A., Italy), while carbohydrates and mineral nutrients were 22% and 3%, respectively (Di Mola et al., 2020; Lucini et al., 2020). In lamb's lettuce and spinach, the foliar application of Trainer produced a significantly enhanced nitrogen uptake in comparison to untreated plants (Di Mola et al., 2020). Collagen has high assimilation due to its abundant amounts of glycine, proline, and lysine (Colla et al., 2015), whereas feather keratins have low bioavailability and are deficient in the amino acids histidine, lysine, methionine, and tryptophan (Callegaro et al., 2019), with distinct functions in plant development and metabolism. Some PH may have reduced biostimulant action because of a lack of certain amino acids (Martín et al., 2022). The major components of PH-based biostimulants obtained from various sources have been documented in Table 4.1.

4.3 Impact of protein hydrolysates on plant growth and developmental traits

The agricultural sector is always on the lookout for new ways to improve plant quality, high yield, and growth (Martín et al., 2022). Biostimulants were produced from an array of seafood and agro sources; these are chemical substances or macromolecules that can promote resistance to abiotic and biotic stress, increase nutrient uptake and plant growth, alter the physiological processes of plants, and improve the overall quality and quantity of their harvests (Shukla et al., 2019; Martín et al., 2022). The use of biostimulants such as microorganisms, humic acid, algal extract, chitosan, or protein hydrolysate in combination with conventional fertilizers may also mitigate some of the negative effects of agrochemicals and pesticides (Aktsoglou et al., 2021).

4.3.1 Root and foliar applications of protein hydrolysates

The physiological reactions of plants to PH treatment appear to be influenced by various factors such as sources of proteins, amino acid content of hydrolysates and dose, and application manner (Aktsoglou et al., 2021). When PHs (granular, soluble powder, or liquid form) are applied as a root or foliar spray, it stimulates plant metabolism and may influence the production and quality of plants (Colla et al., 2017) (Fig. 4.1). The stability of PH used in agricultural nutritional programs relies heavily on its amino acid content. To avoid insoluble aggregates or unwanted interactions with other nutrients, notably minerals due to their chelating features, it has been suggested that PH added to fertigation solutions needs a high balance of hydrophobic or hydrophilic oligopeptides (Ertani et al., 2013a). Furthermore, the appropriate use of peptides via the root system is dependent on several factors, including the coefficient of root absorption, the biochemistry of soil, secretions of

TABLE 4.1 Major components of PH-based biostimulants obtained from various sources.

Source	Formulation	Hydrolysis method	Major Compositions	References
Sunflower defatted seed meal	Liquid	Two-step enzymatic hydrolysis (Alcalase-Flavourzyme)	Most peptides consist of low molecular weight (<25 kDa), free amino acid contents: 164.4 g/kg, carbon content: 41.7%, organic matter: 75.8%, protein content: 67.7%	Ugolini et al. (2015)
Alfalfa	Liquid	Not reported	Total amino acid contents: 5.1%, free amino acids contents: 1.5%, organic matter contents: 23%.	Soppelsa et al. (2018)
Fish (heads and tails)	Liquid	Enzymatic hydrolysis (Alcalase)	Total organic matter contents: 87.2%, total amino acid contents: 99.75%.	Al-Malieky and Jerry (2019)
		Enzymatic hydrolysis (Flavourzyme)	Total organic matter contents: 84.07%, total amino acid contents: 96.73%.	
Chickpea	Liquid	Enzymatic hydrolysis	Organic nitrogen contents: 5.11%, total nitrogen contents: 2.7%, total carbon content: 23.6%, 17 free amino acids were identified and the highest content was reported for arginine (0.223%).	Ertani et al. (2019)
Chicken feather	Liquid	Carbon dioxide-assisted pressure hydrolysis	Free amino acids contents: 0.039 g/L; peptides contents: 2.81 g/L	Schmidt et al. (2020)
Legume seeds (Trainer)	Liquid	Enzymatic hydrolysis	Amino acids contents: 27%, total nitrogen contents: 5%, organic matter contents: 35.5%, and also contains soluble peptides.	Choi et al. (2022)
Chicken feather	Soluble power	Alkaline hydrolysis	Amino acids contents: 165.5 mg/g, total nitrogen contents: 3.06%, total carbon contents: 9.42%.	Raguraj et al. (2022)

hydrolytic enzymes from the root, and so on (Moreno-Hernández et al., 2020). Additionally, proper amino acid or peptide acquisition during foliar application of PHs depends on several elements, including stomata activity of leaf porosity, leaf area to promote diffusion, evaporation parameters, temperature, relative humidity, and peptide sequence or size (Koukounararas et al., 2013). Sestili et al. (2018) reported that the addition of PH to the soil is more successful than using foliar sprays in enhancing the growth of plants and total nitrogen absorption as nitrate transporters are activated by the free amino acids of PHs. When used as a nutrient hydroponic solution, amino acids in PH may be just as effective as inorganic nitrogen fertilizers (Aktsoglou et al., 2021). The addition of PH such as Amino16® (Evyp LLP, Greece) to the nutrient solution for hydroponically grown spearmint and peppermint did not show negative or positive plant development, which was ascribed to either the enhanced root growth or low rate (lower than 0.5%) of PHs administered (Aktsoglou et al., 2021). There is evidence to show that amino acid transport proteins in plants absorb and translocate PH-derived short peptides and amino acids from root or leaf tissues, which are involved in intracellular transport, import into seed, xylem phloem transfer, and phloem unloading and loading (Yang et al., 2020).

The utilization of PHs as foliar spray makes more peptides and amino acids available to plants for absorption by lowering microbial competition (Colla et al., 2015). The swift

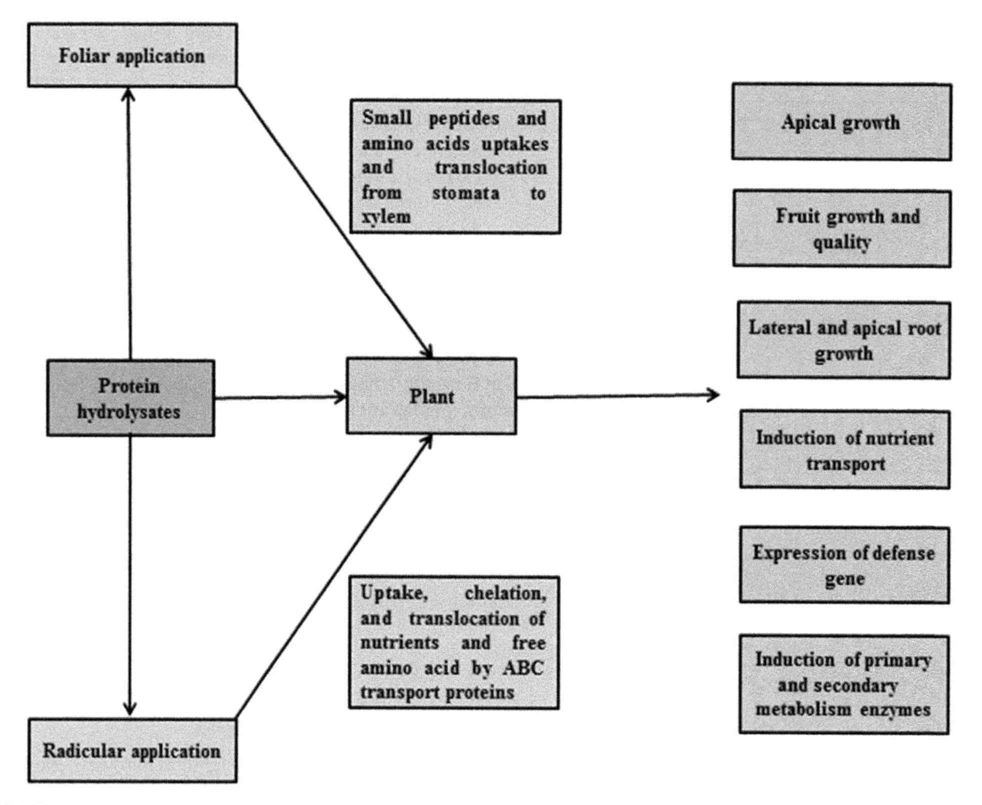

FIGURE 4.1 Schematic representation of the impact of radicular and foliar application of protein hydrolysate on tomato plants.

uptake of glutamic acid through the bentgrass (creeper) foliage serves as a direct precursor for the biosynthesis of proline and gamma-aminobutyric acid. These two crucial metabolites play well-established roles in enhancing plant stress tolerance (Rouphael and Colla, 2020). For the production of indoleacetic acid (IAA), a hormone with significant effects on plant development, tryptophan is regarded as a key amino acid. When used alone, however, its efficacy may suffer (Martín et al., 2022). L-methionine was shown to increase lettuce growth characteristics in research; however, L-tryptophan and L-glycine had different effects when administered to the root of hydroponically grown butterhead lettuce (Rouphael and Colla, 2020). According to Paul et al. (2019), when PH is applied to the foliar, it is absorbed through the stomata, epidermal cells, and cuticle to reach the mesophyll, whereas when applied to the root system hydroponically it is absorbed through the root epidermal cells via ABC membrane transport and redistributed through the xylem. Most PH-based biostimulants induce beneficial physiological effects like growth and development, and they also improve nutrient uptake from soil or rhizosphere microorganisms. However, PH must be able to penetrate the plant tissue at low dosages, which in turn depends on the vegetative stage, cultivars, and species, as well as environmental factors, stomata, and the cuticle, which act as obstacles (Pecha et al., 2012).

4.3.2 Effects of protein hydrolysates applications in plant growth and quality

The biostimulants impact significantly on various traits of plants in terms of quality and productivity, which include enhancement of microelements and macronutrients in vegetative cells, antioxidant activity, stimulation of growth, phenolic contents, protein, increased pigment content, photosynthetic rate, changes in the endogenous phytohormone levels, and root architecture (Ambrosini et al., 2021; Martín et al., 2022). A recent investigation suggested that PH-based biostimulants include compounds that exhibit plant hormone-like actions (Moreno-Hernández et al., 2020). The peptides associated with PHs may perform gibberellin (GA)-like and auxin-like functions, which activate signaling pathways and promote vegetal growth of crops and early maturing of fruits (Drobek et al., 2019), effects activated by few endogenous controlling protein-like hormones (such as phytosulfokine) and signaling peptides affect productive traits for instance thickness of stem, root length, and fruit maturation or encouraging secondary and primary metabolism biosynthesis via the initiation of numerous signaling pathways that comprise second messengers the encourage the nitrate assimilation pathway enzymes, such as glutamine synthetase and nitrate reductase which catalyze a rate-limiting step in the absorption of nitrogen (Ertani et al., 2013b). The PH prepared from *Spirulina platensis* and *Cicer arietinum* L. exhibited GA-like and indole-3-acetic acid (IAA)-like activities and stimulated the growth and accretion of nitrogenous compounds such as phenols, chlorophylls, and proteins in a hydroponically grown *Zea mays* L. culture (Ertani et al., 2019). Additionally, it was observed that the PH also augmented the activities of esterase and peroxidase associated with the differentiation of organogenesis and plant growth. Similarly, in tomato plants, the improvement in root growth facilitated by chorismate-derived phytohormones such as salicylic acid has been reported by Casadesús et al. (2020). The foliar application of vegetable-derived PH on cuttings of tomato plants was found to stimulate rooting, length, and quantity of lateral root branching and increase biomass density. This led to increased plant development and growth, which was attributed to the stimulation of IAA precursors or GA, cytokinin, and auxins biosynthesis (Ceccarelli et al., 2021). The foliar application of fish by-products derived from PH on lettuce displays noteworthy effects on the shoot-fresh weight, total yield, total soluble solids, dry matter, proline, carbohydrate, and the total number and area of leaves (Al-Malieky and Jerry, 2019). The animal-PH (bovine epithelium hydrolysate), Trainer (legume-derived PH), and plant-PH Tyson obtained from soy protein extract, whether applied radicular or foliar enhanced growth and nutrient balance in celery (Consentino et al., 2020). In a study conducted, the biostimulant efficacy of five plant-derived PHs was tested on tomatoes to see whether or not they might stimulate roots in tomato cuttings after a brief dipping procedure. All of the PHs resulted in an increase in root number (between 37% and 56%) and root length (between 45% and 93%) (Ceccarelli et al., 2021).

In addition to its action as a biostimulant, PH also facilitates plant acclimation to several different stress situations, which include salinity, drought, heat, cold, and mineral depletion. Moreover, to mitigate the deleterious effects of saline stress on horticultural crops and fruit trees, StresSal and Trainer have been applied as osmoregulators (Visconti et al., 2015; Luziatelli et al., 2019). Ambrosini et al. (2021) examined in hydroponic culture the ability of a commercially obtained bovine collagen-derived PH to mitigate Fe deficiency stress in the roots of Zea mays. It was reported that PH revealed improved growth and absorption of the roots by chelation of the Fe area in comparison to the control condition. These investigations demonstrate the positive effect of PH in hydroponic or foliar culture. The applications of PH drench on *Solanum lycopersicum* L. plants improve transpiration use efficiency and transpiration rate, with a progressive effect on the metabolic profile and biomass (Paul et al., 2019).

Proline or proline-precursor (ornithine and/or glutamate) in hydrolysates have been shown to have protective benefits in several PH-based biostimulants owing to their chemical chaperone and osmolyte activities under diverse stressful environmental conditions during plant growth. The exact method by which peptides present in PH work is still unclear, but the majority of the data points to the fact that these compounds not only directly provide nutrients to plants but also encourage their uptake and serve as an alternative to artificial fertilizers. The PHs have a beneficial impact on plant crops because they include free amino acids and signaling peptides that improve crop yield, the quality of fruits and vegetables, seedling development, and germination (Rouphael and Colla, 2020). In Table 4.2, various examples of PH are provided whose potential as plant biostimulants has been examined for a variety of plant attributes.

TABLE 4.2 Role of PH-based bio-stimulant on plant growth and developments during last one decade.

Plant	Protein hydrolysate	Mode of application	Enhancement of plant traits	References
Chickpea	Chicken feathers PH	Radicular	Increased root phenology maintaining surface area, mass, and primary specific growth rate. Stimulated secondary root growth and development, germination rate, and seedling vigor,	Paul et al. (2013)
Lettuce	Amino 16®	Radicular and foliar	Increases the production of antioxidants, leaves chemistry quality, yield, leaf morphology uniformity, and nutrient uptake.	Tsouvaltzis et al. (2014)
Lettuce	Legume seeds PH	Radicular and foliar	Osmoregulates under saline conditions, increase stress response to reactive oxygen species indicators, root physiology, chlorophyll and protein production, and stem growth rate.	Lucini et al. (2015)
Wheat	Terra-Sorb complex	Foliar	Increases yield and seed sugar content, photosynthetic rate, mass, and leaves area.	Martinez-Esteso et al. (2016)
Lettuce	Fish PH	Radicular	Increases quality traits such as nutraceutical and succulence, shoot fresh and dry weight, stem diameter, and leaves number.	Xu and Mou (2017)
Apple	Alfalfa PH	Foliar	Enhances biotic post-harvest resistance, anthocyanin content of fruits, and quality traits such as sugar content and color index.	Soppelsa et al. (2018)
Soybean	Terra-Sorb complex	Foliar	Improves oil content, seed quality and yield, and pod numbers.	Kocira (2019)

(*Continued*)

TABLE 4.2 (Continued)

Plant	Protein hydrolysate	Mode of application	Enhancement of plant traits	References
Celery	Bovine animal epithelium and soy extracted PHs	Foliar	Increases yield, nutrient uptake, weight, and plant length.	Consentino et al. (2020)
Tomato	CycoFlow	Radicular	Improves pollen viability, water status, yield, and antioxidant contents under drought stress.	Francesca et al. (2021)
Wheat	AGROMOREE (PH of Rainbow Trout (*Oncorhynchus mykiss*))	Radicular	Promotes germination, length and number of roots, seed weight, quality of grains in the ear, length and quality of ears, protein, and gluten content of seeds.	Mironenko et al. (2022)
Maize	Chicken feather PH and *Bacillus pumilus* AR57	Radicular	Enhances carbohydrate, protein, and chlorophyll contents.	Jagadeesan et al. (2023)

4.4 Conclusion and future prospects

New forms of agriculture must be implemented to achieve sustainable development and the goal of food security. Because of the constraints of conventional crop production methods and the difficulties posed by the climate crisis, biostimulants are a matter of strategic importance. Biostimulants based on PH might pave the way to more sustainable and productive agricultural practices. These molecules may be synthesized from a wide range of protein substrates, including those found in food scraps and industrial offcuts, which has the dual benefit of repurposing these residues and reducing their negative impacts on the environment. The goal of biostimulant creation and study should be the isolation of a functioning PH that can be tested in the laboratory and put on the market for use in the agriculture sector. The PHs have the ability to enhance plant growth and development, mostly during biotic and abiotic stress conditions. It has been shown by many investigators that the radicular application of PHs to plants improves their nutrient assimilation and absorption, root development, nutrient availability, and nutrient usage efficiency across an array of crops. The hormone-like actions (particularly GA-like and auxin-like activities) are shown by PHs when applied to the foliar (leaf) and root (substrate drench) levels, resulting in increased fruit set and enlargement, plant growth, and seed germination. The presence of antioxidants, the concentration of soluble solids, the color of the fruit skin, and the size of the fruit are just a few of the quality criteria that are enhanced by the application of animal or plant-derived PHs. Leafy plants like spinach and lettuce might benefit greatly from the use of PHs because of their ability to decrease nitrate buildup. However, the precise processes that control the positive effects of PHs on plants remain unclear. A deeper systematic knowledge about the composition, timing, and rate of application of PHs, which specifically and directly alter the physiological processes of plants, is required to further maximize the positive effects of PHs. Further, there is growing agreement that peptides of small size contribute significantly to the biological action of PHs. The majority of bioactive peptides, however, remain uncharacterized. More research is needed to identify the signaling peptides that give PHs their biostimulant properties. These results may have the potential to improve the effectiveness of the PH manufacturing process in creating bioactive peptides.

References

Aktsoglou, D.-C., Kasampalis, D.S., Sarrou, E., Tsouvaltzis, P., Chatzopoulou, P., Martens, S., et al., 2021. Protein hydrolysates supplement in the nutrient solution of soilless grown fresh peppermint and spearmint as a tool for improving product quality. Agronomy 11, 1–13.

Al-Malieky, H.M., Jerry, A.N., 2019. Preparation protein hydrolysates from fish by-product and study effected on lettuce (*Lactuca sativa* L.) growth, yield, quality and enhanced salt tolerance. Basrah J. Agric. Sci. 32, 246–255.

Ambrosini, S., Sega, D., Santi, C., Zamboni, A., Varanini, Z., Pandolfini, T., 2021. Evaluation of the potential use of a collagen-based protein hydrolysate as a plant multi-stress protectant. Front. Plant Sci. 12, 600623.

Baglieri, A., Cadili, V., Mozzetti Monterumici, C., Gennari, M., Tabasso, S., Montoneri, E., et al., 2014. Fertilization of bean plants with tomato plants hydrolysates. Effect on biomass production, chlorophyll content and N assimilation. Sci. Hortic. 176, 194–199.

Bonner, C.A., Jensen, R.A., 1997. Recognition of specific patterns of amino acid inhibition of growth in higher plants, uncomplicated by glutamine reversible 'general amino acid inhibition. Plant Sci. 130, 133–143.

Callegaro, K., Brandelli, A., Daroit, D.J., 2019. Beyond plucking: feathers bioprocessing into valuable protein hydrolysates. Waste Manag. 95, 399–415.

Calvo, P., Nelson, L., Kloepper, J.W., 2014. Agricultural uses of plant biostimulants. Plant Soil 383, 3–41.

Casadesús, A., Pérez-Llorca, M., Munné-Bosch, S., Polo, J., 2020. An enzymatically hydrolyzed animal protein-based biostimulant (Pepton) increases salicylic acid and promotes growth of tomato roots under temperature and nutrient stress. Front. Plant Sci. 2020 (11), 953.

Ceccarelli, A.V., Miras-Moreno, B., Buffagni, V., Senizza, B., Pii, Y., Cardarelli, M., et al., 2021. Foliar application of different vegetal derived protein hydrolysates distinctively modulates tomato root development and metabolism. Plants 10, 326.

Choi, S., Colla, G., Cardarelli, M., Kim, H.-J., 2022. Effects of plant-derived protein hydrolysates on yield, quality, and nitrogen use efficiency of greenhouse grown lettuce and tomato. Agronomy 12, 1018.

Colla, G., Rouphael, Y., Canaguier, R., Svecova, E., Cardarelli, M., 2014. Biostimulant action of a plant-derived protein hydrolysate produced through enzymatic hydrolysis. Front. Plant Sci. 5, 448.

Colla, G., Nardi, S., Cardarelli, M., Ertani, A., Lucini, L., Canaguier, R., et al., 2015. Protein hydrolysates as biostimulants in horticulture. Sci. Hortic. 96, 28–38.

Colla, G., Hoagland, L., Ruzzi, M., Cardarelli, M., Bonini, P., Canaguier, R., et al., 2017. Biostimulant action of protein hydrolysates: unraveling their effects on plant physiology and microbiome. Front. Plant Sci. 8, 2202.

Consentino, B.B., Virga, G., Placa, G.G.P., Sabatino, L., Rouphael, Y., Ntatsi, G., et al., 2020. Celery (*Apium graveolens* L.) performances as subjected to different sources of protein hydrolysates. Plants 9, 1633.

Di Mola, I., Cozzolino, E., Ottaiano, L., Nocerino, S., Rouphael, Y., Colla, G., et al., 2020. Nitrogen use and uptake efficiency and crop performance of baby spinach (*Spinacia oleracea* L.) and lamb's lettuce (*Valerianella locusta* L.) grown under variable sub-optimal n regimes combined with plant-based biostimulant application. Agronomy 10, 278.

Drobek, M., Frąc, M., Cybulska, J., 2019. Plant biostimulants: importance of the quality and yield of horticultural crops and the improvement of plant tolerance to abiotic stress-a review. Agronomy 9, 335.

du Jardin, P., 2015. Plant biostimulants: definition, concept, main categories and regulation. Sci. Hortic. 196, 3–14.

Ertani, A., Cavani, L., Pizzeghello, D., Brandellero, E., Altissimo, A., Ciavatta, C., et al., 2009. Biostimulant activities of two protein hydrolysates on the growth and nitrogen metabolism in maize seedlings. J. Plant Nutr. Soil Sci. 172, 237–244.

Ertani, A., Schiavon, M., Muscolo, A., Nardi, S., 2013a. Alfalfa plant-derived biostimulant stimulate short-term growth of salt stressed *Zea mays* L. plants. Plant Soil 364, 145–158.

Ertani, A., Pizzeghello, D., Altissimo, A., Nardi, S., 2013b. Use of meat hydrolyzate derived from tanning residues as plant biostimulant for hydroponically grown maize. J. Plant Nutr. Soil Sci. 176, 287–295.

Ertani, A., Pizzeghello, D., Francioso, O., Sambo, P., Sanchez-Cortes, S., Nardi, S., 2014. *Capsicum chinensis* L. growth and nutraceutical properties are enhanced by biostimulants in a long-term period: chemical and metabolomic approaches. Front. Plant Sci. 5, 375.

Ertani, A., Nardi, S., Francioso, O., Sanchez-Cortes, S., Foggia, M.D., Schiavon, M., 2019. Effects of two protein hydrolysates obtained from chickpea (*Cicer arietinum* L.) and *Spirulina platensis* on *Zea mays* (L.) plants. Front. Plant Sci. 10, 954.

Francesca, S., Cirillo, V., Raimondi, G., Maggio, A., Barone, A., Rigano, M.M., 2021. A novel protein hydrolysate-based biostimulant improves tomato performances under drought stress. Plants 10, 783.

Halpern, M., Bar-Tal, A., Ofek, M., Minz, D., Muller, T., Yermiyahu, U., 2015. The use of biostimulants for enhancing nutrient uptake. Adv. Agron. 130, 141–174.

Jagadeesan, Y., Meenakshisundaram, S., Raja, K., Balaiah, A., 2023. Sustainable and efficient-recycling approach of chicken feather waste into liquid protein hydrolysate with biostimulant efficacy on plant, soil fertility and soil microbial consortium: a perspective to promote the circular economy. Process. Saf. Environ. Prot. 170, 573–583.

Kocira, S., 2019. Effect of amino acid biostimulant on the yield and nutraceutical potential of soybean. Chil. J. Agric. Res. 79, 17–25.

Koukounararas, A., Tsouvaltzis, P., Siomos, A.S., 2013. Effect of root and foliar application of amino acids on the growth and yield of greenhouse tomato in different fertilization levels. J. Food Agric. Environ. 11, 644–648.

Lucini, L., Rouphael, Y., Cardarelli, M., Canguier, R., Kumar, P., Colla, G., 2015. The effect of a plant-derived biostimulant on metabolic profiling and crop performance of lettuce grown under saline conditions. Sci. Hortic. 182, 124–133.

Lucini, L., Miras-Moreno, B., Rouphael, Y., Cardarelli, M., Colla, G., 2020. Combining molecular weight fractionation and metabolomics to elucidate the bioactivity of vegetal protein hydrolysates in tomato plants. Front. Plant Sci. 11, 976.

Luziatelli, F., Ficca, A.G., Colla, G., Švecová, E.B., Ruzzi, M., 2019. Foliar application of vegetal-derived bioactive compounds stimulates the growth of beneficial bacteria and enhances microbiome biodiversity in lettuce. Front. Plant Sci. 10, 601–616.

Martín, M.-H.J., Ángel, M.-M.M., Aarón, S.-L.J., Israel, B.-G., 2022. Protein hydrolysates as biostimulants of plant growth and development. In: Ramawat, N., Bhardwaj, V. (Eds.), Biostimulants: Exploring Sources and Applications. Plant Life and Environment Dynamics, 2022. Springer, Singapore, pp. 141–176.

Martinez-Esteso, M.J., Vilella-Anton, M.T., Selles-Marchart, S., Martínez Márquez, A., Botta-Català, A., Piñol-Dastis, R., et al., 2016. A DIGE proteomic analysis of wheat flag leaf treated with TERRA-SORB® foliar, a free amino acid high content biostimulant. J. Integr. OMICS 6, 9–17.

Matsumiya, Y., Kubo, M., 2011. Soybean peptide: novel plant growth promoting peptide from soybean. In: El-Shemy, H. (Ed.), Soybean and Nutrition. Tech Europe Publisher, Rijeka, pp. 215–230.

Mironenko, G.A., Zagorskii, I.A., Bystrova, N.A., Kochetkov, K.A., 2022. The effect of a biostimulant based on a protein hydrolysate of rainbow trout (*Oncorhynchus mykiss*) on the growth and yield of wheat (*Triticum aestivum* L.). Molecules 27, 6663.

Moreno-Hernández, J.M., Benítez-García, I., Mazorra-Manzano, M.A., Ramírez-Suárez, J.C., Sánchez, E., 2020. Strategies for production, characterization and application of protein-based biostimulants in agriculture: a review. Chil. J. Agric. Res. 80, 274–289.

Nardi, S., Pizzeghello, D., Schiavon, M., Ertani, A., 2016. Plant biostimulants: physiological responses induced by protein hydrolyzed-based products and humic substances in plant metabolism. Sci. Agric. 73, 18–23.

Paleckiene, R., Sviklas, A., Šlinkšiene, R., 2007. Physicochemical properties of a microelement fertilizer with amino acids. Russ. J. Appl. Chem. 80, 352–357.

Paul, T., Halder, S.K., Das, A., Bera, S., Maity, C., Mandal, A., et al., 2013. Exploitation of chicken feather waste as a plant growth promoting agent using keratinase producing novel isolate *Paenibacillus woosongensis* TKB2. Biocatal. Agric. Biotechnol. 2, 50–57.

Paul, K., Sorrentino, M., Lucini, L., Rouphael, Y., Cardarelli, M., Bonini, P., et al., 2019. A combined phenotypic and metabolomic approach for elucidating the biostimulant action of a plant-derived protein hydrolysate on tomato grown under limited water availability. Front. Plant Sci. 10, 493.

Pecha, J., Fürst, T., Kolomazník, K., Friebrová, V., Svoboda, P., 2012. Protein biostimulant foliar uptake modeling: the impact of climatic conditions. AIChE J. 58, 2010–2019.

Philippot, L., Raaijimakers, J.M., Lemanceau, P., van der Putten, W.J., 2013. Going back to the roots: the microbial ecology of the rhizosphere. Nat. Rev. Microbiol. 11, 789–799.

Popko, M., Michalak, I., Wilk, R., Gramza, M., Chojnacka, K., Górecki, H., 2018. Effect of the new plant growth biostimulants based on amino acids on yield and grain quality of winter wheat. Molecules (Basel, Switz.) 23, 470.

Quartieri, M., Lucchi, A., Marangoni, B., Tagliavini, M., Cavani, L., 2002. Effects of the rate of protein hydrolysis and spray concentration on growth of potted kiwifruit (*Actinidia deliciosa*) plants. Acta Hortic. 594, 341–347.

Raguraj, S., Kasim, S., Md Jaafar, N., Nazli, M.H., 2022. Growth of tea nursery plants as influenced by different rates of protein hydrolysate derived from chicken feathers. Agronomy 12, 299.

Rouphael, Y., Colla, G., 2020. Toward a sustainable agriculture through plant biostimulants: from experimental data to practical applications. Agronomy 10, 1461.

Rouphael, Y., Cardarelli, M., Bonini, P., Colla, G., 2017. Synergistic action of a microbial-based biostimulant and a plant derived-protein hydrolysate enhances lettuce tolerance to alkalinity and salinity. Front. Plant Sci. 8, 131.

Ryan, C.A., Pearce, G., Scheer, J., Moura, D.S., 2002. Polypeptide hormones. Plant Cell 14, S251–S264.

Schaafsma, G., 2009. Safety of protein hydrolysates, fractions thereof and bioactive peptides in human nutrition. Eur. J. Clin. Nutr. 63, 1161–1168.

Schmidt, C.S., Mrnka, L., Frantík, T., Barnet, M., Vosatka, M., Svecova, E.B., 2020. Impact of protein hydrolysate biostimulants on growth of barley and wheat and their interaction with symbionts and pathogens. Agric. Food Sci. 29, 222–238.

Sestili, F., Rouphael, Y., Cardarelli, M., Pucci, A., Bonini, P., Canaguier, R., et al., 2018. Protein hydrolysate stimulates growth in tomato coupled with N-dependent gene expression involved in N assimilation. Front. Plant Sci. 9, 1233.

Shukla, P.S., Mantin, E.G., Adil, M., Bajpai, S., Critchley, A.T., Prithiviraj, B., 2019. *Ascophyllum nodosum*-based biostimulants: sustainable applications in agriculture for the stimulation of plant growth, stress tolerance, and disease management. Front. Plant Sci. 10, 1–22.

Soppelsa, S., Kelderer, M., Casera, C., Bassi, M., Robatscher, P., Andreotti, C., 2018. Use of biostimulants for organic apple production: effects on tree growth, yield, and fruit quality at harvest and during storage. Front. Plant Sci. 9, 1342.

Tsouvaltzis, P., Koukounaras, A., Siomos, A.S., 2014. Application of amino acids improves lettuce crop uniformity and inhibits nitrate accumulation induced by the supplemental inorganic nitrogen fertilization. Int. J. Agric. Biol. 16, 951–955.

Ugolini, L., Cinti, S., Righetti, L., Stefan, A., Matteo, R., D'Avino, L., et al., 2015. Production of an enzymatic protein hydrolyzate from defatted sunflower seed meal for potential application as a plant biostimulant. Ind. Crop. Prod. 75, 15–23.

Visconti, F., de Paz, J.M., Bonet, L., Jorda, M., Quinones, A., Intrigliolo, D.S., 2015. Effects of a commercial calcium protein hydrolysate on the salt tolerance of *Diospyros kaki* L. cv. "Rojo Brillante" grafted on *Diospyros lotus* L. Sci. Hortic. 185, 129–138.

Xu, C., Mou, B., 2017. Drench application of fish-derived protein hydrolysates affects lettuce growth, chlorophyll content, and gas exchange. Horttechnology 27, 539–543.

Yakhin, O.I., Lubyanov, A.A., Yakhin, I.A., Brown, P.H., 2017. Biostimulants in plant science: a global perspective. Front. Plant Sci. 7, 2049.

Yang, G., Wei, Q., Huang, H., Xia, J., 2020. Amino acid transporters in plant cells: a brief review. Plants 9, 967.

CHAPTER

5

Paramylon marvels: enhancing plant productivity and resilience

Sharfa Naaz[1], *Abdul Raheem*[1], *Mohammad Amir*[1], *Puhpanjali Yadav*[1], *Ragib Husain*[2], *Vinay Kumar Singh*[2] *and Mohammad Israil Ansari*[1]

[1]Department of Botany, Lucknow University, Lucknow, Uttar Pradesh, India [2]Department of Botany, Bareilly College, Bareilly, Uttar Pradesh, India

5.1 Introduction

Plants are stationary and subject to a variety of harmful stressors (biotic or abiotic). To defend themselves, plants must be able to detect and identify potential threats. Plants have two major types of defenses: preformed defenses (cell wall and cuticle) and their innate immune system (composed of surveillance systems that perceive several general microbe elicitors) against microbial pathogens (Newman et al., 2013). Plants recognize the microbe-specific molecules, referred to as pathogen-associated molecular patterns (PAMPs), by the plant innate immune system pattern recognition receptors (PRRs) (Jones and Dangl., 2006; Howe and Jander., 2008; Newman et al., 2013). The heterogeneous glucose polymers known as β-glucans are naturally occurring structural elements of algae, bacteria, fungi, and plants that act as PAMPs.

A type of β-glucan known as "Paramylon" is also found in the single-celled asexual algae known as *Euglena* (sometimes also treated as a protist), in the form of membrane-bound cytoplasmic storage granules. Paramylon was first described in detail by Gottlieb (1850). It serves as the primary energy store in these algae and can be used as an alternate carbon source to glucose, glutamate, malate, pyruvate, etc. Moreover, this specialized complex carbohydrate, "Paramylon," also allows these organisms to thrive in low-light environments. Studies have shown that in this algal group, paramylon accumulates in larger amounts in heterotrophic (HT) and mixotrophic (MT) conditions in the presence of organic carbon than under phototrophic (PT) conditions (Fig. 5.1) (Grimm et al., 2015).

In *Euglena*, about 90% of paramylon is crystallized, and it is entirely composed of D-glucose (Fig. 5.2) (Barsanti et al., 2011). Contrary to practically all other β (1, 3) glucans,

Biostimulants in Plant Protection and Performance
DOI: https://doi.org/10.1016/B978-0-443-15884-1.00001-4

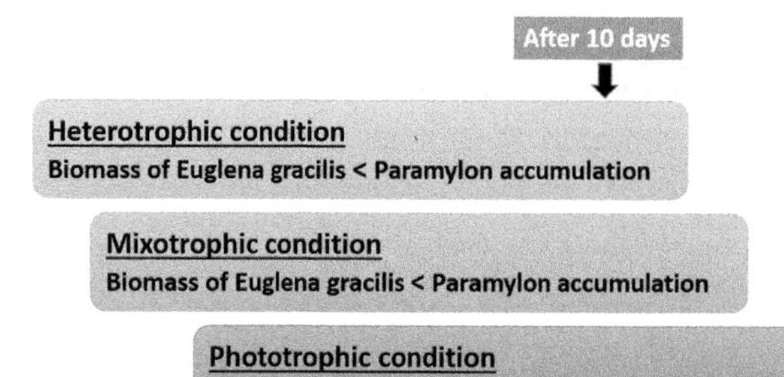

FIGURE 5.1 Cultivation of *Euglena gracilis* and paramylon accumulation during heterotrophic, mixotrophic, and phototrophic condition after 10 days. Source: *Adopted from Grimm et al. (2015).*

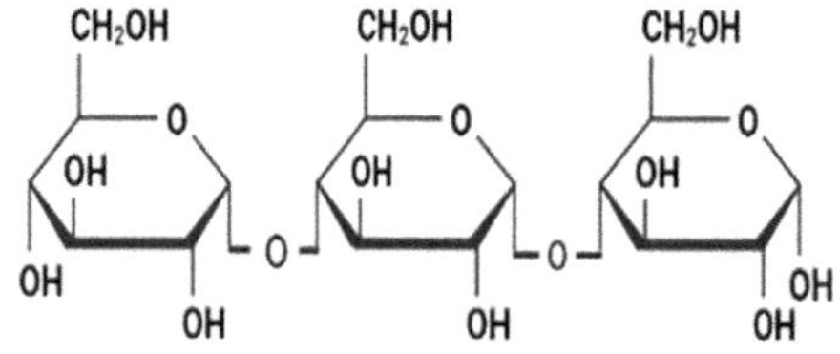

FIGURE 5.2 Molecular structure of paramylon, β-1,3-glucan chain. Source: *Adopted from Barsanti et al. (2011).*

which have side chains, paramylon is a linear polysaccharide (Shibakami et al., 2013). Since the effects of β (1, 3) glucans on signal transduction depend on both the molecular structure and the degree of polymerization, the processing of paramylon from *Euglena* enables the production of linear fibers of β (1, 3) glucans, a trait that significantly improves the interaction with PRRs present on the cell membranes of plants (Fu et al., 2011; Russo et al., 2017). *Euglenoid* paramylons come in a wide variety of shapes and quantities; they can be both big and scarce as well as small and plentiful. Combinations of tiny and large paramylon carbohydrate grains are frequently found. It is also known as "dimorphic" when there are two size classes of paramylon grains present in a single cell (Milanowski et al., 2006). These paramylon granules most likely remain unbound in the cytoplasm of these algae, where they may also be accompanied by pyrenoids that may cap or cover them (but are external to the chloroplast). Large grains are the most certain way to identify paramylon characteristics. The shape of the paramylons is species-specific and constitutes an efficient taxonomic tool (Gojdics, 1953). Numerous euglenoid species have been identified by the number, shape, and exterior appearance of paramylon grains, thus resolving the taxonomic issues as well (Ciugulea et al., 2008).

According to Rodríguez-Zavala et al. (2006), *Euglena gracilis* (a freshwater member of *Euglena*) has achieved specialization in accumulating paramylon (Fig. 5.3). *Euglena gracilis* is known to adapt to harsh conditions and live in places with high concentrations of heavy metals, acidic rivers, and other pollutants (Casiot et al., 2004).

The synthetic form of paramylon derived from *Euglena gracilis* may be used to produce products with additional value in plants. Research on plants revealed that when nutrients are scarce, this paramylon aids in the development and survival of higher plants. To make crops that are toxin-free and have greater potential, the modern agriculture system promotes an alternative to synthetic chemical fertilizers and pesticides. As a successful and sustainable alternative to be used in plant growth, microbial biostimulants such as *Azotobacter and*

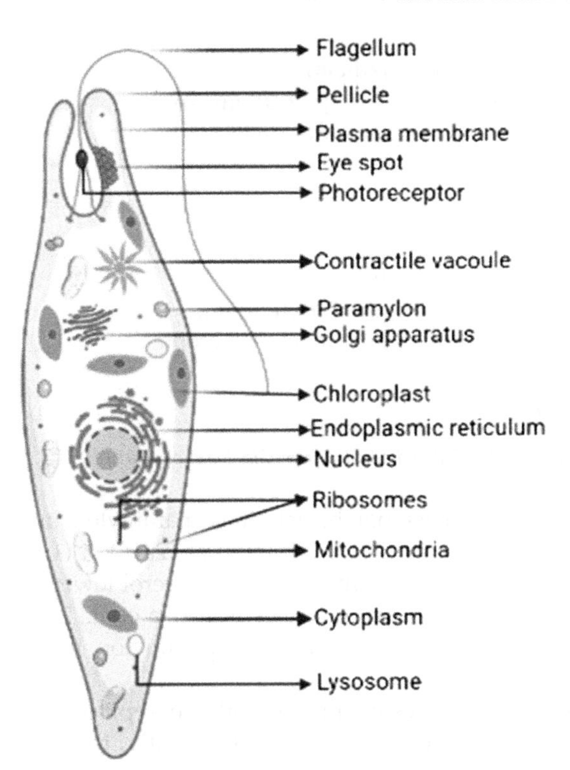

FIGURE 5.3 Euglena gracilis—a fresh water, single-cell alga.

Rhizobium and nonmicrobial biostimulants such as silicon, animal and plant-based protein hydrolysate, humic substances, and algal extracts have positive effects on biodiversity, environment, economy, and human health. Plants and crops will be more productive if extracts from these sustainable alternatives are used in their growth. The paramylon extract may act as a plant biostimulant, especially with sustainable farming methods. The role of paramylon as an elicitor must be interesting for plant and crop growth to facilitate performance under biotic and abiotic stresses (salinity, high temperature, drought, nutrient deficiency, etc.) by augmenting nutrient use efficiency and stress resilience, promoting phytohormonal levels and physiological responses, or by naturally increasing crop plant production. Several studies have shown that β (1, 3) glucan from other algal sources, such as *Laminaria digitata*, acts as an elicitor in dicots (tobacco, tomato, bean, grapevine, alfalfa, and *Arabidopsis*), as well as monocots (wheat and rice), despite the general lack of information on the effects of paramylon from *Euglena* on plant metabolism (Fu et al., 2011; Gauthier et al., 2014).

5.2 Possible functions of *Euglena gracilis*

Some *euglenoids* are significant because they possess characteristics of both plants and animals. *Euglenoids* have the potential to be consumed by both humans and animals. *E. gracili's* potential advantages as food, medicine, crop biostimulant, and biofuels are noted in Fig. 5.4 (Gissible et al., 2019).

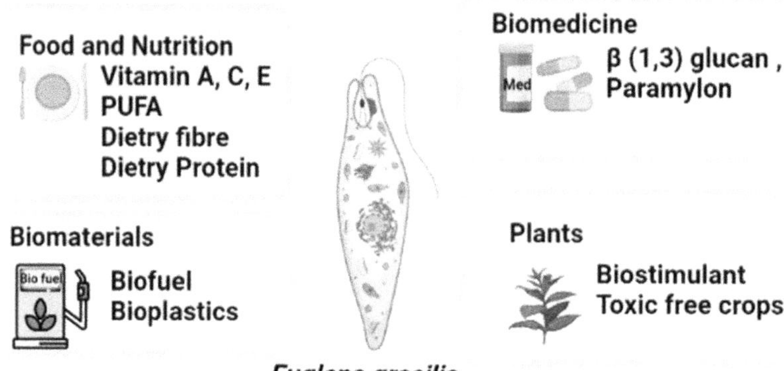

FIGURE 5.4 Possible function of *Euglena gracilis*. Source: *Modified and adopted from Gissible et at. (2019).*

Due to their alleged immunostimulatory and antibacterial bioactivities, paramylon and other β (1, 3) glucans are of particular interest (Russo et al., 2017; Gissibl et al., 2018). β (1, 3) glucans act as hepato-protective, antidiabetic, antihyperglycemic, and cholesterol-lowering agents. They have also been used to treat gastric and colon malignancies (Kataoka et al., 2002; Barsanti et al., 2011; Gissible et al., 2019). Drug delivery is made possible by paramylon, found in *euglenoids*, which also have immunostimulatory effects.

Pressure-sensitive adhesives and petroleum-based resins might both be replaced by paramylon. One possibility is to employ *euglenoids* to make bioethanol (currently, they are not competitive with fossil fuels).

The versatility of paramylon makes it a great fertilizer alternative for plants. According to the studies of Barsanti et al. (2001) and Monfils et al. (2011), paramylon is deposited as granules in the cytosol of *euglenoids* and is quickly broken down and used as a source of carbon when there is a lack of oxygen in plants. Furthermore, *Euglena gracilis* can sequester heavy metals and withstand a variety of external stimuli, including acidic growth conditions and ionizing radiation (García-García et al., 2018).

Euglena gracilis competes for a good source of dietary protein, lipids, vitamins (A, C, and E), and paramylon (a starch-like storage product). When compared to other algae, paramylon yields are relatively high. Vitamin A molecules that act as antioxidants, including retinol, retinal, retinoic acids, and retinyl esters, are abundant in *Euglena*. Vitamin C (ascorbate), which is crucial for human health, is also produced by *Euglena*. Because they contain a lot of lipids, microalgae have been suggested as a source of feedstock for the creation of biofuels. Additionally, lipids can be utilized as a lubricant or as a base ingredient in candles and cosmetics. The biggest difficulties lie in expanding the procedures for industrial applications and future growth.

5.3 Paramylon biosynthesis

Paramylon (Fig. 5.2) is a water-insoluble storage polysaccharide of 100–500 kDa molecular weight and is made up of β (1, 3) linked glucose subunits (Barsanti et al., 2011; Gissibl et al., 2018). The intermolecular triple helix arrangement of paramylon molecules results in

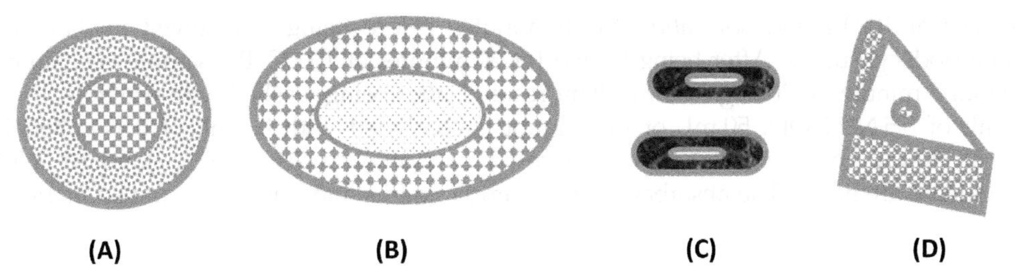

FIGURE 5.5 Different shapes of paramylon granules (A) disk, (B) ellipticle, (C) rod and (D) wedge shape.

microfibrils, which in turn form fibers. Granules made of these fibers are rectangular and wedge-shaped segments that can be synthesized by *E. gracilis* in different shapes like ellipses or rods (Fig. 5.5) (Gissibl et al., 2018). The granules stand out from other carbohydrate storage products seen in plants and algae because they are 1–6 m long, enclosed by a biomembrane, and exhibit an unusually high degree of crystallinity (Miyatake and Kitaoka, 1983; Koizumi et al., 1993; Bäumer et al., 2001; Monfils et al., 2011).

E. gracilis accumulates paramylon during PT, MT, and HT growth (Fig. 5.1) (Grimm et al., 2015). While light is detrimental to the accumulation and conservation of paramylon under MT growth, it is likely because the metabolic switch to paramylon degradation is under the influence of a photoreceptor. Cultivation under strict HT conditions increases paramylon levels during the exponential growth phase (Kiss et al., 1986; Barsanti et al., 2001). Using a medium supplemented with potato liquor, vitamins, and a high concentration of glucose (30 g/L), one of the highest paramylon titers has ever been recorded (16 g/L culture) in a repeated-batch cultivation under HT conditions in the dark (Grimm et al., 2015; Gissibl et al., 2018).

In different sites inside the cell, linear β (1,3) glucan, known as paramylon, exists in granular form. The paramylon granules are dispersed throughout the cytoplasm in some species or are clumped together in the other. Granules are scarce but massive and positioned in a predictable manner in some other species. Granules vary significantly in size and shape, but their distribution pattern in cells serves as a taxonomic trait. As a macromolecule that occurs naturally, paramylon possesses an unusually high crystallinity that confers the benefit of enabling the controlled axenic cultivation of paramylon granules from Euglena cells in fermenters. The process is quick and inexpensive, and it only requires breaking up the cells and refining the granules through repeated washings with a weak detergent (Barsanti et al., 2021). The higher-order aggregates of nanofibrils, measuring 4–10 nm, made up of unbranched triple helices of β (1,3)–D-glucan chains are what give paramylon its crystallinity (Barsanti et al., 2019). In order to evaluate the potential health benefits of β -glucans in various experimental models involving plants, animals, and humans as well as to look into the structure-function relationship and mechanism of action, paramylon has been used, either in the form of granules or as nanofibers.

5.4 Biomass and paramylon production analysis

The content of paramylon is measured according to Suzuki et al. (2015). Approx. 50 mg of the lyophilized material is dissolved in 3 mL of acetone, followed by centrifugation for

5 minutes at 1000 g and sonication for 10 minutes before being redissolved in 3 mL of 1% sodium dodecyl sulfate. After being heated for 30 minutes at 100°C, the sample is again centrifuged for 5 minutes at 2000 g. The resultant residue, after being washed with water, is dissolved in 2 mL of 0.5 N NaOH. 50 mL of the aliquot is combined with 0.5 mL of 0.5% phenol and 5 mL of H_2SO_4. To draw a calibration curve for paramylon measurement, standard solutions with glucose are made. The absorbance of all samples and standards is determined at 480 nm.

5.5 Possibility of paramylon acting as a biostimulant

Hormone signaling is a crucial aspect of basal resistance in plants; hormones like salicylic acid (SA), jasmonic acid (JA), and abscisic acid (ABA) can operate independently or interact through cross-talk among signaling pathways in an antagonistic or cooperative manner (Mundy et al., 2006; Jaillais and Chory 2010). JA is widely believed to have a larger role in defense against necrotrophs, herbivores, and wound responses, while SA is typically involved in defense against biotrophs and initiates systemic acquired resistance (Furch et al., 2014). According to Sauter et al. (2001) and Allègre et al. (2009), root-produced ABA is typically viewed as a root-to-shoot stress signal that is carried in the xylem sap and involved in the modulation of stomatal response.

Research on tomatoes under water stress, however, indicates that root ABA may not be required for either early or late stomatal responses (Christmann et al., 2007; Manzi et al., 2015), leaving an unanswered question regarding the nature of the systemic stress signal. Alterations in the pH of the xylem sap as well as other phytohormones such as strigolactones may also have a role in the process (Holbrook et al., 2002, Visentin et al., 2016).

Furthermore, there are indications that ABA can go from shoots to roots in tomato plants that have been dehydrated and that basipetal transport is primarily responsible for the root's ABA concentration (Manzi et al., 2015). The complicated interaction among ABA, JA, and SA has been studied in tomato plants subjected to soil drying (Muñoz-Espinoza et al., 2015). ABA synergizes with JA and displays an antagonistic relationship against SA (Thaler and Bostock 2004; Ton and Mauch-Mani 2004; Ton et al., 2009).

In investigations by Allègre et al. (2009) and Fu et al. (2011), the effect of β (1, 3) glucans on stomata response has been investigated with foliar spray treatments of grapevine plants with β (1, 3) glucans, resulting in high production of reactive oxygen species (ROS) (Allègre et al., 2009). Tobacco leaf epidermal peels treated with (1, 3) glucans exhibit notable effects on stomata motions when exposed to light, causing both restriction of opening and stimulation of closure in a way similar to ABA (Fu et al., 2011). There is, however, a complete dearth of knowledge regarding how root treatments with β (1, 3) glucans alter xylem hormonal levels and photosynthetic parameters, impacting the plant's ability to react to biotic and abiotic stress. Inadequate water availability is thought to be the primary environmental factor limiting photosynthesis and, as a result, plant growth and crop yield globally (Morison et al., 2008).

Research on the role of β (1, 3) glucan as elicitors of plant response to water stress may be crucial in revealing the physiological and photosynthetic responses of the plant under drought, thereby supporting the use of *Euglena* paramylon as a novel treatment for the enhancement of plant water-use efficiency (*WUE*) and hence encouraging the use of paramylon in plant production and growth.

5.5.1 Improved quality and increased drought resistance in tomato plant

A significant environmental factor that inhibits photosynthesis and speeds up the salinization of arable land is drought (low water availability) (Leng et al., 2019). One of the most economically and nutritionally significant crops in the world, the tomato, is adversely impacted by water shortages, which lower fruit yield and quality. Research mostly focuses on choosing drought-tolerant genotypes with lower water demands to mitigate its effects. It has been demonstrated that paramylon nanofibers applied to tomato plant roots can modify hormone levels, photosynthesis, and water use efficiency (Scartazza et al., 2017).

5.5.1.1 Tomato plant hormone response to paramylon treatment for drought stress

Several plant hormones have been recognized as crucial elements of signaling cascades in plant development and response to diverse biotic stressors or when plants are exposed to synthetic and biological substances. Scartazza et al. (2017) have examined the presence of stress-related phytohormones (SA, JA, and ABA) in tomato xylem exudates following the application of various doses of paramylon to the roots to determine whether some plant hormones are components of the root signals obtained from paramylon-root cell contact. Paramylon microfibrils tightly interact with the plant and gradually cover the radical hairs' surface (Russo et al., 2017). By examining the xylem exudates, it is possible to better understand how phytohormones are transferred from the root to the shoot in the plant vascular system. When compared to untreated control plants, hormone analysis of the xylem sap of tomato plants treated with paramylon exhibits a dose-dependent reduction of the SA in all treatments.

Applying paramylon to roots could mimic a pathogen attack, especially since numerous pathogenic fungi and bacteria release glucans, which are important virulence factors (El Oirdi et al., 2011). According to Muñoz-Espinoza et al. (2015), ABA exerts a negative regulation of SA via modifying PAL1 expression in tomatoes under water stress. The decrease in SA levels in the xylem sap may potentially be caused by an increase in ABA levels. It is generally known that when a plant recognizes a pathogen attack, an oxidative burst occurs. It can therefore be summarized that in tomato plants after paramylon treatment, root SA may be involved in preserving redox equilibrium, hence reducing the amount of SA in the xylem sap transmitted to the aerial portions.

Analysis of these stress-related hormones has previously been done in the xylem exudates of *Brassica napus* L. infected with *Verticillium longisporum* and *Cucurbita maxima* after root treatment with flagellin 22 (Furch et al., 2014).

In tobacco and grapevine, treatments with β (1, 3) glucans cause a range of defensive and resistance reactions, including the emission of NO and an H_2O_2 burst (Fu et al., 2011; Gauthier et al., 2014). Horváth et al. (2007) have demonstrated that pretreatment with modest concentrations of SA boosts tolerance toward the majority of abiotic stressors, mostly because of increased antioxidant ability.

JA or its derivatives significantly regulate plant responses to biotic and abiotic stressors, as well as plant development (Wasternack and Hause, 2013). In tomatoes, additional signals (such as strigolactones) seem to be implicated in the stomatal responses.

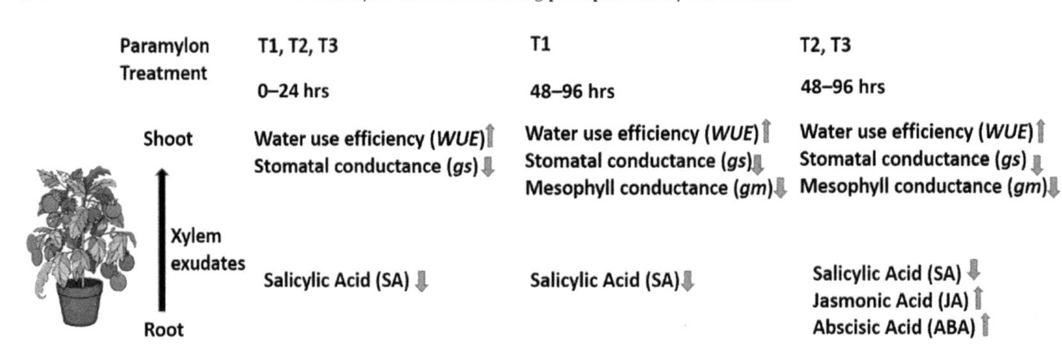

FIGURE 5.6 Changes of xylem hormones level and gas exchange in tomato shoots part in response to paramylon root treatments. Source: *Adopted from Scartazza et al. (2017), where T1–90 mg/L, T2–189 mg/L and T3–500 mg/L.*

The drought stress typically generates ABA in root tissues that subsequently gets distributed throughout the plant in distant organs to boost resistance (Manzi et al., 2015; Visentin et al., 2016). ABA is also involved in defensive reactions (Ma and Ma, 2016). A study by Scartazza et al. (2017) depicts that paramylon treatments do not produce any effect on the amount of ABA in the xylem from 0 to 48 hours, but a noticeable surge in ABA occurs between 48 and 96 hours in response to the dose of elicitor, which results in a threefold increase of the hormone (Fig. 5.6).

The observed hormonal profiles point to a detrimental interaction between the JA and SA pathways during the 48–96 hour period. In various plants, including tomato, rice, tobacco, and *Arabidopsis*, JA and SA signaling have been found to interact antagonistically. JA responses in tomatoes have also been connected to the inhibition of SA induction (Thaler et al., 2002; Thaler et al., 2012). It is well known that ABA interacts strongly with both the JA and SA pathways. Numerous studies suggest that ABA has an antagonistic relationship with SA-based resistance and a favorable synergistic interaction with JA-induced defense (Thaler and Bostock, 2004; Ton et al., 2009). Additionally, under biotic stress, ABA negatively regulates tomato PAL activity and the defense responses of SA signaling (Audenaert et al., 2002).

Muñoz-Espinoza et al. (2015) have examined tomato plants' responses to drought by looking at the hormone profile and gene expression of important enzymes in ABA, JA, and SA production. In contrast, in situations of water stress, the study concludes that ABA controls SA levels negatively in roots and leaves, demonstrating a clear link with JA. Together, the hormonal profiles of SA, JA, and ABA in the study support the results, particularly in the 48–96 hour time period.

5.5.1.2 Tomato plant photosynthetic parameters response to paramylon treatment for drought stress

To ascertain whether the altered xylem hormone composition affects the tomato response at the leaf level, stomatal and nonstomatal photosynthetic parameters have been measured in the tomato plant treated with paramylon (Scartazza et al., 2017). Even though the ABA content in the xylem sap does not vary considerably, the paramylon doses, merely 24 hours after treatment, significantly reduce stomatal conductance (*gs)* (Scartazza et al., 2017).

Thus although there exist considerable differences in the xylem ABA concentration and paramylon dosage during the first 24 hours of treatment, all tomato plants respond by partially closing their stomata. According to Poór and Tari (2012), the tomato's stomatal responses could be influenced by the observed decrease in xylem SA content.

However, additional signals like strigolactones and ROS may have an impact on stomatal motions. According to Allègre et al. (2009) and Fu et al. (2011), spraying the β-glucan laminarin and curdlan onto leaves causes rapid stomatal movements that are mediated by ROS generation. Furthermore, strigolactones control stomata sensitivity to ABA and are part of the systemic signal of drought stress in tomatoes, as studied by Visentin et al. (2016).

Paramylon elicits partial stomatal closure, which decreases the CO_2 assimilation rate (*A*) and transpiration rate (*E*). While intercellular CO_2 concentration (*Ci*) decreases and intrinsic and instantaneous *WUE* increase in both treatments, *E* exhibits a greater drop than *A* at 24 hours. A dose-dependent response to paramylon is discernible after 24 hours. Compared to Control (C), T1, T2, and T3 display further stomatal closure and a reduction in CO_2 diffusion within the mesophyll (*gm*) from 48 to 96 hours.

The increase in ABA and JA levels in the xylem sap that accompanies this impact raises the possibility that both hormones may function as secondary root-to-shoot hormonal signals that modify the CO_2 diffusional limitations. Numerous JA derivatives have been linked to the endogenous ABA level and stomatal movements in the literature (Melotto et al., 2017). Therefore, paramylon may cause a coordinated control of stomatal and mesophyll conductance in tomatoes, similar to what occurs in response to exogenous ABA treatment and under drought or other environmental changes (Fig. 5.6) (Sorrentino et al., 2016; Scartazza et al., 2017).

5.5.2 Paramylon treatment improves the quality profile and drought resistance in tomatoes in the aeroponic system

Solanum lycopersicum was subjected to drought stress with or without root treatment with paramylon nanofibers to test the viability of using paramylon as a novel root treatment to help tomato plants cope with low water availability (Barsanti et al., 2019). In an aeroponics system, which is a soil-free air-water system ideal for studying plant roots since the nutrient solution is sprayed directly into the roots by atomizers, plants were grown under regulated conditions (Fig. 5.7). Stressed plants treated with paramylon exhibited around a 2-week head start in flowering and fruit ripening compared to untreated, well-watered plants, whereas the fruits of stressed, untreated plants do not ripen (Fig. 5.7).

Barsanti et al. (2019) discovered that the ideal water regimen can be reduced because of paramylon activity. Water stress had a significant impact on the eco-physiological measures (such as leaf water potential, stomatal conductance, and photosynthetic yield), which

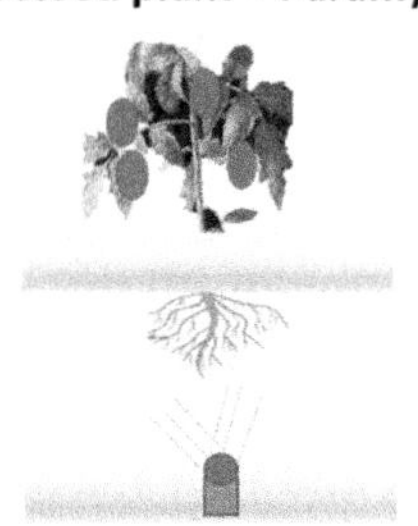

FIGURE 5.7 Late flowering and fruit ripening in well-watered plants, while early flowering and fruit ripening in stress plants treated with paramylon in an aeroponic condition. Source: *Adopted from Barasnti et al. (2019).*

all decreased continuously until saturation. Water-stressed plants administered root treatment with paramylon nanofibers exhibited a return of their physio-morphological parameters to the levels of those of well-watered plants. Paramylon nanofibers have an impact on stomatal behavior that is helpful in improving water use efficiency and preventing dehydration (Barsanti et al., 2019).

The fruits of paramylon-treated plants have higher levels of carbohydrates (glucose, fructose, and sucrose) and antioxidant components (vitamin A/C/E, lycopene, beta-carotene, and phenols) (Dumas et al., 2003; Slimestad et al., 2009; Raiola et al., 2015; Scartazza et al., 2017, Barsanti et al., 2019), which enhance their nutritional value and quality (Malundo et al., 1995; Sivakumar et al., 2002). Additionally, the greater postharvest storage capacity made possible by the increased dry matter content (i.e., lower moisture) lengthens the shelf life and boosts the value of the commercial product.

These findings support the role of paramylon nanofibers in stimulating plant adaptation to abiotic stress and in enhancing plant capacity for environmental sensitization.

5.5.3 Impact of paramylon on tomato plants cultivated to resist off *fusarium* wilt

Fusarium oxysporum, one of the most significant economic illnesses in the world's major tomato-growing regions, is the culprit behind *fusarium* wilt in tomatoes. The disease affects tomatoes grown in warm environments (greenhouses or fields) and reduces fruit yield in wilted plants by 15% − 50%. Because of their unfavorable impacts on nontarget organisms and their possible negative consequences on the environment and human health, the widespread use of chemical fungicides has been a source of public worry and security.

In the soil, *F. oxysporum* can persist for a very long time and become challenging to get rid of. Through direct penetration or wounds, the pathogen infects the roots. As a result of fungal mycelia blocking xylem channels, leaves experience a water deficit with poor gaseous exchange capabilities (stomatal conductance, transpiration, and photosynthesis). Because of this, infected plants exhibit symptoms such as stunted plant growth, reduced leaf area, and dry matter accumulation, which is followed by withering of the vascular tissue and eventually plant death (Gordon, 2017). Stomatal closure brought on by disease-induced water stress or disturbance of the metabolic pathways of photosynthesis, such as the diminished activity of Rubisco, has been linked to a drop in the net photosynthetic rate of infected leaves (Santos et al., 2000).

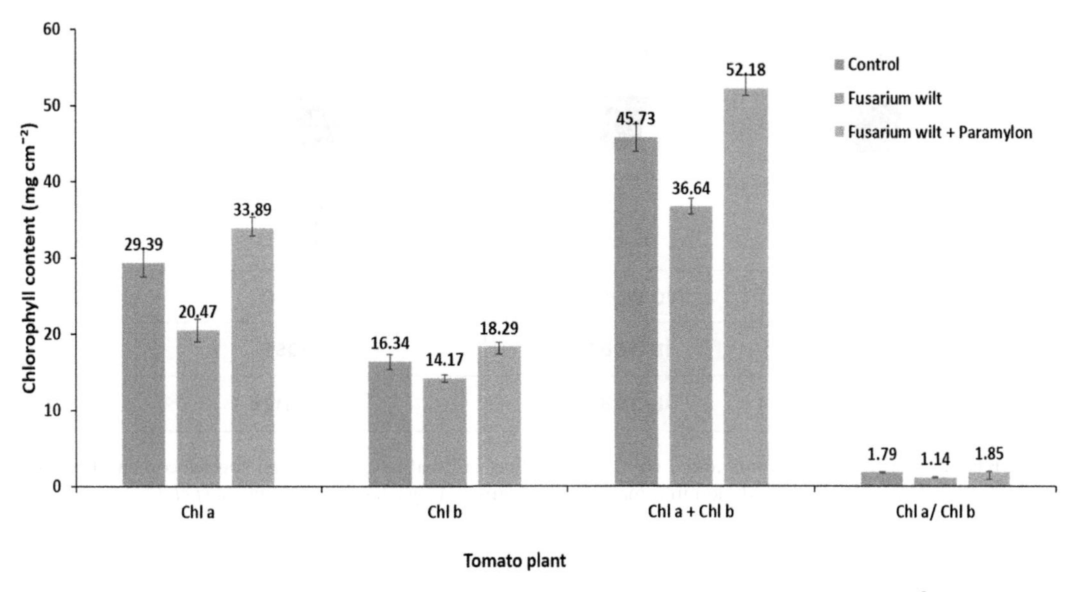

FIGURE 5.8 *Fusarium oxysporum* and β (1, 3) glucan effect on the chlorophyll content (mg/cm^2) in the tomato leaves under *fusarium* wilt. Source: *Adopted from Kabashnikova et al. (2020).*

According to Kabashnikova et al. (2020), *F. oxysporum* infection in tomato seedlings diminishes several morphological and biochemical parameters encompassing a decrease in the growth of tomato shoots and roots, a rise in LPO levels, and a noticeably increased activity of the antioxidant defense enzymes peroxidase, catalase, and superoxide dismutase (Hanaa et al., 2011). Since malondialdehyde and the amount of chlorophylls and carotenoids in leaves are biochemical indicators of the severity of stress conditions, both water deficiency and *Fusarium* wilt can affect the synthesis of these metabolites (Rasheed et al., 2018).

The impact of β (1, 3) glucan and *F. oxysporum* on tomato leaf pigment content is summarized in Fig. 5.8 (Kabashnikova et al., 2020). Infected leaves have a decrease in the content of chlorophyll pigments relative to healthy ones. Chl a content in infected leaves declines to a greater extent than Chl b content, resulting in a decreased Chl a/Chl b ratio. In comparison to infected plants, the treatment with β (1, 3) glucan has a protective impact on the level of chlorophyll pigments during pathogenesis, augmenting the content of Chl a, Chl b, Chl (a + b), and Chl a/Chl b ratio.

When compared to healthy plants, the infected tomato leaves with *fusarium* wilt (Fig. 5.9) display the same overall ROS concentration. However, the treated tomato plants dramatically increased hydrogen peroxide (H_2O_2) concentration and LPO activity compared to the untreated, infected control plants. The amount of ROS increases significantly as a result of the β (1, 3) glucan pretreatment, although H_2O_2 levels marginally drop in comparison to the infected leaves but remain high in contrast to the control (Kabashnikova et al., 2020).

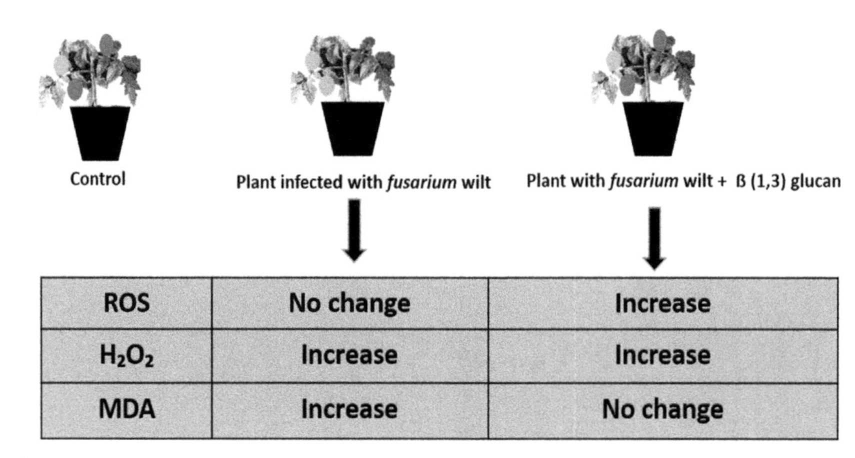

	Plant infected with *fusarium* wilt	Plant with *fusarium* wilt + ß (1,3) glucan
ROS	No change	Increase
H_2O_2	Increase	Increase
MDA	Increase	No change

FIGURE 5.9 *Fusarium oxysporum* and β (1, 3) glucan effect on total reactive oxygen species content, hydrogen peroxide content and lipid peroxidation in tomato leaves. Source: *Adopted from Kabashnikova et al. (2020).*

5.6 Conclusion

Paramylon nanofibers successfully reduce biotic and abiotic stress in tomato plants. As a result, paramylon can be used as a biobased stimulant to boost plants against any stresses, increase nutrients, and ensure stable, high yields. Paramylon activity has the potential to reduce the optimal plant watering schedule. Plants exposed to paramylon show early fruit ripening and flowering compared to the control.

The physical-chemical quality of the plant, including antioxidant compounds and nutrient content, significantly improves in the fruits of paramylon-treated plants. Additionally, because of the higher dry matter content (i.e., reduced moisture) and increased postharvest storage capacity, the commercial product's shelf life gets extended, and its value increases.

Paramylon treatment of *F. oxysporum*-infected tomato seedlings displays a variety of morphological and biochemical alterations, such as a reduction in the growth of shoots and roots, an increase in LPO, and a markedly enhanced activity of the antioxidant defense enzymes peroxidase, catalase, and superoxide dismutase. The findings indicate that physiological, biochemical, and defense hormone parameters exhibit a dose-dependent response to paramylon treatments.

Following root treatment with paramylon nanofibers, stressed plants regain their pre-stressed condition. Defense hormone levels, stomatal behavior, and balancing oxidative stress are all significantly influenced by paramylon. Breeding disease-resistant cultivars is widely regarded as the most environmentally friendly method of plant protection. Resistance-enhancing cultivars can be created by crossing resistant wild-type plants with existing cultivars for traits like color, shape, and flavor. Paramylon can be utilized to make plants resistant to all biotic and abiotic stresses. Paramylon therapy acts as an immunostimulatory agent and enhances the content of nutraceuticals. As a result of past research on the tomato, we may conclude that additional rigorous study on the effects of paramylon therapy on other crops is required.

References

Allègre, M., Héloir, M.C., Trouvelot, S., Daire, X., Pugin, A., Wendehenne, D., et al., 2009. Are grapevine stomata involved in the elicitor-induced protection against downy mildew. Mol. Plant-Microbe Interact. 22 (8), 977–986.

Audenaert, K., Pattery, T., Cornelis, P., Höfte, M., 2002. Induction of systemic resistance to Botrytis cinerea in tomato by *Pseudomonas aeruginosa* 7NSK2: role of salicylic acid, pyochelin, and pyocyanin. Mol. Plant-Microbe Interact. 15 (11), 1147–1156.

Barsanti, L., Gualtieri, P., 2019. Paramylon, a potent immunomodulator from WZSL mutant of *Euglena gracilis*. Molecules 24 (17), 3114.

Barsanti, L., Vismara, R., Passarelli, V., Gualtieri, P., 2001. Paramylon (β-1, 3-glucan) content in wild type and WZSL mutant of *Euglena gracilis*. Effects of growth conditions. J. Appl. Phycol. 13 (1), 59–65.

Barsanti, L., Passarelli, V., Evangelista, V., Frassanito, A.M., Gualtieri, P., 2011. Chemistry, physico-chemistry and applications linked to biological activities of β-glucans. Nat. Prod. Rep. 28 (3), 457–466.

Barsanti, L., Ciurli, A., Birindelli, L., Gualtieri, P., 2021. Remediation of dairy wastewater by *Euglena gracilis* WZSL mutant and β-glucan production. J. Appl. Phycol. 33, 431–441.

Bäumer, D., Preisfeld, A., Ruppel, H.G., 2001. Isolation and characterization of paramylon synthase from *Euglena gracilis* (Euglenophyceae) 1. J. Phycol. 37 (1), 38–46.

Casiot, C., Bruneel, O., Personné, J.C., Leblanc, M., Elbaz-Poulichet, F., 2004. Arsenic oxidation and bioaccumulation by the acidophilic protozoan, *Euglena mutabilis*, in acid mine drainage (Carnoules, France). Sci. Total. Environ. 320 (2–3), 259–267.

Christmann, A., Weiler, E.W., Steudle, E., Grill, E., 2007. A hydraulic signal in root-to-shoot signalling of water shortage. Plant J. 52 (1), 167–174.

Ciugulea, I., Nudelman, M.A., Brosnan, S., Triemer, R.E., 2008. Phylogeny of the euglenoid loricate genera *Trachelomonas* and *Dtrombomonas* (Euglenophyta) inferred from nuclear SSU and LSU rDNA (1). J. Phycol. 44 (2), 406–418.

Dumas, Y., Dadomo, M., Di Lucca, G., Grolier, P., 2003. Effects of environmental factors and agricultural techniques on antioxidantcontent of tomatoes. J. Sci. Food Agric. 83 (5), 369–382.

El Oirdi, M., El Rahman, T.A., Rigano, L., El Hadrami, A., Rodriguez, M.C., Daayf, F., et al., 2011. *Botrytis cinerea* manipulates the antagonistic effects between immune pathways to promote disease development in tomato. Plant Cell 23 (6), 2405–2421.

Fu, Y., Yin, H., Wang, W., Wang, M., Zhang, H., Zhao, X., et al., 2011. β-1, 3-Glucan with different degree of polymerization induced different defense responses in tobacco. Carbohydr. Polym. 86 (2), 774–782.

Furch, A.C., Zimmermann, M.R., Kogel, K.H., Reichelt, M., Mithöfer, A., 2014. Direct and individual analysis of stress-related phytohormone dispersion in the vascular system of *Cucurbita maxima* after flagellin 22 treatment. N. Phytol. 201 (4), 1176–1182.

García-García, J.D., Peña-Sanabria, K.A., Sánchez-Thomas, R., Moreno-Sánchez, R., 2018. Nickel accumulation by the green algae-like *Euglena gracilis*. J. Hazard. Mater. 343, 10–18.

Gauthier, A., Trouvelot, S., Kelloniemi, J., Frettinger, P., Wendehenne, D., Daire, X., et al., 2014. The sulfated laminarin triggers a stress transcriptome before priming the SA-and ROS-dependent defenses during grapevine's induced resistance against *Plasmopara viticola*. PLoS One 9 (2), e88145.

Gissibl, A., Care, A., Parker, L.M., Iqbal, S., Hobba, G., Nevalainen, H., et al., 2018. Microwave pretreatment of paramylon enhances the enzymatic production of soluble β-1, 3-glucans with immunostimulatory activity. Carbohydr. Polym. 196, 339–347.

Gissible, A., Sun, A., Care, A., Nevalainen, H., Sunna, A., 2019. Bioproducts from *Euglena gracilis*: synthesis and applications. Front. Bioeng. Biotechnol. 7, 108.

Gojdics, M., 1953. The Genus Euglena. Univ. Wisc. Press, Madison, pp. 1–268.

Gordon, T.R., 2017. *Fusarium oxysporum* and the fusarium wilt syndrome. Annu. Rev. Phytopathol. 55, 23–39.

Gottlieb, J., 1850. Ueber eine neue, mit Stärkmehl isomere Substanz. Justus Liebigs Annalen der Chem. 75 (1), 51–61.

Grimm, P., Risse, J.M., Cholewa, D., Müller, J.M., Beshay, U., Friehs, K., et al., 2015. Applicability of *Euglena gracilis* for biorefineries demonstrated by the production of α-tocopherol and paramylon followed by anaerobic digestion. J. Biotechnol. 215, 72–79. Available from: https://doi.org/10.1016/j.jbiotec.2015.04.004.

Hanaa, R.F., Abdou, Z.A., Salama, D.A., Ibrahim, M.A., Sror, H.A.M., 2011. Effect of Neem and willow aqueous extracts on fusarium wilt disease in tomato seedlings: induction of antioxidant defensive enzymes. Ann. Agric. Sci. 56 (1), 1–7.

Holbrook, N.M., Shashidhar, V.R., James, R.A., Munns, R., 2002. Stomatal control in tomato with ABA-deficient roots: response of grafted plants to soil drying. J. Exp. Bot. 53 (373), 1503–1514.

Horváth, E., Szalai, G., Janda, T., 2007. Induction of abiotic stress tolerance by salicylic acid signaling. J. Plant Growth Regul. 26, 290–300.

Howe, G.A., Jander, G., 2008. Plant immunity to insect herbivores. Annu. Rev. Plant Biol. 59 (1), 41–66.

Jaillais, Y., Chory, J., 2010. Unraveling the paradoxes of plant hormone signaling integration. Nat. Struct. Mol. Biol. 17 (6), 642–645.

Jones, J.D., Dangl, J.L., 2006. The plant immune system. Nature 444 (7117), 323–329.

Kabashnikova, L., Abramchik, L., Domanskaya, I., Savchenko, G., Shpileuski, S., 2020. β-1, 3-glucan effect on the photosynthetic apparatus and oxidative stress parameters of tomato leaves under fusarium wilt. Funct. Plant Biol. 47 (11), 988–997.

Kataoka, K., Muta, T., Yamazaki, S., Takeshige, K., 2002. Activation of macrophages by linear (1→ 3)-β-d-glucans: implications for the recognition of fungi by innate immunity. J. Biol. Chem. 277 (39), 36825–36831.

Kiss, J.Z., Vasoconcelos, A.C., Triemer, R.E., 1986. Paramylon synthesis and chloroplast structure associated with nutrient levels in *Euglena* (Euglenophyceae) 1. J. Phycol. 22 (3), 327–333.

Koizumi, N., Sakagami, H., Utsumi, A., Fujinaga, S., Takeda, M., Asano, K., et al., 1993. Anti-HIV (human immunodeficiency virus) activity of sulfated paramylon. Antivir. Res. 21 (1), 1–14.

Leng, G., Hall, J., 2019. Crop yield sensitivity of global major agricultural countries to droughts and the projected changes in the future. Sci. Total Environ. 654, 811–821.

Ma, K.W., Ma, W., 2016. Phytohormone pathways as targets of pathogens to facilitate infection. Plant Mol. Biol. 91, 713–725.

Malundo, T.M.M., Shewfelt, R.L., Scott, J.W., 1995. Flavor quality of fresh tomato (*Lycopersicon esculentum* Mill.) as affected by sugar and acid levels. Postharvest Biol. Technol. 6 (1–2), 103–110.

Manzi, M., Lado, J., Rodrigo, M.J., Zacarías, L., Arbona, V., Gómez-Cadenas, A., 2015. Root ABA accumulation in long-term water-stressed plants is sustained by hormone transport from aerial organs. Plant Cell Physiol. 56 (12), 2457–2466.

Melotto, M., Zhang, L., Oblessuc, P.R., He, S.Y., 2017. Stomatal defense a decade later. Plant Physiol. 174 (2), 561–571.

Milanowski, R., Kosmala, S., Zakrys, B., Kwiatowski, J., 2006. Phylogeny of photosynthetic euglenophytes based on combined chloroplast and cytoplasmic SSU rDNA sequence analysis. J. Phycol. 42, 721–730.

Miyatake, K., Kitaoka, S., 1983. Comparison of the methods for determining degree of polymerization of water-insoluble β-1, 3-glucans. Bull. Univ. Osaka Prefect. Ser. B Agric. Biol. 35, 55–58.

Monfils, A.K., Triemer, R.E., Bellairs, E.F., 2011. Characterization of paramylon morphological diversity in photosynthetic euglenoids (Euglenales, Euglenophyta). Phycologia 50 (2), 156–169.

Morison, J.I.L., Baker, N.R., Mullineaux, P.M., Davies, W.J., 2008. Improving water use in crop production. Philos. Trans. R. Soc. B: Biol. Sci. 363 (1491), 639–658.

Mundy, J., Nielsen, H.B., Brodersen, P., 2006. Crosstalk. Trends Plant Sci. 11 (2), 63–64.

Muñoz-Espinoza, V.A., López-Climent, M.F., Casaretto, J.A., Gómez-Cadenas, A., 2015. Water stress responses of tomato mutants impaired in hormone biosynthesis reveal abscisic acid, jasmonic acid and salicylic acid interactions. Front. Plant Sci. 6, 997.

Newman, M.A., Sundelin, T., Nielsen, J.T., Erbs, G., 2013. MAMP (microbe-associated molecular pattern) triggered immunity in plants. Front. Plant Sci. 4, 139. Available from: https://doi.org/10.3389/fpls.2013.00139.

Poór, P., Tari, I., 2012. Regulation of stomatal movement and photosynthetic activity in guard cells of tomato abaxial epidermal peels by salicylic acid. Funct. Plant Biol. 39 (12), 1028–1037.

Raiola, A., Tenore, G.C., Barone, A., Frusciante, L., Rigano, M.M., 2015. Vitamin E content and composition in tomato fruits: beneficial roles and bio-fortification. Int. J. Mol. Sci. 16 (12), 29250–29264.

Rasheed, R., Iqbal, M., Ashraf, M.A., Hussain, I., Shafiq, F., Yousaf, A., et al., 2018. Glycine betaine counteracts the inhibitory effects of waterlogging on growth, photosynthetic pigments, oxidative defence system, nutrient composition, and fruit quality in tomato. J. Hortic. Sci. Biotechnol. 93 (4), 385–391.

Rodríguez-Zavala, J.S., Ortiz-Cruz, M.A., Moreno-Sanchez, R.A.F.A.E.L., 2006. Characterization of an aldehyde dehydrogenase from *Euglena gracilis*. J. Eukaryot. Microbiol. 53 (1), 36–42.

Russo, R., Barsanti, L., Evangelista, V., Frassanito, A.M., Longo, V., Pucci, L., et al., 2017. *Euglena gracilis* paramylon activates human lymphocytes by upregulating pro-inflammatory factors. Food Sci. Nutr. 5 (2), 205–214.

Santos, L., Lucio, J., Odair, J., Carneiro, M.L., Alberto, C., 2000. Symptomless infection of banana and maize by endophytic fungi impairs photosynthetic efficiency. N. Phytol. 147, 609–615. Available from: https://doi.org/10.1046/j.1469-8137.2000.00722.

Sauter, A., Davies, W.J., Hartung, W., 2001. The long-distance abscisic acid signal in the droughted plant: the fate of the hormone on its way from root to shoot. J. Exp. Bot. 52 (363), 1991–1997.

Scartazza, A., Picciarelli, P., Mariotti, L., Curadi, M., Barsanti, L., Gualtieri, P., 2017. The role of *Euglena gracilis* paramylon in modulating xylem hormone levels, photosynthesis and water-use efficiency in Solanum lycopersicum L. Physiol. Plant 161 (4), 486–501.

Shibakami, M., Tsubouchi, G., Nakamura, M., Hayashi, M., 2013. Polysaccharide nanofiber made from euglenoid alga. Carbohydr. Polym. 93 (2), 499–505.

Sivakumar, P., Sharmila, P., Jain, V., Pardha-Saradhi, P., 2002. Sugars have potential to curtail oxygease acitivity of Rubisco. Biochem. Biophys. Res. Commun. 298, 247–250.

Slimestad, R., Verheul, M., 2009. Review of flavonoids and other phenolics from fruits of different tomato (*Lycopersicon esculentum* Mill.) cultivars. J. Sci. Food Agric. 89 (8), 1255–1270.

Sorrentino, G., Haworth, M., Wahbi, S., Mahmood, T., Zuomin, S., Centritto, M., 2016. Abscisic acid induces rapid reductions in mesophyll conductance to carbon dioxide. PLoS One 11 (2), e0148554.

Suzuki, K., Mitra, S., Iwata, O., Ishikawa, T., Kato, S., Yamada, K., 2015. Selection and characterization of *Euglena anabaena* var. *minor* as a new candidate *Euglena* species for industrial application. Biosci. Biotechnol. Biochem. 79 (10), 1730–1736.

Thaler, J.S., Bostock, R.M., 2004. Interactions between abscisic-acid-mediated responses and plant resistance to pathogens and insects. Ecology 85 (1), 48–58.

Thaler, J.S., Karban, R., Ullman, D.E., Boege, K., Bostock, R.M., 2002. Cross-talk between jasmonate and salicylate plant defense pathways: effects on several plant parasites. Oecologia 227–235.

Thaler, J.S., Humphrey, P.T., Whiteman, N.K., 2012. Evolution of jasmonate and salicylate signal crosstalk. Trends Plant Sci. 17 (5), 260–270.

Ton, J., Mauch-Mani, B., 2004. β-amino-butyric acid-induced resistance against necrotrophic pathogens is based on ABA-dependent priming for callose. Plant J. 38 (1), 119–130.

Ton, J., Flors, V., Mauch-Mani, B., 2009. The multifaceted role of ABA in disease resistance. Trends Plant Sci. 14 (6), 310–317.

Visentin, I., Vitali, M., Ferrero, M., Zhang, Y., Ruyter-Spira, C., Novák, O., et al., 2016. Low levels of strigolactones in roots as a component of the systemic signal of drought stress in tomato. N. Phytol. 212 (4), 954–963.

Wasternack, C., Hause, B., 2013. Jasmonates: biosynthesis, perception, signal transduction and action in plant stress response, growth and development. An update to the 2007 review in Annals of Botany. Ann. Bot. 111 (6), 1021–1058.

CHAPTER

6

Humic substances-based products for plants growth and abiotic stress tolerance

Santiago Atero-Calvo, Eloy Navarro-León, Juan Jose Rios, Begoña Blasco and Juan Manuel Ruiz

Department of Plant Physiology, Faculty of Sciences, University of Granada, Granada, Spain

6.1 Introduction

Increased world population and climate change are the two major challenges for modern agriculture. It is estimated that the human population will reach 10 billion people in 2050, and more than 11 billion by 2100, which means that 60% more food will need to be produced (Tian et al., 2021). At the same time, the negative impact of climate change on the environment is steadily increasing, with crops facing different adverse growing conditions such as drought and salinity (Rouphael and Colla, 2020). For this reason, the main objectives of agricultural researchers are to enhance crop production and tolerance to biotic and abiotic stress through different agronomic techniques while reducing agrochemical use (Malik et al., 2020). Breeding programs have historically been used to increase crop production and performance. However, the long periods to breed more efficient varieties (10–15 years), greatly limits the immediate application of this technique (Deolu-Ajayi et al., 2022). An interesting alternative, which has recently attracted farmers' attention, is soil or plant application of products known as biostimulants (Deolu-Ajayi et al., 2022; Rouphael and Colla, 2020).

Plant biostimulants (PBs) were defined by the Fertilizing Products Regulation (EU 2019/1009) of the European Parliament and the Council as: "*A plant biostimulant shall be an EU fertilizing product the function of which is to stimulate plant nutrition processes independently of the product's nutrient content with the sole aim of improving one or more of the following characteristics of the plant or the plant rhizosphere: (a) nutrient use efficiency, (b) tolerance to abiotic stress, (c) quality traits, and (d) availability of confined nutrients in the soil or rhizosphere*" (EU., 2019). Different substances that increase agricultural production and improve tolerance to abiotic stresses while

Biostimulants in Plant Protection and Performance
DOI: https://doi.org/10.1016/B978-0-443-15884-1.00025-7

reducing the use of pesticides and fertilizers can be included in this definition. In this regard, Du Jardin (2015), in his literature review "Plant Biostimulants: Definition, Concept, Main Categories, and Regulation," established six categories of PBs of diverse nature, which are as follows: (1) humic substances; (2) protein hydrolysates from crop and animal wastes; (3) seaweed and plant extracts; (4) chitosan; (5) inorganic elements such as selenium (Se) and inorganic salts, for example, phosphite; (6) beneficial fungi such as arbuscule-forming mycorrhiza; and (7) beneficial bacteria like plant growth-promoting rhizobacteria.

Humic substances (HS) constitute up to 80% of the organic matter on the earth's surface and play a central role in various chemical reactions at the soil level (Canellas and Olivares, 2014). HS is generated by the biodegradation of soil organic matter, including plants, compost, and peat, in a process known as "humification," where soil microorganisms are the main actors (Cha et al., 2021). It is difficult to define the chemical structure of HS because it will depend on its age and origin. Even so, they could be considered heterogenous small molecules associated with supramolecular structures by van der Waals forces and hydrogen bonds (Lipczynska-Kochany, 2018a). According to their aqueous solubility, HS has been classified by soil researchers into three groups: humin, humic acids (HA), and fulvic acids (FA). Using 0.1 M NaOH, HA and FA are extracted from soils, while the fraction remaining in the sample belongs to the insoluble organic matter, humin. Then, the solution containing HA and FA is acidified with HCl to pH 1–2, so that HA precipitates (due to its insolubility at acidic pH) and FA remains in solution (because of its solubility at all pHs) (Islam et al., 2020).

The beneficial effects of HS on plant growth and development, as well as its potential use as PBs, have been extensively investigated in recent decades. The indirect effects of HS on plants are related to their impact on the physical properties of soils. Thus, HS has the ability to form aggregates with soil particles, such as the Si-OH and Al-OH groups of clays, which improves the water-holding capacity and drainage of the soil as well as its porosity. HS also improves cation exchange capacity and carbon sequestration (Islam et al., 2020; Yang et al., 2021b). On the other hand, the direct effects of HS involve their influence on essential nutrient uptake, effect on photosynthesis, tricarboxylic acid cycle (TCA), and hormone balance, as well as protection against abiotic stresses (Canellas et al., 2015; Canellas and Olivares, 2014; Shah et al., 2018; Trevisan et al., 2010). These effects depend on the type of HS, their molecular weight, dosage, the source from which they have been extracted, the mode of application (foliar or root), and on the plant species (Nardi et al., 2021).

This book chapter aims to give the reader an insight into the potential use of HS as an organic PB for plant performance and protection. First, we will analyze the main functions of HS in plant physiology. Then, we will also review the implications of HS for the improvement of plant tolerance to different abiotic stress conditions such as drought, salinity, heavy metal toxicity, and organic pollutants.

6.2 Humic substances and plant growth

The main pathways by which HS improves plant growth and production are summarized in Fig. 6.1. Despite the effect of HS on soil properties, we will only focus on the direct effect of HS on plant physiology.

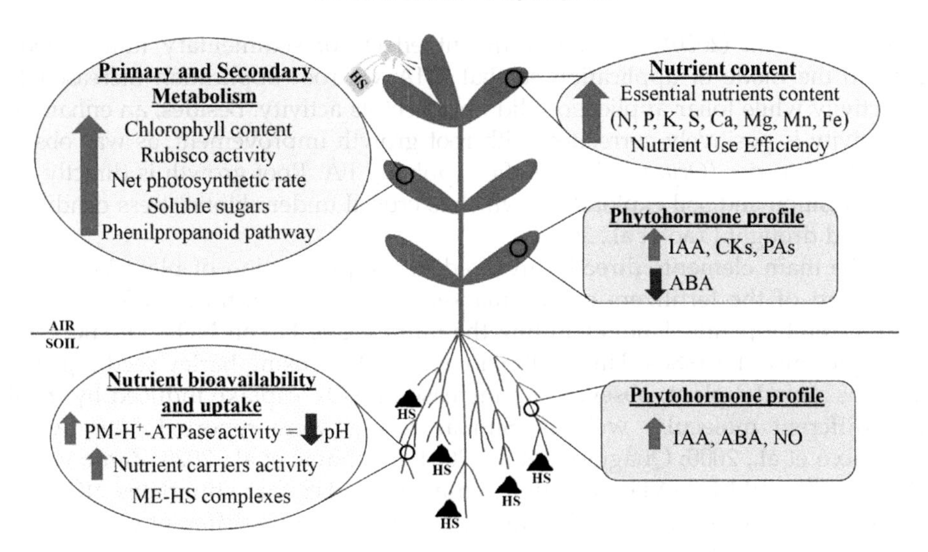

FIGURE 6.1 Main pathways by which HS increase plant growth and production. *HS*, humic substances; *N*, nitrogen; *P*, phosphorus; *K*, potassium; *S*, sulfur; *Ca*, calcium; *Mg*, magnesium; *Mn*, manganese; *Fe*, iron; *IAA*, indole acetic acid; *CKs*, cytokinins; *PAs*, polyamines; *ABA*, abscisic acid; *NO*, nitric oxide; *PM-H$^+$-ATPase*, plasma membrane-H$^+$-ATPase; *ME-HS*, mineral element-humic substances.

6.2.1 Effects of humic substances on nutrient availability and uptake

There is no question that nutrient availability is essential for plant growth and development. For this reason, understanding the key processes that directly affect nutrient availability, their uptake, and assimilation is crucial for enhancing crop production (Kumar et al., 2021). Carbon (C), hydrogen (H), oxygen (O), nitrogen (N), phosphorus (P), sulfur (S), potassium (K), calcium (Ca), magnesium (Mg), manganese (Mn), iron (Fe), boron (B), molybdenum (Mo), zinc (Zn), and copper (Cu) are known to be essential for plants (Kumar et al., 2021; Yang et al., 2021b). The well-known Green Revolution was characterized by the massive use of N and P fertilizers to feed the world's growing population by increasing agricultural production (Kumar et al., 2021; Tian et al., 2021). However, these fertilizers negatively affect the environment, which is why different PBs, such as HS, are presented as an alternative to increase nutrient availability and their uptake by plants.

Modifying rhizospheric pH and secreting organic acids that form complexes with nutrients are some of the plants' strategies to increase nutrient availability and uptake (Liu et al., 2022). The plasma membrane (PM)-H$^+$-ATPase located in root epidermal cells plays a key role in rhizosphere acidification by H + excretion. At the same time, H$^+$ accumulation in the extracellular space and the consequent creation of an electrochemical gradient allow nutrient uptake by secondary active transport as well as organic acids (i.e., citrate and malate) excretion from roots to soil solution (Li et al., 2022). Different studies have shown that HS enhances PM-H$^+$-ATPase activity via the indoleacetic acid (IAA) pathway and the synthesis of new PM-H$^+$-ATPase proteins (Canellas et al., 2002; Canellas and Olivares, 2014; De Hita et al., 2020; Olaetxea et al., 2019). Thus, Canellas et al. (2002) revealed an increase in PM-H$^+$-ATPase activity induced by HA derived from earthworm compost in maize (*Zea mays* L.)

roots. De Hita et al. (2020) observed different effects of sedimentary humic acid (SHA) according to the mode of application so that SHA via root application increased PM-H^+-ATPase activity while foliar application did not affect its activity. Besides, an enhancement in ATPase activity is positively correlated with root growth improvement, as was observed by Yao et al. (2019) in rice (*Oryza sativa* L.) after applying FA. Root growth is directly linked to nutrient acquisition and soil exploration, which is crucial under abiotic stress conditions such as salinity and drought (Yao et al., 2019).

One of the main elements directly involved in the production of plant biomass, which is part of most of the fertilizers on the market, is nitrogen. Nitrate (NO_3^-) is the major form of N taken by plants. Understanding the influence of HS on NO_3^- uptake has been a research topic since the 1980s. Thus, Albuzio et al. (1986), using barley seedlings (*Hordeum vulgare* L.) as a model plant, observed an increase in NO_3^--uptake induced by fractions of HS with different molecular weights. Similar results were reported later by different authors (Cacco et al., 2000; Quaggiotti et al., 2004; Vujinović et al., 2020; Zanin et al., 2019). In addition to the PM-H^+-ATPase activity induced by HS that stimulates NO_3^--symport (2:1 H^+: NO_3^-) (Canellas et al., 2015), HS appears to modulate different genes involved in NO_3^--uptake, such as the high-affinity transport of NO_3^- (*NRT*) (Zanin et al., 2019).

Induction of specific nutrient transporters and enhanced root PM-H^+-ATPase activity, with a consequent increase in root length, are not the only mechanisms by which HS increases nutrient uptake. This is because HS can also act at the rhizospheric level. The presence of functional groups containing O, such as carboxyl, carbonyl, alcoholic, and phenolic hydroxyl groups, on the HS surface allows HS to interact with soil particles and soil minerals, which improves nutrient bioavailability (Yang et al., 2021a). P is an essential plant nutrient whose concentration in the soil does not correlate with its bioavailability because it can be found interacting with soil minerals, which limits its uptake by plants. The competition between HS and P for P adsorption sites allows the release of P (Fu et al., 2013). Thus, the application of HA does not change P concentration in soil, but it does increase the amount of P bioavailable (Yang et al., 2019). Likewise, the interaction of HS with metals such as Fe and aluminum (Al) prevents the formation of Fe-P and Al-P salts, which keeps P available for plants (Yang et al., 2021a).

On the other hand, like citrate and malate, HS can act as chelators by adsorbing mineral elements (ME) to HS functional groups. The resulting ME-HS complex is more bioavailable for plants (Zanin et al., 2019). This mechanism has been extensively studied for Fe, where the formation of Fe-HS complexes increases Fe bioavailability, making it easier for Fe to reach roots and be uptaken by plants (Abros'kin et al., 2016; Chen et al., 2004; Zanin et al., 2019). In this way, it is said that Fe-HS could be a better Fe chelate than others such as EDTA (ethylenediaminetetraacetic acid) (Olaetxea et al., 2018). Finally, Table 6.1 shows examples of different mechanisms by which HS improves the bioavailability and uptake of essential nutrients by plants.

6.2.2 Effects of humic substances on primary and secondary metabolism

Primary metabolism is the set of biochemical processes that directly affect plant physiology, including growth, reproduction, and survival. Those compounds involved in primary

TABLE 6.1 Examples of HS mechanisms of action to increase bioavailability and uptake of essential nutrients.

Essential nutrient	Plant species	HS mechanism of action	References
Cu, Zn, and Fe	*Triticum aestivum* L. *Medicago sativa* L.	Formation of Cu-, Zn-, and Fe-SHE (selected humic extract) increases micronutrient bioavailability.	García-Mina et al. (2004)
N	*Zea mays* L.	Up-regulation of H + -ATPase gene (*Mha2*) by low molecular HS which increases NO3- transport.	Quaggiotti et al. (2004)
Fe	*Helianthus annuus* L.	Fe complexed by HA of low and high molecular weight, that enhanced Fe uptake by plants.	Bocanegra et al. (2006)
N	*Cucumis sativus* L.	HA improves PM-H + -ATPase activity and increases NO3- concentration in shoots.	Mora et al. (2010)
N and S	*Brassica napus* L.	HS increase expression levels of NO3- transporters (*BnNRT1.1* and *BnNRT2.1*) and sulfate (SO_4^{2-}) transporters (*BnSultr1.1* and *BnSultr1.2*)	Jannin et al. (2012)
Fe	*Solanum lycopersicum* L.	Fe complexed with water-extractable humic substances (Fe-WHS) induced Fe(III)-chelate reductase (*LeFRO1*) and Fe transporters (*LeIRT1* and *LeIRT2*).	Tomasi et al. (2013)
N, Cu, Fe, and Zn	*Brassica napus* L.	HA increases expression levels of NO_3^- transporters (*BnNRT1.1* and *BnNRT2.1*), Cu transporter (*COPT2*), Fe transporter (*IRT1*), and Fe or Zn transporter (*NRAMP3*).	Billard et al. (2014)
P	*Solanum lycopersicum* L.	HS-induced high-affinity P transporter (*LePT2*) at high and low P concentrations.	Jindo et al. (2016)

metabolism are called "primary metabolites" and are classified into three groups: carbohydrates, lipids, and proteins (Canellas and Olivares, 2014). In addition to affecting the bioavailability and uptake of essential nutrients, HS also seems to influence primary metabolism-enhancing plant growth. The changes in cellular metabolism induced by HS appear to be due to functional groups (alcohols, phenols, and carboxyls) present in these organic compounds (Schiavon et al., 2010). Thus, one of the most significant results at the photosynthetic level is the increase in chlorophyll concentration after the application of HS, which has been observed by different authors (Ertani et al., 2011; Fan et al., 2015, 2014; Turan et al., 2022). Besides, Fan et al. (2014) analyzed different chlorophyll fluorescence parameters in *Chrysanthemum morifolium* L. Φ_{PSII} indicate the efficiency with which photosystem II (PSII) captures primary light energy. Plants treated with HA showed a significant increase in Φ_{PSII}, indicating that most of the light energy captured by PSII in these plants was used in the photosynthetic process (Fan et al., 2014). Moreover, rubisco activity is stimulated by HS, and soluble sugars (glucose, fructose, and sucrose) increase after HS treatment while starch content decreases (Ertani et al., 2011; Merlo et al., 1991). Therefore, the application of HS results in an increase in the net photosynthetic rate, which is correlated with enhanced crop production (Azcona et al., 2011; Canellas et al., 2013; Fan et al., 2014).

On the other hand, the influence of HS on glycolysis and TCA, or Krebs cycle, was extensively studied by Nardi et al. (2007) in maize leaves after applying HA and three HA fractions (I, II, and III) of different sizes and molecular compositions. They analyzed the activity of different enzymes from both metabolic pathways: glucokinase, phosphoglucose isomerase, PPi-dependent phosphofructokinase, and pyruvate kinase from glycolysis and citrate synthase, malate dehydrogenase, and $NADP^+$-isocitrate dehydrogenase from TCA. HA and fraction III positively affected the activity of most of the enzymes studied, which could lead to increased plant growth (Nardi et al., 2007). In addition, HS can also stimulate other metabolic processes such as nitrogen and sulfur assimilation. Nitrate reductase (NR) and nitrite reductase are responsible for NO_3^- reduction to ammonium (NH_4^+). The assimilation of NH_4^+ to glutamine (Gln) and glutamate (Glu) is driven by Gln synthetase (GS) and Glu synthase (GOGAT) (Rubio-Wilhelmi et al., 2012). NR, GS, and GOGAT activity can be enhanced by HS (Ertani et al., 2011; Hernandez et al., 2015), which is associated with an increase in N compounds such as proteins (Ertani et al., 2011). Similarly, genes coding for the enzymes ATP sulfurylase and serine acetyltransferase, which are key in S assimilation, appear to be upregulated by HS (Jannin et al., 2012). All of this allows HS to improve nutrient use efficiency.

Secondary metabolism is not directly related to plant growth and development, but it plays a critical role in plant defense against adverse growth conditions (Erb and Kliebenstein, 2020). It involves different "secondary metabolites," among which phenolic compounds (flavonoids and phenylpropanoids) are the most relevant. The synthesis of these compounds is carried out by the phenylpropanoid pathway, where phenylalanine (tyrosine) ammonia-lyase (PAL/TAL), which catalyzes the conversion of phenylalanine to *trans*-cinnamic acid and tyrosine to *p*-coumaric acid with the release of NH_4^+, are key enzymes (Sharma et al., 2022). The first study that related HS to the phenylpropanoid pathway was performed by Schiavon et al. (2010). These authors found that HS derived from earthworms enhanced PAL/TAL gene expression and increased total phenolic acids and flavonoids content in maize (Schiavon et al., 2010), which has subsequently been verified in different studies (Conselvan et al., 2017; Cruz et al., 2014; Savarese et al., 2022). All these results suggest that HS induces changes in plant secondary metabolism, which could be positive for the interaction of plants with the environment.

6.2.3 Effects of humic substances on phytohormone profile

Phytohormones can be defined as chemical messengers that modulate different plants' physiological processes. Phytohormone-like activity is considered one of the main physiological effects induced by HS in plants that indirectly affect the mechanisms of action discussed above: increased PM-H^+-ATPase activity, enhanced nutrient uptake, and changes in primary and secondary metabolism (De Hita et al., 2020; Souza et al., 2022). Pizzeghello et al. (2001) reported the presence of hormones such as IAA in the HS structure. If we stop to think about it, it is logical to find hormones in fertile soil because soil microorganisms, as well as plant roots, can produce hormones such as auxins and gibberellins (Nardi et al., 2017). Thus, HS derived from soils rich in organic matter are likely to have hormones embedded in their structure. On the other hand, purified HS, such as those derived from leonardite, have no measurable hormone concentrations. Even so, such HS can increase the concentration of hormones (i.e., IAA, nitric oxide, and ethylene) in plant tissues as well as bind to cell walls and act as a physiological signal similar to a plant hormone. For this reason, it is said that HS has "hormone-like activity" (Mora et al., 2012; Schiavon et al., 2010).

Root growth is important for plants to survive adverse environmental conditions, increasing the soil exploration zone and thus improving water and nutrient uptake (Zandonadi et al., 2010). As discussed in Section 6.2.1., increased root PM-H^+-ATPase activity by HS can be guided by auxins (IAA), acidifying the apoplast and increasing cell wall plasticity, which results in increased cell length and greater root growth (Bettoni et al., 2016). Similarly, other hormones such as nitric oxide (NO) and ethylene are involved in root growth promotion by HS activating PM-H^+-ATPase (Lucena et al., 2006; Zandonadi et al., 2010). Olaetxea et al. (2019), using specific inhibitors, found that abscisic acid (ABA) plays a critical role in cucumber (*Cucumis sativus* L.) root growth after applying HA. Besides, the effects of HS on lateral root proliferation are stimulated by auxins and NO via the induction of H^+ pumps. All of this is translated into an increase in the contact surface between plant roots and soil solution, which results in enhanced nutrient uptake in plants treated with HS (Yang et al., 2021b).

The promotion of root growth through different hormonal pathways ultimately stimulates shoot growth, mainly by improving the absorption and translocation of essential nutrients from the soil to the leaves. Similarly, the translocation of hormones from roots to leaves promotes shoot growth. After applying HA to cucumber, Mora et al. (2010) observed an increase in shoot growth due to changes in the root-shoot distribution of cytokinins and polyamines, which were found in higher concentrations in shoots. Moreover, HS-induced changes in hormone concentration at the root level seem to affect shoot growth. Thus, Mora et al. (2014) found that purified SHA promoted shoot growth in cucumber plants by increasing IAA, NO, ethylene, and ABA concentrations in roots. Likewise, the application of HS to the leaves can increase IAA in roots and shoots, as well as decrease ABA concentration in shoots, which is relevant to promoting shoot growth (De Hita et al., 2020).

On the other hand, the connection between HS and secondary metabolism could also be mediated by hormones. This is due to the correlation between auxin transport and flavonoid accumulation. Because different types of HS contain auxins in their structure, the promotion of phenylpropanoid metabolism, discussed in the previous section, could be auxin-dependent (Schiavon et al., 2010). Therefore, all the changes in hormonal concentrations at root and leaf levels induced by HS are directly connected with plant growth promotion due to the effect of these phytohormones on different physiological functions, including primary and secondary metabolism.

6.3 Crops protection against abiotic stress by humic substances

Abiotic stress conditions such as drought, salinity, heavy metals, and organic pollutants are one of the main challenges for agriculture. An increase in reactive oxygen species (ROS), including superoxide radicals (O_2^-) and hydrogen peroxide (H_2O_2), can be observed in plant cells exposed to abiotic stress. These ROS react with essential biomolecules such as lipids, proteins, and DNA, leading to peroxidation of the lipid bilayer, protein denaturation, and mutations. This negatively affects essential physiological processes such as photosynthesis, resulting in reduced plant growth or even plant death, causing a significant decrease in agricultural productivity. To cope with ROS, the plant induces its enzymatic (catalase, superoxide dismutase, ascorbate peroxidase, etc.) and nonenzymatic (ascorbate, glutathione, carotenoids, etc.) antioxidant systems (Aguiar et al., 2016; Bulgari et al., 2019).

Increased root and leaf growth, improved uptake of essential nutrients, and the induction of primary and secondary metabolism by HS as well as their hormone-like activity could improve crop yield under abiotic stress conditions through HS application. Several studies have highlighted the use of HS as a potential plant biostimulant in the protection of crops against stress conditions. Some of the effects by which HS improves plant tolerance to abiotic stresses are shown in Fig. 6.2. However, the physiological mechanisms by which HS improves plant tolerance are not completely clear, and this field of study requires further research.

6.3.1 Drought and salinity

Drought or water deficits are considered the main environmental factors that limit agricultural growth and production. By 2100, the earth's average temperature is expected to increase by 1.8°C–4.0°C, which translates into an extension of drought, making agriculture more difficult in most parts of the globe (Ozturk et al., 2021). Lack of water in the soil has

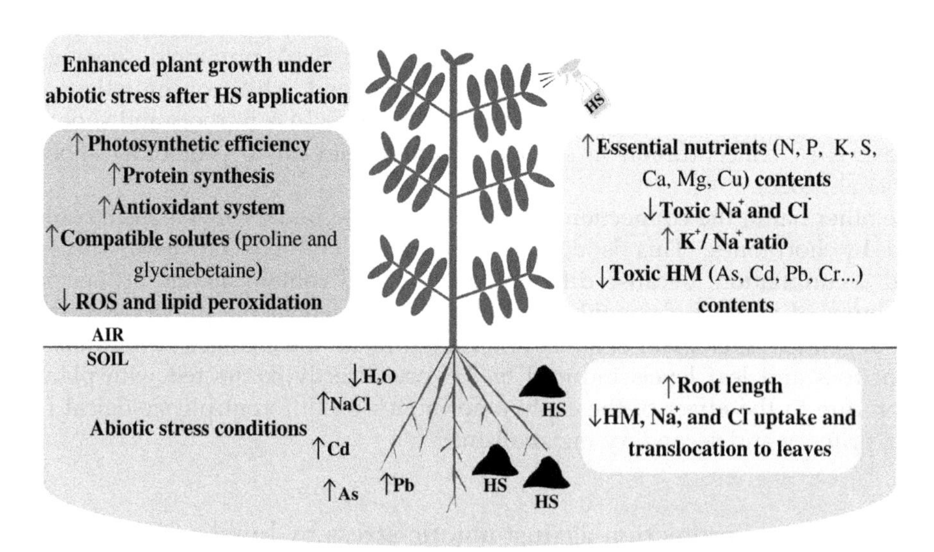

FIGURE 6.2 Some of the mechanisms by which HS improves plant tolerance to abiotic stress conditions such as drought, salinity, and heavy metals. *HS*, humic substances; *HM*, heavy metals; *N*, nitrogen; *P*, phosphorus; *K*, potassium; *S*, sulfur; *Ca*, calcium; *Mg*, magnesium; *Cu*, copper; *Na*, sodium; *Cl*, chloride; *As*, arsenic; *Cd*, cadmium; *Pb*, lead; *Cr*, chromium; *ROS*, reactive oxygen species.

negative consequences for plants, such as a decrease in cell growth, which leads to a reduction in the size of roots, stems, and leaves. In addition, the closure of stomata due to drought decreases the transport of essential nutrients, and the lack of CO_2 leads to a disruption of photosynthesis as well as an increase in ROS (Farooq et al., 2012).

On the other hand, soil water evaporation causes an increase in the concentration of salts in the soil. As a result, drought is often followed by salt stress. Currently, 20%–50% of irrigated cropland is affected by salinity, and it is estimated that by 2050, half of all agricultural cropland will become salinized (Angon et al., 2022). It is considered that under salt stress, the plant cell experiences three types of stress simultaneously: osmotic stress, ionic stress, and oxidative stress due to the decrease of soil water potential, the toxicity of sodium (Na^+) and chloride (Cl^-) ions, and the increase of ROS, respectively (Ahmad et al., 2022; Navarro-León et al., 2020).

HS can improve plant tolerance to drought and salinity, increasing plant growth and yield, as has been verified by many researchers. Similar effects of HS have been described in plants grown under water stress and salinity conditions. The increase in root elongation by HS via root PM-H^+-ATPase activation enhances deep soil exploration, which is crucial in dry environments (Aguiar et al., 2016; Canellas et al., 2002). The stimulation of nutrient uptake by HS is also relevant for plants grown under stressful conditions. Thus, Çimrin et al. (2010) observed that HA increased N, P, K, Ca, S, Fe, Cu, Mn, and Zn in roots and N, P, K, Ca, S, Mg, Cu, and Mn in shoots, and decreased Na in roots and shoots of pepper (*Capsicum annuum* L.) plants subjected to salt stress which was correlated with increased plant tolerance to salinity. It is important to note that it seems that HS can contribute to maintaining a high K^+/Na^+ ratio, which is crucial to salt tolerance.

Regarding photosynthetic efficiency, Shen et al. (2020) found that HS enhanced the photosynthetic status of millet (*Setaria italica* L.) grown under water deficit. Thus, these authors observed that HA enhanced net photosynthetic rate, stomatal conductance, and electron transport rate at the PSII level. Finally, in addition to the physiological effects discussed so far, HS appears to stimulate the activity of antioxidant enzymes such as superoxide dismutase (SOD), peroxidase (POD), catalase (CAT), and ascorbate peroxidase (APX) (Elmongy et al., 2018). Thus, plants treated with HS and exposed to drought or salinity show a decrease in O_2^-, H_2O_2, and malondialdehyde (MDA) contents and, consequently, a lower lipid peroxidation compared to plants not treated with HS (García et al., 2014; Hatami et al., 2018; Shen et al., 2020). This results in increased tolerance of HS-treated crops to abiotic stress conditions.

6.3.2 Heavy metals and organic pollutants

Another challenge for agricultural researchers is to cope with heavy metal (HM) toxicity using inexpensive and environmentally friendly techniques. The presence of HM in soil and water is mainly due to anthropogenic activities such as industries, mining, sewage sludge, and the use of pesticides. Examples of these HMs are cadmium (Cd), lead (Pb), mercury (Hg), arsenic (As), chromium (Cr), and nickel (Ni) (Paniagua-López et al., 2021; Rehman et al., 2021). Similarly, human activities have led to an increase in organic pollutants (OPs) such as polychlorinated biphenyls, polychlorinated dibenzo-p-dioxins, polychlorinated dibenzofurans, pharmaceutical compounds, and herbicides in the environment (Zhang et al., 2017). HM and OPs are highly toxic to plants and animals because, among other reasons, they increase ROS content and affect essential processes such as photosynthesis and the synthesis of essential biomolecules. As a result, high concentrations of HM and OPs cause significant reductions in plant growth and crop production (Hassan et al., 2022; Zhang et al., 2017).

Different examples of HM stress mitigation by HS are summarized in Table 6.2. HS can reduce the uptake and translocation of HM from the soil solution to the leaves, thus reducing the toxic effect of HM on plants. As described by Yang et al. (2021b) in their detailed review, there are two main mechanisms by which HS can deal with HM in soil: binding, by which HM precipitates in soil and stays fixed, and redox reactions through which HM toxicity is reduced. Complexation of HM by HS is made possible by the presence of phenolic and carboxylic groups in the HS structure, thus allowing the formation of HM-HS complexes, which are usually immobile and remain immobilized in soil without being taken up to a lesser extent by plants (Canellas et al., 2015; Yang et al., 2021b). However, HS is usually more efficient in the fixation of certain metals such as Pb and Cu compared to others such as Ni, Zn, and Cd (Weng et al., 2002). In addition to its chelation properties,

TABLE 6.2 Examples of studies of heavy metal stress mitigation in plants by HS application.

Heavy metal(s)	Plant species	Study results	References
Cd and Zn	*Ceratophyllum demersum* L.	HA forms complexes with Cd and Zn which reduces its uptake and toxicity.	Bunluesin et al. (2006)
Pb	*Vicia faba* L.	High levels of FA complex Pb, decrease its uptake by the plant and consequently alleviate Pb stress.	Shahid et al. (2012)
Cr	*Triticum aestivum* L.	Foliar FA application reduces Cr uptake and translocation, increases plant growth, chlorophyll content, and antioxidant system.	Ali et al. (2015)
Pb	*Camellia sinensis* L.	HA reduces Pb stress by three mechanisms: enhancement of the antioxidant system, increase in cell wall synthesis, and stabilization of protein metabolism.	Duan et al. (2020)
Cd	*Pennisetum* L.	Foliar application of HA reduces Cd toxicity by increasing plant biomass, photosynthetic pigments, net photosynthetic rate, and decreasing MDA content.	Song et al. (2020)
As	*Lactuca sativa* L.	HA reduces As toxicity by enhancing the antioxidant system, increasing nutrient uptake and synthesis of carbohydrates and amino acids.	Wang et al. (2021)
Cd	*Oryza sativa* L.	HS together with calcareous substances, modify the pH and precipitate Cd, reducing its translocation to the leaves and alleviating Cd stress.	Liu et al. (2022)
Pb and Cd	*Vallisneria natans* L.	HA reduces Pb and Cd toxicity by metal binding, enhances protein synthesis and antioxidant system, decreases lipid peroxidation, changes the biofilm microbial community, and regulates gene expression.	Huang et al. (2023)

HS can directly act as a heavy metal-reducing agent. In this way, HS can mediate the reduction of HM to a less toxic chemical form (Yang et al., 2021b). For example, the reduction of Cr (VI) (toxic) to Cr (III) (nontoxic) can be mediated by HS derived from peat and leonardite, as has been previously described (Zhilin et al., 2004). Besides, HS can reduce HM using electrons from microorganisms or ME (Yang et al., 2021b). However, some studies show that HM uptake may increase rather than decrease (Canellas et al., 2015), which would be interesting for the development of phytoextraction programs using HS. In this case, the combination of HS and plants may provide good results in the removal of HM from contaminated soils and waters.

Regarding OPs, the mechanisms of remediation are very similar to those discussed for HM. It is said that natural organic matter is the main sorbent for OPs present in the environment. The aliphatic fractions of the HS structure have the properties to bind OPs through hydrogen bonds as well as van der Waals forces. In this way, soil HS can regulate the mobility of OPs as well as their bioavailability and be used in the remediation of these pollutants in soil and water (Van Stempvoort and Lesage, 2002). Besides, HS can accelerate the OP remediation by the Fenton reaction (Lipczynska-Kochany, 2018b). On the other hand, HS can act as a redox agent through electron transfer from microorganisms to OPs, thus promoting its chemical transformation (Yang et al., 2021b). Likewise, the presence of HS can stimulate microorganisms' growth through different mechanisms. HS can form complexes with xenobiotics and decrease the toxicity of these toxic compounds to microorganisms. HS can also serve as a source of energy and carbon for microorganisms (Lipczynska-Kochany, 2018b). These microorganisms, whose populations are increased by the presence of HS, can stimulate the degradation of OPs. Thus, Chen et al. (2016) found that a humic substance analog increased *Geobacter* sp. Populations, which was critical in accelerating the degradation of pentachlorophenol. Although many studies demonstrate the beneficial effects of HS in the remediation of OPs, there is scarcely any research on the physiological implications in plants of HS application under organic compound contamination.

6.4 Co-application of humic substances with other biostimulants

The co-application of different types of biostimulants is a common strategy in agriculture to increase plant growth, development, and tolerance to abiotic and biotic stress conditions. Many studies apply ME, organic compounds, or plant growth-promoting bacteria (PGPB) and, in addition, HS to enhance the stimulatory effect or reduce the possible toxicity of ME (Xiao et al., 2021). Thus, Xiao et al. (2021) observed that the application of a biostimulant containing HA together with Se alleviated Se (VI) stress in wheat and increased wheat biomass without affecting Se concentration. Similarly, foliar application of a mixture of HA with amino acids increased growth and nutritional quality in beans (*Vicia faba* L.) (El-Ghamry, 2009). On the other hand, the combined application of HS and PGPB has beneficial effects on plant growth, as has been demonstrated in different studies. HS increases the PGPB activity, and consequently, the stimulatory effect on the plant is also increased. HS induces the release of organic acids from the roots into the soil solution, and these organic acids can serve as nutrients for PGPB. In addition, the chemical structure of HS allows higher adsorption between PGPB and plant roots (da Silva et al., 2021). Thus, Galambos et al. (2020) applied endophytic bacteria alone and in combination with HA to tomato (*Solanum lycopersicum* L.) plants and found that, in the presence of HA, plant growth was higher for strains *Paraburkholderia phytofirmans* and *Pantoea agglomerans* compared to plants inoculated with these strains alone. These authors found that, in the presence of HA, genes related to protein and hormone metabolism, signal transduction, and transport were upregulated (Galambos et al., 2020).

Moreover, the co-application of HS with other biostimulants also improves plant tolerance to environmental stresses. Most of these research studies have been carried out under drought or salinity conditions. The application of HA and Ca (as calcium nitrate) increases growth, antioxidant capacity, and tolerance to salt stress in pepper (Akladious and Mohamed, 2018). Salt stress is also alleviated after HA and proline application by increasing compatible solute concentrations and reducing Na^+ uptake in pomegranate plants (*Punica granatum* L.) (Rashedy et al., 2022). Similarly, the combined application of HA and glycine betaine improves the tolerance of apple (*Malus robusta* L.) seedlings to drought by increasing antioxidant enzyme activity (SOD, POD, and CAT) and decreasing lipid peroxidation (Zhang et al., 2013). Finally, the physiological changes generated after HS and PGPB application also improve plant tolerance to different types of abiotic stresses (da Silva et al., 2021). Thus, Khan et al. (2020) showed that the co-application of HS and *Bacillus cereus* increases heat tolerance in tomato plants through induction of the antioxidant system, increase in K, P, and Fe contents, and induction of the high-affinity potassium transporter (*SlHKT1*) and transcription factor response to heat stress (*SlHsfA1a*). Therefore, all these studies indicate that HS can also be applied together with other biostimulants to increase crop productivity and stress tolerance.

6.5 Conclusions and remarks

HS (humic and fulvic acids) continue to be studied by agricultural researchers because of their potential use as plant biostimulants. The results presented in this review show the different direct effects of HS on different plant species from a physiological point of view. Uptake and assimilation of essential nutrients, photosynthesis, and secondary metabolism are some of the processes stimulated by HS. In addition, HS can change the phytohormone profile as well as act as a hormone-like signal, with physiological implications for plant growth. Furthermore, different abiotic stress conditions, such as drought, salinity, HM toxicity, and OPs, can be alleviated by HS. However, the physiological, metabolic, transcriptomic, and genetic mechanisms through which HS affects plant growth are not completely clear and need to be further studied. Finally, HS can be applied to plants together with other types of bioestimulants, which shows the complementarity of HS with other compounds of biological nature, enhancing their stimulatory effect, which is appealing to farmers. Therefore, HS-based products are a promising tool to achieve the goal of sustainable agriculture by increasing agricultural productivity, crop quality, and crop tolerance to environmental stress conditions while decreasing the application of synthesized chemicals as fertilizers.

References

Abros'kin, D.P., Fuentes, M., Garcia-Mina, J.M., Klyain, O.I., Senik, S.V., Volkov, D.S., et al., 2016. The effect of humic acids and their complexes with iron on the functional status of plants grown under iron deficiency. Euras. Soil Sci. 49, 1167–1177. Available from: https://doi.org/10.1134/S1064229316100021.

Aguiar, N.O., Medici, L.O., Olivares, F.L., Dobbss, L.B., Torres-Netto, A., Silva, S.F., et al., 2016. Metabolic profile and antioxidant responses during drought stress recovery in sugarcane treated with humic acids and endophytic diazotrophic bacteria. Ann. Appl. Biol. 168, 203–213. Available from: https://doi.org/10.1111/aab.12256.

Ahmad, A., Blasco, B., Martos, V., 2022. Combating salinity through natural plant extracts based biostimulants: a review. Front. Plant Sci. 13, 1665. Available from: https://doi.org/10.3389/fpls.2022.862034.

Akladious, S.A., Mohamed, H.I., 2018. Ameliorative effects of calcium nitrate and humic acid on the growth, yield component and biochemical attribute of pepper (*Capsicum annuum*) plants grown under salt stress. Sci. Hortic. 236, 244–250. Available from: https://doi.org/10.1016/j.scienta.2018.03.047.

Albuzio, A., Ferrari, G., Nardi, S., 1986. Effects of humic substances on nitrate uptake and assimilation in barley seedlings. Can. J. Soil Sci. 66, 731–736. Available from: https://doi.org/10.4141/cjss86-07.

Ali, S., Bharwana, A., Rizwan, S., Farid, M., Kanwal, M., Ali, S., et al., 2015. Fulvic acid mediates chromium (Cr) tolerance in wheat (*Triticum aestivum* L.) through lowering of Cr uptake and improved antioxidant defense system. Environ. Sci. Pollut. Res. 22, 10601–10609. Available from: https://doi.org/10.1007/s11356-015-4271-7.

Angon, P.B., Tahjib-Ul-Arif, M., Samin, S.I., Habiba, U., Hossain, M.A., Brestic, M., 2022. How do plants respond to combined drought and salinity stress?—a systematic review. Plants 11, 2884. Available from: https://doi.org/10.3390/plants11212884.

Azcona, I., Pascual, I., Aguirreolea, J., Fuentes, M., García-Mina, J.M., Sánchez-Díaz, M., 2011. Growth and development of pepper are affected by humic substances derived from composted sludge. J. Plant Nutr. Soil Sci. 174, 916–924. Available from: https://doi.org/10.1002/jpln.201000264.

Bettoni, M.M., Mogor, Á.F., Pauletti, V., Goicoechea, N., Aranjuelo, I., Garmendia, I., 2016. Nutritional quality and yield of onion as affected by different application methods and doses of humic substances. J. Food Compos. Anal. 51, 37–44. Available from: https://doi.org/10.1016/j.jfca.2016.06.008.

Billard, V., Etienne, P., Jannin, L., Garnica, M., Cruz, F., Garcia-Mina, J.M., et al., 2014. Two biostimulants derived from algae or humic acid induce similar responses in the mineral content and gene expression of winter oilseed rape (*Brassica napus* L.). J. Plant Growth Regul. 33, 305–316. Available from: https://doi.org/10.1007/s00344-013-9372-2.

Bocanegra, M.P., Lobartini, J.C., Orioli, G.A., 2006. Plant uptake of iron chelated by humic acids of different molecular weights. Commun. Soil Sci. Plant Anal. 37, 239–248. Available from: https://doi.org/10.1080/00103620500408779.

Bulgari, R., Franzoni, G., Ferrante, A., 2019. Biostimulants application in horticultural crops under abiotic stress conditions. Agronomy 9, 306. Available from: https://doi.org/10.3390/agronomy9060306.

Bunluesin, S., Pokethitiyook, P., Lanza, G.R., Tyson, J.F., Kruatrachue, M., Xing, B., et al., 2006. Influences of cadmium and zinc interaction and humic acid on metal accumulation in *Ceratophyllum Demersum*. Water Air Soil Pollut. 180, 225–235. Available from: https://doi.org/10.1007/s11270-006-9265-0.

Cacco, G., Attinà, E., Gelsomino, A., Sidari, M., 2000. Effect of nitrate and humic substances of different molecular size on kinetic parameters of nitrate uptake in wheat seedlings. J. Plant Nutr. Soil Sci. 163, 313–320. https://doi.org/10.1002/1522–2624(200006)163:3<313::AID-JPLN313>3.0.CO;2-U.

Canellas, L.P., Balmori, D.M., Médici, L.O., Aguiar, N.O., Campostrini, E., Rosa, R.C.C., et al., 2013. A combination of humic substances and *Herbaspirillum seropedicae* inoculation enhances the growth of maize (*Zea mays* L.). Plant Soil 366, 119–132. Available from: https://doi.org/10.1007/s11104-012-1382-5.

Canellas, L.P., Olivares, F.L., 2014. Physiological responses to humic substances as plant growth promoter. Chem. Biol. Technol. Agric. 1, 1–11. Available from: https://doi.org/10.1186/2196-5641-1-3.

Canellas, L.P., Olivares, F.L., Aguiar, N.O., Jones, D.L., Nebbioso, A., Mazzei, P., et al., 2015. Humic and fulvic acids as biostimulants in horticulture. Sci. Hortic. 196, 15–27. Available from: https://doi.org/10.1016/j.scienta.2015.09.013.

Canellas, L.P., Olivares, F.L., Okorokova-Façanha, A.L., Façanha, A.R., 2002. Humic acids isolated from earthworm compost enhance root elongation, lateral root emergence, and plasma membrane H^+-ATPase activity in maize roots. Plant Physiol. 130, 1951–1957. Available from: https://doi.org/10.1104/PP.007088.

Cha, J.Y., Kang, S.H., Geun Ji, M., Shin, G.I., Jeong, S.Y., Ahn, G., et al., 2021. Transcriptome changes reveal the molecular mechanisms of humic acid-induced salt stress tolerance in *Arabidopsis*. Molecules 26, 782. Available from: https://doi.org/10.3390/molecules26040782.

Chen, M., Tong, H., Liu, C., Chen, D., Li, F., Qiao, J., 2016. A humic substance analogue AQDS stimulates *Geobacter* sp. abundance and enhances pentachlorophenol transformation in a paddy soil. Chemosphere 160, 141–148. Available from: https://doi.org/10.1016/j.chemosphere.2016.06.061.

Chen, Y., Magen, H., Clapp, C.E., 2004. Mechanisms of plant growth stimulation by humic substances: the role of organo-iron complexes. Soil Sci. Plant Nutr. 50, 1089–1095. Available from: https://doi.org/10.1080/00380768.2004.10408579.

Çimrin, K.M., Türkmen, Ö., Turan, M., Tuncer, B., 2010. Phosphorus and humic acid application alleviate salinity stress of pepper seedling. Afr. J. Biotechnol. 9, 5845–5851.

Conselvan, G.B., Pizzeghello, D., Francioso, O., Di Foggia, M., Nardi, S., Carletti, P., 2017. Biostimulant activity of humic substances extracted from leonardites. Plant Soil 420, 119–134. Available from: https://doi.org/10.1007/s11104-017-3373-z.

Cruz, R., Gomes, T., Ferreira, A., Mendes, E., Baptista, P., Cunha, S., et al., 2014. Antioxidant activity and bioactive compounds of lettuce improved by espresso coffee residues. Food Chem. 145, 95–101. Available from: https://doi.org/10.1016/j.foodchem.2013.08.038.

da Silva, M.S.R.D.A., dos Santos, B.D.M.S., da Silva, C., Santos, R.D.A., da Silva, C., Santos, R.D.A., et al., 2021. Humic substances in combination with plant growth-promoting bacteria as an alternative for sustainable agriculture. Front. Microbiol. 12719653. Available from: https://doi.org/10.3389/fmicb.2021.719653.

De Hita, D., Fuentes, M., Fernández, V., Zamarreño, A.M., Olaetxea, M., García-Mina, J.M., 2020. Discriminating the short-term action of root and foliar application of humic acids on plant growth: emerging role of jasmonic acid. Front. Plant Sci. 11493. Available from: https://doi.org/10.3389/fpls.2020.00493.

Deolu-Ajayi, A.O., van der Meer, I.M., van der Werf, A., Karlova, R., 2022. The power of seaweeds as plant biostimulants to boost crop production under abiotic stress. Plant Cell Environ. 45, 2537–2553. Available from: https://doi.org/10.1111/pce.14391.

Du Jardin, P., 2015. Plant biostimulants: definition, concept, main categories and regulation. Sci. Hortic. 196, 3–14. Available from: https://doi.org/10.1016/j.scienta.2015.09.021.

Duan, D., Tong, J., Xu, Q., Dai, L., Ye, J., Wu, H., et al., 2020. Regulation mechanisms of humic acid on Pb stress in tea plant (*Camellia sinensis* L.). Environ. Pollut. 267115546. Available from: https://doi.org/10.1016/J.envpol.2020.115546.

El-Ghamry, A.M., 2009. Amino and humic acids promote growth, yield and disease resistance of Faba bean cultivated in clayey soil. Aust. J. Basic. Appl. Sci. 3, 731–739.

Elmongy, M.S., Zhou, H., Cao, Y., Liu, B., Xia, Y., 2018. The effect of humic acid on endogenous hormone levels and antioxidant enzyme activity during in vitro rooting of evergreen azalea. Sci. Hortic. 227, 234–243. Available from: https://doi.org/10.1016/J.scienta.2017.09.027.

Erb, M., Kliebenstein, D.J., 2020. Topical review plant secondary metabolites as defenses, regulators, and primary metabolites: the blurred functional trichotomy. Plant Physiol. 184, 39–52. Available from: https://doi.org/10.1104/pp.20.00433.

Ertani, A., Francioso, O., Tugnoli, V., Righi, V., Nardi, S., 2011. Effect of commercial lignosulfonate-humate on *Zea mays* L. metabolism. J. Agric. Food Chem. 59, 11940–11948. Available from: https://doi.org/10.1021/jf202473e.

EU., 2019. Regulation of the European Parliament and the Council laying down rules on the making available on the market of EU fertilising products and amending Regulations (EC) No 1069/2009 and (EC) No 2003/2003. https://eur-lex.europa.eu/legal-content/EN/TXT/?uri = OJ:L:2019:170:TOC.

Fan, H.M., Li, T., Sun, X., Sun, X.Z., Zheng, C.S., 2015. Effects of humic acid derived from sediments on the postharvest vase life extension in cut chrysanthemum flowers. Postharvest Biol. Technol. 101, 82–87. Available from: https://doi.org/10.1016/j.postharvbio.2014.09.019.

Fan, H.M., Wang, X.W., Sun, X., Li, Y.Y., Sun, X.Z., Zheng, C.S., 2014. Effects of humic acid derived from sediments on growth, photosynthesis and chloroplast ultrastructure in chrysanthemum. Sci. Hortic. 177, 118–123. Available from: https://doi.org/10.1016/j.scienta.2014.05.010.

Farooq, M., Hussain, M., Wahid, A., Siddique, K.H.M., 2012. Drought stress in plants: an overview. In: Aroca, R. (Ed.), Plant Responses to Drought Stress. Springer, Berlin, Heidelberg, pp. 1–33. Available from: https://doi.org/10.1007/978-3-642-32653-0_1.

Fu, Z., Wu, F., Song, K., Lin, Y., Bai, Y., Zhu, Y., et al., 2013. Competitive interaction between soil-derived humic acid and phosphate on goethite. Appl. Geochem. 36, 125–131. Available from: https://doi.org/10.1016/j.apgeochem.2013.05.015.

Galambos, N., Compant, S., Moretto, M., Sicher, C., Puopolo, G., Wäckers, F., et al., 2020. Humic acid enhances the growth of tomato promoted by endophytic bacterial strains through the activation of hormone-, growth-, and transcription-related processes. Front. Plant Sci. 11, 1–18. Available from: https://doi.org/10.3389/fpls.2020.582267.

García-Mina, J.M., Antolín, M.C., Sanchez-Diaz, M., 2004. Metal-humic complexes and plant micronutrient uptake: a study based on different plant species cultivated in diverse soil types. Plant. Soil. 258, 57–68. Available from: https://doi.org/10.1023/B:PLSO.0000016509.56780.40.

García, A.C., Santos, L.A., Izquierdo, F.G., Rumjanek, V.M., Castro, R.N., dos Santos, F.S., et al., 2014. Potentialities of vermicompost humic acids to alleviate water stress in rice plants (*Oryza sativa* L.). J. Geochem. Explor. 136, 48–54. Available from: https://doi.org/10.1016/j.gexplo.2013.10.005.

Hassan, M.U., Nawaz, M., Mahmood, A., Shah, A.A., Shah, A.N., Muhammad, F., et al., 2022. The role of zinc to mitigate heavy metals toxicity in crops. Front. Environ. Sci. 10, 1712. Available from: https://doi.org/10.3389/fenvs.2022.990223.

Hatami, E., Shokouhian, A.A., Ghanbari, A.R., Naseri, L.A., 2018. Alleviating salt stress in almond rootstocks using of humic acid. Sci. Hortic. 237, 296–302. Available from: https://doi.org/10.1016/j.scienta.2018.03.034.

Hernandez, O.L., Calderín, A., Huelva, R., Martínez-Balmori, D., Guridi, F., Aguiar, N.O., et al., 2015. Humic substances from vermicompost enhance urban lettuce production. Agron. Sustain. Dev. 35, 225–232. Available from: https://doi.org/10.1007/s13593-014-0221-x.

Huang, S., Wang, Z., Song, Q., Hong, J., Jin, T., Huang, H., et al., 2023. Potential mechanism of humic acid attenuating toxicity of Pb^{2+} and Cd^{2+} in *Vallisneria natans*. Sci. Total. Environ. 864160974. Available from: https://doi.org/10.1016/j.scitotenv.2022.160974.

Islam, M.A., Morton, D.W., Johnson, B.B., Angove, M.J., 2020. Adsorption of humic and fulvic acids onto a range of adsorbents in aqueous systems, and their effect on the adsorption of other species: a review. Sep. Purif. Technol. 247116949. Available from: https://doi.org/10.1016/j.seppur.2020.116949.

Jannin, L., Arkoun, M., Ourry, A., Laîné, P., Goux, D., Garnica, M., et al., 2012. Microarray analysis of humic acid effects on *Brassica napus* growth: involvement of N, C and S metabolisms. Plant Soil 359, 297–319. Available from: https://doi.org/10.1007/s11104-012-1191-x.

Jindo, K., Soares Soares, T., Lazaro Eustáquio, P.P., Golçalvez Azevedo, I., Oliveira Aguiar, N., Mazzei, P., et al., 2016. Phosphorus speciation and high-affinity transporters are influenced by humic substances. J. Plant Nutr. Soil Sci. 179, 206–214. Available from: https://doi.org/10.1002/jpln.201500228.

Khan, M.A., Asaf, S., Khan, A.L., Jan, R., Kang, S.M., Kim, K.M., et al., 2020. Extending thermotolerance to tomato seedlings by inoculation with SA1 isolate of *Bacillus cereus* and comparison with exogenous humic acid application. PLoS One 15, e0232228. Available from: https://doi.org/10.1371/journal.pone.0232228.

Kumar, S., Kumar, S., Mohapatra, T., 2021. Interaction between macro- and micro-nutrients in plants. Front. Plant Sci. 12, 753. Available from: https://doi.org/10.3389/fpls.2021.665583.

Li, Y., Zeng, H., Xu, F., Yan, F., Xu, W., 2022. H^+-ATPases in plant growth and stress responses. Annu. Rev. Plant Biol. 73, 495–521. Available from: https://doi.org/10.1146/annurev-arplant-102820-114551.

Lipczynska-Kochany, E., 2018a. Effect of climate change on humic substances and associated impacts on the quality of surface water and groundwater: a review. Sci. Total. Environ. 640–641, 1548–1565. Available from: https://doi.org/10.1016/j.scitotenv.2018.05.376.

Lipczynska-Kochany, E., 2018b. Humic substances, their microbial interactions and effects on biological transformations of organic pollutants in water and soil: a review. Chemosphere 202, 420–437. Available from: https://doi.org/10.1016/j.chemosphere.2018.03.104.

Liu, H., Zhang, T., Tong, Y., Zhu, Q., Huang, D., Zeng, X., 2022. Effect of humic and calcareous substance amendments on the availability of cadmium in paddy soil and its accumulation in rice. Ecotoxicol. Environ. Saf. 231113186. Available from: https://doi.org/10.1016/j.ecoenv.2022.113186.

Liu, S., He, F., Kuzyakov, Y., Xiao, H., Hoang, D.T.T., Pu, S., et al., 2022. Nutrients in the rhizosphere: a meta-analysis of content, availability, and influencing factors. Sci. Total. Environ. 826153908. Available from: https://doi.org/10.1016/j.scitotenv.2022.153908.

Lucena, C., Waters, B.M., Romera, F.J., García, M.J., Morales, M., Alcántara, E., et al., 2006. Ethylene could influence ferric reductase, iron transporter, and H^+-ATPase gene expression by affecting FER (or FER-like) gene activity. J. Exp. Bot. 57, 4145–4154. Available from: https://doi.org/10.1093/jxb/erl189.

Malik, A., Mor, V.S., Tokas, J., Punia, H., Malik, S., Malik, K., et al., 2020. Biostimulant-treated seedlings under sustainable agriculture: a global perspective facing climate change. Agronomy 11, 14. Available from: https://doi.org/10.3390/agronomy11010014.

Merlo, L., Ghisi, R., Passera, C., Rascio, N., 1991. Effects of humic substances on carbohydrate metabolism of maize leaves. Can. J. Plant Sci. 71, 419–425. Available from: https://doi.org/10.4141/cjps91-058. 71, 419–425.

Mora, V., Bacaicoa, E., Baigorri, R., Zamarreno, A.M., García-Mina, J.M., 2014. NO and IAA key regulators in the shoot growth promoting action of humic acid in *Cucumis sativus* L. J. Plant Growth Regul. 33, 430–439. Available from: https://doi.org/10.1007/s00344-013-9394-9.

Mora, V., Bacaicoa, E., Zamarreño, A.M., Aguirre, E., Garnica, M., Fuentes, M., et al., 2010. Action of humic acid on promotion of cucumber shoot growth involves nitrate-related changes associated with the root-to-shoot distribution of cytokinins, polyamines and mineral nutrients. J. Plant Physiol. 167, 633–642. Available from: https://doi.org/10.1016/j.jplph.2009.11.018.

Mora, V., Baigorri, R., Bacaicoa, E., Zamarreño, A.M., García-Mina, J.M., 2012. The humic acid-induced changes in the root concentration of nitric oxide, IAA and ethylene do not explain the changes in root architecture caused by humic acid in cucumber. Environ. Exp. Bot. 76, 24–32. Available from: https://doi.org/10.1016/j.envexpbot.2011.10.001.

Nardi, S., Muscolo, A., Vaccaro, S., Baiano, S., Spaccini, R., Piccolo, A., 2007. Relationship between molecular characteristics of soil humic fractions and glycolytic pathway and krebs cycle in maize seedlings. Soil Biol. Biochem. 39, 3138–3146. Available from: https://doi.org/10.1016/j.soilbio.2007.07.006.

Nardi, S., Pizzeghello, D., Ertani, A., di Agronomia, D., Naturali Ambiente, R., 2017. Hormone-like activity of the soil organic matter. Appl, Soil Ecol. 123, 517–520. Available from: https://doi.org/10.1016/j.apsoil.2017.04.020.

Nardi, S., Schiavon, M., Francioso, O., 2021. Chemical structure and biological activity of humic substances define their role as plant growth promoters. Molecules 2021 (26), 2256. Available from: https://doi.org/10.3390/molecules26082256.

Navarro-León, E., López-Moreno, F.J., de la Torre-González, A., Ruiz, J.M., Esposito, S., Blasco, B., 2020. Study of salt-stress tolerance and defensive mechanisms in *Brassica rapa* CAX1a TILLING mutants. Environ. Exp. Bot. 175104061. Available from: https://doi.org/10.1016/j.envexpbot.2020.104061.

Olaetxea, M., De Hita, D., Garcia, C.A., Fuentes, M., Baigorri, R., Mora, V., et al., 2018. Hypothetical framework integrating the main mechanisms involved in the promoting action of rhizospheric humic substances on plant root- and shoot- growth. Appl. Soil Ecol. 123, 521–537. Available from: https://doi.org/10.1016/j.apsoil.2017.06.007.

Olaetxea, M., Mora, V., Bacaicoa, E., Baigorri, R., Garnica, M., Fuentes, M., et al., 2019. Root ABA and H^+-ATPase are key players in the root and shoot growth-promoting action of humic acids. Plant Direct 3e00175. Available from: https://doi.org/10.1002/PLD3.175.

Ozturk, M., Turkyilmaz Unal, B., García-Caparrós, P., Khursheed, A., Gul, A., Hasanuzzaman, M., 2021. Osmoregulation and its actions during the drought stress in plants. Physiol. Plant 172, 1321–1335. Available from: https://doi.org/10.1111/ppl.13297.

Paniagua-López, M., Vela-Cano, M., Correa-Galeote, D., Martín-Peinado, F., Garzón, F.J.M., Pozo, C., et al., 2021. Soil remediation approach and bacterial community structure in a long-term contaminated soil by a mining spill (Aznalcóllar, Spain). Sci. Total. Environ. 777145128. Available from: https://doi.org/10.1016/j.scitotenv.2021.145128.

Pizzeghello, D., Nicolini, G., Nardi, S., 2001. Hormone-like activity of humic substances in *Fagus sylvaticae* forests. N. Phytol. 151, 647–657. Available from: https://doi.org/10.1046/j.0028-646x.2001.00223.x.

Quaggiotti, S., Ruperti, B., Pizzeghello, D., Francioso, O., Tugnoli, V., Nardi, S., 2004. Effect of low molecular size humic substances on nitrate uptake and expression of genes involved in nitrate transport in maize (*Zea mays* L.). J. Exp. Bot. 55, 803–813. Available from: https://doi.org/10.1093/jxb/erh085.

Rashedy, A.A.H., Abd-ElNafea, M.H., Khedr, E.H., 2022. Co-application of proline or calcium and humic acid enhances productivity of salt stressed pomegranate by improving nutritional status and osmoregulation mechanisms. Sci. Rep. 12 (1), 10. Available from: https://doi.org/10.1038/s41598-022-17824-6.

Rehman, A., Masood, S., Ullah, N., Abbasi, E., Hussain, Z., Ali, I., 2021. Molecular basis of Iron Biofortification in crop plants; A step towards sustainability 12–22. https://doi.org/10.1111/pbr.12886.

Rouphael, Y., Colla, G., 2020. Editorial: biostimulants in agriculture. Front. Plant Sci. 1140. Available from: https://doi.org/10.3389/fpls.2020.00040.

Rubio-Wilhelmi, M.del M., Sanchez-Rodriguez, E., Rosales, M.A., Blasco, B., Rios, J.J., Romero, L., et al., 2012. Ammonium formation and assimilation in PSARK::IPT tobacco transgenic plants under low N. J. Plant Physiol. 169, 157–162. Available from: https://doi.org/10.1016/j.jplph.2011.09.011.

Savarese, C., Cozzolino, V., Verrillo, M., Vinci, G., De Martino, A., Scopa, A., et al., 2022. Combination of humic biostimulants with a microbial inoculum improves lettuce productivity, nutrient uptake, and primary and secondary metabolism. Plant Soil 1–30. Available from: https://doi.org/10.1007/s11104-022-05634-8.

Schiavon, M., Pizzeghello, D., Muscolo, A., 2010. High molecular size humic substances enhance phenylpropanoid metabolism in maize (*Zea mays* L.) 36, 662–669. https://doi.org/10.1007/s10886-010-9790-6.

Shah, Z.H., Rehman, H.M., Akhtar, T., Alsamadany, H., Hamooh, B.T., Mujtaba, T., et al., 2018. Humic substances: determining potential molecular regulatory processes in plants. Front. Plant Sci. 9263. Available from: https://doi.org/10.3389/fpls.2018.00263.

Shahid, M., Dumat, C., Silvestre, J., Pinelli, E., 2012. Effect of fulvic acids on lead-induced oxidative stress to metal sensitive *Vicia faba* L. plant. Biol. Fertil. Soils 48, 689–697. Available from: https://doi.org/10.1007/s00374-012-0662-9.

Sharma, H., Chawla, N., Ajmer, S.D., 2022. Role of phenylalanine/tyrosine ammonia lyase and anthocyanidin synthase enzymes for anthocyanin biosynthesis in developing *Solanum melongena* L. genotypes. Physiol. Plant 174e13756. Available from: https://doi.org/10.1111/ppl.13756.

Shen, J., Guo, M.J., Wang, Y.G., Yuan, X.Y., Wen, Y.Y., Song, X.E., et al., 2020. Humic acid improves the physiological and photosynthetic characteristics of millet seedlings under drought stress. Plant Signal. Behav. 15, 1774212. Available from: https://doi.org/10.1080/15592324.2020.1774212.

Song, X., Chen, M., Chen, W., Jiang, H., Yue, X., 2020. Foliar application of humic acid decreased hazard of cadmium toxicity on the growth of Hybrid *Pennisetum*. Acta Physiol. Plant 42, 1–11. Available from: https://doi.org/10.1007/s11738-020-03118-9.

Souza, A.C., Lopes Olivares, F., Eustáquio, L.P.P., Piccolo, A., Canellas, L.P., 2022. Plant hormone crosstalk mediated by humic acids. Chem. Biol. Technol. Agric. 9, 29. Available from: https://doi.org/10.1186/s40538-022-00295-2.

Tian, Z., Wang, J.W., Li, J., Han, B., 2021. Designing future crops: challenges and strategies for sustainable agriculture. Plant J. 105, 1165–1178. Available from: https://doi.org/10.1111/tpj.15107.

Tomasi, N., De Nobili, M., Gottardi, S., Zanin, L., Mimmo, T., Varanini, Z., et al., 2013. Physiological and molecular characterization of Fe acquisition by tomato plants from natural Fe complexes. Biol. Fertil. Soils 49, 187–200. Available from: https://doi.org/10.1007/s00374-012-0706-1.

Trevisan, S., Francioso, O., Quaggiotti, S., Nardi, S., 2010. Humic substances biological activity at the plant-soil interface. Plant Signal. Behav. 5, 635–643. Available from: https://doi.org/10.4161/psb.5.6.11211.

Turan, M., Ekinci, M., Kul, R., Kocaman, A., Argin, S., Zhirkova, A.M., et al., 2022. Foliar applications of humic substances together with Fe/nano Fe to increase the iron content and growth parameters of spinach (*Spinacia oleracea* L.). Agronomy 12, 2044. Available from: https://doi.org/10.3390/agronomy12092044.

Van Stempvoort, D.R., Lesage, S., 2002. Binding of methylated naphthalenes to concentrated aqueous humic acid. Adv. Environ. Res. 6, 495–504. Available from: https://doi.org/10.1016/S1093-0191(01)00076-4.

Vujinović, T., Zanin, L., Venuti, S., Contin, M., Ceccon, P., Tomasi, N., et al., 2020. Biostimulant action of dissolved humic substances from a conventionally and an organically managed soil on nitrate acquisition in maize plants. Front. Plant Sci. 10, 1–14. Available from: https://doi.org/10.3389/fpls.2019.01652.

Wang, Q., Wen, J., Zheng, J., Zhao, J., Qiu, C., Xiao, D., et al., 2021. Arsenate phytotoxicity regulation by humic acid and related metabolic mechanisms. Ecotoxicol. Environ. Saf. 207, 111379. Available from: https://doi.org/10.1016/j.ecoenv.2020.111379.

Weng, L., Temminghoff, E.J.M., Lofts, S., Tipping, E., Van Riemsdijk, W.H., 2002. Complexation with dissolved organic matter and solubility control of heavy metals in a sandy soil. Environ. Sci. Technol. 36, 4804–4810. Available from: https://doi.org/10.1021/es0200084.

Xiao, T., Boada, R., Llugany, M., Valiente, M., 2021. Co-application of Se and a biostimulant at different wheat growth stages: Influence on grain development. Plant Physiol. Biochem. 160, 184–192. Available from: https://doi.org/10.1016/j.plaphy.2021.01.025.

Yang, F., Sui, L., Tang, C., Li, J., Cheng, K., Xue, Q., 2021a. Sustainable advances on phosphorus utilization in soil via addition of biochar and humic substances. Sci. Total. Environ. 768, 145106. Available from: https://doi.org/10.1016/j.scitotenv.2021.145106.

Yang, F., Tang, C., Antonietti, M., 2021b. Natural and artificial humic substances to manage minerals, ions, water, and soil microorganisms. Chem. Soc. Rev. 50, 6221–6239. Available from: https://doi.org/10.1039/d0cs01363c.

Yang, X., Chen, X., Yang, X., 2019. Effect of organic matter on phosphorus adsorption and desorption in a black soil from Northeast China. Soil. Tillage Res. 187, 85–91. Available from: https://doi.org/10.1016/j.still.2018.11.016.

Yao, Y., Wang, X., Yang, Y., Shen, T., Wang, C., Tang, Y., et al., 2019. Molecular composition of size-fractionated fulvic acid-like substances extracted from spent cooking liquor and its relationship with biological activity. Environ. Sci. Technol. 53, 14752–14760. Available from: https://doi.org/10.1021/acs.est.9b02359.

Zandonadi, D.B., Santos, M.P., Dobbss, L.B., Olivares, F.L., Canellas, L.P., Binzel, M.L., et al., 2010. Nitric oxide mediates humic acids-induced root development and plasma membrane H^+-ATPase activation. Planta 231, 1025–1036. Available from: https://doi.org/10.1007/s00425-010-1106-0.

Zanin, L., Tomasi, N., Cesco, S., Varanini, Z., Pinton, R., 2019. Humic substances contribute to plant iron nutrition acting as chelators and biostimulants. Front. Plant Sci. 10675. Available from: https://doi.org/10.3389/fpls.2019.00675.

Zhang, C., Feng, Y., Liu, Y.W., Chang, H.Q., Li, Z.J., Xue, J.M., 2017. Uptake and translocation of organic pollutants in plants: a review. J. Integr. Agric. 16, 1659–1668. Available from: https://doi.org/10.1016/S2095-3119(16)61590-3.

Zhang, L., Gao, M., Zhang, L., Li, B., Han, M., Alva, A.K., et al., 2013. Role of exogenous glycinebetaine and humic acid in mitigating drought stress-induced adverse effects in *Malus robusta* seedlings. Turk. J. Botany. 37, 920–929. Available from: https://doi.org/10.3906/bot-1212-21.

Zhilin, D.M., Schmitt-Kopplin, P., Perminova, I.V., 2004. Reduction of Cr(VI) by peat and coal humic substances. Environ. Chem. Lett. 2, 141–145. Available from: https://doi.org/10.1007/s10311-004-0085-4.

CHAPTER

7

Use of melatonin in plants' growth and productions

Noureddine Chaachouay[1], Abdelhamid Azeroual[1], Bouchaib Bencherki[1], Allal Douira[2] and Lahcen Zidane[2]

[1]Agri-Food and Health Laboratory (AFHL), Faculty of Sciences and Techniques of Settat, Hassan First University, Settat, Morocco [2]Plant, Animal Productions and Agro-industry Laboratory, Department of Biology, Faculty of Sciences, Ibn Tofail University, Kenitra, Morocco

7.1 Introduction

Plants are subjected to various unfavorable circumstances that restrict their growth and development in the environments in which they are cultivated. Stress refers to the situations that impede average plant growth, maturation, and metabolism (Chaachouay et al., 2023b; Rhodes and Nadolska-Orczyk, 2001). In many places around the globe, the detrimental effects of environmental stress on plant growth have intensified due to continuing climate change and a growing human population (Denby and Gehring, 2005). It has been argued that environmental stress factors are the key limiting factors of agricultural productivity around the globe, resulting in more than 50% net reductions for most derivatives (Mahajan and Tuteja, 2005). However, various biostimulants are employed to reduce crop failure and counteract the negative impact of environmental conditions. Recently, phytomelatonin, a novel chemical found in plants that helps them survive in stressful conditions, has been revealed to assist plants in dealing with these challenges (Reiter et al., 2015).

Several recent investigations have demonstrated that melatonin has a role in various plant stressors, including nutritional insufficiency, salt stress, drought stress, fruit maturation, cold stress, biomass production, seed germination, circadian rhythm, leaf senescence, membrane integrity, root development, redox network, osmoregulation, plant development and growth, photosynthesis, heavy metals, flowering, and senescence of plants (Chaachouay et al., 2022a; Erland et al., 2015; Kobylińska et al., 2018; Li et al., 2019;

Biostimulants in Plant Protection and Performance
DOI: https://doi.org/10.1016/B978-0-443-15884-1.00004-X

Shi et al., 2015; Yu et al., 2018). Consequently, it may be essential to use melatonin as a biofertilizer for sustainable agricultural output with little environmental effect.

N-acetyl-5-methoxytryptamine (melatonin) is a tryptophan-derived, inherently occurring, low-molecular-weight chemical produced by all lifeforms (Quarles et al., 1974). The molecular structure of melatonin is classified as an indoleamine molecule, and its manufacture in both plants and mammals occurs through a tryptophan-dependent mechanism (Dubocovich et al., 2003). It is often synthesized in the chloroplasts and mitochondria of plant roots and leaves, then transmitted to the meristem, flowers, and fruits (Wang et al., 2016). In vertebrates, melatonin is generated predominantly by the pineal gland and released periodically into the circulation following synthesis (Macias et al., 1999). In animals, it is formed at night because its content in the blood steadily decreases throughout the day (Nawaz et al., 2020). This chapter methodically and comprehensively discussed the research development on melatonin and its plant growth and production. It looked ahead to prospective investigation to give some theoretical foundation for using melatonin in horticulture crop production.

7.2 Discovery of melatonin

In 1958, Lerner Aron and colleagues isolated and described melatonin from the extract of the bovine pineal gland. This organ was referred to as the soul seat by philosopher and French scientist René Descartes in the 16th century (Lerner et al., 1958). In 1959, N-acetyl-5-methoxy-tryptamine was recognized as melatonin, and shortly afterward, its biosynthesis from tryptophan with serotonin as an intermediate was established (Axelrod and Weissbach, 1960). However, melatonin was first found in humans in 1959 (Lerner et al., 1959), and its occurrence in numerous vertebrates such as amphibians, fish, and birds was recorded in the 1960s and 1970s (Fenwick, 1970; Lauber et al., 1968). Therefore, melatonin was identified in invertebrates such as insects, annelids, crustaceans, mollusks, and planarians in the 1980s (Balzer et al., 1997; Morita et al., 1987; Vivien-Roels et al., 1984). In 1993, O'Neill and Vantassel reported the first discovery of endogenous melatonin in herbs in a conference presentation (Vantassel et al., 1993). The botanists identified melatonin in the *Ipomoea nil* (L.) Roth and fruits of tomato using radioimmunoassay (RIA) and gas chromatography with mass spectrometry (HPLC-MS) (Arnao, 2014). In addition, two articles published simultaneously in 1995 confirmed the presence of melatonin in higher herbs (Vantassel et al., 1995). Dubbels and colleagues measured melatonin classes in extracts of cultivated tobacco and five other plant species (*Beta vulgaris* L., *Musa acuminata* Colla, *Solanum pimpinellifolium* L., *Solanum lycopersicum* L., and *Capsicum annuum* L.) using HPLC-MS and RIA (Dubbels et al., 1995). Since then, melatonin has been discovered and measured in several plant parts, such as roots, shoots, leaves, fruits, and seeds.

7.3 Chemistry of melatonin

Physiochemically, isolated melatonin appears as an off-white powder with a density of 1.17 g/cm^3 and a molecular mass of 232.28 g/mol. The boiling point is 512.8°C, while the melting temperature varies from 116.5°C to 118°C (Allegra et al., 2003; Hugel and Kennaway, 1995; Quarles et al., 1974). Melatonin is recognized from a chemical standpoint by the formula $C_{13}H_{16}N_2O_2$. The indole chemical scaffold is designed and synthesized with 3-amide and 5-alkoxy groups. Furthermore, since it is derived from a tryptophan molecule, it is an indolamine substance (Omer et al., 2021). This specific chemical form is very stable due to its strong resonance mesomerism (Fig. 7.1). Additionally, the 5-alkoxy group and 3-amide group are mainly accountable for this molecule's amphiphilicity.

7.4 Biosynthesis of melatonin in plants

Melatonin biosynthesis in plants varies from that in animals (Fig. 7.2). Several factors impact its formation in plants, with sunlight being among the most important. Melatonin

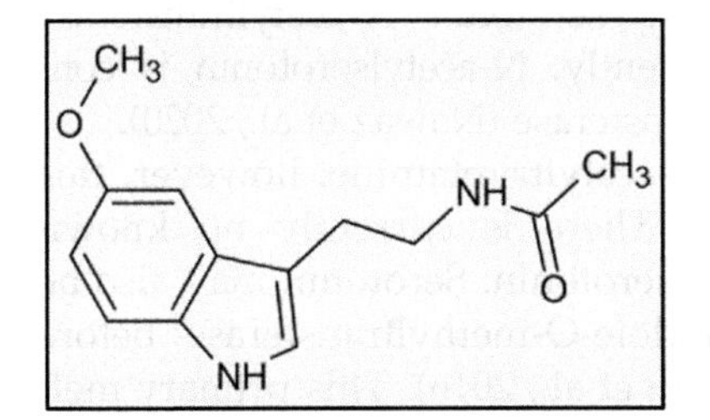

FIGURE 7.1 Melatonin's chemical structure.

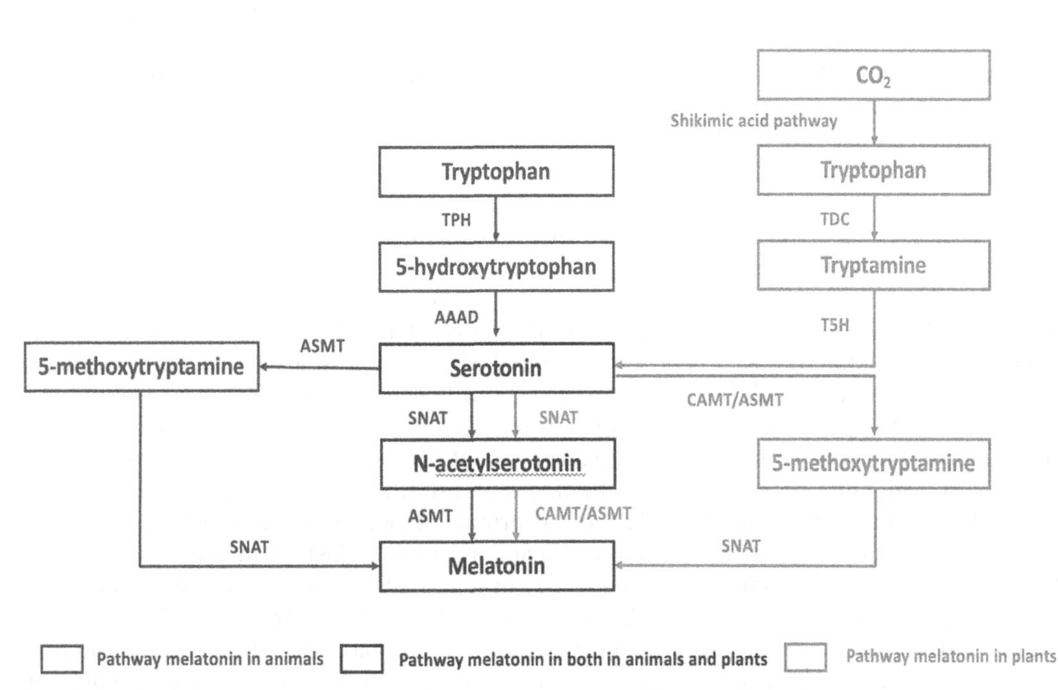

FIGURE 7.2 A comparison of melatonin production processes in plants and animals.

production appears mainly in mitochondria and chloroplasts in plants. Various enzyme groups are found in these compartments to generate melatonin through biosynthetic processes (Acharya et al., 2021). Melatonin synthesis is a two-step process; hence, if melatonin synthesis is inhibited in mitochondria, it will begin in chloroplasts (Nawaz et al., 2020). Melatonin production in plants starts with the endogenous synthesis of tryptophan in plant cells through the chorismate route. N-acetyltransferase, N-acetylserotonin, tryptophan decarboxylase, tryptamine 5-hydroxylase, caffeic acid O-methyltransferase, methyltransferase, serotonin, and tryptophan hydroxylase (TPH) are the six enzymes responsible for converting tryptophan to melatonin (Back et al., 2016; Kang et al., 2008; Tan et al., 2016).

Melatonin production begins with the synthesis of tryptophan through tryptophan decarboxylase. Tryptophan decarboxylase is converted to tryptamine, which is subsequently converted to tryptamine-5-hydroxylase. The enzyme T5H converts tryptamine into serotonin. Serotonin is transformed into melatonin, catalyzed by TDC, and turned into tryptamine (Zhao et al., 2019). 5-hydroxytryptophan is then converted into serotonin by AAAD. After 2 days, serotonin is converted to N-acetylserotonin by serotonin N-acetyltransferase (SNAT)/aryl alkylamine N-acetyltransferase, and subsequently, N-acetylserotonin is converted to melatonin by ASMT /hydroxy-indole-O-methyltransferase (Nawaz et al., 2020).

Tryptamine may also be produced from SNAT into N-acetyltryptamine; however, this cannot be converted into N-acetyl-serotonin by T5H. There is currently no known mechanism for converting N-acetyltryptamine to N-acetylserotonin. Serotonin may also be transformed into 5-methoxy-tryptamine by hydroxy-indole-O-methyltransferase before being turned into melatonin by SNAT (Fan et al., 2018; Tan et al., 2016). This primary melatonin biosynthetic route may provide choices, such as serotonin production through 5-hydroxytryptophan. However, this potential seems limited to mammals since the necessary enzyme (TPH) has not been identified in plants. Even if its synthesis process and metabolic mechanisms are uncertain, the existence of melatonin in plants is a broad phenomenon (Kolár et al., 2002).

7.5 Biosynthesis of melatonin in animals

1960 marked the discovery by Axelrod's team of the mammalian melatonin biosynthesis route, which is today well understood (Mannino et al., 2021; Reiter, 1991). Nonetheless, the conventional melatonin biosynthetic route has been broadened in all animals and applies to additional species, especially insects (Hardeland and Poeggeler, 2003; Zhao et al., 2019). The availability of the precursor tryptophan is a distinguishing factor between animal and plant melatonin production. Animals, unlike plants, cannot synthesize L-tryptophan from scratch; the amino acid must be consumed in food. As a result, animals have a lower melatonin metabolism than plants. Mitochondria are the primary biosynthesis locations and the organelles with the most significant melatonin levels in animals, just as in plants (Fernández-Ortiz et al., 2020). The production of melatonin in mammals involves four steps (Hardeland and Poeggeler, 2003; Mannino et al., 2021; Zhao et al., 2019). Initially, TPH generates 5-hydroxytryptophan. The cofactors oxygen and Tetrahydrobiopterin (BH4) are required for this reaction (Fig. 7.2). TDC transforms 5-hydroxytryptophan into serotonin, the primary substrate for melatonin production (Back et al., 2016). This process results in the emission of carbon dioxide. Following that, aralkyl-amine N-acetyltransferase generates N-acetyl-serotonin at the cost of acetyl-CoA. Finally, N-acetylserotonin methyl-transferase produces melatonin while converting the cofactor S-Adenosylmethionine (SAM) into S-Adenosyl-$_L$-homocysteine (Agathokleous et al., 2019). In addition to animals and plants, several microorganisms possess the ability to synthesize melatonin (Germann et al., 2016; Ma et al., 2017). Utilizing modified microbes to produce melatonin is a possible commercial manufacturing technique. The process through which bacteria produce melatonin is comparable to that used by mammals. It involves sequential enzymatic processes involving coenzymes or cofactors (Fernandez-Cruz et al., 2019; Germann et al., 2016).

7.6 Role of melatonin in plants' growth and productions

Many plant hormones, including auxin, were engaged in plant development and ripening. Melatonin, as an indoleamine, has the same starting constituent, tryptophan, as IAA, suggesting that it may function in controlling plant development and growth (Arnao and Hernández-Ruiz, 2014; Fan et al., 2018; Murch and Saxena, 2002). It was demonstrated that melatonin is spontaneously generated endogenously in several plants, including medicinal species, cereals, and fruits, despite melatonin concentrations varying across plant species (Fan et al., 2018). Melatonin is a growth-promoting chemical because of its dual nature in plant growth promotion. It improves numerous growth characteristics in vivo and in vitro and boosts plant products by controlling ion homeostasis. It is a crucial mediator of gene expression related to plant development factors and may alter their effects (Chaachouay et al., 2023a). Several investigations have shown that melatonin's growth-promoting actions on herbs are similar to those of other herb development stimulants. Much investigation has concentrated on the auxin-like action of melatonin, which may promote development in plant roots and shoots and increase root formation.

Melatonin is recognized as a development chemical, similar to auxin, in hypocotyls of *Lupinus albus* L., monocot taxa, such as common wheat, canary grass, barley, and oat, and dicot species, such as *Arabidopsis thaliana* (L.) Heynh. It is thus an auxinic hormone in plants (Arnao and Hernández-Ruiz, 2007; Chen et al., 2017). Moreover, the melatonin levels utilized in most of these investigations reached physiological norms in plants, raising suspicion about the phytohormone's development characteristics.

Over the last several decades, several investigations have highlighted melatonin as a critical chemical in root development and growth (Arnao and Hernández-Ruiz, 2007; Chen et al., 2017; Zhao et al., 2019). The influence of melatonin on root evolution and growth has gained a lot of attention because of its molecular similarity to auxin. Additionally, mounting data shows that melatonin plays a vital function in root-microorganism interactions in the rhizosphere (Asif et al., 2019; Jiao et al., 2016).

Melatonin exogenously enhances root organogenesis in different plants, such as adventitious and lateral root formation, notably Arabidopsis, cucumber, Lupinus, and rice (Arnao and Hernández-Ruiz, 2007; Chen et al., 2017; Pelagio-Flores et al., 2012; Zhang et al., 2014). Melatonin enhanced the quantity of lateral and adventitious roots in Arabidopsis by up to twice and threefold, respectively (Pelagio-Flores et al., 2012). Melatonin-treated cucumbers exhibited dramatically enhanced lateral root growth (Zhang et al., 2014). Melatonin's significance in stimulating rhizogenesis was identified for the first time by (Arnao and Hernández-Ruiz, 2007), who showed that melatonin stimulated root primordia production from pericycle cells in lupine, leading to lateral and adventitious root synthesis. Later, the same researchers claimed that melatonin might play a vital role in roots' gravitropic action (Arnao and Hernández-Ruiz, 2017).

Melatonin plays a crucial role in agricultural development, as it is a possible biostimulant to boost crop yields, safety, and sustainability, and it is also eco-friendly. It is safer to use in organic agriculture and is biodegradable. Due to its antioxidant properties, increased melatonin consumption by plants may prevent the peroxidation of plant products, extending the stand vitality of herbs and growing agricultural yields (Asif et al., 2019; Chaachouay et al., 2022b; Hardeland, 2009).

Because chlorophyll is an essential element of plants, preserving and stimulating photosynthetic activity is critical to crop development (Chaachouay et al., 2023c). Melatonin acts as an antioxidant in plants, preventing chlorophyll breakdown. Various trials and investigations have developed its chlorophyll conservation action in numerous species. An inquiry into *Hordeum vulgare* L. plants fed a melatonin-containing solution for 48 hours indicated a 2X higher chlorophyll content in melatonin-treated leaves than in regulated leaves (Arnao and Hernández-Ruiz, 2009). Melatonin increases the quantity of soluble carbohydrates, mainly sucrose and sorbitol, throughout the ripening process. This might be due to melatonin promoting starch accumulation and accelerating pear fruit growth, which significantly influences output (Yan et al., 2020). Besides, the action of melatonin protects cucumber plants that experience heat stress (Xu, 2010).

Melatonin may also prevent aging by preserving photosynthetic mechanisms and associated subcellular structures and activities, such as cytokinins. In one investigation, scientists boosted the synthesis of endogenous melatonin in the chloroplasts of several *Arabidopsis thaliana* (L.) Heynh. transgenic lines. Since melatonin is employed in agricultural development, another critical consideration is whether melatonin can enhance plant biomass. For the development of transgenic crops, the indoleamine metabolic enzyme of melatonin was modified in several ways (Tan et al., 2016). Melatonin can potentially increase plant biomass by improving the effectiveness of plant development, establishing resistance to stress responses, and promoting the germination of seeds and seedlings. It is believed that increasing melatonin synthesis has enormous possibilities for improving agricultural productivity.

7.7 Conclusion and future perspectives

Recently, considerable advancement has been achieved in plant melatonin investigations. This development expands our understanding of melatonin's existence, chemistry, metabolism, and potential effects on plant growth and production. As highlighted in this chapter, melatonin is discovered in animals, microorganisms, plants, and organs, although melatonin levels in various plants and organ systems are unstable. Because plants lack a pineal gland, their melatonin biosynthesis route differs from that of animals. Despite certain particular enzymes, the melatonin production route is identical to that of auxin in plants. Several investigations have revealed that melatonin is crucial for plant growth and production.

As a plant development promoter, melatonin may decrease pesticide usage by improving plant growth and preventing disease. Therefore, as a novel form of growth-regulating agent, it offers excellent performance and environmental safety properties. Its production and use may significantly promote the healthy development of horticultural plants, increase the fertilizer usage ratio, and reduce the occurrence of plant diseases, insect pests, and pesticide dosages. It may play a crucial role in minimizing the usage of chemical fertilizers and pesticides and has enormous development potential for promoting the agricultural industry's healthy expansion. Furthermore, other melatonin features in plants, such as the metabolic process and regulation under stressful circumstances, remain unidentified. Finally, the patterns of melatonin administration in response to individual and combination stressors in field settings should be explored.

References

Acharya, D., Satapathy, S., Somu, P., Parida, U.K., Mishra, G., 2021. Apoptotic effect and anticancer activity of biosynthesized silver nanoparticles from marine algae Chaetomorpha linum extract against human colon cancer cell HCT-116. Biol. Trace Elem. Res. 199 (5), 1812–1822.

Agathokleous, E., Kitao, M., Calabrese, E.J., 2019. New insights into the role of melatonin in plants and animals. Chem. Biol. Interact. 299, 163–167.

Allegra, M., Reiter, R.J., Tan, D.-X., Gentile, C., Tesoriere, L., Livrea, M.A., 2003. The chemistry of melatonin's interaction with reactive species. J. Pineal Res. 34 (1), 1–10.

Arnao, M.B., 2014. Phytomelatonin: discovery, content, and role in plants. Adv. Botany 2014.

Arnao, M.B., Hernández-Ruiz, J., 2007. Melatonin promotes adventitious-and lateral root regeneration in etiolated hypocotyls of Lupinus albus L. J. Pineal Res. 42 (2), 147–152.

Arnao, M.B., Hernández-Ruiz, J., 2009. Protective effect of melatonin against chlorophyll degradation during the senescence of barley leaves. J. Pineal Res. 46 (1), 58–63.

Arnao, M.B., Hernández-Ruiz, J., 2014. Melatonin: plant growth regulator and/or biostimulator during stress? Trends plant. Sci. 19 (12), 789–797.

Arnao, M.B., Hernández-Ruiz, J., 2017. Growth activity, rooting capacity, and tropism: three auxinic precepts fulfilled by melatonin. Acta Physiol. Plant. 39 (6), 1–9.

Asif, M., Pervez, A., Ahmad, R., 2019. Role of melatonin and plant-growth-promoting rhizobacteria in the growth and development of plants. CLEAN–Soil Air Water 47 (6), 1800459.

Axelrod, J., Weissbach, H., 1960. Enzymatic O-methylation of N-acetylserotonin to melatonin. Science 131 (3409), 1312.

Back, K., Tan, D.-X., Reiter, R.J., 2016. Melatonin biosynthesis in plants: multiple pathways catalyze tryptophan to melatonin in the cytoplasm or chloroplasts. J. Pineal Res. 61 (4), 426–437.

Balzer, I., Espínola, I.R., Fuentes-Pardo, B., 1997. Daily variations of immunoreactive melatonin in the visual system of crayfish. Biol. Cell 89 (8), 539–543.

Chaachouay, N., Azeroual, A., Bencharki, B., Douira, A., Zidane, L., 2023a. Hormonal interactions during fruit development and ripening. In: Husen, A., Zhang, W. (Eds.), Hormonal Cross-Talk, Plant Defense and DevelopmentPlant Biology, Sustainability and Climate Change. Academic Press, pp. 37–346. Available from: https://doi.org/10.1016/B978-0-323-95375-7.00014-8.

Chaachouay, N., Azeroual, A., Bencharki, B., Douira, A., Zidane, L., 2023b. Various metabolites and or bioactive compounds from vegetables, and their use nanoparticles synthesis, and applications. Nanomaterials from Agricultural and Horticultural Products. In: Husen, A. (Ed.), 1st ed. Springer, Singapore, pp. 187–1209. Available from https://doi.org/10.1007/978-981-99-3435-5_10

Chaachouay, N., Azeroual, A., Bencharki, B., Zidane, L., 2022a. Ethnoveterinary practices of medicinal plants among the Zemmour and Zayane tribes, Middle Atlas, Morocco. S. Afr. J. Bot. 151 (1Part A), 826–840. Available from: https://doi.org/10.1016/j.sajb.2022.11.009.

Chaachouay, N., Azeroual, A., Zidane, L., 2023c. Taxonomy, ethnobotany, phytochemistry and biological activities of thymus saturejoides: a review. Acta Bot. Hung. 65 (1–2), 35–51. Available from: https://doi.org/10.1556/034.65.2023.1-2.2.

Chaachouay, N., Douira, A., Zidane, L., 2022b. Herbal medicine used in the treatment of human diseases in the Rif, Northern Morocco. Arab. J. Sci. Eng. 47, 131–153. Available from: https://doi.org/10.1007/s13369-021-05501-1.

Chen, L., Fan, J., Hu, Z., Huang, X., Amombo, E., Liu, A., et al., 2017. Melatonin is involved in regulation of bermudagrass growth and development and response to low K+ stress. Front. Plant Sci. 8, 2038.

Denby, K., Gehring, C., 2005. Engineering drought and salinity tolerance in plants: lessons from genome-wide expression profiling in Arabidopsis. Trends Biotechnol. 23 (11), 547–552.

Dubbels, R., Reiter, R.J., Klenke, E., Goebel, A., Schnakenberg, E., Ehlers, C., et al., 1995. Melatonin in edible plants identified by radioimmunoassay and by high performance liquid chromatography-mass spectrometry. J. Pineal Res. 18 (1), 28–31.

Dubocovich, M.L., Rivera-Bermudez, M.A., Gerdin, M.J., Masana, M.I., 2003. Molecular pharmacology, regulation and function of mammalian melatonin receptors. Front. Biosci. 8 (4), 1093–1108.

Erland, L.A., Murch, S.J., Reiter, R.J., Saxena, P.K., 2015. A new balancing act: the many roles of melatonin and serotonin in plant growth and development. Plant Signal. Behav. 10 (11), e1096469.

Fan, J., Xie, Y., Zhang, Z., Chen, L., 2018. Melatonin: a multifunctional factor in plants. Int. J. Mol. Sci. 19 (5), 1528.

Fenwick, J.C., 1970. Demonstration and effect of melatonin in fish. Gen. Comp. Endocrinol. 14 (1), 86–97.

Fernandez-Cruz, E., González, B., Muñiz-Calvo, S., Morcillo-Parra, M.Á., Bisquert, R., Troncoso, A.M., et al., 2019. Intracellular biosynthesis of melatonin and other indolic compounds in Saccharomyces and non-Saccharomyces wine yeasts. Eur. Food Res. Technol. 245 (8), 1553–1560.

Fernández-Ortiz, M., Sayed, R.K., Fernández-Martínez, J., Cionfrini, A., Aranda-Martínez, P., Escames, G., et al., 2020. Melatonin/Nrf2/NLRP3 connection in mouse heart mitochondria during aging. Antioxidants 9 (12), 1187.

Germann, S.M., Baallal Jacobsen, S.A., Schneider, K., Harrison, S.J., Jensen, N.B., Chen, X., et al., 2016. Glucose-based microbial production of the hormone melatonin in yeast Saccharomyces cerevisiae. Biotechnol. J. 11 (5), 717–724.

Hardeland, R., 2009. Melatonin: Signaling mechanisms of a pleiotropic agent. Biofactors 35 (2), 183–192.

Hardeland, R., Poeggeler, B., 2003. Non-vertebrate melatonin. J. Pineal Res. 34 (4), 233–241.

Hugel, H.M., Kennaway, D.J., 1995. Synthesis and chemistry of melatonin and of related compounds. A review. Org. Prep. Proced. Int. 27 (1), 1–31.

Jiao, J., Ma, Y., Chen, S., Liu, C., Song, Y., Qin, Y., et al., 2016. Melatonin-producing endophytic bacteria from grapevine roots promote the abiotic stress-induced production of endogenous melatonin in their hosts. Front. Plant Sci. 7, 1387.

Kang, K., Kang, S., Lee, K., Park, M., Back, K., 2008. Enzymatic features of serotonin biosynthetic enzymes and serotonin biosynthesis in plants. Plant Signal. Behav. 3 (6), 389–390.

Kobylińska, A., Borek, S., Posmyk, M.M., 2018. Melatonin redirects carbohydrates metabolism during sugar starvation in plant cells. J. Pineal Res. 64 (4), e12466.

Kolár, J., Johnson, C.H., Machácková, I., 2002. Presence and possible role of melatonin in a short-day flowering plant, Chenopodium rubrum. Melatonin Four Decades. Springer, pp. 391–393.

Lauber, J.K., Boyd, J.E., Axelrod, J., 1968. Enzymatic synthesis of melatonin in avian pineal body: extraretinal response to light. Science 161 (3840), 489–490.

Lerner, A.B., Case, J.D., Mori, W., Wright, M.R., 1959. Melatonin in peripheral nerve. Nature 183 (4678), 1821.

Lerner, A.B., Case, J.D., Takahashi, Y., Lee, T.H., Mori, W., 1958. Isolation of melatonin, the pineal gland factor that lightens melanocyteS1. J. Am. Chem. Soc. 80 (10), 2587.

Li, J., Liu, J., Zhu, T., Zhao, C., Li, L., Chen, M., 2019. The role of melatonin in salt stress responses. Int. J. Mol. Sci. 20 (7), 1735.

Ma, Y., Jiao, J., Fan, X., Sun, H., Zhang, Y., Jiang, J., et al., 2017. Endophytic bacterium Pseudomonas fluorescens RG11 may transform tryptophan to melatonin and promote endogenous melatonin levels in the roots of four grape cultivars. Front. Plant Sci. 7, 2068.

Macias, M., Rodrigueez-Cabezas, M.N., Reiter, R.J., Osuna, A., Acuña-Castrovejo, D., 1999. Presence and effects of melatonin in Trypanosoma cruzi. J. Pineal Res. 27 (2), 86–94.

Mahajan, S., Tuteja, N., 2005. Cold, salinity and drought stresses: an overview. Arch. Biochem. Biophys. 444 (2), 139–158.

Mannino, G., Pernici, C., Serio, G., Gentile, C., Bertea, C.M., 2021. Melatonin and phytomelatonin: chemistry, biosynthesis, metabolism, distribution and bioactivity in plants and animals—an overview. Int. J. Mol. Sci. 22 (18), 9996.

Morita, M., Hall, F., Best, J.B., Gern, W., 1987. Photoperiodic modulation of cephalic melatonin in planarians. J. Exp. Zool. 241 (3), 383–388.

Murch, S.J., Saxena, P.K., 2002. Melatonin: a potential regulator of plant growth and development? Vitro Cell. Dev. Biol. Plant 38 (6), 531–536.

Nawaz, K., Chaudhary, R., Sarwar, A., Ahmad, B., Gul, A., Hano, C., et al., 2020. Melatonin as master regulator in plant growth, development and stress alleviator for sustainable agricultural production: current status and future perspectives. Sustainability 13 (1), 294.

Omer, R.A., Koparir, P., Ahmed, L., Koparir, M., 2021. Computational and spectroscopy study of melatonin. Indian J. Chem. Sect. B (IJC-B) 60 (5), 732–741.

Pelagio-Flores, R., Muñoz-Parra, E., Ortiz-Castro, R., López-Bucio, J., 2012. Melatonin regulates Arabidopsis root system architecture likely acting independently of auxin signaling. J. Pineal Res. 53 (3), 279–288.

Quarles, W.G., Templeton, D.H., Zalkin, A., 1974. The crystal and molecular structure of melatonin. Acta Crystallogr. Sect. B: Struct. Crystallogr. Cryst. Chem. 30 (1), 99–103.

Reiter, R.J., 1991. Pineal melatonin: cell biology of its synthesis and of its physiological interactions. Endocr. Rev. 12 (2), 151–180.

Reiter, R.J., Tan, D.-X., Zhou, Z., Cruz, M.H.C., Fuentes-Broto, L., Galano, A., 2015. Phytomelatonin: assisting plants to survive and thrive. Molecules 20 (4), 7396–7437.

Rhodes, D., Nadolska-Orczyk, A., 2001. Plant stress physiology. eLS.

Shi, H., Jiang, C., Ye, T., Tan, D.-X., Reiter, R.J., Zhang, H., et al., 2015. Comparative physiological, metabolomic, and transcriptomic analyses reveal mechanisms of improved abiotic stress resistance in bermudagrass [Cynodon dactylon (L). Pers.] by exogenous melatonin. J. Exp. Botany 66 (3), 681–694.

Tan, D.-X., Hardeland, R., Back, K., Manchester, L.C., Alatorre-Jimenez, M.A., Reiter, R.J., 2016. On the significance of an alternate pathway of melatonin synthesis via 5-methoxytryptamine: comparisons across species. J. Pineal Res. 61 (1), 27–40.

Vantassel, D.L., Li, J.A., Oneill, S.D., 1993. Melatonin-identification of a potential dark signal in plants. Plant Physiol. 102 (1), 117.

Vantassel, D.L., Roberts, N.J., Oenill, S.D., 1995. Melatonin from higher-plants-isolation and identification of N-acetyl 5-methoxytryptamine. Plant Physiol. 108 (2), 101.

Vivien-Roels, B., Pevet, P., Beck, O., Fevre-Montange, M., 1984. Identification of melatonin in the compound eyes of an insect, the locust (Locusta migratoria), by radioimmunoassay and gas chromatography-mass spectrometry. Neurosci. Lett. 49 (1-2), 153–157.

Wang, R., Yang, X., Xu, H., Li, T., 2016. Research progress of melatonin biosynthesis and metabolism in higher plants. Plant Physiol. J. 52, 615–627.

Xu, X.D., 2010. Effects of exogenous melatonin on physiological response of cucumber seedlings under high temperature stress. Master's degree thesis, Northwest A and F University.

Yan, Y., Shi, Q., Gong, B., 2020. Review of melatonin in horticultural crops. Melatonin-The Hormone of Darkness and its Therapeutic Potential and Perspectives. IntechOpen.

Yu, Y., Wang, A., Li, X., Kou, M., Wang, W., Chen, X., et al., 2018. Melatonin-stimulated triacylglycerol breakdown and energy turnover under salinity stress contributes to the maintenance of plasma membrane H^+-ATPase activity and K^+/Na^+ homeostasis in sweet potato. Front. Plant Sci. 9, 256.

Zhang, N., Zhang, H.-J., Zhao, B., Sun, Q.-Q., Cao, Y.-Y., Li, R., et al., 2014. The RNA-seq approach to discriminate gene expression profiles in response to melatonin on cucumber lateral root formation. J. Pineal Res. 56 (1), 39–50.

Zhao, D., Yu, Y., Shen, Y., Liu, Q., Zhao, Z., Sharma, R., et al., 2019. Melatonin synthesis and function: evolutionary history in animals and plants. Front. Endocrinol. 10, 249.

CHAPTER

8

Use of amino acids in plant growth, photosynthetic assimilation, and nutrient availability

Shakeelur Rahman[1], *Sahil Mehta*[2] *and Azamal Husen*[3,4,5]

[1]Prakriti Bachao Foundation, Ranchi, Jharkhand, India [2]Department of Botany, Hansraj College, University of Delhi, New Delhi, India [3]Department of Biotechnology, Smt. S.S. Patel Nootan Science & Commerce College, Sankalchand Patel University, Visnagar, Gujarat, India [4]Department of Biotechnology, Graphic Era (Deemed to be University), Dehradun, Uttarakhand, India [5]Wolaita Sodo University, Wolaita, Ethiopia

Abbreviations

BCAA Branched-Chain Amino Acids
ECM Extra Cellular Matrix
GB Glycine Betaine
IAA Indole Acetic Acid
LPR Leucine-Rich Repeat
TCA Tricarboxylic Acid

8.1 Introduction

In plants, amino acids play several crucial roles in metabolism (Liao et al., 2022; Bocso and Butnariu, 2022). In various metabolic processes, amino acids serve as intermediates for final metabolites. Amino acids help in the regulation of physiological, biochemical, and various metabolic processes (Yang et al., 2018; Liao et al., 2022). The production of stored proteins depends on amino acids, which are also catabolized when energy is needed and fed into the tricarboxylic acid cycle (TCA cycle) to provide energy (Galili, 2011; Zhang et al., 2015). Many amino acids serve important functions in plant

Biostimulants in Plant Protection and Performance
DOI: https://doi.org/10.1016/B978-0-443-15884-1.00016-6

development and response to environmental stimuli in addition to protein synthesis. Moreover, amino acids have crucial functions in human nutrition, either as a source of nutraceutical substances or as crucial dietary components. They serve as precursors for a variety of primary and secondary metabolites (Moormann et al., 2022).

Undoubtedly, nine out of the twenty-one proteinogenic amino acids cannot be produced in animals, and three or more of the remaining amino acids cannot be synthesized in enough amounts to meet metabolic requirements. On the contrary, plants manufacture all twenty-one proteinogenic amino acids independently, in contrast to humans and animals (Hou and Wu, 2018). As a result, amino acids are crucial for the overall health and function of all living things, including plants. An overview of the function of amino acids in plant growth, development, and metabolism is given in this book chapter.

8.2 Amino acids in plants

Proteins called amino acids are bio-activators that provide plants with energy to make up for losses brought on by respiration and degradation (Baqir et al., 2019). Amino acids are ionic molecules that are colorless and soluble to unstable degrees in both hot and cold water as well as alcohol. Because of their high melting point, they are referred to as hybrid ions. In addition to synthesizing all twenty-one amino acids necessary for protein synthesis, amino acids also serve as precursors for several primary and secondary metabolites. Plants' non-proteinogenic amino acids can play a key role in how they react to environmental challenges (Trovato et al., 2021). Although amino acids can be found in plants in two different forms—freely or in combination to form proteins and peptide compounds—the free form is more prevalent because it decomposes the tiny bonds and leaves the amino acids free, solitary, and easily penetrable (Abd El Hafez, 2011; Baqir et al., 2019).

Among the many important roles, amino acids and their derivatives play an important role in protein synthesis, growth and development, nutrition, and stress responses (Hildebrandt et al., 2015; Baqir et al., 2019). Amino acids are crucial for managing nutrients and photosynthesis, as has become obvious. In addition to serving as a fundamental component of protein synthesis, amino acids also play important functions in the growth of plants and the control of stress. Moreover, amino acids are effective biostimulants for plants. In particular, under biotic and abiotic stress situations, it increases plant productivity (Yakhin et al., 2016; Shahrajabian et al., 2022). Therefore, amino acids are the subject of extensive research regarding plant physiology, nutrition, and other metabolite biosynthesis (Fig. 8.1).

8.2.1 Plant growth

Amino acids not only act as building blocks for protein synthesis, but many amino acids also play active roles in plant growth and development (Baqir et al., 2019). The importance and effectiveness of amino acids depend on the different stages of plant growth. Amino acids play a significant role in several biotic processes, whether they are

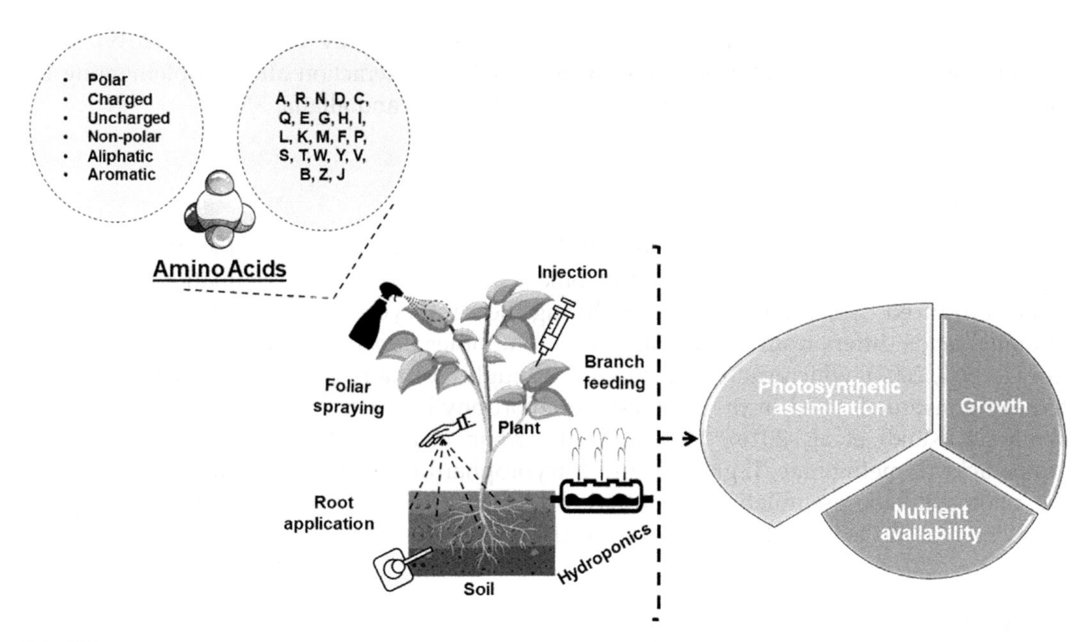

FIGURE 8.1 Amino acids exposure and their role in plant growth and functions.

free or as constituent of proteins. They assist plant cells in the uptake of water and nutrients to increase synthesizing proteins, enhancing plant metabolism and the carbon assimilation rate for proper growth of plants (Sharma-Natu and Ghildiyal 2005). Cell division, protein contents, natural hormones like ethylene and indol acetic acid (IAA), and plant pigments are enhanced by amino acids (Ahmed et al., 2014). Vitamins, nucleotides, and plant growth regulators are synthesized by a substance formed by amino acids. These are important components of protoplasm and living matter that help in the formation of enzymes and conducting enzymatic reactions inside the plant cells (Kamar and Omar, 1987). The positive impact of amino acids on stimulating plant growth and yield performance has been observed in several studies. It was observed that the spraying of amino acids over different crops enhanced root and shoot growth by increasing chlorophyll contents (Lozek and Fecenko, 1996; Zaki et al., 2007; Ashoori et al., 2013).

A study was conducted on wheat crops where it was found that spraying 2000 mg/L of amino acids on wheat plants 55 and 75 days after sowing increased the number of tillers, stimulated plant height, growth, and enhanced chlorophyll and protein contents (El-Naggar and El-Ghamry, 2007; Meijer, 2003). Histidine is one of the most important amino acids in proteins and plays a significant role in plant growth and development. It has a good role in increasing crop growth, early yield, and enhancing phosphorus action inside plants (Denby and Last, 1999). Hydroxylysine repressed the growth of Rhizobium in some leguminous plants. Hydroxylysine plays an important role in the effective symbiotic relationship by regulating the growth of the root nodule bacteria on the root tip or the elongation of the roots of host plants (Ligero et al., 1986). Recently, the exogenous application of amino acids also improved the growth and yield of lettuce (*Lactuca sativa* L.) by Khan et al. (2019). More recently, Smirnova et al. (2022) focused on the metabolic interaction after supplementing the amino acids between plant growth-promoting rhizobacteria and alfalfa.

8.2.2 Photosynthesis

Amino acids in large amounts are found in the mitochondria and chloroplast of plant species as a result of the availability of ketonic acids formed by the assimilation of carbohydrates produced by carbon assimilation during the Kreb cycle (Beavers, 1991). The quantity of amino acids differs from one plant species to another depending on the metabolic procedure (Abed, 2007), which can easily pass on through root hairs and plant vessels. Glycine is a major constituent used in the synthesis of chlorophyll, which enhances the rate of photosynthesis (Kandi et al., 2016). More recently, Khan et al. (2019) studied the potential of amino acid (L-methionine, L-glycine, and L-tryptophan) root supplementation in enhancing the photosynthetic assimilation of butterhead lettuce. In this study, L-methionine significantly increased the growth performance at the lowest concentrations of 0.2 and 0.02 mg/L.

8.2.3 Metabolism and nutrient management

Amino acids as bioinducers boost mineral absorption and improve the efficiency of nutrient elements, which help in the enhancement of grain quality in several crops (Baqir et al., 2019). Amino acids play a significant role as a chelate material for calcium, iron, zinc, copper, and magnesium, as these elements can be absorbed and passed through plants easily (Vernieri et al., 2005). Amino acids help in the balancing of soil microorganisms that enhance the mineralization of the organic matter around the plant roots (Sokol et al., 2022). The roots absorb the organic matter, which provides essential nutrients, including nitrogen, to the plants (Ashmead, 1986; Persson and Nasholm, 2002; Rushimisha et al., 2023). It has been reported that amino acids influence directly or indirectly the physiological functions of plants; as a result, protein synthesis and the formation of carbohydrates take place by increasing photosynthesis and enzymatic activities (Neri et al., 2002; Ruta et al., 2011; Shafeek et al., 2012; Khan et al., 2019) (Table 8.1).

Aspartic acid, or aspartate (Asp), is an essential amino acid for protein synthesis and works as a central building block in carbon and nitrogen metabolism for various metabolic processes, including the biosynthesis of other amino acids, nucleotides, organic acids in the TCA cycle, sugars in glycolysis, and hormones, which are all essential for plant growth and stress resistance (Han et al., 2021). Asp plays an important role in the aspartate-glutamate pathway of amino acid metabolism, functioning as a common precursor for basic amino acids such as isoleucine, lysine, methionine, and threonine. It can be changed into glutamate and asparagine (Azevedo et al., 2006). Furthermore, Asp absorbs inorganic nitrogen, providing a nitrogen source for synthesizing other nitrogen-containing compounds in plants (Buchanan, 1980). Asp can also be converted to oxaloacetate, a kind of organic acid, through the catalytic activities of enzymes, including Asp aminotransferase and prephenate aminotransferase (Holland et al., 2018). Asp is an important amino acid for cell propagation and participates in vegetation reactions to abiotic threats (Jander and

TABLE 8.1 Amino acids and their various role in plant growth and physiological processes.

Name of amino acids	Role of amino acids	Key references
Hyl	Increasing crop growth and early yield	Ligero et al. (1986)
Leu	Leucine-rich repeat proteins have a significant role in plant defenses	Jones and Jones (1997)
His	Increasing crop growth, early yield, and enhancing phosphorus action inside the plant	Denby and Last (1999)
The	Increasing plant tolerance to diseases	Hardie (1999)
Hyp	Hydroxyproline provides structural integrity to mediate cell-cell interactions and communication	Wu et al. (2001)
Tyr	Increases plant disease tolerance	Blume-Jensen and Hunter (2001)
Ile	Increasing the shoot system, growth, and early yield	Diebold et al. (2002)
Ala	Affecting plant growth velocity and activating chlorophyll formation	Broeckling et al. (2005)
Lys	Increasing the shoot system, growth, and early yield	Azevedo et al. (2006)
Met	Accelerates the fruit ripening as it enters the cycle of ethylene formation and has a role in the root activation	Azevedo et al. (2006)
Val	Affecting the velocity of growth, root formation, and seed production	Song et al. (2007)
Asp	Asp is an important amino acid for cell propagation and participates in vegetation reactions to abiotic threats	Jander and Joshi (2010)
Glu	Glutamic acid increased plant height, number of leaves, number of branches disk diameter, root length, plant dry weight, fresh weight, and the plant content of carbohydrates and N, P, and K	Haroun et al. (2010)
Gly	Activating photosynthesis and raising its efficiency as it enhances chlorophyll formation and encourages vegetative growth as well as has a role in pollination and fruitfulness	Aldesuquy et al. (2012)
Ser	Increasing plant tolerance to diseases, activating chlorophyll, and has a role in hormone balance inside the plant	Ros et al. (2012)
Pro	Increasing tolerance to hard conditions, activating pollen grain germination, organizing osmosis potential, maintaining the colloidal properties of the cell protoplasm, and removing the negative effect of free radicals	Talat et al. (2013)
Cys	Cysteine led to growth and increased total photosynthesis pigments in plants	El-Hosary et al. (2013)
Phe	Improving plant cells and embryo formation	Yoo et al. (2013)
Try	Helping to the formation of active auxins IAA necessary for plant growth and has a role in the early yield	Kandil and Marie (2017)
Arg	Increasing tolerance to hard conditions such as heat, frost, drought, and salinity. It has a role in chlorophyll formation and enhancing root formation as well as cell division and poly amid formation	Trovato et al. (2021)

Joshi, 2010). Asp is considered a drought-responsive biomarker because of its rapid changes in content in plants exposed to stress (Han et al., 2021). Spraying of Asp can enhance plant tolerance to salinity (Sadak and Abdelhamid 2015) and cadmium toxicity (Rizwan et al., 2017) based on plant growth and physiological responses. Lei et al. (2022) focused on the metabolic and physiological regulation of perennial ryegrass growing under heat-stress conditions upon supplementation with Asp.

Moreover, cysteine and methionine are two important amino acids containing sulfur that help in the synthesis of many plant proteins (Yassen, 2001). Sulfur forms disulfide bonds in proteins between polypeptide chains to keep the protein folded and provide proper shape to the proteins (Havlin et al., 2005). The non-proteinogenic amino acid β-alanine has significant roles in plant physiology and metabolism, both directly as a defense compound that allows plants to hold out various stresses such as hypoxia, water logging, and drought and indirectly as a precursor to the compounds pantothenate and CoA, which are involved in various functions. In addition, the amino acid is transformed into β-alanine betaine, which has supplementary protective roles such as salt tolerance, and homoglutathione, which may be critical for nitrogen fixation (Broeckling et al., 2005). Various reports regarding the importance of sulfur-containing amino acids in modulating plant metabolism are very well documented in a recently published article by Narayan et al. (2022).

Hydroxyproline-rich glycoproteins released by plant cells are supposed to play a vast range of roles, ranging from providing structural integrity to mediating cell–cell interactions and communication. The pistil and pollen tube extracellular matrix is enriched in these highly glycosylated proteins (Wu et al., 2001). Similarly, isoleucine is one of the branched-chain amino acids that are essential substrates for protein synthesis in all organisms. Although the metabolic pathway for isoleucine has been well characterized in higher plants, it is not known whether it plays a specific role in plant development (Diebold et al., 2002).

Phenylalanine is an aromatic amino acid that serves as a vital component of proteins in all living organisms, and in plants is a precursor for thousands of additional metabolites. Animals are incapable of synthesizing phenylalanine and must primarily obtain it directly or indirectly from plants. While plants can synthesize phenylalanine in plastids through arogenate, the contribution of an alternative pathway via phenylpyruvate, as it occurs in most microbes, has not been demonstrated. Here we establish that plants also utilize the microbial-like phenylpyruvate pathway to produce phenylalanine, and flux through this route is increased when the entry point to the arogenate pathway is limiting. Unexpectedly, we find the plant phenylpyruvate pathway utilizes a cytosolic aminotransferase that links the coordinated catabolism of tyrosine to serve as the amino donor, thus interconnecting the extra-plastidial metabolism of these amino acids. This discovery unlocks another level of complexity in the plant aromatic amino acid regulatory network, unveiling new targets for metabolic engineering (Yoo et al., 2013). Atteya and the group supplemented both proline and phenylalanine to salted moringa and observed enhancement in growth, productivity, and oil composition (Atteya et al., 2022).

Lysine is one of the essential amino acids present in very low amounts in plants as a degradation factor of the metabolism pathway. It degrades starch metabolism, tryptophan metabolism, the tricarboxylic acid cycle, and unfolded protein response. It has been reported that the biosynthesis of lysine focuses on improving the nutritional values of plants. It was observed that the impacts of lysine catabolism on stress tolerance are because of the production of proline and pipecolate from glutamate and α-aminoadipate-δ-semialdehyde, respectively, which is mediated by the saccharopine pathway (Yang et al., 2014). Tryptophan is the main origin of IAA in the majority of living organisms (Kandil and Marie, 2017). Spraying Tryptophan over the wheat crop after 40 days of sowing increased the plant height (Martens and Fankenberger 1994). Similarly, the spraying of 150 mg/L Cysteine also recorded a significance of growth higher than the treatment of 100 mg/L of this amino acid. It was explained that foliar spray of the used concentrations of Tryptophan and Cysteine led to growth and increased total photosynthesis pigments in all wheat growth varieties compared to the control one (El-Hosary et al., 2013). Spraying 50 mg/L Tryptophan improved the root and vegetative growth of various crops (Zahir et al., 2005). In another instance, Gondek and Mierzwa-Hersztek (2021) analyzed the positive effect of soil-applied L-tryptophan on the amount of biomass and nitrogen and sulfur utilization by maize.

Spraying glycine betaine on a standing wheat crop has an impact on the yield performance and its mechanism positively, with an important variation compared to the control treatment, as well as significantly better in most growth traits such as plant height, spike length, and root system length (Aldesuquy et al., 2012). Arginine plays an important role in seed germination, germination stimulation indicator, germination velocity, seed vigor, seedling dry weight, and root system (AL-Qaisi et al., 2016). Arginine has an impact on increasing the chlorophyll index, plant height, and leaf area (Dawood, and Glaim, 2018). It was reported that glutamic acid increased all growth criteria of plants, including plant height, number of leaves, number of branches, disk diameter, root length, plant dry weight, fresh weight, and the plant content of carbohydrates and N, P, and K (Haroun et al., 2010; Mazher et al., 2011). The effect of proline was observed by Talat et al. (2013) on wheat plants, where they observed enhanced performance of plant height, fresh weight, total chlorophyll content, potassium concentration, sodium concentration, and transpiration rate. Leucine-rich repeat (LPR) proteins have an important role in plant protection from different pathogens. It develops resistance to a wide range of pathogens, including bacteria, nematodes, fungi, and viruses. LRR proteins work as proteins required for resistance proteins to function. Pollen self-incompatibility and species incompatibility may also be understood as forms of defense (Jones and Jones, 1997).

Serine is a polar amino acid that plays a fundamental role in plant metabolism, plant development, and cell signaling. In addition to being a building block for proteins, serine participates in the biosynthesis of biomolecules such as amino acids, nucleotides, phospholipids, and sphingolipids (Ros et al., 2012). Recently, Zimmermann et al. (2021) confirmed the role of a phosphorylated pathway of serine biosynthesis in linking plant growth with nitrogen metabolism. Like the earlier discussed amino acids, the role of valine, tyrosine, and threonine supplementation in modulating plant metabolism and nutrient levels was confirmed by several coworkers over the years (Hardie, 1999; Blume-Jensen and Hunter, 2001; Song et al., 2007; Diedhiou et al., 2008; Feduraev et al., 2020).

8.3 Conclusion

One of the most important and complex networks in plants' biological systems is their amino acid metabolism. Plants use amino acids and their derivatives for several crucial processes, including protein synthesis, growth and development, nourishment, and stress. Whether plants grow in hills or fields, the importance of amino acid pools remains the same across different domains. After synthesis, amino acids enter the metabolic maze in the plant cell. However, the functions of all amino acids as a metabolic signal to regulate metabolism have yet to be established in plants. Unlike unicellular systems such as *E. coli* and yeast, plants consist of different cell types/tissues with varying metabolic statuses. Moreover, the intra- and intercellular trafficking of metabolites and the flux of metabolites in the symplast, apoplast, and vascular systems further complicate the study of metabolic regulation in plants. With a reliable technique for monitoring the dynamic cellular and subcellular levels of amino acids, we can clearly understand the role of multiple amino acids in conjunction. Nevertheless, the signaling role and its function in plant immunity and defense are also new avenues of research. The journey to search for receptors mediating amino acid-induced plant immunity promises to give more surprises to plant scientists in the future. Thus, it is critical to do extensive research to comprehend how amino acids operate in plants.

References

Abd EL-hafez, A.A.Y., 2011. Use of amino acids in improving the quantity and performance of horticultural crops under Egyptian conditions. Acad. Sci. Res. J. Sci. 413.

Abed, A.K.M., 2007. Study of amino acids and fatty acids in date plant fruit *Phoenix dactylifera* L. Cultivars ALdehin and brain of three male date plant pollinators. J. Basrah Res. Sci. 31 (3), 31–37.

Ahmed, F.F., Abdelaal Salah, A.H.M., El-Masry, E.M.A., Farag, W.B.M.M., 2014. Response of superior grapevines to foliar application of some micronutrients, calcium, amino acids, and salicylic acids. World Rural. Observer 6 (3), 57–64.

AL-Qaisi, F.W.A., Mahdi, T.S., Mahmud, R.W., 2016. Effect of hydrogen peroxide and arginine in seeds germination and seedling growth of *Zea mays* and Surface growth of *Fusarium moxysporum*. J. AL Mustansiriyah Sci. 27 (4), 17–21.

Aldesuquy, S.H., Abbas, M.A., Abo-Hamed, S.A., Elhakem, A.H., Alsokari, S.S., 2012. Glycine betaine and salicylic acid-induced modification in productivity of two different cultivars of wheat grown under water stress. J. Stress. Physiol. Biochem. 8 (2), 72–89.

Ashmead, H.D., 1986. The Absorption Mechanism of amino acid chelates by plant cells. In: Ashmead, H.D., Ashmead, H.H., Miller, G.W., Hsu, H.H. (Eds.), Foliar feeding of plants with amino acid chelates. Noyes Publications, Park Ridge, New Jersey, USA, pp. 219–235.

Ashoori, M., Esfehani, M., Abdollahi, S., Rabiei, B., 2013. Effects of organic fertilizer complements application on grain yield, nitrogen use efficiency and milling properties in two rice cultivars *Oryza sativa* L. Iran. J. Field Crop Sci. 43, 701–713.

Atteya, A.K., El-Serafy, R.S., El-Zabalawy, K.M., Elhakem, A., Genaidy, E.A., 2022. Exogenously supplemented proline and phenylalanine improve growth, productivity, and oil composition of salted moringa by upregulating osmoprotectants and stimulating antioxidant machinery. Plants 11 (12), 1553.

Azevedo, R.A., Lancien, M., Lea, P.J., 2006. The aspartic acid metabolic pathway, an exciting and essential pathway in plants. Amino Acids 30 (2), 143–162.

Baqir, H.A., Zeboon, N.H., Al-behadili, A.A.J., 2019. The role and importance of amino acids within plants: a review. Plant Arch. 19 (2), 1402–1410.

Beavers, L., 1991. Nitrogen metabolism in plant. Hand Book (Translator) Ministry of Higher Education and Scientific Research. University of Baghdad, p. 477.

Blume-Jensen, P., Hunter, T., 2001. Oncogenic kinase signalling. Nature 411, 355–365.

Bocso, N.S., Butnariu, M., 2022. The biological role of primary and secondary plants metabolites. J. Nutr. Food Process. 5 (3), 1–7.

Broeckling, C.D., Huhman, D.V., Farag, M.A., Smith, J.T., May, G.D., Mendes, P., 2005. Metabolic profiling of *Medicago truncatula* cell cultures reveals the effects of biotic and abiotic elicitors on metabolism. J. Exp. Botany 56, 323–336.

Buchanan, B.B., 1980. Role of light in the regulation of chloroplast enzymes. Annu. Rev. Plant Physiol. 31, 341–374.

Dawood, W.M., Glaim, A.A., 2018. Response of growth traits and grain yield of *Zea mays* L. to spray by proline and Arginine Diyalay of Agriculture Science 2(10):129–136.

Denby, K.J., Last, R.L., 1999. Diverse regulatory mechanisms of amino acid biosynthesis in plants. Genet. Eng. 21, 173–189.

Diebold, R., Schuster, J., Daschner, K., Binder, S., 2002. The branched-chain amino acid transaminase gene family in *Arabidopsis* encodes plastid and mitochondrial proteins. Plant Physiol. 129, 540–550.

Diedhiou, C.J., Popova, O.V., Dietz, K.J., Golldack, D., 2008. The SNF1-type serine-threonine protein kinase SAPK4 regulates stress-responsive gene expression in rice. BMC Plant Biol. 28, 49.

El-Hosary, A.A.G.Y., El-Morsi, A., Hassan, E.A., El-Awadi, M.E., Abdel-Baky, Y.R., 2013. Effect of some bio regulators on growth and yield of some wheat varieties under newly cultivated land conditions. N. Y. Sci. J. 6 (6), 310–334.

El-Naggar, E.M., El-Ghamry, A.M., 2007. Effect of bio and chemical nitrogen fertilizers with foliar of humic and amino acids on wheat. J. Agric. Sci. 32 (5), 4029–4043.

Feduraev, P., Skrypnik, L., Riabova, A., Pungin, A., Tokupova, E., Maslennikov, P., et al., 2020. Phenylalanine and tyrosine as exogenous precursors of wheat (*Triticum aestivum* L.) secondary metabolism through PAL-associated pathways. Plants 9 (4), 476.

Galili, G., 2011. The aspartate-family pathway of plants: linking production of essential amino acids with energy and stress regulation. Plant Signal. Behav. 6, 192–195.

Gondek, K., Mierzwa-Hersztek, M., 2021. Effect of soil-applied L-tryptophan on the amount of biomass and nitrogen and sulfur utilization by maize. Agronomy 11 (12), 2582.

Han, M., Zhang, C., Suglo, P., Sun, S., Wang, M., Su, T., 2021. L-Aspartate: an essential metabolite for plant growth and stress acclimation. Molecules 26, 1887.

Hardie, D.G., 1999. Plant protein serine/threonine kinases: classification and functions. Annu. Rev. Plant Physiol. Plant Mol. Biol. 50, 97–131.

Haroun, S.A., Shukry, W.M., El-Sawy, O., 2010. Effect of asparagine or glutamine on growth and metabolic changes in *Phaseolus vulgaris* under in vitro conditions. Biosci. Res. 7 (1), 1–21.

Havlin, J.L., Beaton, J.D., Tisdale, S.L., Nelson, W.L., 2005. Soil fertility and fertilizers: an introduction to nutrient management. Seventh Edition, p. 515.

Hildebrandt, T.M., Nesi, A.N., Araujo, W.L., Braun, H.P., 2015. Amino acid catabolism in plants. Mol. Plant 8, 1563–1579.

Holland, C.K., Berkovich, D.A., Kohn, M.L., Maeda, H., Jez, J.M., 2018. Structural basis for substrate recognition and inhibition of prephenate aminotransferase from *Arabidopsis*. Plant J. 94, 304–314.

Hou, Y., Wu, G., 2018. Nutritionally essential amino acids. Adv. Nutr. 9, 849–851.

Jander, G., Joshi, V., 2010. Recent progress in deciphering the biosynthesis of aspartate-derived amino acids in plants. Mol. Plant 3, 54–65.

Jones, D.A., Jones, J.D.G., 1997. The role of leucine-rich repeat proteins in plant defense. Adv. Botanical Res. 24, 89–167.

Kamar, M.E., Omar, A., 1987. Effect of nitrogen levels and spraying wih aminal-forte (amino acids salvation) on yield of and potatoes. J. Agric. Sci. 12 (4), 900–907.

Kandil, E.E., Marie, E.A.O., 2017. Response of some wheat cultivars to nano-mineral fertilizers and amino acids foliar application. Alex. Sci. Exch. J. 38 (1), 53–68.

Kandi, A.A., Sharief, A.E.M., Seadh, S.E., Altai, D.S.K., 2016. Role of humic acid and amino acids in limiting loss of nitrogen fertilizer and increasing productivity of some wheat cultivars grown under newly reclaimed sandy soil. Int. J. Adv. Res. Biol. Sci. 3 (4), 123–136.

Khan, S., Yu, H., Li, Q., Gao, Y., Sallam, B.N., Wang, H., et al., 2019. Exogenous application of amino acids improves the growth and yield of lettuce by enhancing photosynthetic assimilation and nutrient availability. Agronomy 9 (5), 266.

Lei, S., Rossi, S., Huang, B., 2022. Metabolic and physiological regulation of aspartic acid-mediated enhancement of heat stress tolerance in perennial ryegrass. Plants 11 (2), 199.

Liao, H.S., Chung, Y.H., Hsieh, M.H., 2022. Glutamate: a multifunctional amino acid in plants. Plant Sci. 318, 111238.

Ligero, F., Lluch, C., Olivares, J., 1986. Evolution of ethylene from roots of *Medicago sativa* plants inoculated with *Rhizobium meliloti*. J. Plant Physiol. 125, 361–365.

Lozek, O., Fecenko, J., 1996. Effect of foliar application of manganese, boron and sodium humate on the potato production. Microelementy Wrolinctwie 1, 169–172.

Martens, D.A., Fankenberger, W.T.J., 1994. Assimilation of exogenous 2–14C-indole-3-acetic acid and 3–14C tryptophan exposed to the roots of three wheat varieties. Plant Soil 166, 281–290.

Mazher, A.A.M., Sahar, M.Z., Safaa, A.M., Hanan, S.S., 2011. Stimulatory effect of kinetin, ascorbic acid and glutamic acid on growth and chemical constituents of *Codiaeum variegatum* L. plants. Am.-Euras. J. Agric. Environ. Sci. 10 (3), 318–323.

Meijer, A.J., 2003. Amino acids as regulators and components of non proteinogenic pathways. J. Nutr. 133, 2057–2062.

Moormann, J., Heinemann, B., Hildebrandt, T.M., 2022. News about amino acid metabolism in plant–microbe interactions. Trends Biochem. Sci. 47, 839–850.

Narayan, O.P., Kumar, P., Yadav, B., Dua, M., Johri, A.K., 2022. Sulfur nutrition and its role in plant growth and development. Plant Signal. Behav. 2030082. Available from: https://doi.org/10.1080/15592324.2022.2030082.

Neri, D., Lodolini, E.M., Chelian, K., Bonanomi, G., Zucconi, F., 2002. Physiological responses to several organic compounds applied to primary leaves of cowpea *Vigna sinensis* L. Acta Hortic. 594, 309–314.

Persson, J., Nasholm, T., 2002. Regulation of amino acid uptake in conifers by exogenous and endogenous nitrogen. Planta 215, 639–644.

Rizwan, M., Ali, S., Zaheer, A.M., Shakoor, M.B., Mahmood, A., Ishaque, W., et al., 2017. Foliar application of aspartic acid lowers cadmium uptake and Cd-induced oxidative stress in rice under Cd stress. Environ. Sci. Pollut. Res. 24, 21938–21947.

Ros, R., Cascales-Minana, B., Segura, J., Anoman, A.D., Toujani, W., Flores-Tornero, M., et al., 2012. Serine biosynthesis by photorespiratory and non-photorespiratory pathways: an interesting interplay with unknown regulatory networks. Plant Physiol. 15, 707–712.

Rushimisha, I.E., Wang, W., Li, Y., Li, X., 2023. Translocation of nitrate in rice rhizosphere and total nitrogen uptake improvement under interactive effect of water and nitrogen supply. Commun. Soil Sci. Plant Anal. 54 (3), 378–391.

Ruta, D., Irena, P., Gvidas, S., Viktoras, P., Sandra, G., 2011. The effect of amino acids fertilizers on winter wheat grain productivity and quality. Proc. Int. Sci. Conf. Rural Dev. 5 (2), 178.

Sadak, M.S., Abdelhamid, M.T., 2015. Influence of amino acids mixture application on some biochemical aspects, antioxidant enzymes and endogenous polyamines of *Vicia faba* Plant grown under seawater salinity stress. Gesunde Pflanz. 67, 119–129.

Shafeek, M.R., Helmy, Y.I., Shalaby, M.A.F., Omer, N.M., 2012. Response of onion plants to foliar application of sources and levels of some amino acid under sandy soil conditions. J. Appl. Sci. Res. 8 (11), 5521–5527.

Shahrajabian, M.H., Cheng, Q., Sun, W., 2022. The effects of amino acids, phenols and protein hydrolysates as biostimulants on sustainable crop production and alleviated stress. Recent Pat. Biotechnol. 16 (4), 319–328.

Sharma-Natu, P., Ghildiyal, M., 2005. Potential targets for improving photosynthesis and crop yield. Curr. Sci. 88 (12), 1918–1928.

Smirnova, I., Sadanov, A., Baimakhanova, G., Faizulina, E., Tatarkina, L., 2022. Metabolic interaction at the level of extracellular amino acids between plant growth-promoting rhizobacteria and plants of alfalfa (*Medicago sativa* L.). Rhizosphere 21, 100477.

Sokol, N.W., Slessarev, E., Marschmann, G.L., Nicolas, A., Blazewicz, S.J., Brodie, E.L., et al., 2022. Life and death in the soil microbiome: how ecological processes influence biogeochemistry. Nat. Rev. Microbiol. 20, 415–430.

Song, P.P., Jia, Y.H., Qu, X., Cao, L., Li, Y.Y., 2007. Detection of free amino acids in plant under the stress by exogenous amino acids. Lett. Biotechnol. 18, 964–967.

Talat, A., Nawaz, K., Hussian, K., Bhatti, K.H., 2013. Foliar application of proline for salt tolerance of two wheat (*Triticum aestivum* L.) cultivars. World Appl. Sci. J. 22 (4), 547–554.

Trovato, M., Funck, D., Forlani, G., Okumoto, S., Amir, R., 2021. Editorial: amino acids in plants: regulation and functions in development and stress defense. Front. Plant Sci. 12, 772810. Available from: https://doi.org/10.3389/fpls.2021.772810.

Vernieri, P., Borghesi, E., Ferrante, A., Magnani, G., 2005. Application of biostimulants in flating system for improving rocket quality. J. Food Agric. Environ. 3, 86–88.

Wu, H., de Graaf, B., Mariani, C., 2001. Hydroxyproline-rich glycoproteins in plant reproductive tissues: structure, functions and regulation. Cell. Mol. Life Sci. 58, 1418–1429.

Yakhin, O.I., Lubyanov, A.A., Yakhin, I.A., 2016. Biostimulants in agrotechnologies: problems, solutions, outlook. Agrochem. Her. 1, 15–21.

Yang, H., Postel, S., Kemmerling, B., Ludewig, U., 2014. Altered growth and improved resistance of *Arabidopsis* against *Pseudomonas syringae* by over expression of the basic amino acid transporter AtCAT1. Plant Cell Environ. 37, 1404–1414.

Yang, Q., Zhao, D., Zhang, C., Wu, H., Li, Q., Gu, M., 2018. A connection between lysine and serotonin metabolism in rice endosperm. Plant Physiol. 176, 1965–1980.

Yassen, B.T., 2001. Basics of Plant Physiologe DrAsharq Printing. University of Qatar.

Yoo, H., Widhalm, J., Qian, Y., 2013. An alternative pathway contributes to phenylalanine biosynthesis in plants via a cytosolic tyrosine:phenylpyruvate aminotransferase. Nat. Commun. 4, 2833.

Zahir, Z.A., Asghar, H.N., Akhtar, M.J., Arshad, M., 2005. Precursor (L-tryptophan)-inoculum (*Azotobacter*) interaction for improving yields and nitrogen uptake of maize. J. Plant Nutr. 28 (5), 805–817.

Zaki, N.M., Hassanein, M.S., Gamal El-Din, K.M., 2007. Growth and yield of some wheat cultivars irrigated with saline water in newly cultivated land as affected by biofertilization. J. Appl. Sci. Res. 3 (10), 112–1126.

Zhang, L., Garneau, M.G., Majumdar, R., Grant, J., Tegeder, M., 2015. Improvement of pea biomass and seed productivity by simultaneous increase of phloem and embryo loading with amino acids. Plant J. 81 (1), 134–146.

Zimmermann, S.E., Benstein, R.M., Flores-Tornero, M., Blau, S., Anoman, A.D., Rosa-Téllez, S., et al., 2021. The phosphorylated pathway of serine biosynthesis links plant growth with nitrogen metabolism. Plant Physiol. 186 (3), 1487–1506.

CHAPTER

9

Use of seaweed extract-based biostimulants in plant growth, biochemical constituents, and productions

Aarushi Gautam[1], Akansha Chauhan[1], Arundhati Singh[2], Shreya Mundepi[1], Manu Pant[1] and Azamal Husen[1,3,4]

[1]Department of Biotechnology, Graphic Era (Deemed to be University), Dehradun, Uttarakhand, India [2]School of Agriculture and Environment, Bentley Perth Campus, Curtin University, Perth, Western Australia, Australia [3]Department of Biotechnology, Smt. S. S. Patel Nootan Science & Commerce College, Sankalchand Patel University, Visnagar, Gujarat, India [4]Wolaita Sodo University, Wolaita, Ethiopia

9.1 Introduction

Biostimulants are best known as compounds that promote plant growth when used even at low concentrations (Goatley and SchmidtMc, 1991). In 2012, the European Biostimulant Industries re-defined biostimulants as substances that promote natural processes when applied to plant parts like the rhizosphere by stimulating nutrient uptake, improving crop quality by making it more tolerant to abiotic stress, improving nutrient efficiency, etc. (Ricci et al., 2019). The increase in demand for natural and eco-friendly products for plant growth has resulted in increasing interest in the usage of biostimulants (Bulgari et al., 2015). These compounds stand apart from agrochemicals in that they neither provide disease resistance nor offer a direct nutrient to the plant as other fertilizers (Drobek et al., 2019). These are instead used as supplements with the involvement of nonliving products like seaweed (SW), nitrobenzene, and gibberellic acid, or with living products like plant growth-promoting rhizobacteria (Mire et al., 2016). With the world's population increasing by an estimated 10 billion by 2050, the demand for crop yield and agriculture footprint has increased tremendously. Biostimulant, therefore, is the best

Biostimulants in Plant Protection and Performance
DOI: https://doi.org/10.1016/B978-0-443-15884-1.00022-1

alternative to enhance crop productivity with minimal chemical interference in plant physiology (EU, 2019; Rouphael and Colla, 2020).

In this scenario, SW extracts have emerged as an important category of biostimulants by showing effective and positive responses in many plants. The benefits include enhanced plant growth, germination rate, chlorophyll synthesis, fruit quality, and postharvest shelf life (Calvo et al., 2014; Goni et al., 2016). SWs are available in red (Rhodophyta), green (Chlorophyta), and brown (Phaeophyta) types. *Phaeophyceae* brown SW extracts have been known to have immense agricultural applications (Goni et al., 2016), while SWs with essential usage in food and industries are *Ascophyllum nodosum, Macrocystis pyrifera,* and *Durvillea potatorum* (Khan et al., 2009). Polyphenols, polysaccharides, betaines, amino acids, and vitamins have been recognized as SW compounds that act toward plant defense mechanisms and as growth promoters (MacKinnon et al., 2010; Rengasamy et al., 2015). It is envisioned that more benefits of SW as a biostimulant can be assessed for the betterment of real-world producers (Samuels et al., 2022). This chapter provides a general overview of the uses of SW extract-based biostimulants in plant growth, their biochemical constituents, production, and characterization. The information provided will enhance our understanding of the concept and optimal utilization of SW as biostimulants for plants.

9.2 Role of biostimulants in plant growth

Biostimulants are biologically derived substances or microorganisms that promote plant growth and development. They enhance nutrient uptake efficiency, increase crop yield, and also help plants cope with abiotic challenges like heat, cold, salinity, and drought. They can either be derived from natural sources, like SW or certain types of bacteria, or they can be synthetic. Biostimulants are available in a range of formulations, made from natural raw materials such as seeds, leaves, and roots, to facilitate their storage, transportation, and ease of handling, maximizing their effectiveness. A biostimulant can be applied in different forms, such as powders, granules, or liquids, and can be used either alone or in conjunction with other products, like pesticides or fertilizers. Proper delivery and application of biostimulant products are essential to ensuring optimal results. Without formulation, biostimulants in plant extracts can deteriorate and volatilize quickly under environmental conditions (Borges et al., 2018). The biostimulants containing humic substances and nitrogen compounds are applied directly to the soil, while various plants and SW extracts are generally applied in foliar form due to their physical properties, such as being sticky, oily, gummy, greasy, or solid, and their low solubility in water (Drobek et al., 2019).

Unlike traditional fertilizers, which directly supply nutrients to plants, biostimulants improve plants' capacity to use the nutrients they already contain (Drobek et al., 2019). Ideally, a biostimulant formulation should be biologically effective, cost-friendly, and have manufacturing convenience. Conventional biostimulants are available in dustable and wettable powder, emulsifiable concrete, and soluble liquids, which are risky to both human health and the environment. New formulations are being developed worldwide to improve safety, reduce pollution and toxicity, increase effectiveness, and be labor-saving and cost-friendly. Research and development in this field include creating water-dispersible granules, concentrated emulsions, microemulsions, controlled release, etc. (Hazra et al., 2019).

9.3 Seaweed as a biostimulant

9.3.1 Introduction to seaweed

SW falls under the kingdom of Protista and is a macroscopic multicellular algae that is devoid of roots, stems, flowers, etc. SWs grow by attaching themselves to any hard substrata or rocky mountains (McHugh, 2003). The whole plant consists of the stipe, holdfast, and blade and is generally termed a thallus. SWs are found all over the globe's coastline in climatic zones such as warm, polar, and temperate zones. Approximately 10,000 species of SWs are known to humankind, but they account for only 10% of marine productivity globally. Some species, like brown algae of the *Sargassum* genus, float freely, while other are found in the benthic region of the water bodies, tethered to solid substrates using holdfasts. Kelps and fucoids can be easily recognized and are well-known SWs to casual onlookers on the shoreline. SWs prove to be a critical part of marine ecosystems as they feed and house various marine species. Out of all the species known of SWs, only some hold great importance in agricultural uses such as mulches, manures, and animal and human nutrition (Rönnbäck et al., 2007; Khan et al., 2009; Rayorath et al., 2009; Craigie, 2011).

SWs are categorized based on the pigmentation discovered in them. The three categories are *Rhodophyta* (red), *Ochrophyta* (previously known as *Phaeophyta*) (brown), and *Chlorophyta* (green). Presently, 4000 red, 900 green, and 1500 brown SW species are known to mankind. Around 221 species of SW (32 Chlorophytes, 64 *Phaeophyceae*, and 125 Rhodophytes) are currently produced worldwide; 145 of these species are used to generate various foods, while hydrocolloid is synthesized from around 101 species (Nayar and Bott, 2014; Sultana et al., 2012). Numerous SWs are claimed to have attributes that boost plant growth; as a result, they have a wide range of long-lasting benefits in horticulture and agriculture as ecological manures and fertilizers (Craigie, 2011). Although SW has been utilized in agriculture since the dawn of time, SW extracts have only recently gained popularity and acceptance as "Plant Biostimulants" (Battacharyya et al., 2015).

Algal formulations are used as biostimulants as they are rich in a variety of beneficial compounds, including polysaccharides, proteins, PUFAs, polyphenols, pigments, minerals, PGRs, and antioxidants (Chojnacka et al., 2012). SW extract-based biostimulants also contain growth-promoting hormones and trace elements that stimulate plant growth (Sivasankari et al., 2006). They are one of the most promising groups of biostimulants that have been shown to improve nutrient uptake, higher yield, strengthen plant roots, germination rate, the quality of fruit, tolerance against stresses, and reduce seed dormancy (Ali et al., 2019). Chickpea plants that were given SW extract before being exposed to salinity conditions had a noticeable boost in vegetative growth as compared to untreated plants (Abdel et al., 2017). The growth of wheat and rice is also enhanced in a saline environment when SW extracts are applied (Zou et al., 2018; Liu et al., 2019).

Brown, green, and red algae are widely used for SW extract preparation. *Ascophyllum nodosum*, *Ecklonia maxima*, and *Macrocystis pyrifera* extracts are mostly used to prepare bioproducts (Gupta et al., 2011). The most extensively studied SW for plant biostimulants is brown SW, *A. nodosum* (Chojnacka et al., 2012). According to a recent study reported by Santaniello et al. (2017), it has been seen that the extract of *A. nodosum* reduces drought stress in soybeans by changing the physiology and expression of genes related to stress. Besides *A. nodosum*, *F. serratus* extracts are also known to contain a higher phenolic content (Audibert et al., 2010; Balboa et al., 2013). Other SWs like *Phaeophyceae* and *Durvillea potatorum* are also known to have numerous agricultural and industrial uses (Khan et al., 2009; Goni et al., 2016).

9.3.2 Characteristics of seaweed to be used as a biostimulant

SWs are abundant with compounds such as nitrogen, phosphorus, potassium, calcium, phenolic compounds, polysaccharides, and several organic and inorganic compounds, which make them highly suitable to be used as a plant biostimulant and help in enhanced plant growth. Some of the explored properties of SW extracts are described below.

9.3.2.1 Enhanced plant nutrient uptake

The nutrient uptake of plants occurs mainly through roots or the foliage surface, and SW extracts are recognized to amend the soil properties as well as the root structure for better uptake of nutrients. A strong bridge polymeric matrix is found in SW extracts, which are proven to ameliorate the water-retaining ability of the soil (Verkleij, 1992; Lattner et al., 2003; Mzibra et al., 2021), thereby prompting root development and microbial activity in the soil (Chen et al., 2003).

This was validated by different studies on the effect of SW *E. maxima* extract on mung bean and tomato, wherein SW acted as a biostimulant in enhancing root growth (Finnie and Van Staden, 1985; Crouch and Van Staden, 1991). SW isolates also have an impact on genes that regulate the nutrient uptake of the plant. Isolates of *A. nodosum* have been shown to amplify the growth in lateral shoots and boost nitrogen assimilation by enhancing the activity of the NRT1.1 nitrate transporter gene, which further upgrades the auxin transport and nitrogen sensing capacity of the plants (Krouk et al., 2010; Castaings et al., 2011). Brown SW isolates have been successfully applied to the foliage of the grapevine for improved copper intake (Turan and Köse, 2004), to the lettuce for iron uptake (Crouch et al., 1990), and to the cabbage for calcium uptake (Kotze and Joubert, 1980). It has been reported that a commercial product produced from brown SW, when applied to tomatoes, enhanced the mineral nutrient content of P, K, Ca, Zn, N, and Fe. (Dobromilska et al., 2008). As reported by Spinelli et al. (2010), *Cajanus cajan* was given the treatment of *Sargassum wightii* in solid and liquid fertilizer form, and the content of total sugar, chlorophyll, amino acid, carotenoid, lipid, and protein increased significantly.

Billard et al. (2014) reported that rapeseed plants treated with *A. nodosum* extract displayed an improved ability to uptake nitrogen and sulfur. The transcription experiments showed that enhanced nutrient uptake was the result of overexpression of the BnNRT 1.1/BnNRT 2.1 and BnSultr 4.1/BnSultr 4.2 genes that code for root transporters involved in the acquisition of nitrogen, sulfur, and iron. It was reported that SW extracts have an impact on important hormonal biosynthesis genes. The effect was seen on pepper and tomato plants, and the results showed that genes involved in the biosynthesis of gibberellins *(Ga_2 O_x)*, auxins *(IAA)*, and cytokinins *(IPT)* were upregulated, and these genes are said to be linked to increased plant growth (Ali et al., 2019, 2020).

9.3.2.2 Abiotic stress tolerance

Extreme temperatures, water logging, salinity, drought, wind, etc. are some of the abiotic strains that affect the yield and productivity of crops (Kobayashi et al., 2008). Intensive agriculture practices also have a significant influence on the development of crops. Among all the abiotic stresses, salinity alone can affect half the irrigated land (Zhu, 2000). The availability of strategies that can reverse the impact of abiotic strains is limited, and the genetic intricacy of abiotic stress in plants is one of the major drawbacks. This drawback is overcome by the use of SW extracts to lower the impact of abiotic stress on crops. In the case of grapes, freezing tolerance has been reported to be enhanced by the application of an isolate of *A. nodosum*. The extract helped in the reduction of osmotic potential in the leaves. The osmotic capacity of treated plants increased to 1.57 MPa after nine days of treatment with SW extract, while the untreated plants had a 1.51 MPa osmotic potential (Wilson, 2001). Freezing tolerance reportedly increased in barley (Dalal et al., 2019) and *A. thaliana* (Ganesan et al., 2015) with the application of SW extract.

According to Patel et al. (2018), the extract of *Kappaphycus alvarezzi*, when applied to several wheat varieties under drought and salinity stress, resulted in the enhancement of chlorophyll content, water content in the tissues, increased carotenoids content, and longer shoots and roots. The extract also significantly decreased electrolyte loss, and lipid peroxidation decreased the $K+/Na+$ ratio while also raising the Ca concentration, which minimized ionic inequality. Moreover, osmoprotectants such as amino acids and total protein were stored by treated wheat plants. It was reported that polysaccharides extracted from SW *Grateloupia filicina* were able to alleviate salt stress from the rice plant during the seed germination stage and were able to produce seed stimulation when subjected to a saline environment (Liu et al., 2019).

9.3.2.3 Biotic stress tolerance

SW extracts are known to be very effective in shielding plants from biotic stress-like attacks by fungi, bacteria, and viral pathogens. As signal molecules from pathogens or the host plant, referred to as elicitors, are detected, a defense response is triggered (Hernández-Herrera et al., 2014). According to Lizzi et al. (1998), an isolate of *A. nodosum*, when used as a spray on the leaves of capsicum and grape, reduced the infection by *Phytophthora capsica* and *Plasmopara viticola*, respectively. It has been reported (Radwan et al., 2012) that adding SW extracts to the soil decreased the quantity of root-knot nematodes on the roots of the tomato plant as well as lowered the amount of *M. incognita* larvae and galls. Extracts from *Halimeda tuna*, *Spatoglossum variabile*, and *Melanothamnus afaqhusainii* are capable of controlling nematode galls on roots, nematode's entry into roots, and fungi *Fusarium solani* and *Rhizoctonia solani* in the roots of tomatoes (Sultana et al., 2012). It was observed in pepper plants that the drenching of soil in SW extract lowered the *Verticillium* wilt of pepper by enhancing pathogen resistance and improving plant health (Rekanovic et al., 2010). Baloch et al. (2013) observed that extracts from *Melanothamnus afaqhusainii*, *Spatoglossum variabile*, and *Stokeyia indica* subdued the impact of root-rotting nematodes and fungi *Meloidogyne incognita* and *Fusarium solani* in eggplant. SW extracts are now considered an alternative for pest and disease control in the *Solanaceae* family due to their impact as plant protectants (Pohl et al., 2019).

9.3.2.4 Physiology and metabolism of plants

Recent studies have reported that SW extracts affect the physiology of plants and their metabolism, changing the metabolome and global transcriptome profile of the plant (Nair et al., 2012; Jannin et al., 2013). Secondary metabolites such as flavonoids are beneficial for plant growth and their relationship with environmental constituents like biotic stress, abiotic stress, and response to UV light. SW extract is used to increase the production of *CHI* (Chalcone isomerase), a major enzyme that produces phenylpropanoid plant defense compounds and flavanol precursors (Battacharyya et al., 2015). It has been reported that the use of SW extract results in an increase in chlorophyll content in strawberries, grapevine, and many other crops (Blunden et al., 1997; Mancuso et al., 2006; Sivasankari et al., 2006; Spinelli et al., 2010; Fan et al., 2013; Jannin et al., 2013). Commercial SW extract has also been applied to the apple trees (Fuji variety), which resulted in tackling the alternate bearing of fruits in nutrient-lacking conditions but not in the typical nutrient management conditions (Spinelli et al., 2009).

9.3.2.5 Phytohormone resembling activity

Phytohormones are crucial for plant development and growth. The phytohormone-like activity of SW extracts has been explored in plants. It was observed that at lower concentrations, SW extracts enhanced the growth of the plant, while at higher concentrations, the extracts proved to be inhibitory for growth (Provasoli and Carlucci, 1974; Khan et al., 2009). It has been reported (Provasoli and Carlucci, 1974; Stirk et al., 2003; Khan et al., 2009; Wally et al., 2013) that brown SW extracts consist of phytohormones like cytokinins, indole acetic acid, gibberellic acid, abscisic acid, and polyamines. Currently, the known phytohormones discovered in SW extracts might have a direct or indirect link to the growth-promoting capabilities of the extract. Depending on the extraction method employed and processing methods, the particular phytohormone-like function in the SW isolate can vary. For instance, extracts prepared fresh from *E. maxima* possessed auxin-like as well as cytokinin-like properties, while after a long period of storage, the auxin-like property was eliminated and the cytokinin-like property altered significantly (Stirk et al., 2004).

9.3.2.6 Plant-microbe interaction

The SW extract applied to the roots and leaves of the plant has shown a positive effect on microbial interaction with plants. Plant-microbe interaction plays an important role in plant growth and development (Ali et al., 2021). Wang et al. (2016) showed that the implementation of a biostimulant extracted from *Lessonia flavicans* and *Lessonia nigrescens* in the transplanted soil of *Malus hupehensis* seeds considerably enhanced the soil in urease, invertase, phosphatase, and proteinase enzymes in comparison to the control. It was also observed by the T-RFLP studies that extracts drastically alter the fungus/bacterial community of the soil by increasing the ratio of fungi/bacteria to the rhizosphere of *Malus* spp.

9.3.2.7 Improved production and standard

SW extracts have been used to raise the standard and lifespan of the product. According to Fan et al. (2011, 2013), the marketed extracts of *A. nodosum* raised the nutritional value of the leaf, preservation quality, and flavonoid content of spinach. The content of fats in olive oil was altered by the SW extract treatment; linolenic acid and oleic acid concentrations significantly increased while palmitoleic acid, linoleic acid, and stearic acid concentrations declined. In addition to increasing productivity and quality, using SW isolates prolonged the storage span of pears and avocados (Blunden et al., 1978; Kamel, 2014).

The latest study (Ashour et al., 2023) has reported that the application of SW extracts True-Algae-Max (TAM) along with 50% NPK fertilizers in strawberry plants showed improved yield and enhanced nutritional content of the plant. The TAM was used at different concentrations (0, 50, and 100%) in combination with NPK fertilizer. Results showed that TAM at 50% concentration worked best to improve the growth yield, upgrade chlorophyll content, plant fresh weight, enhance root and shoot length, increase anthocyanin content, etc.

The overall benefits of using SW extracts on crops and fruits are compiled in Tables 9.1 and 9.2 and Fig. 9.1.

9.3.3 Mechanism of action of seaweed biostimulant on plants in abiotic and biotic stress

The effectiveness of SW isolates and their mode of activity in plants remains to be discussed elaborately. Phycocolloids are present in the SW extracts (absent in land plants) which provide them with a specific mode of action (Deolu-Ajayi et al., 2022).

The impact of the application of *A. nodosum* isolate in the form of foliar spray on the *Paspalum vaginatum* cultivar under irregular intervals of irrigation and saline stress

TABLE 9.1 Effect of seaweed isolates on the crops.

	Observed effects of seaweed						
Crops	Enhanced germination rate and seedling vigor	Increased root and shoot development	Increased chlorophyll	Abiotic stress resistance	Increased nutrient uptake	Improved production	Pathogen resistance
	Bean	Lettuce	Tomato	Soybean	Onion	Okra	Carrot
	Onion	Gram mung bean	Wheat	Tomato	Soybean	Cauliflower	Broccoli
	Rice	Sweet pepper	Sweet pepper	Potato	Rice	Cucumber	Cucumber
	Maize	Tomato	Okra	Spinach	Onion	Eggplant	Onion
	Tomato					Pepper	Tomato
						Tomato	Sugarcane
						Lettuce	

TABLE 9.2 Effects of seaweed on fruits.

	Observed effects on the fruit of seaweed application						
Fruits	Improved fruit quality	Improved yield	Abiotic stress resistance	Pathogen resistance	Improved nutrient uptake	Improved nutrient content	Increased chlorophyll content
	Strawberry	Strawberry	Avocado	Strawberry	Kiwifruit	Strawberry	Strawberry
	Pomegranate	Apple	Grapefruit	Citrus	Olive	Peach	Olive
	Newhall navel orange	Date palm	Apple	Cherry		Olive	Peach
		Grapevine	Pomegranate			Mango seedling	Banana
		Kiwifruit	Loquat			Pecan nut	Grapevine
		Pomegranate	Tangerine orange				Apricot

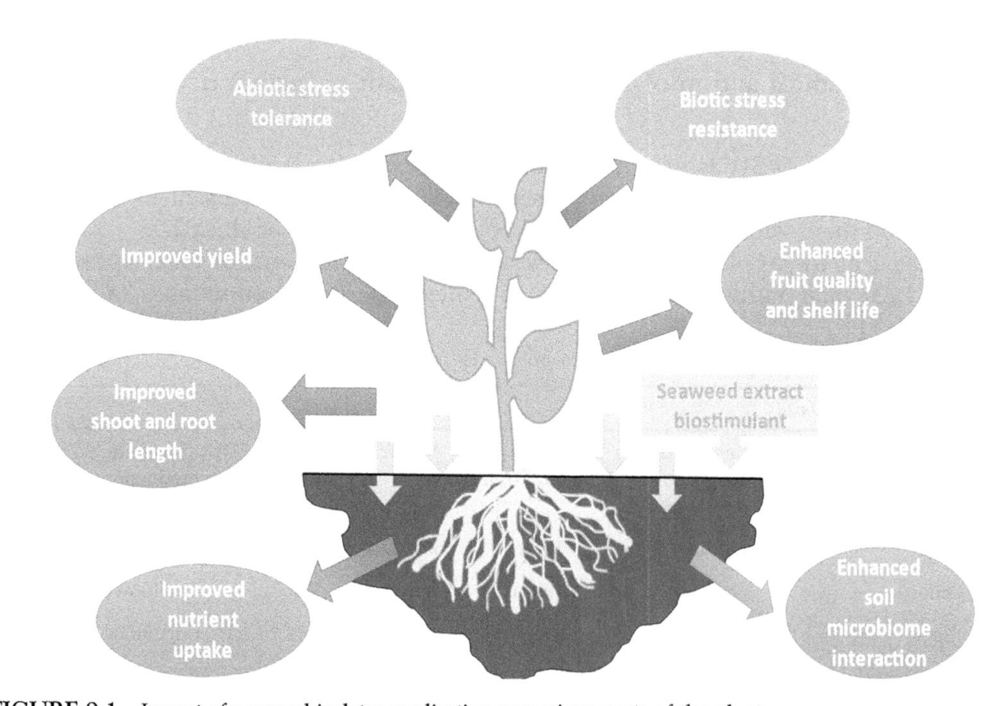

FIGURE 9.1 Impact of seaweed isolates application on various parts of the plant.

(Elansary et al., 2017) showed that there was an enhancement in lipid peroxidation (confirmed by DPPH (2,2′-diphenyl-1-picrylhydrazyl) and linoleic acid analysis). Also, a significant increase in the APX (ascorbate peroxidase), SOD, and CAT (catalase) activities was observed which indicated the boosted antioxidant defense system resulting in the ROS (H_2O_2) reduction in the SW-treated plants.

In a study on *Arabidopsis thaliana*, the impact of treatment with *A. nodosum* was observed in a saline environment. The physiological and transcriptomics assay results showed that plants dosed with subfractions (ethyl acetate) of *A. nodosum* downregulated 91 and 262 genes on days 1 and 5, respectively, and an upregulated 184 and 257 genes on days 1 and 5, respectively. Amongst all the upregulated genes, abiotic strain genes made up 2.2% on first day and rose up to 6% on fifth day (Sharma et al., 2019). The role of the SW isolate of *A. nodosum* on asparagus was observed in a salty environment by Al-Ghamdi and Elansary (2018). The extract was given weekly at a concentration of 7 mL/L under saline conditions (2000 and 400 ppm NaCl). The results showed that ANN1, ANN2, PIP1 aquaporin, water-controlling-related genes, and biologically working molecule-metabolism-related genes CHS and P5CS1 have upregulated redox-related genes. GPX3 and APX1 were found to be upregulated in comparison to the control. The experiment also revealed that salinity stress can be eliminated by improving the antioxidant activities of proline and phenols.

Under drought stress, the treatment of *Kappaphycus alvarezii* sap (K sap) in wheat plants upregulated the ROS-scavenging genes TdCAT, TdSOD, transcript expression of transcription factor TaWRKY10, which is abiotic stress-responsive, and WCK-1, which is a stress signaling cascade gene. The researchers concluded that the treatment of K sap resulted in the induction of phytohormones and also resulted in the upregulation of transcription factors and genes under abiotic stress (Patel et al., 2018).

The result of the application of SW extracts to the plant has a positive effect on controlling the pathogens that attack the plant and helps in upgrading the plant defense system. For example, in a study conducted by Sahayaraj et al. (2011), it was observed that brown SW isolates, particularly benzene and chloroform extracts obtained from *Padina pavonica*, showed the capacity to drastically shorten or lengthen the nymphal development period as well as display nymphicidal activity by interfering with the physiology of cotton pests (*Dysdersus cingulatus*). Also, the anti-juvenile hormone component of the SW was thought to be responsible for the above effect, which prevented pest development at the blastokinesis or germ bund stages. SW extracts obtained from various macroalgae species such as *P. gymnospora*, *O. secundiramea*, *L. dendroidea*, *H. musciformis*, and *P. capillacea* extracted from the Brazilian Manguinhos beach were tested for their antifungal properties. The results displayed that the isolates were successful in restricting the action of anthracnose induction in banana and papaya plants by *Colletotrichum gloeosporioides*. After the GC-MS analysis, it was deduced that the antifungal property of the extracts was imparted due to the presence of quercetin, halogenated terpenes, and fatty acids (Machado et al., 2014). The percentage of galls was reduced significantly (76%) and the overall population of nematodes was also reduced dramatically after the application of the powder extract from *Ulva lactuca* when administered as soil treatment at a rate of 5 g/kg. This substantially reduced root nematode infestations in the banana plants and it was directly linked to the existence of phenolic compounds in the SW extract (El-Ansary and Hamouda, 2014). When the extracts of *Acanthophora spicifera, A. nodosum, and Sargassum vulgare* were sprayed on tomato and sweet pepper in both soil and greenhouse conditions, the result was interpreted a decreased infection by the fungus *Alternaria solani* and *Xanthomonas campestris* pv. *vesicatoria* by upregulating the genes of ET, SA, and JA defense signaling, including *ETR-1*, *PR1-a*, and *PinII*, respectively, along with increasing the enzyme defense system (Ali et al., 2019, 2020).

9.4 Preparation of seaweed extracts

Various commercial methods are used to prepare SW-based biostimulants (Fig. 9.2), where the extracts are either in liquid or powdered form (Kadam et al., 2013; Michalak and Chojnacka, 2015). Some extraction methods include water-based, acid hydrolysis, alkaline hydrolysis, enzyme-assisted extraction, supercritical fluid extraction, and pressurized liquid extraction, which are mentioned in the figure along with the effect of their target application (Shukla et al., 2018).

Water-based extraction involves harvesting the biostimulant compounds by blending, followed by hydration of the dried form of SW meal in water (Sharma et al., 2014). Then the separation is based on filtration, where the biostimulants are rich in activities with phytohormones (Blunden and Wildgoose, 1977; Crouch and van Staden, 1993).

Acid hydrolysis is done using acids like sulfuric acid at 40°C–50°C for 30 minutes (Sharma *et al.*, 2014). This method mainly extracts sulfated polysaccharides as it removes complex phenolic compounds (Ale et al., 2012; Flórez Fernández et al., 2018).

Alkaline hydrolysis is a widely used process in industries, and it gathers the extract from *A. nodosum* (Flórez Fernández et al., 2018). This method uses alkaline solutions like NaOH and KOH at temperatures between 70°C–100°C. This process breaks down the complex polysaccharides into simpler forms. The compounds produced are generally not

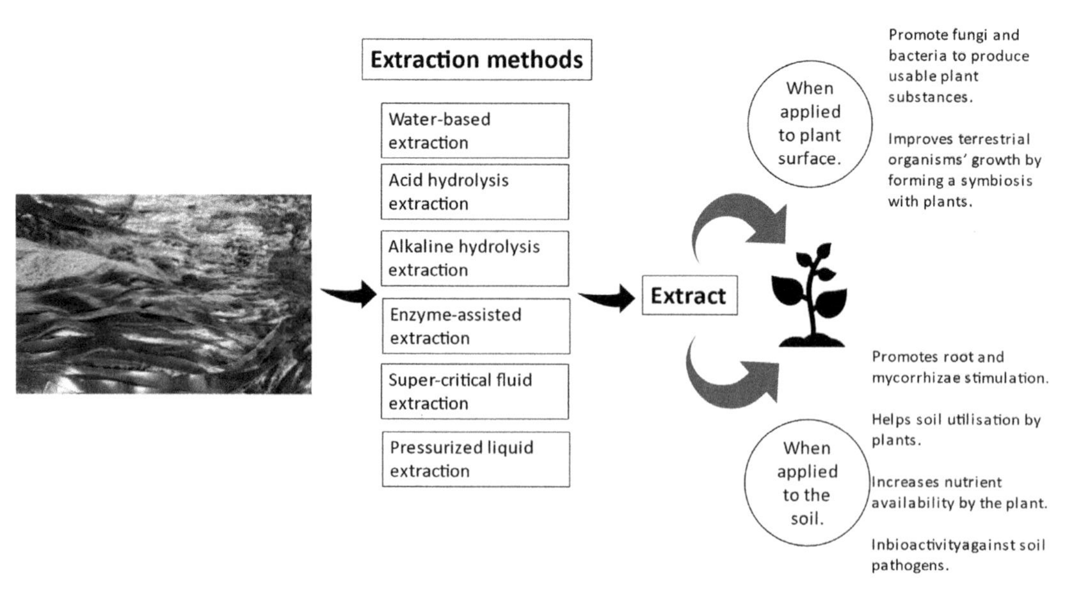

FIGURE 9.2 Methods of seaweed extract preparation.

present in the SW, as the resultant forms a bond between the hydrolysis chemical and the tissues of brown SW (Craigie, 2011).

Supercritical fluid extraction, which extracts the chemicals while preventing the parent SW material from being degraded by biochemical and thermal processes, is an environmentally beneficial form of extraction (Herrero et al., 2010; Michalak et al., 2015; da Silva et al., 2016).

On the other hand, pressurized liquid extraction uses a high pressure of 3.5–20 MPa and a temperature range of 50°C–200°C (Kadam et al., 2013). The high pressure increases the temperature above the boiling point, resulting in the extraction of bioactive compounds (Kadam et al., 2013; Michalak and Chojnacka, 2015).

Last but not least, one environmentally friendly extraction technique that uses no solvents is enzyme-assisted extraction (Kadam et al., 2013). The degradation process determines the efficiency of this process by the enzyme of the complex compounds present in the cell wall of the SW (Wijesinghe and Jeon, 2012; Kadam et al., 2013). The extracts show high antioxidant properties and involve protease and carbohydrate-degradation enzymes.

9.4.1 Characterization of seaweed extracts

Among the prominent substances found in SW extracts are polysaccharides (alginates, laminarin, and galactan), proteins, fatty acids, pigments (chlorophylls, carotenes, and xanthophylls), secondary metabolites (phenolics), and phytohormones (Zhong et al., 2020; Mzibra et al., 2021). These components vary among the three classes of SWs, i.e., phaeophyta, rhodophyta, and chlorophyta.

The characterization of these components can be done based on composition analysis, spectroscopic techniques, and thermal analysis (Table 9.2).

9.4.1.1 Composition analysis

9.4.1.1.1 Proximate analysis

The proximate compositions such as crude protein, moisture content, ash content, and carbohydrate of SW extracts can be characterized by proximate analysis. The quantification of crude fiber content present in the extract can be done by making powder of the freeze-dried plant material and by doing successive hydrolysis with sulfuric acid and sodium hydroxide in a hot plate apparatus. The fat content present in the SW sample can be detected with petroleum ether by a Soxtec system. The moisture content can be detected by drying the powdered plant sample in a hot air oven at 105°C overnight. The sample should be kept at 550°C for 24 hours in a muffle furnace to depict the amount of ash using gravimetric analysis, and the total protein of the sample can be quantified using the Kjeltec system by multiplying the nitrogen content by 6.25.

9.4.1.1.2 Ultimate analysis

The ultimate analysis is a quantitative measurement of various elements such as carbon, nitrogen, hydrogen, and sulfur in SW extracts, which can be done by a CHNS element analyzer.

9.4.1.2 Spectroscopic techniques

The use of spectroscopy to depict the structure of bio-stimulatory compounds and identify particular chemical groups in SW extract has been growing rapidly in recent years. Infrared (IR) and Raman vibrational spectroscopy can provide precise details about the molecular characteristics and structure of materials. The most prominent vibrational method for analyzing plant components until recently was IR spectroscopy. These spectroscopic methods are time-saving, nondestructive, and efficient even with small sample quantities.

9.4.1.2.1 Fourier transform infrared spectroscopy

Fourier transform infrared spectroscopy (FTIR) is a reliable method for the identification and characterization of bio-stimulatory components of SW as a semiquantitative or qualitative technique that is capable of identifying these compounds by stretching and bending bands present in the biomolecule. The spectra of air- or freeze-dried samples can be scanned over a wavelength ranging from 400 to 4000 cm^{-1} at room temperature.

9.4.1.2.2 LC-MS Analysis

The characterization of the phenolic compounds of bio-stimulating plants can be accomplished via LC-MS analysis. A qualitative assessment of the phenolic components such as flavonoids, other polyphenols, and lignans from various SW extracts can be accomplished by LC-MS analysis in both positive and negative ionization modes.

9.4.1.2.3 HPLC Analysis

SW bioactive components can be effectively analyzed using the HPLC method. The method allows for the systematic characterization of the various plant extracts and specifically focuses on the identification of the variable compounds. It can characterize the plant's primary (phytohormones) and secondary (phenolic compounds) metabolites, which help the plant stimulate its defense mechanisms.

9.4.1.3 Thermal analysis

9.4.1.3.1 Thermogravimetric analysis

The thermogravimetric analysis (TGA) can be done in both isothermal and non-isothermal conditions. Thermograms can be used to observe the thermal weight loss features, and various tools can be used to perform quantitative analyses. TGA of SW extracts can be carried out using a thermal analyzer, which evaluates the changes in the specimens in a stable gas environment and at varying cooling and heating rates. To effectively assess the continuous distribution process of different components of SW, the samples should be pyrolyzed at 25°C–800°C at a 10°C/min temperature rating and in a stable nitrogen atmosphere of 100 mL/min (Table 9.3).

TABLE 9.3 Methods and results of previously conducted studies for the characterization of seaweed components.

Seaweed extract	Characterization type	Component identified	Result analysis	Key references
Hypnea japonica, *Hypnea Charoides*, *Ulva lactuca*	Proximate analysis	Protein content	19.0 ± 0.36, 18.4 ± 0.30, and 7.06 ± 0.06, respectively	Wong and Cheung (2000)
Hypnea japonica, *Hypnea Charoides*, *Ulva lactuca*	Proximate analysis	Ash content	22.1 ± 0.72, 22.8 ± 2.23, and 21.3 ± 2.78, respectively	Wong and Cheung (2000)
Sargassum spp.	Ultimate analysis; proximate analysis	C, H, N; moisture, volatile matter, fixed carbon, and ash content	25.35, 3.81, 1.37; 9.8 ± 1.1, 43.9 ± 1.8, 25.1 ± 0.8, and 21.2 ± 1.5, respectively	Ali and Bahadar (2017)
Sargassum spp.	Proximate analysis	Crude protein, total carbohydrate, and crude fiber	10.25, 41.81, and 9.84 ± 0.07, respectively	Borines et al. (2013)
Laminaria, *Ascophyllum nodosum*	FTIR analysis	Alginic acid, fucoidans, mannuronic acid, and glucuronic acid	Strong band at 1612–1547, 1230, 814, 732 cm^{-1} respectively	Ertani et al. (2018)
Sargassum thunbergia, *Mastocarpus stellatus* and *Ulva* sp.	FTIR analysis	Amino acids, aliphatic compound, and fucoidan	Broad bands at 3550–3200, 2930, and 1257–1254 cm^{-1}, respectively	Yang et al. (2021)
Caulerpa lentillifera, *Gracilaria coronopifolia* and *Chaetomorpha linum*	TGA	Carbohydrate and protein	Peak at 180°C–270°C and 320°C–450°C	Ong et al. (2019)
U. fasciata, *U. lactuca*, *U. linza*, *U. reticulata*, *U. taeniata*, *M. oxyspermum*	HPLC	Abscisic acid	19.68 ± 2.15, 34.85 ± 1.75, 39.74 ± 5.68, 68.28 ± 2.09, 38.5 ± 1.29, and 11.35 ± 0.48, respectively	Gupta et al. (2011)

9.5 Commercial products of seaweed-based bioinoculant

The increasing demand for organic products as alternatives to synthetic chemical inputs is boosting the growth of the global biostimulant industry. By 2025, the biostimulants market is anticipated to reach a value of $4.14 to $4.9 billion (Kumar et al., 2020). The SW industry is estimated to be worth $5.5–6 billion a year, with food products accounting for the majority of that amount at $5 billion. Other products, including hydrocolloids, fertilizers, and additives for animal feed, make up the remaining billion dollars. Annually, the industry consumes 7.5–8 million tonnes of wet SW. Both Indian companies and MNCs have increased their involvement in the biostimulant product segment in recent years. One example is the Canadian company Acadian Seaplants Ltd., which developed SW-based plant stimulants to help Indian farmers overcome stress and increase their agricultural outputs. The company is expanding its presence in the Indian market with its wide range of products. In the coming years, the Indian market for biostimulants is anticipated to expand rapidly as farmers become more aware of the benefits of these products.

A commercial SW liquid extract, TAM, is prepared by using three species named *Ulva lactuca*, *Jania rubens*, and *Pterocladia capillaries*. The study has shown that it has enhanced the morpho-agronomic and bioactive qualities of strawberry plants suitable for bioethanol production (Ashour et al., 2023). The extract of brown SW derived from *Fucus* and the *Ascophyllum* genus (containing phenolic compounds) used to stimulate the symbiosis between Mycorrhiza al fungus and Rhizobium is patented and can be used as fertilizer or treatment (Briand and Salamagne, 2018). High concentrations of humic-like polyphenols, primarily derived from brown SW, can be found in the extracts produced by Maxicrop and Acadian Seaplants (Craigie, 2011). For instance, *A. nodosum* from Norway and Nova Scotia contains 15%–25% extractable polyphenols with high molecular weight (Glombitza et al., 1977).

9.6 Methods of application of seaweed-based biostimulants on plants

The method in which SW extracts are applied is crucial to their effectiveness and plant responses. They can be applied through foliar and root applications, or a combination of the two. They can be added to soil or growing medium using methods such as fertigation, drenching, or dripping (Sangeetha et al., 2015). Due to the rapid contact with plant tissues and quick foliar absorption, foliar applications of extracts are more efficient. Furthermore, soil particles can adsorb extracts, reducing their mobility (Ali et al., 2016; Ali et al., 2020). Additionally, the best plant responses have been observed when these extracts are applied every 10 to 14 days (Arioli et al., 2015). Because SW liquid extracts contain growth-stimulating hormones and trace elements that can be added to the soil or sprayed on seeds to encourage plant growth, their significance in agriculture has expanded as foliar sprays (Sivasankari et al., 2006). Biostimulants are absorbed by plants when administered on the leaves via the cuticle, epidermal cells, and stomata (Fernández and Eichert, 2009); however, when applied as a root drench (drenching technique), it is taken through root epidermal cells and spread throughout the plant via the xylem (Subbarao et al., 2015).

It has been reported that foliar sprays with a concentration of less than or equal to 0.05% v/v of the extract are optimal for the crop, resulting in disease control and increased yield (Ali et al., 2016). The quality of tomato fruit and nutrient uptake were both improved by foliar applications (5.0% concentration) of *K. alvarezii* sap (Zodape et al., 2011). Spraying rice crops with SW extracts at a concentration of 2000 ppm resulted in a notable increase in the weight of 100 grains and produced the highest yield per plant under salt-stress conditions. The use of 12.5 kg/ha SW extract in the soil and a foliar spray of SW liquid improved various aspects of rice growth and production (Deepana et al., 2021). Similarly, the application of liquid SW extracts improved soil quality in foxtail millet (Rathinapriya et al., 2020). The foliar spray of liquid biostimulant *K. alvarezii* sap on rice crops improved crop growth traits, including increased height of the plant, more leaves per plant, and higher grain yield, compared to using the recommended amount of fertilizer (Arun et al., 2019). The application of three sprays of *K. alvarezii* and *Gracilaria edulis* at different doses (2.5%, 5%, 7.5%, and 15%) to the maize crop at various stages, along with a water spray as a control, significantly increased grain yield by 18.5% and 26.0%. This was due to an increase in the number of rows of cobs, cob length, the weight of 100 grains, and improved nutrient uptake (Basavaraja et al., 2018). The uptake of N, P, and K was enhanced by 116%, 113%, and 93% in onions grown under water stress when SW extracts were applied (Almaroai and Eissa, 2020), thus reducing the usage of NPK fertilizers. Hence, SW has the potential to significantly enhance the availability of both macro and micronutrients for crops.

9.7 Conclusion

With increasing food demand and low land availability, SW biostimulants are in great demand to increase the production of crops in the limited land area to meet global food demand. The application of SW isolates as plant biostimulants is a good source of nutrients that enhance the overall development of the plant and is a great alternative for sustainable agriculture. Although the SW isolates are used in small concentrations in the plant, they have far more pronounced effects. Among the three categories of SWs, the most popular extracts are made from brown SW, and out of all the SW extracts, the most commercialized extract is prepared from *A. nodosum*. The SW extracts possess the abilities of plant hormones, which trigger different pathways to upregulate or downregulate genes to enable plants to tolerate abiotic and biotic strains. In the future, there is scope to understand the mode of action of SW extracts on plant physiology under different conditions. Although these extracts have gained popularity among many other plant biostimulants, their results are still inconsistent in several cases due to their broad-spectrum applications. Apart from that, the extraction methods of SW extracts are yet to be fully optimized. The agronomic competence of SW extracts has to be increased because the composition and the collection environment of SW biomass can vary drastically. To overcome these challenges, algal biorefineries are being looked up to as a novel, competent idea that will ensure the mass production of algae under controlled conditions and assist in the standard biostimulant formulation.

References

Abdel, L.A.A.H., Srivastava, A.K., Saber, H., Alwaleed, E.A., Tran, L.S.P., 2017. *Sargassum muticum* and *Jania rubens* regulate amino acid metabolism to improve growth and alleviate salinity in chickpea. Sci. Rep. 7 (1), 1–12.

Ale, M.T., Mikkelsen, J.D., Meyer, A.S., 2012. Designed optimization of a single-step extraction of fucose-containing sulfated polysaccharides from *Sargassum* sp. J. Appl. Phycol. 24, 715–723.

Al-Ghamdi, A.A., Elansary, H.O., 2018. Synergetic effects of 5-aminolevulinic acid and *Ascophyllum nodosum* seaweed extracts on *Asparagus* phenolics and stress related genes under saline irrigation. Plant Physiol. Biochem. 129, 273–284.

Ali, I., Bahadar, A., 2017. Red Sea seaweed (*Sargassum spp.*) pyrolysis and its devolatilization kinetics. Algal. Res. 21, 89–97.

Ali, N., Farrell, A., Ramsubhag, A., Jayaraman, J., 2016. The effect of *Ascophyllum nodosum* extract on the growth, yield and fruit quality of tomato grown under tropical conditions. J. Appl. Phycol. 28, 1353–1362.

Ali, O., Ramsubhag, A., Jayaraman, J., 2019. Biostimulatory activities of *Ascophyllum nodosum* extract in tomato and sweet pepper crops in a tropical environment. PLoS One. 14, e0216710.

Ali, O., Ramsubhag, A., Jayaraman, J., 2020. Phytoelicitor activity of *Sargassum vulgare* and *Acanthophora spicifera* extracts and their prospects for use in vegetable crops for sustainable crop production. J. Appl. Phycol. 33, 639–651.

Ali, O., Ramsubhag, A., Jayaraman, J., 2021. Biostimulant properties of seaweed extracts in plants: implications towards sustainable crop production. Plants 10, 531.

Almaroai, Y.A., Eissa, M.A., 2020. Role of marine algae extracts in water stress resistance of onion under semiarid conditions. J. Soil Sci. Plant Nutr. 20, 1092–1101.

Arioli, T., Mattner, S.W., Winberg, P.C., 2015. Applications of seaweed extracts in Australian agriculture: past, present and future. J. Appl. Phycol. 27, 2007–2015.

Arun, M.N., Kumar, R.M., Nori, S., Singh, A., Tuti, M.D., Srinivas, D., et al., 2019. Effect of seaweed extract as biostimulant on crop growth and yield in rice (*Oryza sativa L.*) under transplanted condition. J. Rice Res. 12 (2), 45–49.

Ashour, M., Al-Souti, A.S., Hassan, S.M., Ammar, G.A.G., Goda, A.M.A.-S., El-Shenody, R., et al., 2023. Commercial seaweed liquid extract as strawberry biostimulants and bioethanol production. Life 13, 85.

Audibert, L., Fauchon, M., Blanc, N., Hauchard, D., Ar Gall, E., 2010. Phenolic compounds in the brown seaweed *Ascophyllum nodosum*: distribution and radical-scavenging activities. Phytochem. Anal. 21 (5), 399–405.

Balboa, E.M., Conde, E., Moure, A., Falqué, E., Domínguez, H., 2013. *In vitro* antioxidant properties of crude extracts and compounds from brown algae. Food. Chem. 138 (2–3), 1764–1785.

Baloch, G.N., Tariq, S., Ehteshamul-Haque, S., Athar, M., Sultana, V., Ara, J., 2013. Management of root diseases of eggplant and watermelon with the application of asafoetida and seaweeds. J. App. Bot. Food Qual. 86 (1), 138–142.

Basavaraja, P.K., Yogendra, N.D., Zodape, S.T., Prakash, R., Ghosh, A., 2018. Effect of seaweed sap as foliar spray on growth and yield of hybrid maize. J. Plant Nutr. 41 (14), 1851–1861.

Battacharyya, D., Babgohari, M.Z., Rathor, P., Prithiviraj, B., 2015. Seaweed extracts as biostimulants in horticulture. Sci. Hortic. 196, 39–48.

Billard, V., Etienne, P., Jannin, L., Garnica, M., Cruz, F., Garcia-Mina, J.M., et al., 2014. Two biostimulants derived from algae or humic acid induce similar responses in the mineral content and gene expression of winter oilseed rape (*Brassica napus L.*). J. Plant Growth Regul. 33, 305–316.

Blunden, G., Jenkins, T., Liu, Y., 1997. Enhanced leaf chlorophyll levels in plants treated with seaweed extract. J. Appl. Phycol. 8, 535–543.

Blunden, G., Jones, E.M., Passam, J.C., 1978. Effects of postharvest treatment of fruit and vegetables with cytokinin-active seaweed extract and kinetin solutions. Botanica Marina 21, 237–240.

Blunden, G., Wildgoose, P.B., 1977. The effects of aqueous seaweed extract and kinetin on potato yields. J. Sci. Food Agric 28 (2), 121–125.

Borges, D.F., Lopes, E.A., Moraes, A.R.F., Soares, M.S., Visôtto, L.E., Oliveira, C.R., et al., 2018. Formulation of botanicals for the control of plant-pathogens: a review. Crop Prot. 110, 135–140.

Borines, M.G., de Leon, R.L., Cuello, J.L., 2013. Bioethanol production from the macroalgae *Sargassum spp.* Bioresour. Technol. 138, 22–29.

Briand, X., Salamagne, S.C., 2018. U.S. Patent Application No. 15/754,741.

Bulgari, R., Cocetta, G., Trivellini, A., Vernieri, P., Ferrante, A., 2015. Biostimulants and crop responses: a review. Biol. Agric. Hortic. 31 (1), 1–17.

Calvo, P., Nelson, L., Kloepper, J.W., 2014. Agricultural uses of plant biostimulants. Plant Soil. 383, 3–41.

Castaings, L., Marchive, C., Meyer, C., Krapp, A., 2011. Nitrogen signalling in *Arabidopsis*: how to obtain insights into a complex signalling network. J. Exp. Bot. 62, 1391–1397.

Chen, S.K., Edwards, C.A., Subler, S., 2003. The influence of two agricultural biostimulants on nitrogen transformations, microbial activity, and plant growth in soil microcosms. Soil Biol. Biochem. 35, 9–19.

Chojnacka, K., Saeid, A., Witkowska, Z., Tuhy, L., 2012. Biologically active compounds in seaweed extracts-the prospects for the applicationAugust Open Conf. Proc. J. 3 (1).

Craigie, J.S., 2011. Seaweed extract stimuli in plant science and agriculture. J. Appl. Phycol. 23, 371–393.

Crouch, I.J., Beckett, R.P., Van Staden, J., 1990. Effect of seaweed concentrate on the growth and mineral nutrition of nutrient-stressed lettuce. J. Appl. Phycol. 2, 269–272.

Crouch, I.J., Staden, J., Van, 1991. Evidence for rooting factors in a seaweed concentrate prepared from *Ecklonia maxima*. J. Plant Physiol. 137, 319–322.

Crouch, I.J., Van Staden, J., 1993. Evidence for the presence of plant growth regulators in commercial seaweed products. Plant Growth Regul. 13, 21–29.

Da Silva, R.P., Rocha-Santos, T.A., Duarte, A.C., 2016. Supercritical fluid extraction of bioactive compounds. Trends Anal. Chem. 76, 40–51.

Dalal, A., Bourstein, R., Haish, N., Shenhar, I., Wallach, R., Moshelion, M., 2019. Dynamic physiological phenotyping of drought- stressed pepper plants treated with "productivity-enhancing" and "survivability-enhancing" biostimulants. Front. Plant Sci. 10, 905.

Deepana, P., Bama, K.S., Santhy, P., Devi, T.S., 2021. Effect of seaweed extract on rice (*Oryza sativa* var. ADT53) productivity and soil fertility in Cauvery delta zone of Tamil Nadu, India. J. Appl. Nat. Sci. 13 (3), 1111–1120.

Deolu-Ajayi, A.O., van der Meer, I.M., van der Werf, A., Karlova, R., 2022. The power of seaweeds as plant biostimulants to boost crop production under abiotic stress. Plant Cell Environ 45 (9), 2537–2553.

Dobromilska, R., Mikiciuk, M., Gubarewicz, K., 2008. Evaluation of cherry tomato yielding and fruit mineral composition after using of Bio-algae S-90 preparation. J. Elem. 13, 491–499.

Drobek, M., Frąc, M., Cybulska, J., 2019. Plant biostimulants: importance of the quality and yield of horticultural crops and the improvement of plant tolerance to abiotic stress—a review. Agron. J. 9 (6), 335.

Elansary, H.O., Yessoufou, K., Abdel-Hamid, A.M.E., El-Esawi, M.A., Ali, H.M., Elshikh, M.S., 2017. Seaweed extracts enhance salam turfgrass performance during prolonged irrigation intervals and saline shock. Front. Plant Sci. 8, 830.

El-Ansary, M.S.M., Hamouda, R.A., 2014. Biocontrol of root-knot nematode infected banana plants by some marine algae. Russ. J. Mar. Biol. 40, 140–146.

Ertani, A., Francioso, O., Tinti, A., Schiavon, M., Pizzeghello, D., Nardi, S., 2018. Evaluation of seaweed extracts from *Laminaria* and *Ascophyllum nodosum spp.* as biostimulants in *Zea mays L.* using a combination of chemical, biochemical and morphological approaches. Front. Plant Sci. 9, 428.

EU, 2019. Regulation of the European Parliament and of the council laying down rules on the making available on the market of EU fertilising products and amending Regulations (EC) No 1069/2009 and (EC) No 1107/2009 and repealing Regulation (EC) No 2003/2003. Off. J. Eur. Union 62, 1–114.

Fan, D., Hodges, D.M., Critchley, A.T., Prithiviraj, B., 2013. A commercial extract of brown macroalga (*Ascophyllum nodosum*) affects yield and the nutritional quality of spinach *in vitro*. Commun. Soil Sci. Plant. Anal. 44, 1873–1884.

Fan, D., Hodges, D.M., Zhang, J., Kirby, C.W., Ji, X., Locke, S.J., et al., 2011. Commercial extract of the brown seaweed *Ascophyllum nodosum* enhances phenolic antioxidant content of spinach (*Spinacia oleracea L.*) which protects *Caenorhabditis elegans* against oxidative and thermal stress. Food Chem. 124, 195–202.

Fernández, V., Eichert, T., 2009. Uptake of hydrophilic solutes through plant leaves: current state of knowledge and perspectives of foliar fertilization. Crit. Rev. Plant Sci. 28 (1–2), 36–68.

Finnie, J.F., Van Staden, J., 1985. Effect of seaweed concentrate and applied hormones on *in vitro* cultured tomato roots. J. Plant Physiol. 120, 215–222.

Flórez-Fernández, N., Torres, M.D., González-Muñoz, M.J., Domínguez, H., 2018. Potential of intensification techniques for the extraction and depolymerization of fucoidan. Algal. Res. 30, 128–148.

Ganesan, S., Vadivel, K., Jayaraman, J., 2015. Sustainable Crop Disease Management Using Natural Products. CABI, Wallingford, UK.

Glombitza, K.W., 1977. Marine Natural Products Chemistry. *Plenum Press*, New York, pp. 191–204, by DJ Faulkner and WH Fenical.

Goatley, J.M., Schmidt, R.E., 1991. Biostimulator enhancement of Kentucky bluegrass sod. HortScience 26 (3), 254–255.

Goni, O., Fort, A., Quille, P., McKeown, P.C., Spillane, C., O'Connell, S., 2016. Comparative transcriptome analysis of two *Ascophyllum nodosum* extract biostimulants: same seaweed but different. J. Agric. Food Chem. 64 (14), 2980–2989.

Gupta, V., Kumar, M., Brahmbhatt, H., Reddy, C.R.K., Seth, A., Jha, B., 2011. Simultaneous determination of different endogenetic plant growth regulators in common green seaweeds using dispersive liquid–liquid microextraction method. Plant Physiol. Biochem. 49 (11), 1259–1263.

Hazra, D.K., Purkait, A., 2019. Role of pesticide formulations for sustainable crop protection and environment management: A review. J. Pharmacogn. Phytochem. 8, 686–693.

Hernández-Herrera, R.M., Virgen-Calleros, G., Ruiz-López, M., Zañudo-Hernández, J., Délano-Frier, J.P., Sánchez-Hernández, C., 2014. Extracts from green and brown seaweeds protect tomato (*Solanum lycopersicum*) against the necrotrophic fungus *Alternaria solani*. J. Appl. Phycol. 26 (3), 1607–1614.

Herrero, M., Mendiola, J.A., Cifuentes, A., Ibáñez, E., 2010. Supercritical fluid extraction: Recent advances and applications. J. Chromatogr. 1217 (16), 2495–2511.

Jannin, L., Arkoun, M., Etienne, P., Laîné, P., Goux, D., Garnica, M., et al., 2013. *Brassica napus* growth is promoted by *Ascophyllum nodosum* (L.) Le Jol. seaweed extract: microarray analysis and physiological characterization of N, C, and S metabolisms. J. Plant Growth Regul. 32, 31–52.

Kadam, S.U., Tiwari, B.K., O'Donnell, C.P., 2013. Application of novel extraction technologies for bioactives from marine algae. J. Agric. Food Chem. 61 (20), 4667–4675.

Kamel, H.M., 2014. Impact of garlic oil, seaweed extract and imazalil on keeping quality of valencia orange fruits during cold storage. J. Hortic. Sci. Ornam. Plants 6, 116–125.

Khan, W., Rayirath, U.P., Subramanian, S., Jithesh, M.N., Rayorath, P., Hodges, D.M., et al., 2009. Seaweed extracts as biostimulants of plant growth and development. J. Plant Growth Regul. 28, 386–399.

Kobayashi, F., Maeta, E., Terashima, A., Kawaura, K., Ogihara, Y., Takumi, S., 2008. Development of abiotic stress tolerance via bZIP-type transcription factor LIP19 in common wheat. J. Exp. Bot. 59, 891–905.

Kotze, W.A.G., Joubert, M., 1980. Influence of foliar spraying with seaweed products on the growth and mineral nutrition of rye and cabbage. Elsen. J. 4, 17–20.

Krouk, G., Lacombe, B., Bielach, A., Perrine-Walker, F., Malinska, K., Mounier, E., et al., 2010. Nitrate-regulated auxin transport by NRT1. 1 defines a mechanism for nutrient sensing in plants. Dev. Cell. 18, 927–937.

Kumar, H.D., Aloke, P., 2020. Role of biostimulant formulations in crop production: an overview. Int. J. Appl. Res. Vet. M. 8, 38–46.

Lattner, D., Flemming, H.C., Mayer, C., 2003. 13C-NMR study of the interaction of bacterial alginate with bivalent cations. Int. J. Biol. Macromol. 33 (1–3), 81–88.

Liu, H., Chen, X., Song, L., Li, K., Zhang, X., Liu, S., et al., 2019. Polysaccharides from *Grateloupia filicina* enhance tolerance of rice seeds (*Oryza sativa L.*) under salt stress. Int. J. Biol. Macromol. 124, 1197–1204.

Lizzi, Y., Coulomb, C., Polian, C., Coulomb, P.J., Coulomb, P.O., 1998. Seaweed and mildew: what does the future hold? Laboratory tests have produced encouraging results [L'algue face au mildiou: quel avenir? Des resultats de laboratoire tres encourageants]. Phytoma 508, 29.

Machado, L.P., Matsumoto, S.T., Jamal, C.M., da Silva, M.B., da Cruz Centeno, D., Neto, P.C., et al., 2014. Chemical analysis and toxicity of seaweed extracts with inhibitory activity against tropical fruit anthracnose fungi. J. Sci. Food Agric. 94, 1739–1744.

MacKinnon, S.L., Hiltz, D., Ugarte, R., Craft, C.A., 2010. Improved methods of analysis for betaines in *Ascophyllum nodosum* and its commercial seaweed extracts. J. Appl. Phycol. 22, 489–494.

Mancuso, S., Azzarello, E., Mugnai, S., Briand, X., 2006. Marine bioactive substances (IPA extract) improve foliar ion uptake and water stress tolerance in potted *Vitis vinifera* plants. Adv. Hortic. Sci. 20, 156–161.

McHugh, D.J., 2003. A Guide to the Seaweed Industry. Food Agric. Orga. of Unit. Natio, Rome, p. 441.

Michalak, I., Chojnacka, K., 2015. Algae as production systems of bioactive compounds. Eng. Life Sci. 15 (2), 160–176.

Michalak, I., Dmytryk, A., Wieczorek, P.P., Rój, E., Łęska, B., Górka, B., et al., 2015. Supercritical algal extracts: a source of biologically active compounds from nature. J. Chem. . Available from: https://doi.org/10.1155/2015/597140.

Mire, G.L., Nguyen, M.L., Jardin, P.D., Verheggen, F., Fassotte, B., 2016. Implementing plant biostimulants and biocontrol strategies in the agroecological management of cultivated ecosystems. Biotechnol. Agron. Soc. Environ. 20, 299–313.

Mzibra, A., Aasfar, A., Benhima, R., Khouloud, M., Boulif, R., Douira, A., et al., 2021. Biostimulants derived from Moroccan seaweeds: seed germination metabolomics and growth promotion of tomato plant. J. Plant Growth Regul. 40, 353–370.

Nair, P., Kandasamy, S., Zhang, J., Ji, X., Kirby, C., Benkel, B., et al., 2012. Transcriptional and metabolomic analysis of *Ascophyllum nodosum* mediated freezing tolerance in *Arabidopsis thaliana*. BMC Genom 13, 643.

Nayar, S., Bott, K., 2014. Current status of global cultivated seaweed production and markets. World Aquac. 45 (2), 32–37.

Ong, M.Y., Syahira Abdul Latif, N.I., Leong, H.Y., Salman, B., Show, P.L., Nomanbhay, S., 2019. Characterization and analysis of Malaysian macroalgae biomass as potential feedstock for bio-oil production. Energy 12 (18), 3509.

Patel, K., Agarwal, P., Agarwal, P.K., 2018. Kappaphycus alvarezii sap mitigates abiotic-induced stress in Triticum durum by modulating metabolic coordination and improves growth and yield. J. Appl. Phycol. 30, 2659–2673.

Pohl, A., Andrzej, K., Agnieszka, S., 2019. Seaweed extracts' multifactorial action: influence on physiological and biochemical status of *Solanaceae* plants. Acta Agrobot. 72 (1), 1758.

Provasoli, L., Carlucci, A.F., 1974. Vitamins and growth regulators. Bot. Monogr. 471–487.

Radwan, M.A., Farrag, S.A.A., Abu-Elamayem, M.M., Ahmed, N.S., 2012. Biological control of the root-knot nematode, *Meloidogyne incognita* on tomato using bioproducts of microbial origin. Appl. Soil Ecol. 56, 58–62.

Rathinapriya, P., Satish, L., Pandian, S., Rameshkumar, R., Balasangeetha, M., Rakkammal, K., et al., 2020. Effects of liquid seaweed extracts in improving the agronomic performance of foxtail millet. J. Plant Nutr. 43 (19), 2857–2875.

Rayorath, P., Benkel, B., Hodges, D.M., Allan-Wojtas, P., MacKinnon, S., Critchley, A.T., et al., 2009. Lipophilic components of the brown seaweed, *Ascophyllum nodosum*, enhance freezing tolerance in. Arabidopsis thaliana. Plan. 230, 135–147.

Rekanovic, E., Potocnik, I., Milijasevic-Marcic, S., Stepanovic, M., Todorovic, B., Mihajlovic, M., 2010. Efficacy of seaweed concentrate from *Ecklonia maxima* (Osbeck) and conventional fungicides in the control of Verticillium wilt of pepper. Pestic. Fitomed. 25 (4), 319–324.

Rengasamy, K.R.R., Kulkarni, M.G., Stirk, W.A., Van Staden, J., 2015. A new plant growth stimulant from the brown seaweed *Ecklonia maxima*. J. Appl. Phycol. 27, 581–587.

Ricci, M., Tilbury, L., Daridon, B., Sukalac, K., 2019. General principles to justify plant biostimulant claims. Front. Plant Sci. 10, 494.

Rönnbäck, P., Kautsky, N., Pihl, L., Troell, M., Söderqvist, T., Wennhage, H., 2007. Ecosystem goods and services from Swedish coastal habitats: identification, valuation, and implications of ecosystem shifts. Ambio. 36, 534–544.

Rouphael, Y., Colla, G., 2020. Editorial: biostimulants in agriculture. Front. Plant Sci. 11, 40.

Sahayaraj, K., Kalidas, S., 2011. Evaluation of nymphicidal and ovicidal effect of a seaweed, *Padina pavonica (Linn.)* (*Phaeophyceae*) on cotton pest, *Dysdercus cingulatus* (*Fab.*). Indian J. Geo-Mar. Sci. 40, 125–129.

Samuels, L.J., Setati, M.E., Blancquaert, E.H., 2022. Towards a better understanding of the potential benefits of seaweed based biostimulants in Vitis vinifera L Cultivars. Plants 11 (3), 348.

Sangeetha, G., Vadivel, K., Jayaraman, J., 2015. Sustainable Crop Disease Management Using Natural Products. CABI, Wallingford, UK.

Santaniello, A., Scartazza, A., Gresta, F., Loreti, E., Biasone, A., Di Tommaso, D., et al., 2017. *Ascophyllum nodosum* seaweed extract alleviates drought stress in *Arabidopsis* by affecting photosynthetic performance and related gene expression. Front. Plant sci. 8, 1362.

Sharma, H.S., Fleming, C., Selby, C., Rao, J.R., Martin, T., 2014. Plant biostimulants: a review on the processing of macroalgae and use of extracts for crop management to reduce abiotic and biotic stresses. J. Appl. Phycol. 26, 465–490.

Sharma, S., Chen, C., Khatri, K., Rathore, M.S., Pandey, S.P., 2019. *Gracilaria dura* extract confers drought tolerance in wheat by modulating abscisic acid homeostasis. Plant Physiol. Biochem. 136, 143–154.

Shukla, P.S., Shotton, K., Norman, E., Neily, W., Critchley, A.T., Prithiviraj, B., 2018. Seaweed extract improve drought tolerance of soybean by regulating stress-response genes. AoB Plas 10 (1), plx051.

Sivasankari, S., Venkatesalu, V., Anantharaj, M., Chandrasekaran, M., 2006. Effect of seaweed extracts on the growth and biochemical constituents of *Vigna sinensis*. Bioresour. Technol. 97 (14), 1745–1751.

Spinelli, F., Fiori, G., Noferini, M., Sprocatti, M., Costa, G., 2009. Perspectives on the use of a seaweed extract to moderate the negative effects of alternate bearing in apple trees. J. Hortic. Sci. Biotechnol. 84, 131–137.

Spinelli, F., Fiori, G., Noferini, M., Sprocatti, M., Costa, G., 2010. A novel type of seaweed extract as a natural alternative to the use of iron chelates in strawberry production. Sci. Hortic. 125, 263–269.

Stirk, W.A., Arthur, G.D., Lourens, A.F., Novak, O., Strnad, M., Van Staden, J., 2004. Changes in cytokinin and auxin concentrations in seaweed concentrates when stored at an elevated temperature. J. Appl. Phycol. 16, 31–39.

Stirk, W.A., Novak, M.S., Van Staden, J., 2003. Cytokinins in macroalgae. Plant Growth Regul. 41, 13–24.

Subbarao, S.B., Hussain, I.A., Ganesh, P.T., 2015. Biostimulant activity of protein hydrolysate: influence on plant growth and yield. J. Plant Sci. Res. 2 (2), 1–6.

Sultana, V., Baloch, G.N., Ara, J., Ehteshamul-Haque, S., Tariq, R.M., Athar, M., 2012. Seaweeds as an alternative to chemical pesticides for the management of root diseases of sunflower and tomato. J. Appl. Bot. Food Qual. 84 (2), 162.

Turan, M., Köse, C., 2004. Seaweed extracts improve copper uptake of grapevine. Acta Agric. Scand. Plant Sci. 54, 213–220.

Verkleij, F.N., 1992. Seaweed extracts in agriculture and horticulture: a review. Biol. Agric. Hortic. 8, 309–324.

Wally, O.S.D., Critchley, A.T., Hiltz, D., Craigie, J.S., Han, X., Zaharia, L.I., et al., 2013. Regulation of phytohormone biosynthesis and accumulation in *Arabidopsis* following treatment with commercial extract from the marine macroalga *Ascophyllum nodosum*. J. Plant Growth Regul. 32, 324–339.

Wang, Y., Fu, F., Li, J., Wang, G., Wu, M., Zhan, J., et al., 2016. Effects of seaweed fertilizer on the growth of *Malus hupehensis Rehd.* seedlings, soil enzyme activities and fungal communities under replant condition. Eur. J. Soil Biol. 75, 1–7.

Wijesinghe, W.A.J.P., Jeon, Y.J., 2012. Enzyme-assistant extraction (EAE) of bioactive components: a useful approach for recovery of industrially important metabolites from seaweeds: a review. Fitoterapia 83 (1), 6–12.

Wilson, S., 2001. Frost Management in Cool Climate Vineyards. Final Report to Grape and Wine Research & Development Corporation. Univ. Tasmania, pp. 1–34.

Wong, K.H., Cheung, P.C., 2000. Nutritional evaluation of some subtropical red and green seaweeds: part I—proximate composition, amino acid profiles and some physico-chemical properties. Food chem. 71 (4), 475–482.

Yang, Y., Zhang, M., Alalawy, A.I., Almutairi, F.M., Al-Duais, M.A., Wang, J., et al., 2021. Identification and characterization of marine seaweeds for biocompounds production. Environ. Technol. Innov. 24, 101848.

Zhong, B., Robinson, N.A., Warner, R.D., Barrow, C.J., Dunshea, F.R., Suleria, H.A., 2020. LC-ESI-QTOF-MS/MS characterization of seaweed phenolics and their antioxidant potential. Mar. drugs 18 (6), 331.

Zhu, J.K., 2000. Genetic analysis of plant salt tolerance using Arabidopsis. Plant Physiol. 124, 941–948.

Zodape, S.T., Gupta, A., Bhandari, S.C., Rawat, U.S., Chaudhary, D.R., Eswaran, K., et al., 2011. Foliar application of seaweed sap as biostimulant for enhancement of yield and quality of tomato (*Lycopersicon esculentum* Mill.). J. Sci. Ind. Res. 70, 215–219.

Zou, P., Lu, X., Jing, C., Yuan, Y., Lu, Y., Zhang, C., et al., 2018. Low-molecular-weight polysaccharides from *Pyropia yezoensis* enhance tolerance of wheat seedlings (*Triticum aestivum* L.) to salt stress. Front. Plant Sci. 9, 427.

CHAPTER

10

Role of alfalfa brown juice in plant growth and development

Nóra Bákonyi[1], *Döme Barna*[1], *Miklós Gábor Fári*[1], *Szilvia Veres*[1], *Tarek Alshaal*[1,2], *Éva Domokos-Szabolcsy*[1] *and Péter Makleit*[1]

[1]Department of Applied Plant Biology, Faculty of Agriculture, Food Science and Environmental Management, Debrecen University, Debrecen, Hungary [2]Soil and Water Department, Faculty of Agriculture, Kafrelsheikh University, Kafr El-Sheikh, Egypt

10.1 Introduction: origin of alfalfa brown juice

Brown juice is a by-product obtained during the wet fractionation of fresh green biomass for leaf protein isolation. The fractionation of alfalfa (*Medicago sativa* L.) green biomass was first studied by Károly Ereky. Fári (2018) and Domokos-Szabolcsy et al. (2023) described in detail the history of green biomass as a protein source, the establishment of the very first Green Protein Biorefinery Factory in Hungary and England, and the green feed mill concept of Károly (Karl) Ereky. In Ereky's process, green juice and fiber fraction were obtained after mechanical pressing of green alfalfa biomass. The process was further developed by N.W. Pirie, who invented the first scientifically relevant use of the term brown juice (Pirie, 1956). In Pirie's method, leaf protein concentrate (LPC) was obtained by coagulating soluble proteins from green juice after blowing steam. The remaining liquid fraction after protein coagulation and separation is the brown juice, which is also referred to as deproteinized plant juice (DPJ), deproteinized leaf extracts, deproteinized leaf juice, and deproteinized whey or phytoserum (Jadhav, 2018; Barna et al., 2020). The next important milestone in LPC research was the patents registered by Hungarian researchers (Holló et al., 1969; Holló, 1970, 1972) and the establishment of WEPEX, the first leaf protein processing plant in the world based on these principles, in 1972 in Hungary. The Pro-Xan method is based on the procedure described by Pirie for producing protein extraction from alfalfa (Pirie, 1987). As a result of this method in the main product, the fiber content

Biostimulants in Plant Protection and Performance
DOI: https://doi.org/10.1016/B978-0-443-15884-1.00027-0

was low while the protein and xanthophyll contents were high as the Pro-Xan name indicates (Mayorga et al., 1972; Saunders et al., 1973; Knuckles et al., 1972; Bickoff et al., 1970; Carroad et al., 1981). The above-mentioned methods were the first technologies developed for leaf protein and brown juice production; however, many attempts have been conducted over the last decades to improve alfalfa fractionation and valorization (Table 10.1).

The most novel technology so far was developed by Fári and Szabolcsy in 2019, based on Ereky's and Pirie's method (Fári and Domokos-Szabolcsy, 2019), by producing LPC with microwave-assisted protein coagulation. The brown juice fraction obtained from alfalfa is a dark brownish liquid, and it sometimes appears greenish. Depending on the applied technology to isolate LPC, it looks clear, translucent, or turbid. Alfalfa brown juice is rich in vitamins, pigments, enzymes, minerals, and other vital phytochemicals; hence, it is a promising material for further valorization (Pirie, 1987; Thomsen et al., 2004; Iliyas, 2019). About 50% of the processed green biomass by microwave-assisted protein coagulation methods is brown juice (Manwatkar et al., 2014; Bákonyi et al., 2020). Brown juice contains about 4%–8% (Bákonyi et al., 2020) or 13%–15% dry matter, 16%–20% protein, and 25%–30% cellulose (Zanin, 1998). Nevertheless, these values vary according to the applied separation technique; for instance, Bákonyi et al. (2020) cited 4%–8% dry matter content. Moreover, the composition of alfalfa brown juice depends heavily on many factors, such as plant species, preharvest weather conditions, plant phenophase, harvest time, and processing technology (Thomsen et al., 2004; Bákonyi et al., 2020).

Although alfalfa is the most widely studied raw material for the production of LPC, and thus brown juice, other plant species were also intensively studied. Brown juice could be also obtained from grass, rice, clover, corn, potatoes, dill, fenugreek, spinach, coriander, chicory, etc. (Sinclair, 2009; Sayyed, 2015; Jadhav and Deshmukh, 2018; Santamaría-Fernández et al., 2019). Green biomass of various plants, such as mulberry (Chowdary et al., 2002), ryegrass, clover, and broccoli (Domokos-Szabolcsy et al., 2022) are raw materials in biorefineries, producing LPC and brown juice. Grasses, such as wheat and oats, are used for leaf protein isolation, and the liquid by-product is called grass brown juice, which may also be suitable for L-lysine (Thomsen and Kiel, 2008; Slade and Birkinshaw, 1938; Holló, Görög, and Koch, 1971; Welch et al., 1979; Kromus et al., 2008; Sinclair, 2009; Kiel et al., 2015; Kamm et al., 2016; Jadhav and Deshmukh, 2018) and biogas production (Feng et al., 2021). Also, rice brown juice and its importance for food safety were reported (Zhang et al., 2019).

10.2 Features of alfalfa brown juice

Alfalfa brown juice contains about 40% carbohydrates (mainly monosaccharides, like glucose and fructose), lipids, macro- and microelements, and 3% N (on a dry mass basis). Also, it has significant amounts of amino acids and various biologically active compounds. The most common macro- and microelements in brown juice are B, Ca, K, Mg, Mn, Mo, P, S, and Zn content. Fig. 10.1 shows the macro and mesoelement content of fermented alfalfa brown juice.

The different nitrogen forms and the microelement content of alfalfa brown juice are represented in Fig. 10.2.

TABLE 10.1 Summary of literatures related to brown juice term.

Year	Term	Purpose of use	Separation technology	Plant species	Key references
1938	Filtrate, liquid	The brown juice produced during the process is recycled to the next batch of biomass for grinding and pressing. The brown juice concentrated by repeated recycling can then be used for yeast cultivation or alcoholic fermentation.	The plant biomass is mixed with water using a rotary knife grinder and filtered. The resulting filtrate was precipitated by acid precipitation (pH 3.5–5 with the addition of hydrochloric acid) and allowed to stand for 10 min. The resulting coagulum is filtered through a cloth to separate it from the brown.	Freshly harvested grasses	Slade and Birkinshaw (1938)
1956	Fluid, juice after coagulation	It can be used for yeast cultivation or as a medium for other microorganisms; it can be added to feed for nonruminants.	After grinding and pressing, different coagulation methods are cooking, steam blowing, and acidic, organic solvent precipitation.	Different species	Pirie (1956)
1969		LPC production and yeast cultivation on the by-product.	LPC production according to Patent and use of brown juice.	Rye, alfalfa, kale, *Sorghum* sp.	Holló et al. (1969)
1970		LPC production and yeast cultivation on the by-product.	LPC production according to Patent and use of brown juice.	Rye, alfalfa, kale, *Sorghum* sp.	Holló (1970)
1972		LPC production and yeast cultivation on the by-product.	LPC production according to Patent and use of brown juice.	Rye, alfalfa, kale, *Sorghum sp.*	Holló (1972)
1971	Liquid phase	Microbial cell mass, protein production.	The green juice is separated by coagulation into precipitated proteins and brown juice, which is inoculated with microbes to utilize the nitrogen content of the brown juice during aerobic fermentation. The cell mass thus obtained is used as feed when added to the protein concentrate.	Grasses	Holló and László (1971)
1976	Alfalfa solubles	The brown juice, mixed with the pressed fiber can be used to feed ruminants directly, dried or after ensiling.	After pressing, adding ammonia to the green juice (pH 8–8.5) precipitates the proteins, stabilizes the carotenoids, and increases the nitrogen content of the remaining brown juice with ammonia.	Alfalfa	Vosloh et al. (1976)
1977	Deproteinized plant juice (DPJ)	The brown juice was used to fertilize alfalfa, smooth brome, and corn in various concentrations. They mention the uses of	Alfalfa biomass processed by pressing for protein production results brown juice 50% of the biomass.	Alfalfa	Ream et al. (1977)

(*Continued*)

TABLE 10.1 (Continued)

Year	Term	Purpose of use	Separation technology	Plant species	Key references
	deproteinized alfalfa juice (DAJ)	brown juice for other purposes like single cell protein production and mixing with direct animal feed.			
1979	Deproteinized alfalfa juice (DAJ), deproteinized oat juice (DOJ)	Alfalfa fertilization in pot experiments.	After chopping and pressing fresh harvested alfalfa, the protein concentrate was separated from the brown juice by coagulation.	Low saponin containing alfalfa and oats	Welch et al. (1979)
1979	Deproteinized alfalfa juice (DAJ)	Alfalfa fertilization in pot experiments.	The brown juice was obtained by heat coagulation of alfalfa green juice.	Alfalfa	Volenec et al. (1979)
1987	Whey, brown juice, deproteinized juice	Versatile, there is no specific mention.	Heat coagulation and separation by filtration.	Alfalfa	Pirie (1987)
1994	Brown juice (BJ)	Production of enzymes, lactic acid, amino acids and microbes.	The brown juice is concentrated by evaporation and utilized as a medium.	Alfalfa, green fodder	Kiel and Andersen (1994)
1998	Deproteinized leaf juice	Investigation of plant growth inhibitory effect.	Brown juice and protein concentrate obtained by heat precipitation.	Alfalfa	Jadhav and Mungikar (1998)
2004	Brown juice (BJ)	L-Lysine production.	The brown juice produced according to the principle of green biorefinery is preserved by lactic acid fermentation and the resulting medium is supplemented and used for lysine production.	Alfalfa, green fodder	Thomsen et al. (2004)
2005	Deproteinized leaf juice (DPJ)	Utilization as a microbial medium.	Green crop fractionation (GCF).	Alfalfa	Sayyed and Mungikar (2005)
2005	Brown juice (BJ)	Lactic acid fermentation.	In the second fractionation step of the biorefining, the brown juice is formed, which can be further utilized by fermentation.	Grasses	Kromus et al. (2004)
2009	Brown juice (BJ)	Molasses-like use in ruminant feed.	The evaporated brown juice can be mixed back into the coagulated and dried protein concentrate.	Meadow grasses, alfalfa, clovers	Sinclair (2009)

2014	Deproteinized leaf Juice (DPJ)	Media for α-amylase production by fungi.	Deproteinized leaf juice (DPJ), remained after isolating leaf protein concentrate (LPC) from the heated green juice.	Alfalfa, dill, fenugreek, spinach, coriander	Sayyed (2014)
2015	Filtrate, liquid	Modifies Slade-Birkinshaw process (M-SB).	The theoretical basis of the method is the patent filed by E.S. Slade and J.H. Birkinshaw in 1938.	Freshly cut grasses	Kiel et al. (2015)
2015	Deproteinised leaf juice (DPJ)	Media for the production of fungal biomass.	Biomass is mechanically fractionated and pressed. The leaf juice was heated to 95°C for 15 min; than LPC was removed by filtration and brown-colored liquid (DPJ) was collected, dried, and preserved.	Fenugreek	Sayyed. (2015)
2016	Brown juice (BJ)	Lactic acid fermentation.	Brown juice is a by-product of processing based on the Green Biorefinery concept.	Alfalfa, grasses	Kamm et al. (2016)
2018	Deproteinized leaf juice (DPJ)	Plant conditioning and fertilization, effect on germination.	Green crop fractionation (GCF). Green juice from fractionation of plant biomass. The protein coagulum and the brown juice can be separated using heat or acid coagulation and filtration.	Alfalfa	Sukte (2018)
2018	Deproteinized juice (DPJ)	Yeast cultivation.	Green crop fractionation (GCF).	Wheat, corn, potatoes	Jadhav and Deshmukh (2018)
2018	Deproteinized alfalfa juice (DPJ)	Evaluation of effect on plant growth.	Application of fermented alfalfa juice, in hydroponics.	Maize	Makleit et al. (2018a)
2018	Deproteinized alfalfa juice (DPJ)	Evaluation of effect on plant growth.	Application of fermented alfalfa juice, in hydroponics.	Maize	Makleit et al. (2018b)
2019	Deproteinized alfalfa juice (DPJ)	Evaluation of effect on plant growth.	Application of fermented alfalfa juice, in hydroponics.	Maize	Makleit et al. (2019)
2019	Deproteinized leaf Juice (DPJ)	Medium for fungal growth and subsequent production of $\propto$-amylase.	After isolation of LPC from leaves the remaining by-product is deproteinized leaf juice (DPJ).	Alfalfa, dill, fenugreek, spinach, coriander	Iliyas (2019)
2019	Brown juice (BJ)	There is no mention of relevant brown juice recovery.	Lactic acid fermentation of pressed green juice, filtration of precipitated proteins by centrifugation.	Alfalfa, red clovers, chicory	Santamaría-Fernández et al. (2019)

(Continued)

TABLE 10.1 (Continued)

Year	Term	Purpose of use	Separation technology	Plant species	Key references
2020	Brown juice (BJ) Deproteinized leaf juice (DPJ), phytoserum, plant whey	Characteristics of alfalfa brown juice, its lacto-fermentation. Effect of fermentation on the composition of brown juice. Foliar application of lacto-fermented brown juice on celosia plants.	Alfalfa green juice is heated by microwave to 85°C based on a patent (Fári and Domokos-Szabolcsy, 2019), than brown juice is separated by filtering from coagulated LPC.	Alfalfa	Bákonyi et al. (2020)
2020	Brown juice (BJ) Deproteinized leaf Juice (DPJ)	Foliar application of lacto-fermented brown juice on sweet basil plants.	Alfalfa green juice is heated by microwave to 85°C based on a patent (Fári and Domokos-Szabolcsy, 2019), than brown juice is separated by filtering from coagulated LPC.	Alfalfa	Kisvarga et al. (2020)
2021	Brown juice (BJ) Deproteinized leaf juice (DPJ)	Effect of raw and lacto-fermented brown juice on the germination and growth and physiological, anatomical parameters of marigold plants.	Alfalfa green juice is heated by microwave to 85°C based on patent (Fári and Domokos-Szabolcsy, 2019), than brown juice is separated by filtering from coagulated LPC.	Alfalfa	Barna et al. (2021)
2021	Fermented brown juice (DPJ)	Usage of fermented brown juice for culture medium of microorganisms.	Lactic acid fermentation of brown juice.	Different species	Bákonyi et al. (2021)

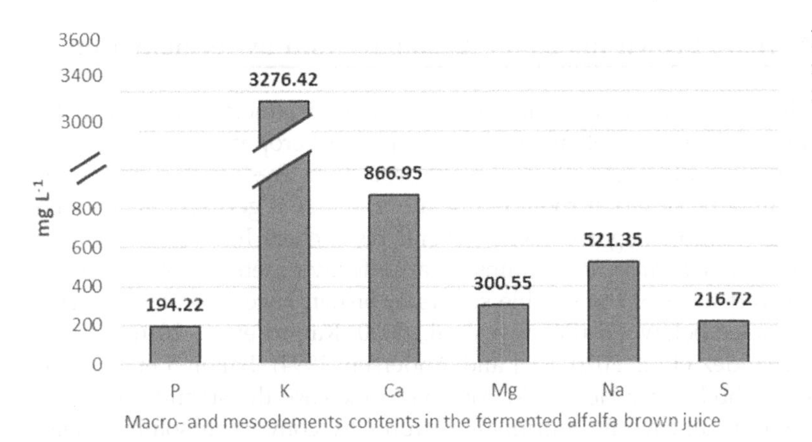

FIGURE 10.1 Macro- and mesoelement content of fermented alfalfa brown juice.

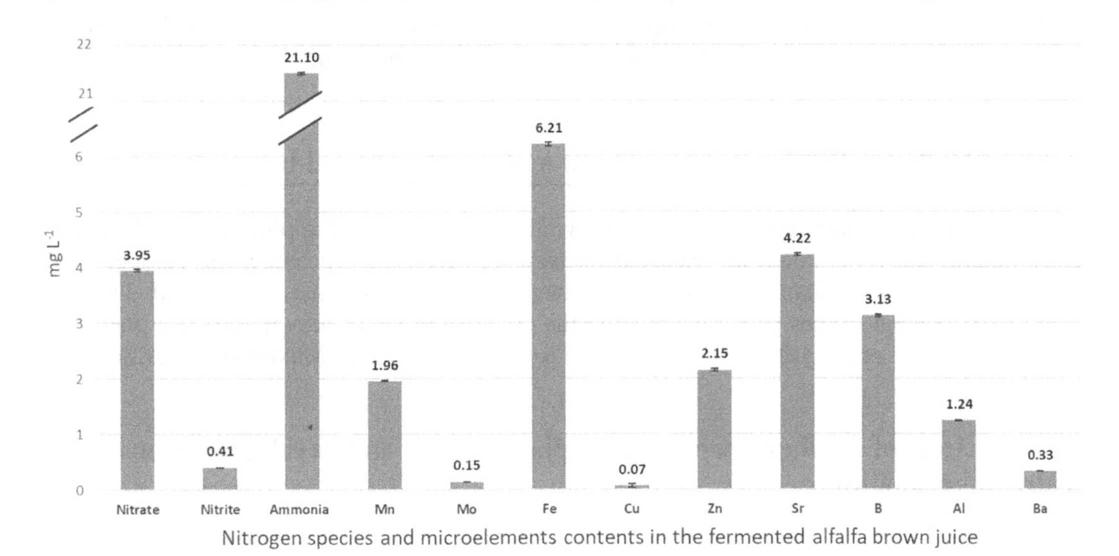

FIGURE 10.2 Nitrogen forms and microelement content of fermented alfalfa brown juice.

Moreover, different organic acids, for example, lactic acid, acetic acid, and citric acid, were measured in brown juice (Thomsen et al., 2004). Brown juice contains compounds with an antioxidant-like effect, such as phenols. Barna et al. (2022) reported that the total phenolic and flavone contents of alfalfa brown juice were in the range of 20–51 μg/mL and 67–253 μg/mL, respectively, depending on the plant variety and harvest time. They also identified several bioactive compounds in alfalfa brown juice, including vitamins (i.e., nicotinamide, biotin, riboflavin, and nicotinic acid), flavones (e.g., genkwanin, apigenin, luteolin, tricin, chrysoeriol, dihydroxyflavone, alfalone, etc.), isoflavones (e.g., coumestrol, formononetin, formononetin, ononin, etc.), flavonoids (e.g., quercetin, isoquercetin, naringenin, etc.), phenolic acids (e.g., ferulic acid, caffeic acid, sinapic acid, etc.), organic acids (e.g., benzoyltartaric acid, phenyllactic acid, etc.), terpenes, and an alkaloid with plant hormone-like effect (trigonelline).

10.3 Instability of alfalfa brown juice and its stabilization by fermentation

Lamont et al. (2017) illustrated the importance of lactic acid–producing bacteria in the preservation of agricultural products, as well as in various aspects of crop production, such as increasing soil fertility, promoting plant growth, and plant protection. Fresh brown juice (with pH 5–6) is unstable and cannot be stored at room temperature, where it gets spoiled within a few days because of its high soluble sugar content. Several researchers have recommended lacto-fermentation using different *Lactobacillus* strains as an effective method for preserving brown juice. *Lactobacilli* convert most of the soluble sugars in brown juice into organic acids, reducing the pH of brown juice below 4.5 (Bákonyi et al., 2020; Kamm et al., 2016; Kromus et al., 2004; Santamaría-Fernández et al., 2020; Kiel and Andersen, 1994). Bákonyi et al. (2020) developed a fermentation method using lactic acid bacteria to increase the stability of brown juice at room temperature and improve its nutritional features by converting soluble sugars into organic acids, which reduce the pH to less than 4. Also, the utilization of lactic acid bacteria brings additional benefits to the nutritional value of brown juice as a result of significant changes in the ratio of organic and mineral components of brown juice.

10.4 Applications of fermented and non-fermented brown juice

Fermented brown juice could be directed toward many significant applications, such as plant biostimulant, soil conditioner, an excellent fermentation feedstock for enzymes, biodegradable plastics, amino acids, vitamins, alcohols, or their precursors (Bákonyi et al., 2020; Thomsen et al., 2004). Several studies have documented the potential usage of brown juice as a plant biostimulant (Bákonyi et al., 2020; Barna et al., 2021; Kisvarga et al., 2020; Makleit et al., 2018b).

10.4.1 Effect of alfalfa brown juice on plant growth and development under hydroponic conditions

Application of fermented alfalfa brown juice at low concentrations (i.e., 0.05–0.1 vol./vol.%) displayed positive effects on the growth dynamics of maize (*Zea mays* L.) seedlings under hydroponic conditions (Makleit et al., 2018a). However, higher rates of alfalfa brown juice (i.e., 1.0–5.0 vol./vol.%) decreased the growth and relative chlorophyll content of maize plants grown in hydroponics. The reduction in plant growth at the high rates of fermented alfalfa brown juice could be attributed to the enhanced uptake of several microelements, particularly toxic ones, due to the low pH of the fermented brown juice, where the solubility of these microelements increases in acidic solutions (Makleit et al., 2018b). The excess uptake of such ions disrupts ion homeostasis in plant tissues, leading to an imbalance in plant growth. However, some certain maize hybrids could tolerate higher brown juice concentrations (up to 0.5 vol./vol.%) (Makleit et al., unpublished data).

Plant cultivation in horticultural perlite and watering plants with a nutrient solution is a type of hydroponics. Under such conditions, the application of the fermented alfalfa brown juice at a rate of 0.5 vol./vol.% increased the growth and photosynthetic capacity of maize plants. Substitution of the nutrient solution by the fermented alfalfa brown juice at a rate of 50% recorded the same plant height, weight, and photosynthetic capacity as those receiving 100% nutrient solution. Furthermore, foliar application of maize plants with the fermented alfalfa brown juice at a rate of 2.5 vol./vol.% completely compensated for the application of nutrient solution (Makleit et al., 2019). However, the same findings could not be recorded in the case of sunflower plants.

Increasing the pH of the fermented alfalfa brown juice could improve its utilization and support its application at high rates of up to 10 vol./vol.%. Wood ash and fly ash are among the promising candidates that could be applied to modify the pH of the fermented alfalfa brown juice. In addition to adjusting pH, wood ash markedly contains a high content of nutrients, especially K. The positive effect of the combined application of the fermented alfalfa brown juice and wood/fly ash was proved using sunflower plants in hydroponics (Makleit et al., unpublished data; Fig. 10.3).

10.4.2 Effect of alfalfa brown juice on plant growth and development under greenhouse conditions

Lacto-fermented alfalfa brown juice could be a potent plant biostimulant; as it contains plant growth–promoting microbes (PGPMs) and plant hormone-like substances (Barna et al., 2022), plant growth regulators (i.e., triacontanol) for plant cell cycle, nodulation,

FIGURE 10.3 Effect of application of fermented alfalfa juice in combination of ash filtrate on sunflower (*Helianthus annuus* L.) plants.

oxidative stress, and plant growth (Minorsky, 2002), and possesses beneficial traits for human and nutritional physiological effects (Garg, 2016; Zhou et al., 2012). Triacontanol—a plant growth regulator—was identified in alfalfa, and it increases plant growth by enhancing the rates of photosynthesis, protein biosynthesis, the transportation of nutrients within plant organs, enzyme activity, and productivity (Naeem et al., 2012).

Application of the fermented brown juice at lower concentrations (below 2.5%) significantly increased plant biometrics, such as the shoot and root weight, length of shoot and root, volume of shoot and root, number of leaves, and content of photosynthetic pigments, of different plant species (i.e., French Marigold, Celosia, Sweet Basil) (Bákonyi et al., 2020; Barna et al., 2021; Kisvarga et al., 2020; Makleit et al., 2018a). The process of fermented alfalfa brown juice preparation and application on celosia is represented in Fig. 10.4.

The effect of the combined application of the fermented alfalfa brown juice and wood/fly ash was examined using sunflower and maize as plant models in different soil types under greenhouse conditions.

Maize and sunflower were consciously chosen because these plant species are cultivated in large areas in the same geographic region, such as alfalfa. Fermented brown juice was applied solely or in combination with wood/fly ash as a foliar spraying or root application via watering. Deionized water and conventional nutrient solutions in different concentrations served as controls. In most cases, there were no differences among treatments concerning growth parameters and relative chlorophyll content. In contrast to hydroponics, the

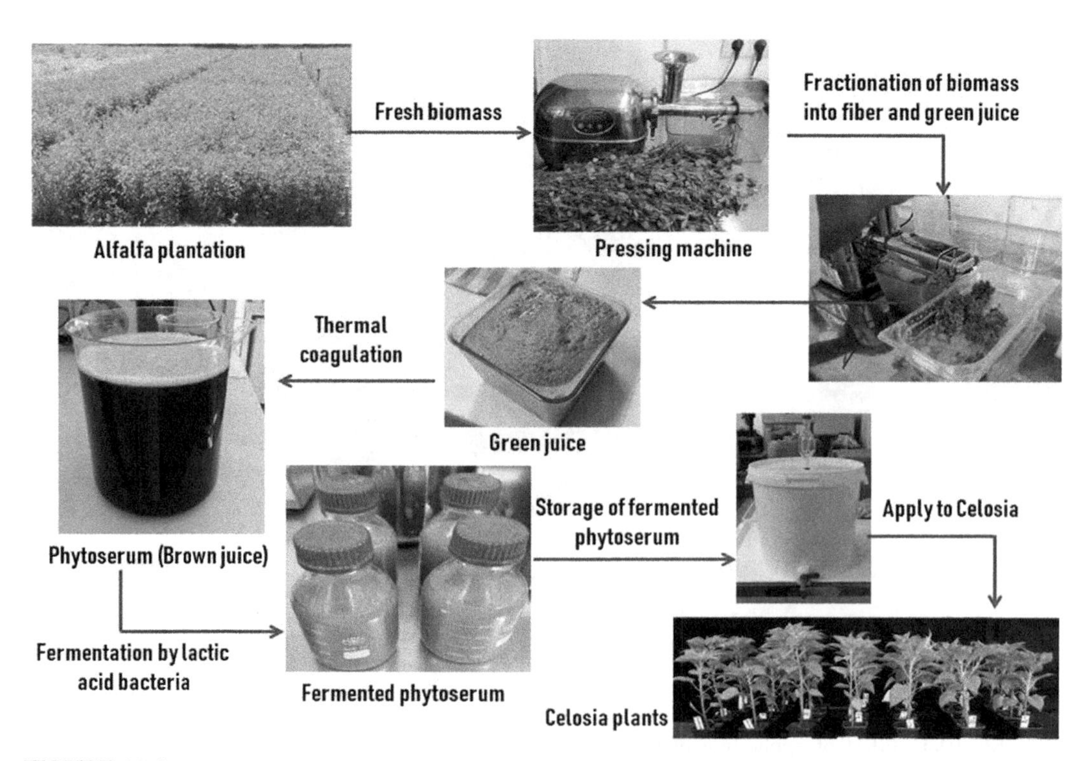

FIGURE 10.4 The process of fermented alfalfa brown juice preparation and application on celosia.

application of fermented brown juice at rates of 5–10 vol./vol.% had no toxicity symptoms on the plants. This could be ascribed to the high adsorption/buffer capacity of soils, which blunts the negative effects of the fermented brown juice, and soils gradually supply grown plants with nutrients. Ca deficiency symptoms developed in maize plants treated with brown juice and wood/fly ash solutions simultaneously, which is also an important observation and must be kept in mind, especially in the case of soils with low Ca levels. The high K content of both ingredients generated Ca deficiency; this phenomenon is known as oppugnance, which originated from an unbalanced K/ Ca ratio (Makleit et al., unpublished data).

The applied rate of fermented brown juice, which has no toxic effects on plant growth in the case of field conditions, is essential information. Compared to the quantity of the used soil in the pots, the total quantity of the fermented brown juice applied batch-wise at different times reached 30% of the soil volume (Makleit et al., unpublished data).

10.4.3 Effect of alfalfa brown juice on plant growth and development under open-field conditions

Lacto-fermented alfalfa brown juice was tested as a potential new generation of organic biostimulant in open-field trials as a foliar application at low concentrations (<2.5%) in the following plant species: cash crops (i.e., corn, winter wheat, and sunflower); vegetable and fruit crops (i.e., tomatoes, apples, grapes, sour cherry); and forestry (i.e., fraxinus). The results showed that the fermented alfalfa brown juice applied at low concentrations (0.5%–2.5%) significantly increased the plant height (by 6.1% in maize, by 2.3% in sunflower, by 3.8% in winter wheat, by 37.5% in fraxinus), the number of leaves (by 20.4% in winter wheat, by 26.3% in fraxinus), shoot mass (by 54.7% in fraxinus), root mass (by 45.9% in fraxinus), the yield (by 3.8% in maize, by 9.7% in sunflower, by 20.6% in apple, by 5.6% in sour cherry), the diameter of the fruit (by 10.2% in apples, by 4.2% in cherries), increased the content of photosynthetic pigments (maize, fraxinus), the plate diameter of sunflower increased by 4.5%, the weight of 1000-grain by 3.5%, and number of grains per spike increased by 20.7% in winter wheat in comparison to the controls (Makleit et al., unpublished data).

10.4.4 Effect of alfalfa brown juice on the growth of cyanobacteria and green microalgae

Bákonyi et al. found that the culture medium containing fermented brown juice improved biomass production of cyanobacteria and green microalgae (Bákonyi et al., 2021; Figs. 10.5 and 10.6).

Based on these results, a line of microorganism-containing plant conditioner products (namely Fotolacto products; License no: 6300/817-2/2021; NÉBIH; nebih.gov.hu) has been developed and marketed in Hungary, in which the culture medium is fermented brown juice. Moreover, Danish researchers planned to focus on the production of microbial protein through the cultivation of methanotrophic bacteria or green microalgae (Hansen, 2021).

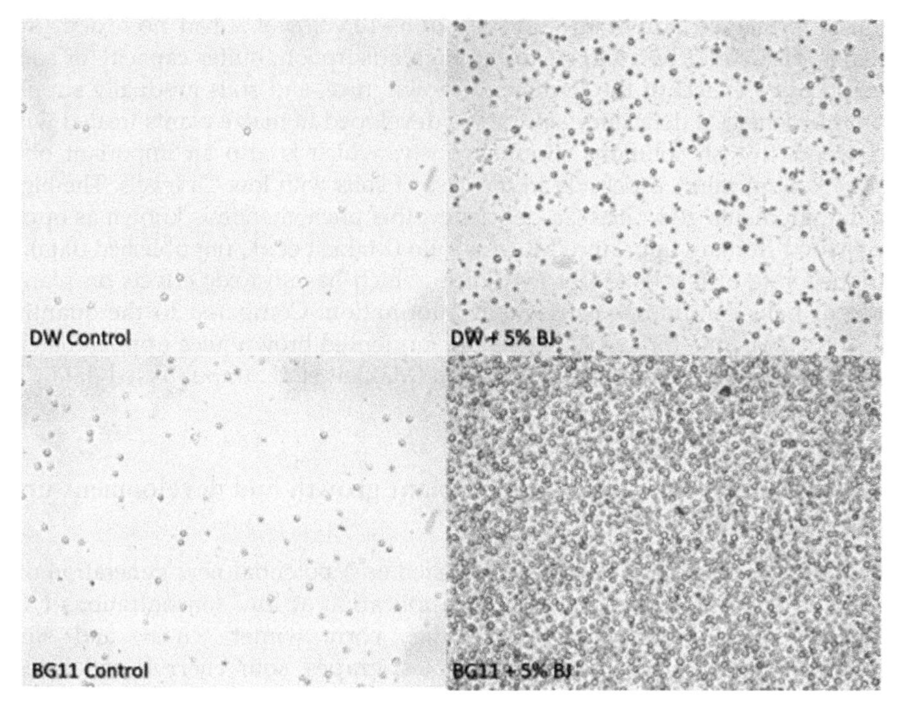

FIGURE 10.5 Micrographs of *Chlorella vulgaris* cultures grown in distilled water (DW) or BG11 medium (BG11) supplemented with 5% alfalfa brown juice.

FIGURE 10.6 *Chlorella vulgaris* cultures grown in distilled water supplemented with alfalfa brown juice in different concentrations.

References

Bákonyi, N., Fári, M., Domokos-Szabolcsy, É., Oláhné, T.I., Barn, D., 2021. Culture medium containing fermented brown juice. Fermentált növényi barnalét tartalmazó tenyészközeg. Patent application no. P2100223, submission no. 132299-1423 KOH/LZS, Hungary.

Bákonyi, N., Kisvarga, S., Barna, D., Tóth, I.O., El-Ramady, H., Abdalla, N., et al., 2020. Chemical traits of fermented alfalfa brown juice: ITS implications on physiological, biochemical, anatomical, and growth parameters of celosia. Agronomy 10 (2), 247. Available from: https://doi.org/10.3390/agronomy10020247.

Barna, D., Tóth, I.O., Fári, M.G., Bákonyi, N., 2020. Deproteinized plant juice as part of circular economy: a short review and brief experimental data. Acta Agrar. Debr. 2 (2), 23–26. Available from: https://doi.org/10.34101/ACTAAGRAR/2/3773.

Barna, D., Kisvarga, S., Kovács, S., Csatári, G., Tóth, I.O., Gábor Fári, M., et al., 2021. Raw and fermented alfalfa brown juice induces changes in the germination and development of French marigold (Tagetes Patula L.) plants. Plants 10 (6), 1076. Available from: https://doi.org/10.3390/plants10061076.

Barna, D., Alshaal, T., Tóth, I.O., Cziáky, Z., Gábor Fári, M., Éva Domokos-Szabolcsy, et al., 2022. Bioactive metabolite profile and antioxidant properties of brown juice, a processed alfalfa (Medicago Sativa) by-product. Heliyon 8 (11). Available from: https://doi.org/10.1016/j.heliyon.2022.e11655.

Bickoff, E.M., Knuckles, B.E., Spencer, R.R., Lazar, M.E., Kohler, G.O., 1970. PRO-XAN process: incorporation and evaluation of sugar cane rolls in wet fractionation of alfalfa. J. Agric. Food Chem. 18 (6), 1086–1089. Available from: https://doi.org/10.1021/jf60172a001.

Carroad, P.A., Anaya-serrano, H., Edwards, R.H., Kohler, G.O., 1981. Optimization of cell disruption for alfalfa leaf protein concentration (Pro-Xan) production. J. Food Sci. 46 (2), 383–386. Available from: https://doi.org/10.1111/j.1365-2621.1981.tb04866.x.

Chowdary, N.B., Naik, V.N., Sharma Govindaiah, D.D., 2002. Deproteinized mulberry leaf juice - a new media for growth of microorganisms. Int. J. Ind. Entomol. 5 (2).

Domokos-Szabolcsy, É., Elhawat, N., Jorge Domingos, G., Kovács, Z., Koroknai, J., Bodó, E., et al., 2022. Comparison of wet fractionation methods for processing broccoli agricultural wastes and evaluation of the nutri-chemical values of obtained products. Foods 11 (16), 2418. Available from: https://doi.org/10.3390/foods11162418.

Domokos-Szabolcsy, É., Reyhan Yavuz, S., Picoli, E., Gabor Fári, M., Kovács, Z., Tóth, C., et al., 2023. 'Green biomass-based protein for sustainable feed and food supply: an overview of current and future prospective. Life 13 (2), 307. Available from: https://doi.org/10.3390/life13020307.

Fári, M.G., Domokos-Szabolcsy, E., 2019. WO2019150144 (A1). Method for producing plant protein coagulum, August 2019.

Fári, M.G., 2018. A zöld fehérjemalom tudományos megalapozása és lehetséges szerepe a fehérjegazdálkodásban (Development of scientific basis of green proteomill concept and its possible influence on protein economy). Állattenyésztés és Takarmányozás 67 (4), 237–253.

Feng, L., James Ward, A., Ambye-Jensen, M., Bjarne Møller, H., 2021. Pilot-scale anaerobic digestion of by-product liquid (brown juice) from grass protein extraction using an un-heated anaerobic filter. Process Saf. Environ. Prot. 146 (February), 886–892. Available from: https://doi.org/10.1016/j.psep.2020.12.026.

Garg, R.C., 2016. Chapter 44 - Fenugreek: multiple health benefits. In: Gupta, R.C. (Ed.), Nutraceuticals. Academic Press, Boston, pp. 599–617. Available from: https://doi.org/10.1016/B978-0-12-802147-7.00044-9.

Hansen, M., 2021. [Project] - expanding the green biorefinery. https://projektbank.dtu.dk/en-us/Pages/BulletinView.aspx?EntityId = 91afab7b-3e59-ea11-8142-005056a057de.

Holló, J., 1970. Process for the production of fibreless green concentrate of full biological value. Danish Patent number: 138873.

Holló, J., 1972. Process for the production of fibreless green concentrate of full biological value. Canadian Patent number: 909063.

Holló, J., Zagyvai, I., Koch, L., 1969. Process for the production of fibreless green concentrate of full biological value. United States Patent number: 3637396; Application number: 846239.

Holló, J., László, E., 1971. The Synthesis of Dextran. I. Importance of Sucrose Concentration in the course of Dextran Fermentation. Period. Polytechn., pp. 35–42.

Iliyas, S., 2019. Use of deproteinised leaf juice of Medicago sativa L. for the production of α-amylase. J. Bio-Sci. 26 (December), 7–14. Available from: https://doi.org/10.3329/jbs.v26i0.44659.

Jadhav, R.K., Deshmukh, S., 2018. Study of yeast quality enhancement by Forage Deproteinised Juice (DPJ) fermentation. J. Exp. Sci. 8, 9–12.

Jadhav, R.K., Mungikar, A.M., 1998. Mitotic inhibition & chomosomal aberrations induced by deproteinised leaf juice of lucerne (Medicago Sativa Linn.) in root tips of onion (Alium Sepa). Int. J. Mendel. 15, 21–22.

Jadhav, R.K., 2018. Yeast utilizing deproteinised leaf juice (DPJ) as a medium for growth and production of metabolites. Plant. Arch. 18 (2), 1716–1720.

Jadhav, R.K., Deshmukh, S., 2018. Study of yeast quality enhancement by forage deproteinised juice (DPJ) fermentation. J. Exp. Sci. 9, 9–12. Available from: https://doi.org/10.25081/jes.2018.v9.3649.

Kamm, B., Schönicke, P., Hille, Ch, 2016. Green biorefinery - industrial implementation. Food Chem. 197 (April), 1341–1345. Available from: https://doi.org/10.1016/j.foodchem.2015.11.088.

Kiel, P., Andersen, M., 1994. Method for obtaining a culture medium from plant sap. Publication of EP0585234A1, Denmark.

Kiel, P., Andersen, M., Lübeck, M., 2015. A method of providing functional proteins from a plant material. December 2015.

Kiel, P, Andersen, M., Lübeck, M., 2015. A method of providing functional proteins from a plant material. PCT/DK2015/050185. Danish Patent and Trademark Office.

Kisvarga, S., Barna, D., Kovács, S., Csatári, G., Tóth, I.O., Gábor Fári, M., et al., 2020. Fermented alfalfa brown juice significantly stimulates the growth and development of sweet basil (Ocimum Basilicum L.) plants. Agronomy 10 (5), 657. Available from: https://doi.org/10.3390/agronomy10050657.

Knuckles, B.E., Bickoff, E.M., Kohler, G.O., 1972. Pro-Xan process: methods for increasing protein recovery from alfalfa. J. Agric. Food Chem. 20 (5), 1055–1057. Available from: https://doi.org/10.1021/jf60183a020.

Kromus, S., Kamm, B., Kamm, M., Fowler, P., Narodoslawsky, M., 2008. Green biorefineries: the green biorefinery concept - fundamentals and potential. Biorefineries-Industrial Processes and Products: Status Quo and Future Directions. Wiley-VCH Verlag GmbH, pp. 253–294Vol. 1. Available from: https://doi.org/10.1002/9783527619849.ch12.

Kromus, S., Wachter, B., Koschuh, W., Mandl, M., Krotscheck, C., Narodoslawsky, M., 2004. The green biorefinery Austria-development of an integrated system for green biomass utilization. Chem. Biochem. Eng. 18, 7–12.

Lamont, J., Wilkins, O., Smith, D.L., Lamont, J.R., Bywater-Ekeg, M., Ard, €, 2017. From yogurt to yield: potential applications of lactic acid bacteria in plant production. Soil Biol. Biochem (111), 1–9. Available from: https://doi.org/10.1016/j.soilbio.2017.03.015.

Makleit, P., Fári, M., Veres, S., 2018a. Proteinmentes Lucerna (Medicago Sativa L.) kivonat (DAJ) növénytáplálási célú felhasználása. In: XXXVII. Óvári Tudományos Napok, 2018. November 9–10 Fenntartható agrárium és környezet, az Óvári Akadémia 200 éve - múlt, jelen, jövő. Szerk.: Szalka Éva, VEAB Agrártudományi Szakbizottság, Széchenyi István Egyetem Mezőgazdaság-és Élelmiszertudományi Kar, Mosonmagyaróvár, 284–290, ISBN: 9786155837159.

Makleit, P., Fári, M.G., Bákonyi, N., Csajbók, J., Veres, S., 2018b. Application of brown juice as a by-product of alfalfa leaf protein concentrate production for plant nutrition. In: Arccal Vagy Háttal a Jövőnek? LX. Georgikon Napok, 240–246.

Makleit, P., Fári, M., Veres, S., (2019). Lucerna-savó növénytáplálási célú alkalmazása. In: Növénynemesítés a 21. század elején: kihívások és válaszok. XXV. Növénynemesítési Tudományos Nap Szerk: Karsai Ildikó, Magyar Tudományos Akadémia Agrártudományok Osztályának Növénynemesítési Tudományos Bizottsága, Budapest, 380–383. ISBN: 9789638351456. https://www.researchgate.net/profile/Peter-Makleit/publications.

Manwatkar, V.G., Gogle D P, Vidya Vikas Arts, 2014. The effect of deproteinised juice (DPJ) on seed germination and seedling growth of different plants (by paper towel method). Biology 65–68.

Mayorga, A.H., Gonzalez, J., Rolz, C., Agrie, J., Washington, D.C., Better, E., 1972. Pilot plant production of an edible white fraction leaf protein concentrate from alfalfa. Can. Inst. Food Sci. Technol. J. 20 (213), 24.

Minorsky, P.V., 2002. The hot and the classic: Trigonelline: a diverse regulator in plants. Plant Phys. 128, 7–8.

Naeem, M., Masroor, M., Khan, A., Moinuddin, 2012. Triacontanol: a potent plant growth regulator in agriculture. J. Plant Interact. 7 (2), 129–142. Available from: https://doi.org/10.1080/17429145.2011.619281.

Pirie, N.W., 1956. Unexploited technological possibilities of making food for man and animals. Proc. Nutr. Soc. 15 (2), 154–160. Available from: https://doi.org/10.1079/pns19560032.

Pirie, N.W., 1987. Leaf Protein and its By-Products in Human and Animal Nutrition. Cambridge University Press, p. 209no. 1 (January). Available from: https://doi.org/10.1017/s0014479700015805.

Ream, H.W., Smith, D., Walgenbach, R.P., 1977. Effects of deproteinized alfalfa juice applied to alfalfa - bromegrass, bromegrass, and corn 1. Agron. J. 69 (4), 685–689. Available from: https://doi.org/10.2134/agronj1977.00021962006900040040x.

Santamaría-Fernández, M., Karkov Ytting, N., Lübeck, M., 2019. Influence of the development stage of perennial forage crops for the recovery yields of extractable proteins using lactic acid fermentation. J. Clean. Prod. 218 (May), 1055–1064. Available from: https://doi.org/10.1016/j.jclepro.2019.01.292.

Santamaría-Fernández, M., Schneider, R., Lübeck, M., Venus, J., 2020. Combining the production of L-lactic acid with the production of feed protein concentrates from alfalfa. J. Biotechnol. 323 (November), 180–188. Available from: https://doi.org/10.1016/j.jbiotec.2020.08.010.

Saunders, R.M., Connor, M.A., Booth, A.N., Bickoff, E.M., Kohler, G.O., 1973. Measurement of digestibility of alfalfa protein concentrates by in vivo and in vitro methods. J. Nutr. 103 (4), 530–535. Available from: https://doi.org/10.1093/jn/103.4.530.

Sayyed, I.U., 2014. Production of A-amylase on deproteinised leaf juice prepared from different plants. Plant. Sci. Feed 4 (4), 31–35.

Sayyed, I.U., 2015. Fungal biomass production using deproteinised leaf juice (DPJ) of Trigonella Foenum-Graecum. Int. J. Bioassays 4.

Sayyed, I.U., Mungikar, A.M., 2005. Use of deproteinised leaf juice (DPJ) in microbial biotechnology. Pollut. Res. 24 (2), 459.

Sinclair, S., 2009. Protein extraction from pasture. Literature review part A: the plant fractionation bio-process and adaptability to farming systems. Milest. Rep. Prep. MAF SFF Grant C.

Slade, R.E., Birkinshaw, J.H.. 1938. Improvements in or relating to the utilization of grass and other green crops, issued on February 18, 1938.

Sukte, S.M., 2018. Physiological Studies on Some Field Crops of Beed District. Lulu Publication. Available from: https://www.lulu.com/it/it/shop/dr-savita-marutirao-sukte/physiological-studies-on-some-field-crops-of-beed-district/paperback/product-15vmjrn4.html?page = 1&pageSize = 4.

Thomsen, M.H., Kiel, P., 2008. Selection of lactic acid bacteria for acidification of brown juice (grass juice), with the aim of making a durable substrate for L-lysine fermentation. J. Sci. Food Agric. 88 (6), 976–983. Available from: https://doi.org/10.1002/jsfa.3176.

Thomsen, M.H., Bech, D., Kiel, P., 2004. Manufacturing of stabilised brown juice for L-lysine production-from university lab scale over pilot scale to industrial production. Chem. Biochem. Eng. Q. 18 (1), 37–46.

Volenec, J., Smith, D., Soberalske, R.M., Ream, H.W., 1979. Greenhouse alfalfa yields with single and split applications of deproteinized alfalfa juice 1. Agron. J. 71 (4), 695–697. Available from: https://doi.org/10.2134/agronj1979.00021962007100040043x.

Vosloh, C.J., Edwards, R.H., Enochian, R.V., Kuzmick, D.D., Kohler, G.O., 1976. Leaf Protein Concentrate (Pro-Xan) From Alfalfa: An Economic Evaluation. Agricultural Economic Reports 307613, United States Department of Agriculture, Economic Research Service.

Welch, D.A., Smith, D., Soberalske, R.M., Ream, H.W., 1979. Growth and composition of alfalfa fertilized in greenhouse trials with deproteinized juice from low and high saponin alfalfa and from oat herbage. J. Plant. Nutr. 1 (2), 151–170. Available from: https://doi.org/10.1080/01904167909362706.

Zanin, V., 1998. A New Nutritional Idea For Man: Lucerne Leaf Concentrate. APEF, Association Pour La Promotion Des Extraits Foliaires En Nutrition.

Welch, D.A., Smith, D., Soberalske, R.M., Ream, H.W., 1979. Growth and composition of alfalfa fertilized in greenhouse trials with deproteinized juice from low and high saponin alfalfa and from oat herbage, Journal of Plant Nutrition, . 2. Taylor and Francis Online, pp. 151–170.

Zhang, L., Yu, R., Yu, Y., Wang, C., Zhang, D., 2019. Determination of four acetanilide herbicides in brown rice juice by ionic liquid/ionic liquid-homogeneous liquid-liquid micro-extraction high performance liquid chromatography. Microchem. J. 146 (May), 115–120. Available from: https://doi.org/10.1016/j.microc.2018.12.062.

Zhou, J., Chan, L., Zhou, S., 2012. Trigonelline: a plant alkaloid with therapeutic potential for diabetes and central nervous system disease. Curr. Med. Chem. 19 (21), 3523–3531. Available from: https://doi.org/10.2174/092986712801323171.

CHAPTER

11

Use of plant water extracts as biostimulants to improve the plant tolerance against abiotic stresses

Muhammad Bilal Hafeez[1], *Asma Hanif*[2], *Sobia Shahzad*[2], *Noreen Zahra*[3,4], *Bilal Ahmad*[5], *Abida Kausar*[4], *Aaliya Batool*[3] *and Muhammad Usman Ibrahim*[1]

[1]Department of Agronomy, University of Agriculture, Faisalabad, Pakistan [2]Department of Botany, Islamia University of Bahawalpur, Bahawalnagar Campus, Pakistan [3]Department of Botany, University of Agriculture, Faisalabad, Pakistan [4]Department of Botany, Government College Women University Faisalabad, Pakistan [5]Agriculture Genomics Institute at Shenzhen, Chinese Academy of Agricultural Sciences, Shenzhen, P.R. China

11.1 Introduction

Food security is seriously threatened due to the burgeoning increase in population and changing climate (Zahra et al., 2022a). Abiotic stresses can disturb plant growth and productivity, which are of significant agricultural importance (Hafeez et al., 2021; Raza et al., 2021). Abiotic stresses such as salinity, drought, cold, heat, water logging, and heavy metals have adversely affected crop production throughout the globe (Mubarik et al., 2021; Zahra et al., 2021a; Shaukat et al., 2022). Many physiological, agronomic, and molecular approaches are used to enhance tolerance to many abiotic stresses using natural and synthetic plant growth regulators (Saddiq et al., 2019; Ahmad et al., 2022; Nizar et al., 2022). The concept of application of hormonal, molecular, and genetic regulations of plant growth in response to various stresses has remained under study for a few decades, while some fruitful and satisfactory achievements have been attained (Rhaman et al., 2020; Ali et al., 2022; Munir et al., 2022). Plants release numerous secondary metabolites (allelochemicals) that boost the tolerance against abiotic stresses when applied in low concentration

Biostimulants in Plant Protection and Performance
DOI: https://doi.org/10.1016/B978-0-443-15884-1.00023-3

(allelopathic hormesis) (Pannacci et al., 2022). Keeping in view, allelopathic hormesis, the least tested but potential approach, could be employed to reduce the adverse effect of abiotic stresses.

Application of allelopathic crop water extracts (ACWE) has been reported as a cost-effective and eco-friendly approach to improving the plant tolerance to various abiotic stresses in field crops (Kamran et al., 2019; Bajwa et al., 2020). The concept of allelopathic hormesis was first proposed by Southam (1943), while experiencing stimulatory response at lower concentrations released from the bark of an oak tree. The term hormesis originate from the Greek word *hormo* means to "excite" also explained this term while describing the stimulatory response of poisonous substance used at lower concentration (Aslam et al., 2017). Many plants naturally produce secondary metabolites and biogrowth regulators having many important ecological and biological functions or providing bases for interaction among neighboring plants and the environment through allelopathic phenomenon (Abbas et al., 2021; Aci et al., 2022). Exogenous use of plant growth regulators showed greater evidence to improve various abiotic stress tolerance in plants (Bulgari et al., 2019; Desoky et al., 2019a). Abbas et al. (2017) have documented a 50% and 42% increase in crop growth and grain yield, respectively, due to plant-released phytotoxins at lower concentrations under controlled conditions and a 42% increase in crop yield under field conditions. Screening of donor plants exhibiting allelopathic hormesis can be done by treating the seed with aqueous extract of such plants through seed priming and foliar application (Maqbool et al., 2013). ACWE of sorghum, neem, carrot, brassica, sugar beet, seaweed, sunflower, and moringa improve crop growth when applied at low concentrations and hence may be used as natural growth enhancer (Farooq et al., 2017; Bulgari et al., 2019). Many natural biostimulants such as moringa leaf extract (MLE), sunflower, sugar beet water extract, seaweed water extract, licorice root extract, carrot root extract and sorghum water extract have been reported for the mitigation of thermal (Farooq et al., 2018; Rashid et al., 2018), chilling (Batool et al., 2019), drought (Alghabari, 2018; Farooq et al., 2018; Khaliq et al., 2022), heavy metals (Khalofah et al., 2020; Ahmed et al., 2021), and salinity stress (Abdel Latef et al., 2019; Aboualhamed and Loutfy, 2020; Alamer et al., 2022; Zahra et al., 2022b).

Various plant-derived extracts contain many plant growth substances such as antioxidants, phytohormones, nutrients, and osmoprotectants that play a significant role in strengthening the defense system of plants to alleviate the adverse effect of environmental stresses. However, a comprehensive study related to this plant-derived water extracts is still missing. This chapter highlights the use of seaweed and seaweed extract, carrot root extract, MLE, brassica water extract, sorghum water extract, licorice root extract (LRE), and *Ocimum* extract against abiotic stresses (Fig. 11.1).

11.2 Application of crop extracts to enhance the tolerance against abiotic stresses

11.2.1 Moringa leaf extract

Moringa oleifera is a valuable tree species, whose leaves having biostimulant potential because they are enriched with mineral nutrients, enzymatic and nonenzymatic antioxidants, essential amino acids, metabolites, phytohormones, and vitamins (Aslam et al., 2020;

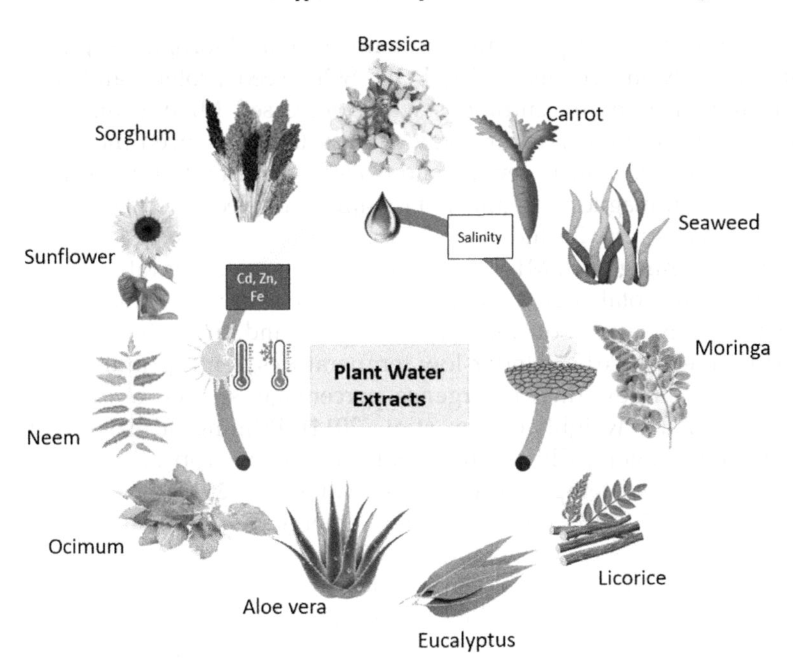

FIGURE 11.1 Effect of different crop water extracts to mitigate the abiotic stresses in various cropping plants.

Faisal et al., 2020; Jahanzaib et al., 2023; Mashamaite et al., 2022). Under drought stress, foliar spray of MLE increased iron, copper, manganese, zinc, SOD, catalase (CAT), and peroxidase (POD) while reduction in the hydrogen peroxide (H_2O_2), superoxide anion ($O_2\bullet^-$) and malondialdehyde (MDA) of wheat crop (Basu et al., 2022). Khan et al. (2022) stated that rice seed priming with MLE (3%) emergence index, final emergence percentage, fresh and dry biomass, shoot and root length, chlorophyll *a* (Chl *a*), chlorophyll *b* (Chl *b*), SOD, and CAT under water stress conditions. In mung bean, the foliar spray of MLE increased the plant fresh and dry weight, pods per plant, and pod dry weight under drought conditions (Alghabari, 2018). Under drought conditions, seed priming with MLE (3%) increased total soluble phenolics, total chlorophyll, grains per spike (GPS), hundred-grain weight, grain yield (GY), and biological yield (BY) of wheat (Farooq et al., 2018). In wheat, under combined terminal heat and drought stress exogenous application of MLE (3%) increased plant height, GPS, BY, GY, harvest index, total soluble phenolics and chlorophylls, and transpiration efficiency (Farooq et al., 2017).

The foliar spray of MLE (3%) increased the total chlorophyll, intrinsic water-use efficiency, photosynthetic rate, ascorbic acid, SOD, CAT, POD, and grain yield under heat stress (Rashid et al., 2018). Under terminal heat stress, the application of MLE (3%) improved number of GPS, GY, thousand-grain weight, SOD, CAT, free proline, and soluble sugars of wheat crop (Afzal et al., 2020). Under heat stress, foliar spray of MLE (3%) increased plant height, leaf weight, leaf area, fresh and dry weight, stem diameter, Chl *a*, Chl *b*, crude fiber and protein, total ash, and forage yield (Ahmad et al., 2016). Under heat stress, priming of MLE (3%) increased Chl *a*, Chl *b*, relative water content, membrane stability index, net assimilation rate, grain yield, biological yield, and grain protein of maize crop (Iqbal et al., 2020). Under terminal heat stress, application of MLE (3%) improved root and shoot lengths and their dry weights, seed and straw yield per plant, carotenoids, total chlorophylls, photosynthesis rate, soluble proteins, glycine betaine, CAT, POD, SOD, seed protein, and root, shoot and seed potassium and calcium concentrations while decreased MDA contents in quinoa crop (Rashid et al., 2022a). In quinoa, foliar spray of MLE (3%) enhanced the Chl *a*, Chl *b*, carotenoids, SOD, CAT, POD, panicle length and weight, thousand-grain weight, grain yield, seed nitrogen, calcium, magnesium, and potassium while decreased H_2O_2 and MDA contents under terminal heat stress (Rashid et al., 2020).

Under chilling stress, wheat priming with MLE (3%) enhanced germination rate, chlorophyll content, amylase activity, and total sugars (Afzal et al., 2012). Priming of maize with MLE (3%) increased sugars, α-amylase, root and shoot lengths, and their dry weights under low temperatures (Imran et al., 2013). Under low temperatures, lentil priming with MLE (3%) increased germination index, final emergence percentage, sugars, α-amylase, root and shoot lengths, and their dry weights (Imran et al., 2014). Priming of maize with MLE (3%) enhanced stand establishment, Chl *a*, Chl *b*, leaf area index, crop growth rate, and relative water content under low temperatures (Bakhtavar et al., 2015). Under low temperatures, maize priming with MLE (3%) increased final emergence percentage, emergence index, number of leaves per plant, plant height, cob diameter, number of grains per cob thousand-grain weight, grain yield, and biological yield (Junaid et al., 2019). Under low temperatures, MLE (3%) foliar application increased number of branches, number of leaves, membrane stability index, total chlorophyll, and phenolics of moringa plants (Batool et al., 2019). Under chilling stress, MLE (3%) leaf area index, crop growth rate, leaf area duration, total dry matter, plant height, number of grains per cob, thousand-grain weight, BY, GY, relative water contents, total phenolics, grain oil, grain protein, and grain starch contents while decreased electrolyte leakage (Waqas et al., 2017).

Under saline conditions, faba bean soil supplementation of MLE increased shoot fresh weight, ascorbate peroxidase (APX), CAT, Chl *a*, Chl *b*, and total phenolics (Ragab et al., 2022) (Table 11.1). MLE (3%) foliar spray increased total chlorophyll, SOD, CAT, POD, spike length, number of GPS, BY, and GY of wheat under saline-sodic conditions (Riaz et al., 2022). Under saline stress, soil supplementation of MLE (3%) increased the root and shoot fresh and dry weights and their lengths, number of leaves, number of roots, carotenoids, Chl *a*, Chl *b*, dehydroascorbate reductase (DHAR), monodehydroascorbate reductase (MDHAR), and glutathione reductase (GR), soluble sugars, free amino acids, free proline, glycine betaine while decreased H_2O_2 of milk thistle (Zahra et al., 2022c). Zahra et al. (2021b) noted that soil drenching of MLE (3%) enhanced the SOD, CAT, POD, soluble protein, alkaloid, anthocyanin, ascorbic acid, and riboflavin while decreased H_2O_2 and MDA of milk thistle. Foliar spray of MLE (3%) improved the plant height, grain protein, 1000 grain weight, yield of straw and grains, grain ions concentrations (iron, nitrogen, zinc, phosphorus, magnesium, and potassium), Chl *a*, Chl *b*, carotenoids contents, CAT, SOD, and POD under saline conditions of wheat (Merwad, 2020).

Foliar application of MLE increased total chlorophyll, carotenoids, proline, shoot and root length, leaves area plant, plant dry mass, pod protein, number of pods, pods yield, membrane stability index, SOD, POD, GR, and CAT under cadmium stress (Howladar, 2014). Similarly, under cadmium stress soil supplementation of MLE enhanced the ascorbic acid, SOD, POD, GR, and CAT while decreased of H_2O_2 and MDA garden cress (Khalofah et al., 2020).

TABLE 11.1 Effect of different crop water extracts to mitigate the abiotic stresses in various cropping plants.

Crop extracts	Abiotic stress and conditions	Treatment time and concentration	Mechanism of stress tolerance	Crop targeted	Reference
Moringa leaf extract	HS, late sown (28°C–35°C)	Seed priming and 1:30 (Moringa:water)	↑Emergence index, ↑grains per cob, ↑BY, ↑GY, ↑CMP, ↑RWC, ↑Chl *a*, Chl *b*, ↓HI	*Zea mays* L.	Mahboob et al. (2015)
Sorghum water extract	HS + DS imposed at booting stage, glass canopy temperature (↑4°C–2°C) than the ambient, 35% WHC till maturity	Spray at booting, anthesis and postanthesis stage and 3% extract	↑PH, ↑GPS, ↑HGW, ↑BY, ↑GY, ↑HI, ↑WUE, ↑TE, ↑TSP, ↑TCC	*Triticum aestivum* L.	Farooq et al. (2017)
Sunflower water extract	HS + DS imposed at booting stage, glass canopy temperature (↑4°C–2°C) than the ambient, 35% WHC till maturity	Spray at booting, anthesis and postanthesis stage and 3% extract	↑PH, ↑GPS, ↑HGW, ↑BY, ↑GY, ↑HI, ↑WUE, ↑TE, ↑TSP, ↑TCC	*T. aestivum* L.	Farooq et al. (2017)
Brassica water extract	HS + DS imposed at booting stage, glass canopy temperature (↑4°C–2°C) than the ambient, 35% WHC till maturity	Spray at booting, anthesis and postanthesis stage and 3% extract	↑PH, ↑GPS, ↑HGW, ↑BY, ↑GY, ↑HI, ↑WUE, ↑TE, ↑TSP, ↑TCC	*T. aestivum* L.	Farooq et al. (2017)
Moringa leaf extract	HS + DS imposed at booting stage, glass canopy temperature (↑4°C–2°C) than the ambient, 35% WHC till maturity	Spray at booting, anthesis and postanthesis stage and 3% extract	↑PH, ↑GPS, ↑HGW, ↑BY, ↑GY, ↑HI, ↑WUE, ↑TE, ↑TSP, ↑TCC	*T. aestivum* L.	Farooq et al. (2017)
Sorghum water extract	DS imposed at booting stage 35% WHC till maturity	Seed priming and 3% extract	↑GPS, ↑HGW, ↑BY, ↑GY, ↑HI, ↑WUE, ↑TE, ↑Chl *b*, ↑proline, ↑TSP, ↑GB, ↓MDA, ↓RMP	*T. aestivum* L.	Farooq et al. (2018)
Sunflower water extract	DS imposed at booting stage 35% WHC till maturity	Seed priming and 3% extract	↑GPS, ↑HGW, ↑BY, ↑GY, ↑HI, ↑WUE, ↑TE, ↑Chl *b*, ↑proline, ↑TSP, ↑GB, ↓MDA, ↓RMP	*T. aestivum* L.	Farooq et al. (2018)
Brassica water extract	DS imposed at booting stage 35% WHC till maturity	Seed priming and 3% extract	↑GPS, ↑HGW, ↑BY, ↑GY, ↑HI, ↑WUE, ↑TE, ↑Chl *b*, ↑proline, ↑TSP, ↑GB, ↓MDA, ↓RMP	*T. aestivum* L.	Farooq et al. (2018)
Moringa leaf extract	DS imposed at booting stage 35% WHC till maturity	Seed priming and 3% extract	↑GPS, ↑HGW, ↑BY, ↑GY, ↑HI, ↑WUE, ↑TE, ↑Chl *a*, ↑Chl *b*, ↑proline, ↑TSP, ↑GB, ↓MDA, ↓RMP	*T. aestivum* L.	Farooq et al. (2018)

(*Continued*)

TABLE 11.1 (Continued)

Crop extracts	Abiotic stress and conditions	Treatment time and concentration	Mechanism of stress tolerance	Crop targeted	Reference
Sorghum water extract	HS imposed at booting stage, glass canopy temperature (↑4°C–2°C) than the ambient	Seed priming and 3% extract	↑GPS, ↑HGW, ↑BY, ↑GY, ↑HI, ↑WUE, ↑TE, ↑Chl *a*, ↑Chl *b*, ↑proline, ↑TSP, ↑GB, ↓MDA, ↓RMP	*T. aestivum* L.	Farooq et al. (2018)
Sunflower water extract	HS imposed at booting stage, glass canopy temperature (↑4°C–2°C) than the ambient	Seed priming and 3% extract	↑GPS, ↑HGW, ↑BY, ↑GY, ↑HI, ↑WUE, ↑TE, ↑Chl *a*, ↑Chl *b*, ↑proline, ↑TSP, ↑GB, ↓MDA, ↓RMP	*T. aestivum* L.	Farooq et al. (2018)
Brassica water extract	HS imposed at booting stage, glass canopy temperature (↑4°C–2°C) than the ambient	Seed priming and 3% extract	↑GPS, ↑HGW, ↑BY, ↑GY, ↑HI, ↑WUE, ↑TE, ↑Chl *a*, ↑Chl *b*, ↑proline, ↑TSP, ↑GB, ↓MDA, ↓RMP	*T. aestivum* L.	Farooq et al. (2018)
Moringa leaf extract	HS imposed at booting stage, glass canopy temperature (↑4°C–2°C) than the ambient	Seed priming and 3% extract	↑GPS, ↑HGW, ↑BY, ↑GY, ↑HI, ↑WUE, ↑TE, ↑Chl *a*, ↑Chl *b*, ↑proline, ↑TSP, ↑GB, ↓MDA, ↓RMP	*T. aestivum* L.	Farooq et al. (2018)
Sorghum water extract	HS imposed at booting stage, glass canopy temperature (↑4°C–2°C) than the ambient	Spray at booting, anthesis and postanthesis stage and 3% extract	↑GPS, ↑HGW, ↑BY, ↑GY, ↑HI, ↑WUE, ↑TE, ↑TSP, ↑TCC	*T. aestivum* L.	Farooq et al. (2018)
Sunflower water extract	HS imposed at booting stage, glass canopy temperature (↑4°C–2°C) than the ambient	Spray at booting, anthesis and postanthesis stage and 3% extract	↑GPS, ↑HGW, ↑BY, ↑GY, ↑HI, ↑WUE, ↑TE, ↑TSP, ↑TCC	*T. aestivum* L.	Farooq et al. (2018)
Brassica water extract	HS imposed at booting stage, glass canopy temperature (↑4°C–2°C) than the ambient	Spray at booting, anthesis and postanthesis stage and 3% extract	↑GPS, ↑HGW, ↑BY, ↑GY, ↑HI, ↑WUE, ↑TE, ↑TSP, ↑TCC	*T. aestivum* L.	Farooq et al. (2018)
Moringa leaf extract	HS imposed at booting stage, glass canopy temperature (↑4°C–2°C) than the ambient	Spray at booting, anthesis and postanthesis stage and 3% extract	↑GPS, ↑HGW, ↑BY, ↑GY, ↑HI, ↑WUE, ↑TE, ↑TSP, ↑TCC	*T. aestivum* L.	Farooq et al. (2018)
Sorghum water extract	DS imposed at booting stage 35% WHC till maturity	Spray at booting, anthesis and postanthesis stage and 3% extract	↑GPS, ↑HGW, ↑BY, ↑GY, ↑HI, ↑WUE, ↑TE, ↑TSP, ↑TCC	*T. aestivum* L.	Farooq et al. (2018)
Sunflower water extract	DS imposed at booting stage 35% WHC till maturity	Spray at booting, anthesis and postanthesis stage and 3% extract	↑GPS, ↑HGW, ↑BY, ↑GY, ↑HI, ↑WUE, ↑TE, ↑TSP, ↑TCC	*T. aestivum* L.	Farooq et al. (2018)

Brassica water extract	DS imposed at booting stage 35% WHC till maturity	Spray at booting, anthesis and postanthesis stage and 3% extract	↑GPS, ↑HGW, ↑BY, ↑GY, ↑HI, ↑WUE, ↑TE, ↑TSP, ↑TCC	*T. aestivum* L.	Farooq et al. (2018)
Moringa leaf extract	DS imposed at booting stage 35% WHC till maturity	Spray at booting, anthesis and postanthesis stage and 3% extract	↑GPS, ↑HGW, ↑BY, ↑GY, ↑HI, ↑WUE, ↑TE, ↑TSP, ↑TCC	*T. aestivum* L.	Farooq et al. (2018)
Sorghum water extract	HS imposed at booting stage, plexi glass canopy temperature (↑7°C–10°C) than the ambient	Spray at anthesis and grain filling stage and 3% extract	↑Chl *a*, ↑Chl *b*, ↑carotenoids, ↑SOD, ↑CAT, ↑POD, ↑panicle length and weight, ↑thousand-grain weight, ↑HI, ↑GY, ↑nitrogen, ↑calcium, ↑magnesium, ↑potassium, ↓H_2O_2 ↓MDA	*Chenopodium quinoa* Willd.	Rashid et al. (2020)
Moringa leaf extract	HS imposed at booting stage, plexi glass canopy temperature (↑7°C–10°C) than the ambient	Spray at anthesis and grain filling stage and 3% extract	↑Chl *a*, ↑Chl *b*, ↑carotenoids, ↑SOD, ↑CAT, ↑POD, ↑panicle length and weight, ↑thousand-grain weight, ↑HI, ↑GY, ↑nitrogen, ↑calcium, ↑magnesium, ↑potassium, ↓H_2O_2 ↓MDA	*C. quinoa* Willd.	Rashid et al. (2020)
Moringa leaf extract	HS imposed at booting stage, plexi glass canopy temperature (↑7°C–10°C) than the ambient	Spray at anthesis and grain filling stage and 3% extract	↑TCC, ↑WUE, ↑Chl *a*, ↑Chl *b*, ↑ascorbic acid, ↑SOD, ↑CAT, ↑POD, ↑GY, ↑BY, ↑inflorescence length, ↑Ca, ↑Mg, ↑Zn, ↑Mg, ↑K, ↓Fe, ↓Na	*C. quinoa* Willd.	Rashid et al. (2018)
Sorghum water extract	HS, late sown	Spray at stem elongation stage and anthesis stage and 3% extract	↑Root length and dry weight, ↑shoot length and dry weight, ↑GY, ↑BY, ↑carotenoids, ↑TCC, ↑photosynthesis rate, ↑TSP, ↑GB, ↑CAT, ↑POD, ↑SOD, ↑seed protein, ↑K, ↑Ca, ↓MDA	*C. quinoa* Willd.	Rashid et al. (2022)
Moringa leaf extract	HS, late sown	Spray at stem elongation stage and anthesis stage and 3% extract	↑RL, ↑RDW, ↑SL, ↑SDW, ↑GY, ↑BY, ↑carotenoids, ↑TCC, ↑photosynthesis rate, ↑TSP, ↑GB, ↑CAT, ↑POD, ↑SOD, ↑seed protein, ↑K, ↑Ca, ↓MDA	*C. quinoa* Willd.	Rashid et al. (2022)
Moringa leaf extract	DS, watering every 30 days of interval	Spray at vegetative growth (50%), flowering, and boll development stage and 3.3% extract	↑Balls harvested, ↑ball weight, ↑seed cotton weight, ↑TSP, ↑phenolics	*Gossypium hirsutum* L.	Hussain et al. (2020)
Sugar beet root extract		Seed priming and 2% extract		*Lens culinaris* L.	Imran et al. (2014)

(*Continued*)

TABLE 11.1 (Continued)

Crop extracts	Abiotic stress and conditions	Treatment time and concentration	Mechanism of stress tolerance	Crop targeted	Reference
	CS, Seed placed incubator (CS; 10°C for 7 days and Normal 25°C for 7 days)		↑Germination index, ↑total sugars, ↑final emergence (%), ↑α-amylase, ↑RL, ↑SL, ↑RDW, ↑SDW		
Moringa leaf extract	CS, Seed placed incubator (CS; 10°C for 7 days and Normal 25°C for 7 days)	Seed priming and 3% extract	↑Germination index, ↑total sugars, final emergence (%), ↑α-amylase, ↑RL, ↑SL, ↑RDW, ↑SDW	*L. culinaris* L.	Imran et al. (2014)
Aloe vera leaf extract	CS, Seed placed incubator (CS; 10°C for 7 days and Normal 25°C for 7 days)	Seed priming and 2% extract	↑Germination index, ↑total sugars, final emergence (%), ↑α-amylase, ↑RL, ↑SL, ↑RDW, ↑SDW	*L. culinaris* L.	Imran et al. (2014)
Sorghum water extract	CS, first week of November 8°C–13°C	Spray at six leave and tasseling stage and 3%	↑Leaf area index, ↑total dry matter, ↑PH, ↑grains per cob, ↑thousand-GW, ↑BY, ↑GY, ↑RWC, ↑TSP, ↑grain oil, ↑grain protein, grain starch, ↓EL	*Z. mays* L.	Waqas et al. (2017)
Moringa leaf extract	CS, first week of November 8°C–13°C	Spray at six leave and tasseling stage and 3%	↑Leaf area index, ↑total dry matter, ↑PH, ↑grains per cob, ↑thousand-GW, ↑BY, ↑GY, ↑RWC, ↑TSP, ↑grain oil, ↑grain protein, ↑grain starch, ↓EL	*Z. mays* L.	Waqas et al. (2017)
Moringa leaf extract	CS, sowing time temperature 14°C–18°C but optimum sowing time 25°C–35°C	Spray first after 1 month and second after 2 months, 3% extract	↑Number of branches, ↑number of leaves, ↑TCC, ↑TSP, ↑membrane stability index	*Moringa oliefera* Lam.	Batool et al. (2019)
Licorice root extract	SS, EC = 14.6 dS/m	Three sprays at 25, 40, and 55 days after sowing; 5 g/L extract	↑APX, ↑SOD, ↑CAT, ↑GR, ↑POD, ↑proline, ↑RWC, ↑membrane stability index, ↑GY, ↑grain protein, ↑K, ↑N, ↓EL, ↓MDA, ↓H_2O_2, ↓$O_2^{\bullet -}$	*T. aestivum* L.	Elrys et al. (2020)
Seaweed extract	SS, EC = 85 mM	Soil drenching after every 14 days to till 90 days after sowing, 2.5 mL/L extract	↑K, ↑Mg, ↑Na, ↑glutamate, ↑glycine, ↑serine, ↑proline, ↑threonine ↓leaves dry and fresh weight, ↓fruit dry and fresh weight, ↓Ca, ↓Cl, ↓starch, hexoses, ↓glucose, ↓sucrose, ↓total amino acids, ↓TSP, ↓glutamine, ↓alanine, ↓GABA, ↓asparagine	*Solanum lycopersicum* L.	Dell'Aversana et al. (2021).

Carrot root extract	SS, seawater was applied	Seed priming and 2% extract	↑Shoot height, ↑RL, ↑RDW, ↑SDW, ↑Chl *a*, ↑Chl *b*, ↑carotenoids ↑total soluble sugars, ↑K, ↑proline, ↑TSP, ↑total free amino acid, ↑SOD, ↑CAT, ↑APX, ↑ascorbic acid, ↓MDA	*Z. mays* L.	Abdel Latef et al. (2019)
Ocimum sanctum leaf extract	Arsenic toxicity, 50 μM sodium arsenite applied to 5-day-old seedling	Apply once to 5-day-old seedling and 3.5 mL of 0.875 mg/mL	↑RL, ↑SL, ↑RW, ↑SW, ↑Oxidized glutathione, ↑RWC, ↓TBARS content, ↓reduced glutathione, ↓ascorbate, ↓DHAR, ↓SOD, ↓APX, ↓MDHAR, ↓CAT, ↓dehydroascorbate, ↓$OH^{\bullet}$, ↓H_2O_2, ↓$O_2^{\bullet-}$	*Oryza sativa L.*	Gautam et al. (2020)
Neem leaf extract	Arsenic toxicity, 50 μM sodium arsenite applied to a 5-day-old seedling	Apply once to a 5-day-old seedling and 3 mL of 0.75 mg/mL	↑RL, ↑SL, ↑RW, ↑SW, ↑oxidized glutathione, ↑RWC, ↓TBARS content, ↓reduced glutathione, ↓ascorbate, ↓DHAR, ↓SOD, ↓APX, ↓MDHAR, ↓CAT, ↓dehydroascorbate, ↓$OH^{\bullet}$, ↓H_2O_2, ↓$O_2^{\bullet-}$	*O. sativa L.*	Gautam et al. (2020)
Licorice extract	Field having EC = 7.75 dS/m and heavy metals contamination	Spray three times (20, 35, and 50 days after transplanting) and 0.5% extract	↑SDW, ↑number of fruits, ↑fruits yield, ↑TCC, ↑proline, ↑soluble sugars, ↑K, ↑P, ↑N	*Capsicum annuum*	Desoky et al. (2019a)

Notes: *BY*, Biological yield; *Chl a*, chlorophyll *a*; *Chl b*, chlorophyll *b*; *CMP*, cell membrane permeability; *CS*, chilling stress; *DS*, drought stress; *EL*, electrolyte leakage; *GB*, glycine betaine; *GPS*, grains per spike; *GY*, grain yield; *HGW*, 100-grain weight; *HI*, harvest index; *HS*, heat stress; *PH*, plant height; *RWCs*, relative water contents; *RDW*, root dry weight; *RL*, root length, *RMP*, relative membrane permeability; *SDW*, shoot dry weight; *SL*, shoot length; *TCC*, total chlorophyll contents; *TE*, transpiration efficiency; *TSP*, total soluble phenolics; *WHC*, water holding capacity; *WUE*, water-use efficiency.

11.2.2 Seaweed extract

Seaweed extract (SWE) contains numerous micro and macro nutrient elements, quaternary ammonium compounds, abscisic acid, auxins, cytokinins, vitamins, and amino acids (Battacharyya et al., 2015; Mukherjee and Patel, 2020). SWE has a positive role in enhancing the abiotic stress tolerance in crops (Guinan et al., 2013). Foliar spray of SWE enhanced flower and fruit number, pollen viability, sucrose, glucose, and fructose contents in tomatoes (Carmody et al., 2020). In maize, the SWE application enhanced the number of seeds per cob, cob diameter, cob length, cob fresh weight, stem diameter, plant height, CAT, APX, GR, SOD, MDHAR, proline while decreased MDA, $^{\circ}O_2^-$, and H_2O_2 under water deficit stress (Trivedi et al., 2018). Foliar treatment of SWE increased the water relative content, net carbon dioxide assimilation, water-use efficiency, SOD, POD, and APX while lower indole-3-acetic acid, abscisic acid, gibberellic acid, lipid peroxidation, and MDA under drought and heat stress conditions (Cabo et al., 2020). In Arabidopsis, priming with SWE enhanced rosette diameter, number of rosette leaves, APX and decreased H_2O_2 under water deficit conditions (Rasul et al., 2021). Under drought stress, soil supplementation of SWE increased the leaf area, PSII effective quantum yield, shoot and root dry weight, root volume, stem diameter, SOD, APX, POD, carotenoids, Chl *a*, and total chlorophyll while lessened MDA contents in soybean crop (do Rosário Rosa et al., 2021).

Under saline conditions, the soil supplementation of SWE increased the potassium magnesium, sodium, glutamate, glycine, serine, proline, and threonine while decreasing leaves' dry and fresh weight, fruit dry and fresh weight, calcium, chloride, nitrate starch, hexoses, glucose, sucrose, total amino acids, soluble proteins, glutamine, alanine, GABA, and asparagine in tomato plants (Dell'Aversana et al., 2021). Moreover, priming with SWE improved the shoot fresh weight, root dry weight, potassium, magnesium, calcium, fruit fresh weight and number of fruits in tomatoes under field saline conditions (Di Stasio et al., 2020). SWE application higher the plant height, leaf area, root length and volume, shoot and root dry weight, total chlorophyll contents, potassium uptake, but decreased sodium uptake and electrolyte leakage under saline conditions in *Calotropis procera* (Bahmani Jafarlou et al., 2022). Under heavy metals contamination, SWE application increased the Chl *a*, Chl *b*, carotenoids contents, root and shoot length, fresh and dry weights, leaf area, root nitrogen, phosphorus, potassium, sodium, calcium, and magnesium while decreasing root cadmium, lead, copper and nickel of radish (Ahmed et al., 2021).

11.2.3 Carrot root extract

Carrot root extract (CRE) is rich in β-carotene, thiamine, riboflavin, pyridoxine, and vitamins C, D, and E (Alasalvar et al., 2001; Mizgier et al., 2016). The seed priming of broad bean with CRE (100%) improved the shoot height, root length, leaf area, root and shoot fresh and dry weights and water contents, photosynthetic pigments, carotenoids contents, photosynthetic activity, CAT, POD, ascorbic acid, sucrose, starch and while decreased glucose, total soluble proteins, MDA contents under drought stress (Kasim et al., 2017).

Application of CRE improved shoot and root lengths, number of leaves, fresh and dry weights of seedlings, photosynthetic pigments, carotenoids contents, total carbohydrates, anthocyanins, flavonoids, ascorbic acid, and POD under saline conditions of cowpea (Abbas and Akladious, 2013). Seed priming of maize with CRE improved the shoot height, root length, root and shoot dry weights, photosynthetic pigments, carotenoids contents, potassium, root and shoot proline, phenolics, total free amino acid, total soluble sugars, SOD, CAT, APX, and ascorbic acid while decreased MDA contents under saline stress (Abdel Latef et al., 2019). Priming of *Lupinus termis* with CRE enhanced root and shoot fresh and dry weights and their lengths, leaf water contents, photosynthetic pigments, and carotenoids contents; however, this decreased proteins, total soluble sugars, MDA, CAT, SOD, and ascorbic acid under saline conditions (Nessim and Kasim, 2019).

11.2.4 Brassica water extract

Brassica water extract (BWE) contains a natural steroid named brassinolide which has been recognized as a plant growth promoter against abiotic stresses (Iqbal, 2014). In wheat, under combined terminal heat and drought stress exogenous application of BWE (3%) increased plant height, GPS, BY, GY, harvest index, total soluble phenolics and chlorophylls, and transpiration efficiency (Farooq et al., 2017). Under drought conditions, seed priming with BWE (3%) increased total soluble phenolics, total chlorophyll, GPS, hundred-grain weight, BY, and GY of wheat (Farooq et al., 2018). In wheat, foliar application of BWE (3%) enhanced the photosynthetic pigments, carotenoids, root and shoot lengths and their fresh and dry weights, photosynthetic rate, POD, SOD, and CAT while decreasing the H_2O_2 under drought stress (Khaliq et al., 2022). Moreover, seed priming with BWE (3%) also increased total soluble phenolics, total chlorophyll, GPS, hundred-grain weight, BY and GY of wheat under heat stress conditions (Farooq et al., 2018).

11.2.5 Sorghum water extract

Sorghum water extract contains numerous phenolics such as p-coumaric acid, p-hydroxybenzoic acid, vanillic acid, p-hydroxybenzaldehyde, and ferulic acid (Hussain et al., 2021). Under terminal heat stress, wheat seed priming with sorghum water extract (3%) increased total soluble phenolics, total chlorophyll, GPS, hundred-grain weight, BY, and GY (Farooq et al., 2018). Under terminal heat stress, the application of sorghum water extract (0.075%) improved GPS, GY, thousand-grain weight, SOD, CAT, free proline, and total soluble sugars in wheat (Afzal et al., 2020). Under heat stress, a foliar spray of sorghum water extract (1%) increased plant height, stem diameter, leaf weight per plant, leaf area, fresh and dry weight per plant, photosynthetic pigments, crude fiber and protein, total ash, and forage yield (Ahmad et al., 2016). Under terminal heat stress, application of sorghum water extract (3%) improved root and shoot lengths and their dry weights, seed and straw yield per plant, carotenoids, total chlorophylls, photosynthesis rate, soluble proteins, glycine betaine, CAT, POD, SOD, seed protein, and root, shoot and seed potassium and calcium concentrations while decreased MDA contents in quinoa crop (Rashid et al., 2022). In quinoa, foliar spray of sorghum water extract (3%) enhanced the Chl *a*, Chl *b*, carotenoids, SOD, CAT, POD, panicle length and weight, thousand-grain weight, grain yield, seed nitrogen, calcium, magnesium, and potassium while decreased H_2O_2 and MDA contents under terminal heat stress (Rashid et al., 2020).

In wheat, under combined terminal heat and drought stress exogenous application of sorghum water extract (3%) increased plant height, GPS, BY, GY, harvest index, total soluble phenolics and chlorophylls, and transpiration efficiency (Farooq et al., 2017). Under drought conditions, seed priming with sorghum water extract (3%) increased total soluble phenolics, total chlorophyll, grains per spike, hundred-grain weight, grain yield, and biological yield of wheat (Farooq et al., 2018). Similarly, foliar spray of sorghum water extract (2%) enhanced the root and shoot lengths and their fresh and dry weights, photosynthetic pigments, carotenoids, photosynthetic rate, POD, SOD, and CAT while decreasing the H_2O_2 under drought stress of wheat crop (Ibrahim et al., 2022).

Under low temperatures, maize priming with sorghum water extract (10 mL/L) increased shoot and root lengths and their dry weights, α-amylase activity, and sugar contents (Imran et al., 2013). Under chilling stress, sorghum water extract (3%) leaf area index, crop growth rate, leaf area duration, total dry matter, plant height, number of grains per cob, thousand-grain weight, grain and biological yield, relative water contents, total phenolics, grain oil, grain protein, and grain starch contents while decreased electrolyte leakage (Waqas et al., 2017).

The sorghum water extract (5%) increased emergence (%), root and shoot lengths and their dry weights, α-amylase activity, Chl *a*, Chl *b*, CAT, POD, SOD activity, and shoot potassium ions while decreased H_2O_2 and MDA contents of camelina under saline conditions (Huang et al., 2021). Under salt stress, sorghum water extract (5%) final germination percent, germination index, leaf expansion, tillers, seedling fresh and dry weight, plant height, total phenolics, Chl *a*, Chl *b*, total chlorophyll, soluble proteins, total sugars, and a-amylase while decreased mean germination time, time to 50% germination of wheat crop (Bajwa et al., 2018). Similarly, foliar application of sorghum water extract (2%) under saline conditions increased the root and shoot fresh and dry weights, photosynthetic pigments, POD, SOD, and CAT, while decreasing the H_2O_2 and MDA contents of maize (Alamer et al., 2022).

11.2.6 Sunflower water extract

Sunflower water extract contains six-membered ring lactams, alcohols, olefins, ester, and aromatic compounds that could also be present in sesquiterpene lactones (Taiwo and Makinde, 2005). In wheat, under combined terminal heat and drought stress exogenous application of sunflower water extract (3%) increased plant height, GPS, BY, GY, harvest index, total soluble phenolics and chlorophylls, and transpiration efficiency (Farooq et al., 2017). Under drought conditions, seed priming with sunflower water extract (3%) increased total soluble phenolics, total chlorophyll, GPS, hundred-grain weight, BY and GY of wheat (Farooq et al., 2018). Seed priming with sunflower water extract (3%) increased total soluble phenolics, total chlorophyll, GPS, hundred-grain weight, BY and GY of wheat under terminal heat stress (Farooq et al., 2018). In *Arabidopsis*, sunflower water extract increased the fresh and dry weight of seedlings, Chl *a*, Chl *b*, carotenoids contents, SOD, APX, POD, and CAT activities while decreased electrical conductivity, H_2O_2 and MDA contents under saline conditions (Li et al., 2022). Sunflower water extract (5%) improved germination (%), number of leaves while decreasing mean germination time in rice crops under saline conditions (Muhammad et al., 2011).

11.2.7 Neem water extract

Neem (*Azadirachta indica*) leaves contain about 300 primary and secondary metabolites, and fats, proteins, and carbohydrate derivatives (Rahal et al., 2019). The exogenous spray of neem leaf extract (5%) enhanced the root and shoot dry and fresh weight, photosynthetic pigments, carotenoids, free proline, phenolics, glycine betaine, CAT, SOD, and POD while reducing the hydrogen peroxide and relative membrane permeability in quinoa under drought stress (Naz et al., 2022). In mung bean, the foliar spray of neem water extract increased the plant's fresh and dry weight, pods per plant, and pod dry weight under drought conditions (Alghabari, 2018). In rice, under arsenic toxicity, the neem application increased root and shoot length and their weights, oxidized glutathione (GSSG), relative water content, however, decreased TBARS content, reduced glutathione (GSH), ascorbate, DHAR, SOD, APX, MDHAR, CAT, dehydroascorbate (DHA), $OH^{\bullet}$, H_2O_2, and $O_2^{\bullet -}$ in rice (Gautam et al., 2020).

11.2.8 Licorice root extract

LRE (*Glycyrrhiza glabra* L.) is rich source of antioxidants, osmoprotectants such as ascorbic acid, α-tocopherol, salicylic acid, soluble sugars, amino acids, glutathione, proline, vitamins (i.e., A, E, and B groups), trihydroxy acid, glycyrrhizic acid, potassium, and calcium (Rady et al., 2019; Desoky et al., 2019a; Adam et al., 2022). Foliar spray of LRE (5%) improved the plant height, grain protein, 1000 grain weight, yield of straw and grains, grain ions concentrations (iron, nitrogen, zinc, phosphorus, magnesium, and potassium), Chl *a*, Chl *b*, carotenoids contents, CAT, SOD, and POD under saline conditions of wheat (Merwad, 2020). In common beans, the foliar application of LRE (0.5%) better the shoot length and dry weight, number of leaves, leaf area, number of pods, pod yield, free proline, total soluble carbohydrates and sugars, selenium, Chl *a*, Chl *b*, carotenoids contents, APX, SOD, CAT, membrane stability index, while decreased electrolyte leakage, MDA, H_2O_2, and $O_2^{\bullet -}$ under saline conditions (Rady et al., 2019). Under salinity stress, the foliar application of LRE (5 g/L) improved the APX, SOD, CAT, GR, POD, proline, relative water content, membrane stability index, grain yield, and yield-related attributes, grain protein, potassium, and nitrogen, however, lower the electrolyte leakage, MDA, H_2O_2, and $O_2^{\bullet -}$ of wheat (Elrys et al., 2020). Seed priming of peas with LRE (0.5%) increased the dry and fresh weights, stomatal conductance, net photosynthetic and transpiration rate, photosynthetic pigments, soluble sugars, CAT, SOD, GR, APX, POD, α-tocopherol, free proline, though, decreased the electrolyte leakage, MDA, H_2O_2, and $O_2{\bullet}^{-}$ under saline conditions (Desoky et al., 2019c).

Under combined heavy metal and saline stress, the foliar treatment of LRE (0.5%) increased the shoot dry weight, number of fruits, fruit yield, chlorophyll contents, proline, soluble sugars, potassium, phosphorus, and nitrogen in pepper plants (Desoky et al.,

2019a). The foliar spray of LRE (0.5%) enhanced the shoot dry weight, number of fruits, fruits yield, chlorophyll contents, proline, soluble sugars, potassium, phosphorus, nitrogen, CAT, SOD, GR, APX, and POD under combined heavy metal and saline stress of pepper (Desoky et al., 2019b).

11.2.9 *Ocimum sanctum* leaf extract

Fresh leaves of *Ocimum sanctum* (*O. sanctum* leaf extract [OLE]) contains rosmarinic acid, apigenin, isothymonin, isothymusin, cirsimaritin, cirsilineol, eugenol, antioxidant compounds, ursolic acid, oleanolic acid, β-caryophyllene, linalool, carvacrol, saponins, flavonoids, tannins, and alkaloids (Mahajan et al., 2013; Singh and Chaudhuri, 2018). In rice, the foliar spray of OLE enhanced the chlorophyll fluorescence, superoxide dismutase (SOD), and number of filled grains per panicle, however, decreased MDA, and proline contents under drought stress (Pandey et al., 2016). Under salinity stress, the OLE treatment increased the plant fresh and dry weights, proline, soluble carbohydrates and proteins, antioxidants activity, mineral elements, and lessened the MDA contents of faba bean (Aboualhamed and Loutfy, 2020). In rice, under arsenic toxicity, the OLE application increased root and shoot length and their weights, GSSG, relative water content, however, decreased TBARS content, GSH, ascorbate, DHAR, SOD, APX, MDHAR, CAT, DHA, $OH^{\bullet}$, H_2O_2, and $O_2^{\bullet -}$ in rice (Gautam et al., 2020).

11.2.10 Other crops extracts

The foliar application of *Eucalyptus globulus* extract under drought conditions enhanced the plant height, plant dry weight, pod per plant, pod length, and pod dry weight of mung bean (Alghabari, 2018). Seed priming of broad bean with garlic extract (100%) improved the shoot height, leaf area, root length, root and shoot dry and fresh weights and water contents, photosynthetic pigments, carotenoids contents, photosynthetic activity, CAT, POD, ascorbic acid, sucrose, starch and while decreasing glucose, total soluble proteins, and MDA contents under drought stress (Kasim et al., 2017). Under low temperatures, lentil priming with aloe vera leaf extract (3%) increased the germination index, final emergence percentage, total sugars, α-amylase, root and shoot lengths, and their dry weights (Imran et al., 2014). Under low temperatures, lentil priming with sugar beet root extract (3%) increased germination index, final emergence percentage, total sugars, α-amylase, root and shoot lengths, and their dry weights (Imran et al., 2014). *Arthrocnemum macrostachyum* extract under saline conditions increased the soybean root and shoot dry and fresh weights and their lengths, number of leaves, photosynthetic pigments, carotenoids contents, soluble sugars and proteins, total free amino acids, proline, total phenolics, and ascorbic acid, however, this decreased H_2O_2 and MDA contents (Osman et al., 2021). Cypress leaf extract (0.5%) fresh and dry weights, photosynthetic quantum yield, total chlorophylls, rubisco activity, carbon dioxide assimilation rate, intercellular carbon dioxide concentration, ascorbate, GSH, GSSH, and DHA while decreased MDA and H_2O_2 in zucchini (ElSayed et al., 2022).

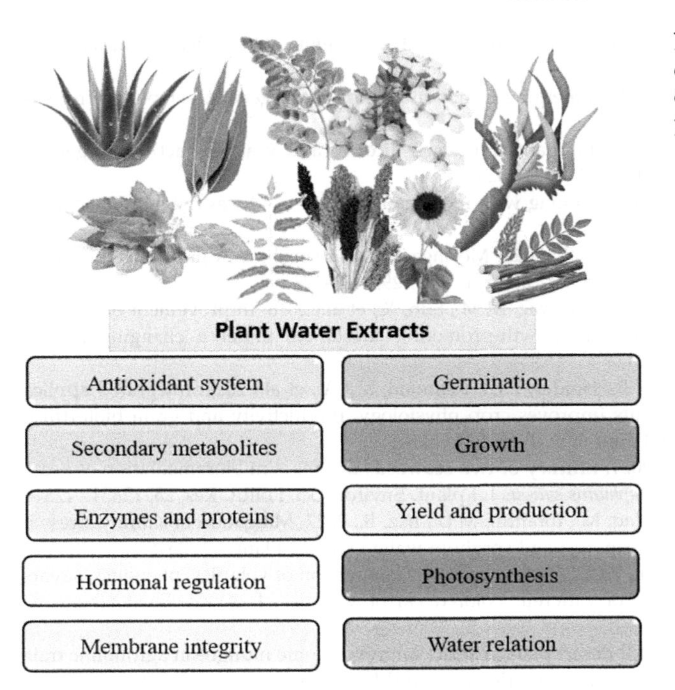

FIGURE 11.2 Effect of plant-derived extracts on different physiological, biochemical pathways and plant growth and yield attributes.

11.3 Conclusions and future perspective

Different crop water extracts induce abiotic stress tolerance in various crop species through modulation of physiological processes including photosynthesis, secondary metabolite biosynthesis, reactive oxygen species sequestration and antioxidant production (Fig. 11.2). Exploitation of these crop water extracts such as seaweed, carrot root water extract, neem, licorice root water extract, brassica and sorghum water extract and moringa and OLE could be helpful to mitigate the adverse effects of abiotic stresses under the changing climate scenario and improve the crop growth and yield production. However, the genetic mechanism to improve the stress-tolerant genes with the application of different water extracts of cropping plants is still unascertained. Moreover, dose-dependent use of other cropping plants may also be tried to improve the pave of biofertilizers.

References

Abbas, S.M., Akladious, S.A., 2013. Application of carrot root extract induced salinity tolerance in cowpea (*Vigna sinensis* L.) seedlings. Pak. J. Bot. 45, 795–806.

Abbas, T., Nadeem, M.A., Tanveer, A., Chauhan, B.S., 2017. Can hormesis of plant-released phytotoxins be used to boost and sustain crop production? Crop. Prot. 93, 69–76.

Abbas, A., Huang, P., Hussain, S., Saqib, M., He, L., Shen, F., et al., 2021. Application of allelopathic phenomena to enhance growth and production of camelina (*Camelina sativa* (L.)). Appl. Ecol. Environ. Res. 19, 453–469.

Abdel Latef, A.A.H., Mostofa, M.G., Rahman, M., Abdel-Farid, I.B., Tran, L.-S.P., 2019. Extracts from yeast and carrot roots enhance maize performance under seawater-induced salt stress by altering physio-biochemical characteristics of stressed plants. J. Plant Growth Regul. 38, 966–979.

Aboualhamed, M.F., Loutfy, N., 2020. *Ocimum basilicum* leaf extract induces salinity stress tolerance in faba bean plants. Egypt. J. Bot. 60, 681–690.

Aci, M.M., Sidari, R., Araniti, F., Lupini, A., 2022. Emerging trends in allelopathy: a genetic perspective for sustainable agriculture. Agronomy 12, 2043.

Adam, A.R., Allafe, M.A., Omar, E.A., 2022. Biostimulants influence (licorice and yeast extract) on vegetative growth of faba bean (*Vicia faba* L.). J. Plant Prod. 13, 321–324.

Afzal, I., Hussain, B., Basra, S., Rehman, H., 2012. Priming with moringa leaf extract reduces imbibitional chilling injury in spring maize. Seed Sci. Technol. 40, 271–276.

Afzal, I., Akram, M., Rehman, H., Rashid, S., Basra, S., 2020. Moringa leaf and sorghum water extracts and salicylic acid to alleviate impacts of heat stress in wheat. S. Afr. J. Bot. 129, 169–174.

Ahmad, W., Noor, M.A., Afzal, I., Bakhtavar, M.A., Nawaz, M.M., Sun, X., et al., 2016. Improvement of sorghum crop through exogenous application of natural growth-promoting substances under a changing climate. Sustainability 8, 1330.

Ahmad, N., Virk, A.L., Hussain, S., Hafeez, M.B., Haider, F.U., Rehmani, M.I.A., et al., 2022. Integrated application of plant bioregulator and micronutrients improves crop physiology, productivity and grain biofortification of delayed sown wheat. Environ. Sci. Pollut. Res. 29, 52534–52543.

Ahmed, D.A.E.-A., Gheda, S.F., Ismail, G.A., 2021. Efficacy of two seaweeds dry mass in bioremediation of heavy metal polluted soil and growth of radish (*Raphanus sativus* L.) plant. Environ. Sci. Pollut. Res. 28, 12831–12846.

Alamer, K.H., Perveen, S., Khaliq, A., Zia Ul Haq, M., Ibrahim, M.U., Ijaz, B., 2022. Mitigation of salinity stress in maize seedlings by the application of vermicompost and sorghum water extracts. Plants 11, 2548.

Alasalvar, C., Grigor, J.M., Zhang, D., Quantick, P.C., Shahidi, F., 2001. Comparison of volatiles, phenolics, sugars, antioxidant vitamins, and sensory quality of different colored carrot varieties. J. Agric. Food Chem. 49, 1410–1416.

Alghabari, F.M., 2018. Foliar application of Saudi desert plants extract improved some mungbean agronomic traits under drought stress. J. King Abdulaziz Univ. Meteorol. Environ. Arid Land Agric. Sci 27, 21–29.

Ali, S., Moon, Y.-S., Hamayun, M., Khan, M.A., Bibi, K., Lee, I.-J., 2022. Pragmatic role of microbial plant biostimulants in abiotic stress relief in crop plants. J. Plant Interact. 17, 705–718.

Aslam, F., Khaliq, A., Matloob, A., Tanveer, A., Hussain, S., Zahir, Z.A., 2017. Allelopathy in agro-ecosystems: a critical review of wheat allelopathy-concepts and implications. Chemoecology 27, 1–24.

Aslam, M.F., Basra, S., Hafeez, M.B., Khan, S., Irshad, S., Iqbal, S., et al., 2020. Inorganic fertilization improves quality and biomass of *Moringa oleifera* L. Agrofor. Syst. 94, 975–983.

Bahmani Jafarlou, M., Pilehvar, B., Modaresi, M., Mohammadi, M., 2022. Seaweed liquid extract as an alternative biostimulant for the amelioration of salt-stress effects in *Calotropis procera* (Aiton) WT. J. Plant Growth Regul. 42, 449–464.

Bajwa, A.A., Farooq, M., Nawaz, A., 2018. Seed priming with sorghum extracts and benzyl aminopurine improves the tolerance against salt stress in wheat (*Triticum aestivum* L.). Physiol. Mol. Biol. Plants 24, 239–249.

Bajwa, A.A., Nawaz, A., Farooq, M., 2020. Allelopathic crop water extracts application improves the wheat productivity under low and high fertilizer inputs in a semi-arid environment. Int. J. Plant Prod. 14, 23–35.

Bakhtavar, M.A., Afzal, I., Basra, S.M.A., Ahmad, A.-U.-H., Noor, M.A., 2015. Physiological strategies to improve the performance of spring maize (*Zea mays* L.) planted under early and optimum sowing conditions. PLoS ONE 10, e0124441.

Basu, S., Prabhakar, A.A., Kumari, S., Kumar, R.R., Shekhar, S., Prakash, K., et al., 2022. Micronutrient and redox homeostasis contribute to *Moringa oleifera*-regulated drought tolerance in wheat. Plant Growth Regul. 1–12.

Batool, S., Khan, S., Basra, S.M., Hussain, M., Saddiq, M.S., Iqbal, S., et al., 2019. Impact of natural and synthetic plant stimulants on Moringa seedlings grown under low-temperature conditions. Int. Lett. Nat. Sci. 76, 50–59.

Battacharyya, D., Babgohari, M.Z., Rathor, P., Prithiviraj, B., 2015. Seaweed extracts as biostimulants in horticulture. Sci. Hortic. 196, 39–48.

Bulgari, R., Franzoni, G., Ferrante, A., 2019. Biostimulants application in horticultural crops under abiotic stress conditions. Agronomy 9, 306.

Cabo, S., Morais, M.C., Aires, A., Carvalho, R., Pascual-Seva, N., Silva, A.P., et al., 2020. Kaolin and seaweed-based extracts can be used as middle and long-term strategy to mitigate negative effects of climate change in physiological performance of hazelnut tree. J. Agron. Crop. Sci. 206, 28–42.

Carmody, N., Goñi, O., Łangowski, Ł., O'Connell, S., 2020. *Ascophyllum nodosum* extract biostimulant processing and its impact on enhancing heat stress tolerance during tomato fruit set. Front. Plant Sci. 11, 807.

Dell'Aversana, E., Cirillo, V., Van Oosten, M.J., Di Stasio, E., Saiano, K., Woodrow, P., et al., 2021. *Ascophyllum nodosum* based extracts counteract salinity stress in tomato by remodeling leaf nitrogen metabolism. Plants 10, 1044.

Desoky, E.-S.M., Elrys, A.S., Rady, M.M., 2019a. Integrative moringa and licorice extracts application improves *Capsicum annuum* fruit yield and declines its contaminant contents on a heavy metals-contaminated saline soil. Ecotoxicol. Environ. Saf. 169, 50–60.

Desoky, E., Elrys, A.S., Rady, M.M., 2019b. Licorice root extract boosts *Capsicum annuum* L. production and reduces fruit contamination on a heavy metals-contaminated saline soil. Int. Lett. Nat. Sci. 73, 1–16.

Desoky, E.-S.M., ElSayed, A.I., Merwad, A.-R.M., Rady, M.M., 2019c. Stimulating antioxidant defenses, antioxidant gene expression, and salt tolerance in *Pisum sativum* seedling by pretreatment using licorice root extract (LRE) as an organic biostimulant. Plant Physiol. Biochem. 142, 292–302.

Di Stasio, E., Cirillo, V., Raimondi, G., Giordano, M., Esposito, M., Maggio, A., 2020. Osmo-priming with seaweed extracts enhances yield of salt-stressed tomato plants. Agronomy 10, 1559.

do Rosário Rosa, V., Dos Santos, A.L.F., da Silva, A.A., Sab, M.P.V., Germino, G.H., Cardoso, F.B., et al., 2021. Increased soybean tolerance to water deficiency through biostimulant based on fulvic acids and *Ascophyllum nodosum* (L.) seaweed extract. Plant Physiol. Biochem. 158, 228–243.

Elrys, A.S., Abdo, A.I., Abdel-Hamed, E.M., Desoky, E.-S.M., 2020. Integrative application of licorice root extract or lipoic acid with fulvic acid improves wheat production and defenses under salt stress conditions. Ecotoxicol. Environ. Saf. 190, 110144.

ElSayed, A.I., Rafudeen, M.S., Ganie, S.A., Hossain, M.S., Gomaa, A.M., 2022. Seed priming with cypress leaf extract enhances photosynthesis and antioxidative defense in zucchini seedlings under salt stress. Sci. Hortic. 293, 110707.

Faisal, M., Iqbal, S., Basra, S., Afzal, I., Saddiq, M., Bakhtavar, M., et al., 2020. Moringa landraces of Pakistan are potential source of premium quality oil. S. Afr. J. Bot. 129, 397–403.

Farooq, M., Rizwan, M., Nawaz, A., Rehman, A., Ahmad, R., 2017. Application of natural plant extracts improves the tolerance against combined terminal heat and drought stresses in bread wheat. J. Agron. Crop. Sci. 203, 528–538.

Farooq, M., Nadeem, F., Arfat, M., Nabeel, M., Musadaq, S., Cheema, S., et al., 2018. Exogenous application of allelopathic water extracts helps improving tolerance against terminal heat and drought stresses in bread wheat (*Triticum aestivum* L. Em. Thell.). J. Agron. Crop. Sci. 204, 298–312.

Gautam, A., Pandey, A.K., Dubey, R.S., 2020. *Azadirachta indica* and *Ocimum sanctum* leaf extracts alleviate arsenic toxicity by reducing arsenic uptake and improving antioxidant system in rice seedlings. Physiol. Mol. Biol. Plants 26, 63–81.

Guinan, K., Sujeeth, N., Copeland, R., Jones, P., O'brien, N., Sharma, H., et al., 2013. Discrete roles for extracts of *Ascophyllum nodosum* in enhancing plant growth and tolerance to abiotic and biotic stresses. Acta Hortic. 1009, 127–135.

Hafeez, M.B., Zahra, N., Zahra, K., Raza, A., Khan, A., Shaukat, K., et al., 2021. Brassinosteroids: molecular and physiological responses in plant growth and abiotic stresses. Plant Stress 2, 100029.

Howladar, S.M., 2014. A novel *Moringa oleifera* leaf extract can mitigate the stress effects of salinity and cadmium in bean (*Phaseolus vulgaris* L.) plants. Ecotoxicol. Environ. Saf. 100, 69–75.

Huang, P., He, L., Abbas, A., Hussain, S., Hussain, S., Du, D., et al., 2021. Seed priming with sorghum water extract improves the performance of camelina (*Camelina sativa* (L.) crantz.) under salt stress. Plants 10, 749.

Hussain, N., Yasmeen, A., Afzal, M.A., 2020. Exogenously applied growth promoters modulate the antioxidant enzyme system to improve the cotton productivity under water stress conditions. Ital. J. Agron. 15, 165–171.

Hussain, M.I., Danish, S., Sánchez-Moreiras, A.M., Vicente, Ó., Jabran, K., Chaudhry, U.K., et al., 2021. Unraveling sorghum allelopathy in agriculture: concepts and implications. Plants 10, 1795.

Ibrahim, M.U., Khaliq, A., Hussain, S., Murtaza, G., 2022. Sorghum water extract application mediates antioxidant defense and confers drought stress tolerance in wheat. J. Plant Growth Regul. 41, 863–874.

Imran, S., Afzal, I., Basra, S., Saqib, M., 2013. Integrated seed priming with growth promoting substances enhances germination and seedling vigour of spring maize at low temperature. Int. J. Agric. Biol. 15, 1251–1257.

Imran, S., Afzal, I., Amjad, M., Akram, A., Khawar, K., Pretorius, S., 2014. Seed priming with aqueous plant extracts improved seed germination and seedling growth under chilling stress in Lentil (*Lens culinaris* Medik). Acta Adv. Agric. Sci. 2, 58–69.

Iqbal, M.A., 2014. Role of moringa, brassica and sorghum water extracts in increasing crops growth and yield: a review. Am. Eurasian J. Agric. Environ. Sci. 14, 1150–1158.

Iqbal, H., Yaning, C., Waqas, M., Ahmed, Z., Raza, S.T., Shareef, M., 2020. Improving heat stress tolerance in late planted spring maize by using different exogenous elicitors. Chil. J. Agric. Res. 80, 30–40.

Jahanzaib, Saddiq, M.S., Basra, S.M., Nawaz, M.F., Iqbal, S., Khan, S., et al., 2023. Inorganic NPK supplementation improves biomass and quality of multicut moringa fodder. J. Plant Nutr. 46, 1519–1532.

Junaid, M., Alderfasi, A., Afzal, I., Wajid, H., Mahmood, A., 2019. Evaluation of different strategies for induction of chilling tolerance in spring maize using moringa leaf extracts. Cercetari Agron. Moldova 52, 228–247.

Kamran, M., Cheema, Z.A., Farooq, M., Ali, Q., Anjum, M.Z., Raza, A., 2019. Allelopathic influence of sorghum aqueous extract on growth, physiology and photosynthetic activity of maize (*Zea mays* L.) seedling. Philipp. Agric. Sci. 102, 33–41.

Kasim, W.A.E.-A., Nessem, A.A., Gaber, A., 2017. Alleviation of drought stress in Vicia faba by seed priming with ascorbic acid or extracts of garlic and carrot. Egypt. J. Bot. 57, 45–59.

Khaliq, A., Ibrahim, M.U., Hussain, S., Zia Ul Haq, M., Al-Huqail, A.A., Nawaz, M., et al., 2022. The hormetic effects of a Brassica water extract triggered wheat growth and antioxidative defense under drought stress. Appl. Sci. 12, 4582.

Khalofah, A., Bokhari, N., Migdadi, H., Alwahibi, M., 2020. Antioxidant responses and the role of *Moringa oleifera* leaf extract for mitigation of cadmium stressed *Lepidium sativum* L. S. Afr. J. Bot. 129, 341–346.

Khan, S., Ibrar, D., Bashir, S., Rashid, N., Hasnain, Z., Nawaz, M., et al., 2022. Application of moringa leaf extract as a seed priming agent enhances growth and physiological attributes of rice seedlings cultivated under water deficit regime. Plants 11, 261.

Li, J., Evon, P., Ballas, S., Trinh, H.K., Xu, L., Van Poucke, C., et al., 2022. Sunflower bark extract as a biostimulant suppresses reactive oxygen species in salt-stressed Arabidopsis. Front. Plant Sci. 13.

Mahajan, N., Rawal, S., Verma, M., Poddar, M., Alok, S., 2013. A phytopharmacological overview on Ocimum species with special emphasis on *Ocimum sanctum*. Biomed. Prev. Nutr. 3, 185–192.

Mahboob, W., ur Rehman, H., Basra, S.M.A., Afzal, I., Abbas, M.A., Naeem, M., et al., 2015. Seed priming improves the performance of late sown spring maize (*Zea mays*) through better crop stand and physiological attributes. Int. J. Agric. Biol. 17.

Maqbool, N., Wahid, A., Farooq, M., Cheema, Z., Siddique, K., 2013. Allelopathy and abiotic stress interaction in crop plants. In: Cheema, Z., Farooq, M., Wahid, A. (Eds.), Allelopathy: Current Trends and Future Applications. Springer, Berlin, Heidelberg, pp. 451–468.

Mashamaite, C.V., Ngcobo, B.L., Manyevere, A., Bertling, I., Fawole, O.A., 2022. Assessing the usefulness of *Moringa oleifera* leaf extract as a biostimulant to supplement synthetic fertilizers: a review. Plants 11, 2214.

Merwad, A.-R.M., 2020. Mitigation of salinity stress effects on growth, yield and nutrient uptake of wheat by application of organic extracts. Comm. Soil. Sci. Plant Anal. 51, 1150–1160.

Mizgier, P., Kucharska, A.Z., Sokół-Łętowska, A., Kolniak-Ostek, J., Kidoń, M., Fecka, I., 2016. Characterization of phenolic compounds and antioxidant and anti-inflammatory properties of red cabbage and purple carrot extracts. J. Funct. Foods 21, 133–146.

Mubarik, M.S., Khan, S.H., Sajjad, M., Raza, A., Hafeez, M.B., Yasmeen, T., et al., 2021. A manipulative interplay between positive and negative regulators of phytohormones: a way forward for improving drought tolerance in plants. Physiol. Plant 172, 1269–1290.

Muhammad, F., Mazhar, H., Abdul, W., Rushna, M., 2011. Employing aqueous allelopathic extracts of sunflower in improving salinity tolerance of rice. J. Agric. Soc. Sci. 7, 75–80.

Mukherjee, A., Patel, J., 2020. Seaweed extract: biostimulator of plant defense and plant productivity. Int. J. Environ. Sci. Technol. 17, 553–558.

Munir, N., Hanif, M., Abideen, Z., Sohail, M., El-Keblawy, A., Radicetti, E., et al., 2022. Mechanisms and strategies of plant microbiome interactions to mitigate abiotic stresses. Agronomy 12, 2069.

Naz, H., Akram, N.A., Ashraf, M., Hefft, D.I., Jan, B.L., 2022. Leaf extract of neem (*Azadirachta indica*) alleviates adverse effects of drought in quinoa (*Chenopodium quinoa* Willd.) plants through alterations in biochemical attributes and antioxidants. Saudi J. Biol. Sci. 29, 1367–1374.

Nessim, A., Kasim, W., 2019. Physiological impact of seed priming with $CaCl_2$ or carrot root extract on Lupinus termis plants fully grown under salinity stress. Egypt. J. Bot. 59, 763–777.

Nizar, M., Shaukat, K., Zahra, N., Hafeez, M.B., Raza, A., Samad, A., et al., 2022. Exogenous application of salicylic acid and hydrogen peroxide ameliorate cadmium stress in milk thistle by enhancing morpho-physiological attributes grown at two different altitudes. Front. Plant Sci. 3157.

Osman, M.S., Badawy, A.A., Osman, A.I., Abdel Latef, A.A.H., 2021. Ameliorative impact of an extract of the halophyte *Arthrocnemum macrostachyum* on growth and biochemical parameters of soybean under salinity stress. J. Plant Growth Regul. 40, 1245–1256.

Pandey, V., Ansari, M., Tula, S., Sahoo, R., Bains, G., Kumar, J., et al., 2016. *Ocimum sanctum* leaf extract induces drought stress tolerance in rice. Plant Signal. Behav. 11, e1150400.

Pannacci, E., Baratta, S., Falcinelli, B., Farneselli, M., Tei, F., 2022. Mugwort (*Artemisia vulgaris* L.) aqueous extract: hormesis and biostimulant activity for seed germination and seedling growth in vegetable crops. Agriculture 12, 1329.

Rady, M., Desoky, E.-S., Elrys, A., Boghdady, M., 2019. Can licorice root extract be used as an effective natural biostimulant for salt-stressed common bean plants? S. Afr. J. Bot. 121, 294–305.

Ragab, S.M., Turoop, L., Runo, S., Nyanjom, S., 2022. The effect of foliar application of zinc oxide nanoparticles and *Moringa oleifera* leaf extract on growth, biochemical parameters and in promoting salt stress tolerance in faba bean. Afr. J. Biotechnol. 21, 252–266.

Rahal, A., Kumar, D., Malik, J.K., 2019. Neem extract. Nutraceuticals in Veterinary Medicine. In: Gupta, R., Srivastava, A., Lall, R. (Eds.). Springer, pp. 37–50.

Rashid, N., Basra, S.M., Shahbaz, M., Iqbal, S., Hafeez, M.B., 2018. Foliar applied moringa leaf extract induces terminal heat tolerance in quinoa. Int. J. Agric. Biol. 20, 157–164.

Rashid, N., Khan, S., Wahid, A., Basra, S.M.A., Alwahibi, M.S., Jacobsen, S.E., 2022. Impact of natural and synthetic growth enhancers on the productivity and yield of quinoa (*Chenopodium quinoa* willd.) cultivated under normal and late sown circumstances. J. Agron. Crop. Sci. 208, 552–566.

Rashid, N., Wahid, A., Basra, S., Arfan, M., 2020. Foliar spray of moringa leaf extract, sorgaab, hydrogen peroxide and ascorbic acid improve leaf physiological and seed quality traits of quinoa (*Chenopodium quinoa*) under terminal heat stress. Int. J. Agric. Biol. 23, 811–819.

Rasul, F., Gupta, S., Olas, J.J., Gechev, T., Sujeeth, N., Mueller-Roeber, B., 2021. Priming with a seaweed extract strongly improves drought tolerance in Arabidopsis. Int. J. Mol. Sci. 22, 1469.

Raza, A., Tabassum, J., Mubarik, M., Anwar, S., Zahra, N., Sharif, Y., et al., 2021. Hydrogen sulfide: an emerging component against abiotic stress in plants. Plant Biol. 24 (4), 540–558.

Rhaman, M.S., Imran, S., Rauf, F., Khatun, M., Baskin, C.C., Murata, Y., et al., 2020. Seed priming with phytohormones: an effective approach for the mitigation of abiotic stress. Plants 10, 37.

Riaz, J., Iqbal, S., Hafeez, M.B., Saddiq, M.S., Khan, S., Jahanzaib, et al., 2022. The potential of moringa leaf extract (MLE) as bio-stimulant for improving biochemical and yield attributes of wheat (*Triticum aestivum* L.) under saline conditions. Fresenius Environ. Bullet 31, 8690–8699.

Saddiq, M.S., Iqbal, S., Afzal, I., Ibrahim, A.M., Bakhtavar, M.A., Hafeez, M.B., et al., 2019. Mitigation of salinity stress in wheat (*Triticum aestivum* L.) seedlings through physiological seed enhancements. J. Plant Nutr. 42, 1192–1204.

Shaukat, K., Zahra, N., Hafeez, M.B., Naseer, R., Batool, A., Batool, H., et al., 2022. Role of salicylic acid–induced abiotic stress tolerance and underlying mechanisms in plants. Emerging Plant Growth Regulators in Agriculture. In: Aftab, A., Naeem, M. (Eds.). Elsevier, pp. 73–98.

Singh, D., Chaudhuri, P.K., 2018. A review on phytochemical and pharmacological properties of Holy basil (*Ocimum sanctum* L.). Indust. Crop. Prod. 118, 367–382.

Southam, C.M., 1943. Effects of extract of western red-cedar heartwood on certain wood-decaying fungi in culture. Phytopathology 33, 517–524.

Taiwo, L., Makinde, J., 2005. Influence of water extract of Mexican sunflower (*Tithonia diversifolia*) on growth of cowpea (*Vigna unguiculata*). Afr. J. Biotechnol. 4, 355–360.

Trivedi, K., Anand, K.V., Vaghela, P., Ghosh, A., 2018. Differential growth, yield and biochemical responses of maize to the exogenous application of *Kappaphycus alvarezii* seaweed extract, at grain-filling stage under normal and drought conditions. Algal Res. 35, 236–244.

Waqas, M.A., Khan, I., Akhter, M.J., Noor, M.A., Ashraf, U., 2017. Exogenous application of plant growth regulators (PGRs) induces chilling tolerance in short-duration hybrid maize. Environ. Sci. Pollut. Res. 24, 11459–11471.

Zahra, N., Hafeez, M.B., Shaukat, K., Wahid, A., Hussain, S., Naseer, R., et al., 2021a. Hypoxia and anoxia stress: plant responses and tolerance mechanisms. J. Agron. Crop. Sci. 207, 249–284.

Zahra, N., Wahid, A., Hafeez, M.B., Alyemeni, M.N., Shah, T., Ahmad, P., 2021b. Plant growth promoters mediated quality and yield attributes of milk thistle (*Silybum marianum* L.) ecotypes under salinity stress. Sci. Rep. 11 (1), 17.

Zahra, N., Hafeez, M.B., Wahid, A., Al Masruri, M.H., Ullah, A., Siddique, K.H., et al., 2022a. Impact of climate change on wheat grain composition and quality. J. Sci. Food Agric. Available from: https://doi.org/10.1002/jsfa.12289.

Zahra, N., Wahid, A., Hafeez, M.B., Lalarukh, I., Batool, A., Uzair, M., et al., 2022b. Effect of salinity and plant growth promoters on secondary metabolism and growth of milk thistle ecotypes. Life 12, 1530.

Zahra, N., Wahid, A., Hafeez, M.B., Shaukat, K., Shahzad, S., Shah, T., et al., 2022c. Plant growth promoters alleviate oxidative damages and improve the growth of milk thistle (*Silybum marianum* L.) under salinity stress. J. Plant Growth Regul. 41, 3091–3116.

CHAPTER

12

Role of biodegradable mulching films and vegetable-derived biostimulant application for enhancing plants performance and nutritive value

Atul Loyal[1], *S.K. Pahuja*[1], *D.S. Duhan*[2], *Naincy Rani*[3], *Divya Kapoor*[6], *Rakesh K. Srivastava*[4], *Gaurav Chahal*[5] *and Pankaj Sharma*[6]

[1]Department of Genetics and Plant Breeding, CCS Haryana Agricultural University, Hisar, Haryana, India [2]Department of Vegetable Sciences, CCS Haryana Agricultural University, Hisar, Haryana, India [3]Department of Chemistry, CCS Haryana Agricultural University, Hisar, Haryana, India [4]International Crops Research Institute for the Semi-Arid Tropics (ICRISAT), Patancheru, Telangana, India [5]College of Agriculture, CCS Haryana Agricultural University, Hisar, Haryana, India [6]Department of Microbiology, CCS Haryana Agricultural University, Hisar, Haryana, India

12.1 Introduction

Humans have been using plants and plant products directly and indirectly in various forms for ages. Plants have also played an important role in the process of human civilization, paving the way for them to live in groups rather than hunter-gatherers. Since then, humans are optimizing their requirements and developing new ways to fulfill their requirements. During this entire process, humans learned various things and started to classify them as plant growth promoters (PGPs; compost, manures, mulches) and growth retardants (weeds, insect pests). With time and the advancement of science, humans gained deeper insights into these growth promoters and growth retardant substances (Kumar et al., 2020). The agricultural practices developed by human beings are

Biostimulants in Plant Protection and Performance
DOI: https://doi.org/10.1016/B978-0-443-15884-1.00020-8

facing a lot of problems like increasing biotic and abiotic stresses, changing climatic conditions, increasing incidence of pathogens, and competition from herbicide-resistant and newly evolved varieties of weeds (Rani et al., 2019, 2022; Sangwan et al., 2023; Singh et al., 2019, 2021; Sharma et al., 2020a, 2020b, 2021a, 2021b, 2021c, 2022a, 2022b, 2023). Therefore agricultural scientists are always on a quest to devise novel practices to ensure global food security along with minimal environmental deterioration. The employment of mulching films and biostimulants appears to be a viable option for combatting these concerns (Cozzolino et al., 2020).

A prevalent farming practice across the globe entails covering the soil around plants with plastic films. This technique was originally introduced in the 1970s. The employment of mulching films aids in suppressing the growth of weeds, shielding the crops from pest and disease infestation, and decreasing the usage of pesticides and herbicides. Mulching films, based on their colors, also absorb and/or reflect sunlight, thus differently changing soil temperature and ultimately altering crop growth and productivity. The major concern associated with the usage of plastic films is that they are never completely removed from a field, leaving remnants that persist in soil for decades. Owing to the inherent limitations of plastic films, there is an urgent need to use compostable and biodegradable materials in modern agriculture. As a potent alternative, biodegradable mulches are also in use and are designed to be tilled into soil. These are made up of bio-based polymers finding their origin from microbes, plants, or fossil-sourced materials, thus can be easily degraded by the soil microflora after tillage. The current chapter will discuss different types and forms of various mulches and biostimulants used throughout the world and their effect on the crop's plants, how they enhance plant performance, and their future prospects.

12.2 What are biodegradable mulching films and biostimulants?

Etymologically "mulch" is a layer that is used to cover the ground/soil and the space in between the adjacent rows of plants to minimize the evaporation losses, to regulate the soil temperature, and to check the growth of weeds. There are at least thousands of years old records of using mulches in crop production in both the old and new world (Lightfoot, 1996). Meanwhile, it is not mentioned in the literature what compelled/motivated humans to use mulching in crop production. The prime advantage offered by mulching offers is restricting the growth of weeds, which eventually enhances the availability and supply of water and nutrients to the crop plants. Initially, only crop residues and tree wastes were used as mulches with the prime purpose of controlling weeds. Nevertheless, the advancement in agricultural sciences also unveiled the other associated aspects of mulching.

However, biostimulants are quite new to use in agriculture as compared to mulches. It was Blagoveshchensky (1955) who first studied the use and effect of biostimulants in agriculture. On being drawn from natural substances he used the term biogenic stimulants which was later known as biostimulants. According to European Union (2019), biostimulants are "a product stimulating plant nutrition processes independently of the product's nutrient content with the sole aim of improving one or more of the following characteristics of the plant or the plant rhizosphere: (a) nutrient use efficiency, (b) tolerance to abiotic stress, (c) quality traits and (d) availability of confined nutrients in soil or rhizosphere."

In Europe, they are only considered as growth enhancers and cannot protect the plants against various biotic/abiotic stresses. Therefore it is not easy to agree on one definition of biostimulants as the norms and criteria change from region to region and country to country. The present agricultural practices rely on heavy usage of pesticides and herbicides that often lead to the contamination of produce with pesticide residues. The increasing environmental and health concerns have put forward the global demand for pesticide-free food, especially in the case of perishables vegetables and fruits, where pesticides and herbicides are used more frequently in order to obtain high yield/returns. The use of pesticides has almost doubled to 4.17 mt (metric tons) in 2019 from 2.30 mt in 1990 (Hannah et al., 2022). The alarmingly increasing population and depleting genetic resources will put more pressure on the farmers and scientists, which ultimately provoke the application of pesticides and fertilizers to enhance production. Here comes the role of biostimulants and biodegradable mulches which will help to meet the humongous demand for food without using health-hazardous pesticides and fertilizers.

12.3 Classification of mulching films and biostimulants

With time and space, the use and forms of biostimulants and mulches have been changed and still, there are new developments in the field pertaining to biostimulants. Therefore it becomes necessary to classify mulches and biostimulants to gain a better sense of understanding. Mulches can be classified into two broad categories: organic and synthetic (inorganic); these two can be further subdivided into various classes which are discussed below.

Organic mulches as the name itself describes the composition as well as nature of these type of mulches. These mulches are either by-products of a plant-based industry or crop residue/postharvest residue of a crop which has either zero or very little economic value, for example, dry coconut shells are used for mulches. So these mulches can be grouped as (1) crop residue based, (2) tree/wood based, and (3) compost. Wheat straw and rice straw mulches are the most commonly used crop residue–based mulches used across the globe (Goel et al., 2019). The major purpose of using crop residue–based mulching is to cover/occupy the ground surface a least 30% (Erenstein, 2002) and limit the growth of the weeds during the critical crop weed competition period, which is generally 15–45 days in most of the field crops. Tree/wood-based mulches are by-product of the timber or wood industry and have no other commercial use. Sawdust, wood chips, leaves, and barks are some tree-based products used in mulching, out of these except leaves rest are by-product of the timber industry, the main by-product of the wood and timber/wood industry. These are used as fuel in most part of the world, and among these, sawdust is preferred to be used as mulch as it stores water and prevents soil erosion. Similarly, compost is also used as mulch despite the fact that the primary objective of compost is to provide nutrition to the plants in addition to its ability to loosen soil (Heckman et al., 2009). In addition to this, other mulches are in use but are considered to be appropriate for a limited period or crop range, for example, blanket application of pebble/gravel mulch finds limitations in its usage in field crops/horticultural/vegetable crops. Although it proves to be an excellent mulch type in terms of its moisture retention trait, it is more beneficial to arid and semiarid regions, where water is scarce and its conservation is the primary concern. In China, gravel mulch has been used extensively for the aforementioned purpose for more than three centuries in semiarid regions (Lv et al., 2019). Meanwhile, the recent focus is on the use of newspaper and cardboard as mulching material (Li et al., 2021). It is a complex as well as cumbersome task to enlist all the mulches and discuss their merits and demerits, so keeping in view of this, only biodegradable mulches and biodegradable mulching films (Fig. 12.1) are discussed in detail in the upcoming segment.

Synthetic mulches are generally two types that are represented by rubber mulch and plastic mulch. Rubber mulch is predominantly used in gardens only for esthetic purposes (Nithisha et al., 2022) and is prepared from old recycled tires and other rubber waste materials. However, it does not provide any nutritional support to the plants at any stage of life, unlike other mulch types. Additionally, it does not help in improving the physical and chemical properties of the soil (Table 12.1). In addition, rubber mulches also contain a wide variety of substances that are defined as toxic to humans on exposure (Chalker-Scott, 2021). Although, such kinds of mulches are also in practice at some places, the number of cons-associated seems to be at par with.

Plastic mulches are a thin sheet of polyethylene that covers the ground space in between the plants. It may be either transparent or colored (generally black plastic mulching films are preferred). Each of them tends to have its own advantages but black plastic mulches are proven to be more beneficial when compared with transparent mulching films (He et al., 2021). Though it has been proven that transparent mulching films have an important role in increasing crop productivity, in recent years, they have also been reported to induce premature senescence and yield decline (Liu et al., 2005; Wang et al., 2009). China is regarded as the global leader in using plastic mulches (Daryanto et al., 2017). Recent years have experienced an exponential increase in the use of plastic film as assessed from the numbers representing 3×10^5 t in 1991 to 14.7×10^5 t in 2017 (Gao et al., 2019). Plastic mulching films are prepared by using low-density polyethylene. Although these are reported as good

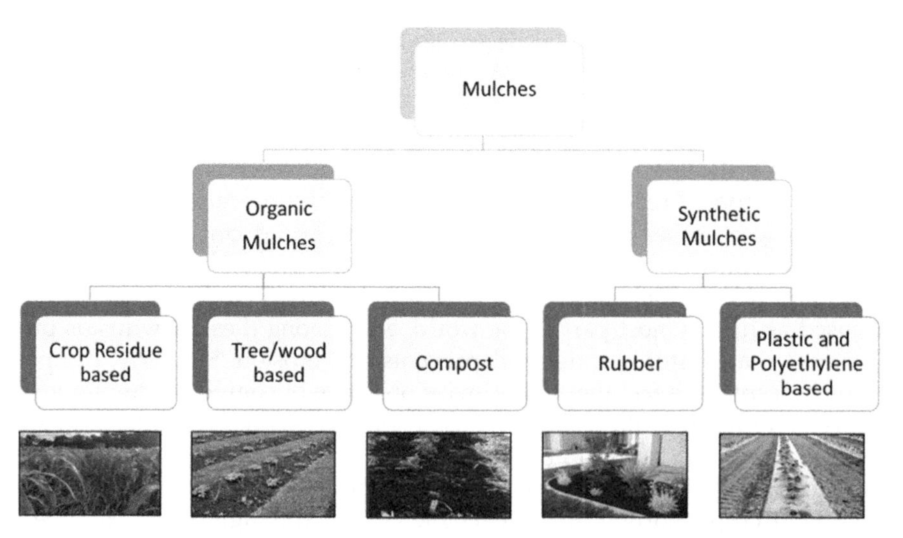

FIGURE 12.1 Types of mulches.

TABLE 12.1 Various kinds of mulches and their effect on crop plants.

Sr. no.	Type of mulches	Crop studied	Effect of mulches	References
1	Crop residue	Corn	13.7% Higher yield, weed management	Erenstein (2003)
2	Tree wood-based	Baby cabbage	57% Higher yield and 18.3% increment in water retention, lower weed infestation	Masarirambi et al. (2013)
3	Compost	Apple	Increased carnivore insect population, pollution remediation	Brown and Tworkoski (2004) Chen and Wu (2005)
4	Rubber	Kitchen and home gardens	Cause Zn toxicity in plants, moisture conservation	Chalker-Scott (2021)
5	Plastic/ polyethylene based	Floricane red raspberry	Suppress weeds, modify soil temperature and moisture, promote earlier and 34% higher yields, and pest control	Zhang et al. (2019) Kasirajan and Ngouajio (2012)
6	Gravel–sand mulch	Spring maize	Improving infiltration and soil temperature, checking wind erosion by 84%–96%, and water erosion also.	Li (2003)

candidates for controlling weeds, their nonbiodegradable nature like other petroleum-derived products should also be taken into consideration before putting them into practice. They persist in the field for a period much longer than the crop cycle and may harm soil health due to the secretion of phthalate and phthalic acid esters (Thompson et al., 2019). Owing to their nonsustainable and environmentally deteriorating nature, biodegradable mulches are projected as a good and potential alternative. Due to their nonenvironment friendly nature, various biodegradable plastic mulches resembling the properties of plastic mulches have been developed which comprises polyesters, polysaccharides, and copolymer or a blend of both. Among these, the most important are poly(3-hydroxybutyrate) and poly(3-hydroxybutyrate-co-3-hydroxyvalerate). Bioplastics (biopolymers): such biopolymers are obtained from genetically engineered plants or microorganisms and these can reduce the sheer dependency on plastic mulches (Bi et al., 2021). Some of the available plastic biodegradable mulches are polyhydroxyalkanoates, polyhydroxy valerate, polyhydroxybutyrate, polyhydroxyalkanoate, polylactic acid, polycaprolactone, polybutylene succinate adipate, polybutylene adipate/terephthalate, aliphatic–aromatic copolyesters, polybutylene succinate (PBS), polyethylene terephthalate, and polymethylene adipate/terephthalate.

Biostimulants are natural occurring substances that are derived from different sources such as vegetal protein extracts, animal protein extracts, protein hydrolyzate (PH), humic substances, and microorganisms (fungi and bacteria) derived biostimulants. Biostimulants when applied in small quantities are capable of boosting the nutrient uptake efficiency of plants and can also supplement the growth and development of plants. Being derived from natural sources they can also ameliorate the soil health conditions and create a better microenvironment for the plant roots (Calvo et al., 2014). They are categorized into five different classes based on the source of extraction: (1) Animal protein hydrolyzate derived,

(2) humic and fulvic substances derived, (3) microbial derived, (4) plant extracts derived/vegetal protein hydrolyzate derived, and (5) seaweed derived (Del Buono, 2021).

Animal protein hydrolyzate-based biostimulants can supplement the growth of roots as these have a high content of aromatic acids (phenylalanine and tryptophan) that act as precursors of the root growth-promoting hormone auxin (Dai et al., 2013). Apart from plant growth-promoting hormones, they have also been shown to provide resistance to various abiotic stresses (salinity, temperature, and drought stress) by means of singling peptides of hydrolyzates (Phelan et al., 2009; Colla et al., 2017).

Humic substances are a crucial component of humans and a major organic section of the soil profile. Similarly, the fulvic substances are naturally occurring substances of the soil, and their extracts are used for the production of the plant biostimulants. These are available in the public domain and widely used for the orchard crops (Canellas et al., 2015). All the positive effect shown by the plants on the application of biostimulants is due to better modification of the root architecture.

Trichoderma ssp. and arbuscular mycorrhizal fungi have been reported to be used as microbial-derived plant biostimulants. Other than these bacteria and fungi both can be used for biostimulant production. Saia et al. (2019) studied the effect of microbial-derived plant biostimulants and found that they showed positive change (20%–97%) in the uptake of various essential nutrients such as P, Mg, Fe, Zn, and Mn (Cozzolino et al., 2020). Meanwhile, Jiao et al. (2011) also suggested the use of plant biostimulants derived from microbial organisms to enhance plant productivity in hostile environmental circumstances. The in vitro production of plant biostimulants production using microorganisms seems feasible and practically possible under in vitro conditions.

Another category of plant biostimulants is plant extracts derived/vegetal protein hydrolyzate derived. These biostimulants are generally derived from plants that possess some antioxidants and protective substances such as *Moringa oleifera* (Lam.), which help plants, especially in coping with abiotic stress/water scarcity (Del Buono, 2021). Biostimulants prepared from moringa have proven their effectiveness in enhancing squash productivity under drought situations (Abd El-Mageed et al., 2017; Zulfiqar et al., 2020). Their effects and implications are discussed in detail in the following section.

Seaweed extracts represent another class of plant biostimulants which are mainly derived from various kinds of macroalgae (brown, red, and green). These are the most commonly used seaweed extracts for the production of biostimulants. They have to be extracted from the sea/oceans which is a major hindrance in the adoption of seaweed extracts derived from plant biostimulants, due to the fact that quality and chemical composition are not so good when it is harvested from the brackish waters of seas and oceans (Rouphael and Colla, 2020a, 2020b). It has a great potential to become a good source form plant biostimulants production if algae are grown in vitro.

12.4 Types and forms of mulching films and biostimulants

Mulching films generally form a thin layer in between the plant spacing. The form of mulching film is largely dependent on the type of mulch used. The other form of synthetic mulches that is in use is polyethylene sheet and more recently polypropylene sheets are also being used. Although the thickness and the color of the mulch in films is a personal choice, it is known to play a significant role and thus are also a subject to change depending on the purpose that needs to be addressed. Black plastic films are found to be superior to transparent mulching films and other colored mulching films (He et al., 2021). In the case of wood/timber-based mulch, it is a very fine powder that is predominantly used in water erosion-prone areas to prevent this. There is slight similarity in the form in which compost-based mulches and sawdust mulch are applied but here major difference is that unlike sawdust mulches size of the compost particles is not uniform and compost mulches also slowly release nutrients and that culminate into improved plant structure as well as soil texture. Pebble/gravel or rubber mulch have uniform-sized particles but they are available in different sizes and generally used for mulch purposes in gardens or esthetic places. Traditionally, the sole purpose of mulches was to reduce the weed growth during the initial period (see Fig. 12.2) and still that is the primary reason for their use, remaining factors (such as temperature regulation, help nutrient uptake, impart resistance to biotic/abiotic resistance) are secondary.

Biostimulants may be used in the form of foliar spray, powder, granules, or solutions. It depends upon the source of their extraction. When extracted or derived from humic and fulvic substances biostimulants are applied directly in the soil in granular form but when prepared from other living entities (microorganisms, plant extracts, and seaweed), they are applied in the foliar application form (Drobek et al., 2019; Kocira et al., 2018).

FIGURE 12.2 Various types of mulches under different environmental conditions: (A) Effect of mulching films on young cucurbits in polyhouse; (B) effect of mulching films on mature cucurbits plants in polyhouse; (C) effect of mulching films on young brinjal plants in polyhouse; (D) effect of mulching films on mature brinjal plants in polyhouse; (E) crop residue mulching in mustard under field conditions; and (F) mixed mulching (paper and straw) in gladiolus under field conditions.

Generally, amount and time are crucial factors in deciding the dose, time, and method of application of these biostimulants. However, the foliar form is the most economical and commonly used in the case of plant biostimulants. With the development in the field of biostimulants, it is revealed that they also provide resistance to plants.

12.5 Benefits and potential of biodegradable mulching films and biostimulants

Plants are most vulnerable to temperature fluctuations at the seedling stage or at the young stage. Mulches provide a shield of protection to the plants by regulating the temperature of microclimate around the plants during these extremes (cold stress and heat stress; Kader et al., 2019). The early exposure of the plants to extreme temperature conditions reduce the plant vigor and also have a negative impact on plant root development and nutrient absorption at a later stage (Chalker-Scott, 2007).

12.5.1 Advantages over traditional mulches and stimulants

Biodegradable mulches and vegetal-derived biostimulants offer numerous advantages over traditional mulches and stimulants. Biodegradable mulch films and vegetal biostimulants are made from natural, renewable materials and break down over time, reducing the need for synthetic fertilizers and pesticides (Fig. 12.3). In comparison, traditional mulches can contain harmful chemicals and take years to break down, potentially leading to environmental pollution (Fig. 12.4).

Thus these mulches are environment friendly and are naturally degraded with the completion of the crop cycle whereas the traditional or plastic mulches are not so easily degradable. They persist for a longer period in the soil and are not even degraded by the action of various microflora present in the soil, ultimately reducing soil fertility and soil health. In contrast to this, biodegradable mulch films and vegetal biostimulants add organic matter to the soil and promote the growth of beneficial soil microorganisms, which can lead to improved soil fertility and reduced soil degradation and in some studies even it has proven that application of biostimulants have promoted the soil microflora growth above ground. These also act as a buffer and prevent abrupt changes in the soil pH due to anthropogenic activities. Yield is the ultimate objective for every crop for a plant biologist

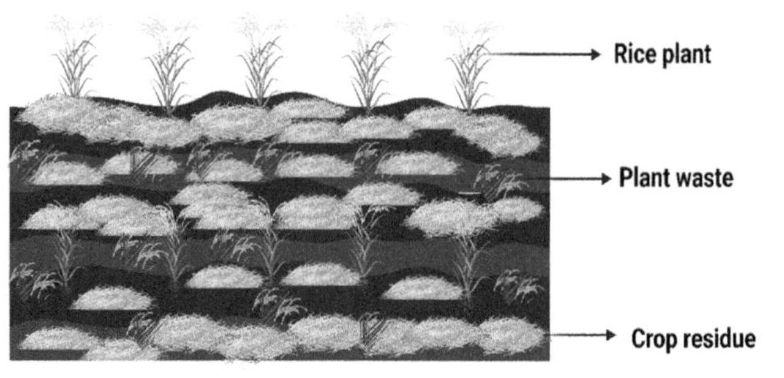

FIGURE 12.3 Plant/crop residue–based mulching.

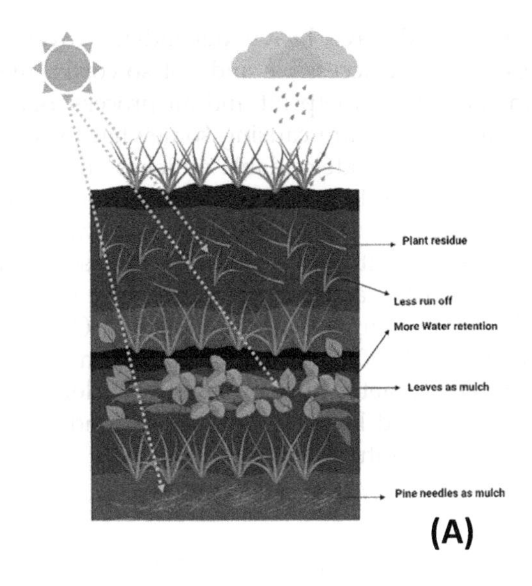

A. Open field with various type of biodegradable mulches regulating temperature, preventing soil erosion, leaching and retaining more water
B. Field without mulch failed to prevent soil erosion, evaporation and leaching

FIGURE 12.4 Contrast between two fields one with biodegradable mulch and another without mulch.

to achieve; at present, we heavily rely on synthetic fertilizers. As we all know, there is limited availability of the core ingredient of these synthetic fertilizers, and they are located in only a few selected countries.. For instance, Morocco has almost 70% of the world's total phosphatic reserves (Walan et al., 2014); similarly, Russia enjoys its monopoly over the production and distribution of potash fertilizers in the world (Al Rawashdeh and Maxwell, 2014).

Biostimulants can replace synthetic fertilizers and have heavy dependency over these. Although synthetic fertilizers and mulches are easily accessible and not so costly, the cost of their disposal is very high and they leave a carbon footprint and all process associated with form their mining to preparation and application are causing somewhat to considerable damage to ecosystem. That is not in the case of biodegradable mulches and biostimulants. Biodegradable mulch films and vegetal biostimulants can improve the quality of the food crops; for example, the change in the amino acid profile of basil plants was observed at the intermediate dose of vegetal protein hydrolyzates 0.15N (nitrogen) dose aspergine (+82%), aspartic acid (+106%), glutamic acid (+50%), and glutamine (+274%) while some showed major improvement at V-PH 0.25N were alanine (+73%), aspergine (+74%), and gamma-aminobutyric acid (+155%) (Rouphael et al., 2021). It is important to note that the effectiveness of biodegradable mulch films and vegetal biostimulants will depend on a number of factors, including soil type, crop type, and local environmental conditions, and they should be used in conjunction with sound agricultural practices

12.5.2 Role of biodegradable mulches and biostimulants in nutrient uptake and enhancing nutrient use efficiency of plants

Plants require various nutrients for the completion of their life cycle. The amount and time of requirement vary from crop to crop but some of the nutrients are common to all plants; for instance, phosphorus, which is part of DNA, is largely required by all plants for root initiation and development at the germination stage.

Some nutrients are required by all plants at a different type of hydrolyzed collagen, including granulated gelatin, gelatin hydrolyzate, and amino acid mixtures simulating gelatin composition, which were evaluated in terms of plant growth on cucumber (Wilson et al., 2018). In their study, the authors reported that gelatin hydrolyzate treatment increased the expression of genes encoding for amino acid permeases (AAP3, AAP6) and transporters of amino acids and nitrogen. Therefore they concluded that gelatin hydrolyzate provided a sustained source of nitrogen and acted as a biostimulant. Furthermore, Luziatelli et al. (2019) conducted a greenhouse experiment on lettuce aiming to assess the effect of three commercial PBs: vegetal-derived PH, vegetal-derived PH enriched with copper and a tropical plant extract on plant growth, and the epiphytic bacterial population. The three commercial PBs boosted the shoot's fresh weight with no significant differences between the three organic PBs. Rouphael and Colla (2020a, 2020b) were also able to demonstrate that PBs can stimulate the growth of epiphytic bacteria (*Pantoea, Pseudomonas, Acinetobacter*, and *Bacillus* genus) with PGP and/or biological control activity against pathogens, thus acting synergistically with organic compounds to increase the marketable fresh yield of lettuce. Biostimulants also improve plants' performance indirectly and humic substances present in biostimulants is its perfect example. These act as a buffer for maintaining soil pH and enhance the cation exchange capacity of the soil (Jardin, 2015). Humic substances also form a complex with the micronutrients/trace elements and prevent their leaching and make them available to the plants (Chen et al., 2004; Garcia-Mina et al., 2004). In addition, the fulvic acids having low molecular size are absorbed only when they combine with micronutrients and directly penetrate through the cell wall to plasmalemma than plant cells but fulvic acids having high molecular sizes neither combine with trace elements nor get absorbed alone. They do have more size and due to this fail to penetrate beyond the call wall (Nardi et al., 2002). Zeng et al. (2002) have also shown that humic and fulvic substances present in the biostimulants binds with the heavy metal ions present in the soil and thus reduce the toxic effect of these metal on plants upon absorption. When small bioactive peptides are isolated from soybean-derived pH, it enhances the numbers of root hairs in *Brassica rapa*, whereas high adventitious root formation was seen in tomatoes. In tomato and creeping bentgrass leaves, nitrogen assimilation enzymes were stimulated when the plants were sprayed with seaweed extracts biostimulants (Zhang et al., 2010; Ramya et al., 2015). Another strange thing was that the effect was higher when plants were grown at low nitrate concentrations, directly indicating that in adverse environmental conditions, biostimulants may enhance nutrient absorption efficiency (Durand et al., 2003). Rouphael et al. (2021) compared the effect of animal-derived biostimulants and vegetal-derived biostimulants on sweet basil (*Ocimum basilicum* L.) under greenhouse conditions at different doses. They reported that mineral contents (N, P, K, S, Ca, Mg, Na, and Cl) were slightly higher in plants treated with vegetal-derived biostimulants. Still, there is a lack of literature that describes the in-depth effect of vegetal-derived biostimulants on plants and soil microflora in sustainable agriculture. Vegetal-derived biostimulants contain varied compounds, simple/complex carbohydrates, and soluble organic nitrogen (Trouvelot et al., 2014), which can be utilized by both plants as well as bacteria for the uptake of nitrogen, carbon, and energy (Farrell et al., 2013).

12.5.3 How biodegradable mulch films and vegetal biostimulants can replace fertilizers?

Biodegradable mulch films and vegetal biostimulants are alternative ways to provide nutrients to crops that can replace or supplement the use of fertilizers. The use of these products can help reduce the dependency on synthetic fertilizers and improve the overall sustainability of agricultural practices. At present (2023), the market for organic food is $294.54 billion which has shown a 13.7% compound annual growth rate (CAGR) from 2022. The demand and market for organic products have been increasing and it is estimated to grow at a CAGR of 14.8% till 2027 and may reach up to $512 billion. (Organic Food Global Market Report, 2023). It is not hidden from scientists that the production sharply drops when we shift to organic agriculture from synthetic fertilizers. In that case, vegetal-derived biostimulants are a good replacement. It is proven that biostimulants provide nutrients in a controlled manner and at different growth phases of plants. It is already mentioned how biostimulants enhance the photosynthetic ability, nutrition uptake (Zhang et al., 2010; Ramya et al., 2015), and nutrition assimilation (Durand et al., 2003) in plants. Although the biostimulants are not as concentrated as the high-grade fertilizers, they help in the modification of the root architecture by promoting the number and growth of root hairs. In addition, the loss of nutrients through leaching and percolation in the case of synthetic fertilizers is very common unlike the case of biostimulants.

12.5.4 How biodegradable mulch films and vegetal biostimulants impart resistance to various stress?

Unfavorable conditions result in decreased crop yields because the plants are under stress and they spend most of their energy to combat the stress rather than increasing yield and biomass. Various factors can cause stress including biotic factors like fungal and bacterial infections and abiotic ones like water, heat, and salinity stresses. Natural treatments like biostimulants and biodegradable mulching can improve crop yield and quality by providing them with resistance to biotic and abiotic stress.

12.5.4.1 *Salinity stress*

Salinity stress, also known as osmotic stress, occurs when the concentration of ions gets increased in soil which results in reduced water absorption by roots. Therefore salinity-stressed plants also come under water stress. Various biostimulants extracted from vegetable sources have been used to protect plants under salinity stress. For example, protein hydrolyzates improved the resistance to salinity stress in various crops like lettuce, hence improving productivity and yield (Sorrentino et al., 2021). Several microbiological biostimulants like *Azospirillum brasilense* have also revealed positive results in crops like *Lactuca sativa*, *Cicer arietinum*, *Vicia faba* under a salinity stress environment.

The indiscriminative use of synthetic fertilizers and other agrochemicals adds salts to the soil profile which increases the soil pH and also contributes to the abiotic stress caused by the salts (Chalker-Scott, 2007). Biodegradable mulching may help solve this problem as it is proven in various studies that mulching enhances water retention in the soil and reduces evaporation up to a certain extent and sometimes they have also been shown to reduce the toxic effects of salts on plants (Ansari et al., 2001; Landis 1988; Yobterik and Timmer, 1994). Protein hydrolyzate-based biostimulants have proven to stimulate phenylalanine ammonia-lyase enzyme and gene expression, stimulate the production of flavonoids under salt stress, and improve plant performance under salt stress (Ertani et al., 2013).

12.5.4.2 *Nutrient stress*

Nutritional impairments can arise from a number of causes, including poor agronomic management and varied soil conditions (low organic matter, nonoptimal pH). By generating an increase in root biomass, biostimulants treatments make higher amounts of soil available providing more nutrition to them. For instance, commercially available biostimulants, that is, Amino Prim and Amino Hort, have been reported to improve nutrition absorption capacities even at trace doses (Rouphael et al., 2021). The nature of biostimulants is complex and variable which makes it difficult to determine the real image behind those mechanisms or which components are playing a crucial role (Van Oosten et al., 2017). Fiorentino et al. (2018) reported that lettuce yield was higher when it was supplemented with two different biostimulant formulations and this happened in the absence of nitrogenic fertilizers. When biotic/abiotic stress is combined with nutritional stress the yield and plant performance are drastically reduced and there are rare chance of recovery from this particular point despite the use of synthetic fertilizers. If things can be predicted and such kind of situations are endemic to a particular region/crop, both biostimulants and biodegradable mulch can only ameliorate the plant from stress under such circumstances. This is because they have a synergic effect on plants and supply water/nutrients in a slow and phased manner. Sometimes soil microflora acting in deeper soil may move to the upper surface of the soil and supplement plant growth by providing the nutrients to the plants in the effective root zone.

12.5.4.3 *Temperature stress*

The arrival of cold temperatures in late spring is one of the most harmful abiotic stimuli that adversely affects productivity. Low temperatures result in the rupturing of cell membranes, leading to ultimate death in some serious cases. Biostimulants, such as BactorS13 or Flortis Micorrize, have been used in plants under cold stress and help them to withstand the stress by increasing the concentration of substances that protect the cell membranes like some osmotically active substances and antioxidants (Micelli et al., 2021). Similar results can also be obtained with microbial biostimulants. Treatments with Azospirillum brasilense have shown positive results on lettuce (*Lactuca sativa* L.), peppers (*Capsicum annuum* L.), chickpeas (*Cicer arietinum* L.), and broad beans (*Vicia faba* L.) grown in a high salinity environment (Ruzzi and Aroca, 2015).

Biodegradable mulches have a unique property to act as a buffer between the plants and the surrounding microenvironment. They prevent the sudden rise and fall in the temperature by regulating the surrounding temperature, it helps the plants to cope up the temperature stress, more specifically from high temperature during the hot summers in arid and semiarid regions. Moore and Wszelaki (2019) and Sekara et al. (2019) also proved that there was a significant difference between the biodegradable mulches and plastic mulches. Moreover, the biodegradable mulches were degraded up to 40%–60% in the same season and rest of it will be degraded in the upcoming season. After use, degradable mulches are tilled back into the soil and supplement the next crop (Somanathan et al., 2022).

In recent times, the duration of heat waves and cold waves have decreased but the frequency/intensity have increased and that lead to damage in cells. So, in this scenario, biostimulants/seaweed extracts can be of prime importance in imparting cold stress resistance. Recently, some scattered work has been done on seaweed extracts to evaluate their ability to impart cold tolerance to plants. Bradáčová et al. (2016) tested their ability to improve the cold tolerance in maize plants and it was observed that extracts having Zn and Fe were only able to enhance the ability of plants to endure cold through reactive oxygen species. In addition, the enduring effects were due to the inclusion of micronutrients in extracts and these micronutrients play a key role in cofactors in antioxidative enzymes. This clearly indicates that a cold amalgamation of biostimulants and micronutrients can be used to overcome cold tolerance. Imran et al. (2013) also found similar results when corn seedlings were exposed to cold/chilling stress conditions and they showed signs of recovery when supplied with micronutrients.

Global warming has also affected the yield and productivity in agriculture by putting the plants under heat stress in various tropical and subtropical regions. Various biostimulants like *Ascophyllum nodosum* extracts induced positive effects in tomato plants under heat stress conditions (Carmody et al., 2020). Similar is the case of biodegradable mulches which keep the microenvironment around the plants slightly cooler than the surrounding atmosphere and helps the plants to maintain turgor pressure, minimizing the evaporation losses.

12.5.4.4 Drought stress

This is the most common abiotic stress which directly affects photosynthesis, and thereby the yield and quality of crops. Commercial biostimulants like Actiwave or Wokozim, procured from extracts of seaweed have been known to increase the accumulation of osmotically active substances which contribute to resisting the changes in plant water status (Abbas et al., 2020). The application of biostimulants obtained from *Ascophyllum nodosum* has improved the green color of leafy vegetables even under water stress conditions, increasing the biosynthesis of chlorophylls in them (Del Buono 2021). Alteration of the soil composition and structure using microorganisms is one of the direct mechanisms for enhancing water uptake. For example, bacterial exopolysaccharides play a key role in the stimulation of plant growth under water stress and form micro and micro aggregates by means of which it improves the soil structure (Grover et al., 2011). In addition, it also recreates a microenvironment around the plants that prevents microorganisms from drying and, also binds to Na^+ ions and reduces the salinity stress effect on plants (Nwodo et al., 2012). Mycorrhizal fungi have the ability to strengthen and enhance the penetration capacity of roots into soils (Bárzana et al., 2012; Khan et al., 2015), resulting in more root biomass, improved soil structure, and improved water retention capacity (Cavagnaro et al., 2015). Glycoprotein (glomaline) is produced by arbuscular mycorrhizae from genus *Glomus* induce proper growth, resistance to water stress, and better growth in orange plants. In addition, to root growth, these induce a better nutrient uptake by means of an aquaporin network. These help in making a balance between root water absorption and turgor pressure recovery. The role of biodegradables mulches in arid and semiarid regions is already discussed.

12.6 Future prospects and limitations

The current world market valuation of biostimulants for the year 2022 is estimated to be $3.5 billion and it is expected to reach $6.2 billion by the end of 2027. It indicates that the acceptance and use of biodegradable mulches and biostimulants will double in the span of 5 years. The main driving force behind the increase of these is the growing concern for sustainable agriculture and the impact of agriculture on the environment. To reduce the use of plastics in agriculture biodegradable mulching film will be promoted. Although these have shown a positive effect on the different plant performance, they still are low performer when compared with fertilizers and synthetic mulches. However, the cost of production of biodegradable mulching films is a limiting factor in their adoption at larger scale (Martin-Closas et al., 2017; Tan et al., 2016). Another limiting factor in their wide-scale adoption is their uncertain mode of action, and it is not crystal clear in some cases how they impart resistance and enhance plant performance. There is a need for in-depth study of mechanisms by which they enhance plant performance under hostile/limiting conditions. This can be realized by means of cautious agronomic experimentation, molecular or biochemical demonstration of positive effects on biological processes or the use of advanced analytical instruments to detect functional constituents (Yakhin et al., 2017). It opens the door to the vast possibilities of use of biostimulants, and can be used for blanket application as well as in combination with natural growth-promoting substances such as various microbial inoculants or plant hormones extracted naturally. They can also be used in combination with nanomaterials as their combined effect has not yet been studied (Loyal et al., 2023). More specifically, these can be phenomenal in enhancing pulses production as pulses are grown on marginal lands and with limited resources and can be viable options for nutritional security and mitigating malnutrition (Kumar et al., 2022). There are still divided opinions on whether biostimulants should be considered plant protectants or not as the European Union does not consider/treat them as pesticides. So, there should be a proper framework for their classification as well as regulation. As they can revolutionize the entire fertilizer industry if some ice-breaking discovery is made in the coming years.

References

Abbas, M., Anwar, J., Zafar-ul-Hye, M., Iqbal Khan, R., Saleem, M., Rahi, A.A., et al., 2020. Effect of seaweed extract on productivity and quality attributes of four onion cultivars. Horticulturae 6, 28.

Abd El-Mageed, T.A., Semida, W.M., Rady, M.M., 2017. Moringa leaf extract as biostimulant improves water use efficiency, physio-biochemical attributes of squash plants under deficit irrigation. Agric. Water Managt. 193, 46–54.

Al Rawashdeh, R., Maxwell, P., 2014. Analysing the world potash industry. Resour. Policy 41, 143–151.

Ansari, R., Marcar, N.E., Khanzada, A.,N., Shirazi, M.U., Crawford, D.F., 2001. Mulch application improves survival but not growth of Acacia ampliceps Maslin, *Acacia nilotica* (L.) Del. and *Conocarpus lancifolius* L. on a saline site in southern Pakistan. Int. For. Rev. 158–163.

Bárzana, G., Aroca, R., Paz, J.A., Chaumont, F., Martinez-Ballesta, M.C., Carvajal, M., et al., 2012. Arbuscular mycorrhizal symbiosis increases relative apoplastic water flow in roots of the host plant under both well-watered and drought stress conditions. Ann. Bot. 109 (5), 1009–1017.

Bi, S., Pan, H., Barinelli, V., Eriksen, B., Ruiz, S., Sobkowicz, M.J., 2021. Biodegradable polyester coated mulch paper for controlled release of fertilizer. J. Clean. Prod. 294, 126348.

Blagoveshchensky, A.V., 1955. Biogenic stimulantss in agriculture. Priroda. 7, 43–47.

Bradáčová, K., Weber, N.F., Morad-Talab, N., Asim, M., Imran, M., Weinmann, M., et al., 2016. Micronutrients (Zn/Mn), seaweed extracts, and plant growth-promoting bacteria as cold-stress protectants in maize. Chem. Bio. Technol. Agric. 3 (1), 1–10.

Brown, M.W., Tworkoski, T., 2004. Pest management benefits of compost mulch in apple orchards. Agric. Ecosyst. Environ. 3, 465–472.

Calvo, P., Nelson, L., Kloeppere, J.W., 2014. Agricultural uses of plant biostimulants. Plant Soil 383 (1–2), 3–41.

Canellas, L.P., Olivares, F.L., Aguiar, N.O., Jones, D.L., Nebbioso, A., Mazzei, P., et al., 2015. Humic and fulvic acids as biostimulants in horticulture. Sci. Hortic. 196, 15–27.

Carmody, N., Goni, O., Langowski, L., O'Connell, S., 2020. Ascophyllum nodosum extract biostimulant processing and its impact on enhancing heat stress tolerance during tomato fruit set. Front. Plant Sci. 11, 807.

Cavagnaro, T.R., Bender, S.F., Asghari, H.R., van der Heijden, M.G., 2015. The role of arbuscular mycorrhizas in reducing soil nutrient loss. Trends Plant Sci. 20 (5), 283–290.

Chalker-Scott, L., 2007. Impact of mulches on landscape plants and the environment—a review. J. Environ. Hortic. 25 (4), 239–249.

Chalker-Scott L., 2021. Rubber mulch use in home gardens and landscapes.

Chen, J.H., Wu, J.T., 2005. Benefits and drawbacks of composting. Food Fertilizer Technology Center.

Chen, Y., Nobili, M.D., Aviad, T., 2004. Stimulatory effects of humic substances on plant growth. In: Magdoff, F., Ray, R.W. (Eds.), Soil Organic Matter in Sustainable Agriculture. pp. 103–129.

Colla, G., Hoagland, L., Ruzzi, M., Cardarelli, M., Bonini, P., Canaguier, R., et al., 2017. Biostimulant action of protein hydrolysates: unraveling their effects on plant physiology and microbiome. Front. Plant Sci. 8, 2202.

Cozzolino, E., Giordano, M., Fiorentino, N., El-Nakhel, C., Pannico, A., Di Mola, I., et al., 2020. Appraisal of biodegradable mulching films and vegetal-derived biostimulant application as eco-sustainable practices for enhancing lettuce crop performance and nutritive value. Agronomy 10 (3), 427.

Dai, X., Mashiguchi, K., Chen, Q., Kasahara, H., Kamiya, Y., Ojha, S., et al., 2013. The biochemical mechanism of auxin biosynthesis by an Arabidopsis YUCCA flavin-containing monooxygenase. J. Biol. Chem. 288 (3), 448–1457.

Daryanto, S., Wang, L., Jacinthe, P.A., 2017. Can ridge-furrow plastic mulching replace irrigation in dryland wheat and maize cropping systems? Agric. Water Manag. 190, 1–5.

Del Buono, D., 2021. Can biostimulants be used to mitigate the effect of anthropogenic climate change on agriculture? It is time to respond. Sci. Total. Environ. 751, 141763.

Drobek, M., Frąc, M., Cybulska, J., 2019. Plant biostimulants: importance of the quality and yield of horticultural crops and the improvement of plant tolerance to abiotic stress—a review. Agronomy 9 (6), 335.

Du Jardin, P., 2015. Bioestimulantes vegetais: definição, conceito, principais categorias e regulacao. Sci. Hortic. 196, 3–14.

Durand, N., Briand, X., Meyer, C., 2003. The effect of marine bioactive substances (N PRO) and, exogenous cytokinins on nitrate reductase activity in Arabidopsis thaliana. Physiol. Plant 119 (4), 489–493.

Erenstein, O., 2002. Crop residue mulching in tropical and semi-tropical countries: an evaluation of residue availability and other technological implications. Soil Tillage Res. 67, 115–133.

Erenstein, O., 2003. Smallholder conservation farming in the tropics and sub-tropics: a guide to the development and dissemination of mulching with crop residues and cover crops. Agric. Ecosyst. Environ. 100 (1), 17–37.

Ertani, A., Schiavon, M., Muscolo, A., Nardi, S., 2013. Alfalfa plant-derived biostimulant stimulate short-term growth of salt stressed *Zea mays* L. plants. Plant Soil 364, 145–158.

European Union, 2019. Regulation of the European parliament and of the council laying down rules on the making available on the market of EU fertilising products and amending regulations (EC) no 1069/2009 and (EC) no 1107/2009 and repealing regulation (EC) no 2003/2003.

Farrell, M., Hill, P.W., Farrar, J., DeLuca, T.H., Roberts, P., Kielland, K., et al., 2013. Oligopeptides represent a preferred source of organic N uptake: a global phenomenon? Ecosystems 16, 133–145.

Fiorentino, N., Ventorino, V., Woo, S.L., Pepe, O., De Rosa, A., Gioia, L., et al., 2018. Trichoderma-based biostimulants modulate rhizosphere microbial populations and improve N uptake efficiency, yield, and nutritional quality of leafy vegetables. Front. Plant Sci. 9, 743.

Gao, H., Yan, C., Liu, Q., Ding, W., Chen, B., Li, Z., 2019. Effects of plastic mulching and plastic residue on agricultural production: a meta-analysis. Sci. Total. Environ. 651, 484–492.

Garcia-Mina, J.M., Antolin, M.C., Sanchez-Diaz, M., 2004. Metal-humic complexes and plant micronutrient uptake: a study based on different plant species cultivated in diverse soil types. Plant Soil 258, 57–68.

Goel, L., Shankar, V., Sharma, R.,K., 2019. Investigations on effectiveness of wheat and rice straw mulches on moisture retention in potato crop (*Solanum tuberosum* L.). Int. J. Recycl. Org. 8, 345–356.

Grover, M., Ali, S.Z., Sandhya, V., Rasul, A., Venkateswarlu, B., 2011. Role of microorganisms in adaptation of agriculture crops to abiotic stresses. World J. Microbiol. Biotechnol. 27, 1231–1240.

Hannah, R., Max, R., Pablo, R., 2022. Pesticides. <OurWorldInData.org>. Retrieved from: <https://ourworldindata.org/pesticides>.

He, G., Wang, Z., Hui, X., Huang, T., Luo, L., 2021. Black film mulching can replace transparent film mulching in crop production. Field Crop. Res. 261, 108026.

Heckman, J.R., Weil, R., Magdoff, F., 2009. Practical steps to soil fertility for organic agriculture. In: Francis, C. (Ed.), Organic Farming: The Ecological System, 54. pp. 139–173.

Imran, M., Mahmood, A., Römheld, V., Neumann, G., 2013. Nutrient seed priming improves seedling development of maize exposed to low root zone temperatures during early growth. Eur. J. Agron. 49, 141–148.

Jiao, G., Yu, G., Zhang, J., Ewart, H.S., 2011. Chemical structures and bioactivities of sulfated polysaccharides from marine algae. Mar. Drugs 9 (2), 196–223.

Kader, M.A., Singha, A., Begum, M.A., Jewel, A., Khan, F.H., Khan, N.I., 2019. Mulching as water-saving technique in dryland agriculture. Bull. Natl. Res. Cent. 43 (1), 1–6.

Kasirajan, S., Ngouajio, M., 2012. Polyethylene and biodegradable mulches for agricultural applications: a review. Agron. Sustain. Dev. 32, 501–529.

Khan, A.L., Hussain, J., Al-Harrasi, A., Al-Rawahi, A., Lee, I.J., 2015. Endophytic fungi: resource for gibberellins and crop abiotic stress resistance. Crit. Rev. Biotechnol. 35 (1), 62–74.

Kocira, A., Swieca, M., Kocira, S., Zlotek, U., Jakubczyk, A., 2018. Enhancement of yield, nutritional and nutraceutical properties of two common bean cultivars following the application of seaweed extract (*Ecklonia maxima*). Saudi J. Biol. Sci. 25 (3), 563–571.

Kumar, R., Sharma, P., Gupta, R.K., Kumar, S., Sharma, M.M.M., Singh, S., et al., 2020. Earthworms for eco-friendly resource efficient agriculture. Resources Use Efficiency in Agriculture. pp. 47–84.

Kumar, S., Bamboriya, S.D., Rani, K., Meena, R.S., Sheoran, S., Loyal, A., et al., 2022. Grain legumes: a diversified diet for sustainable livelihood, food, and nutritional security. Advances in Legumes for Sustainable Intensification. Academic Press, pp. 157–178.

Landis, T.D., 1988. Management of forest nursery soils dominated by calcium salts. N. For. 2, 173–193.

Li, X.Y., 2003. Gravel–sand mulch for soil and water conservation in the semiarid loess region of northwest China. Catena 52 (2), 105–127.

Li, A., Zhang, J., Ren, S., Zhang, Y., Zhang, F., 2021. Research progress on preparation and field application of paper mulch. Environ. Technol. Innov. 24, 101949.

Lightfoot, D.R., 1996. The nature, history, and distribution of lithic mulch agriculture: an ancient technique of dryland agriculture. Agric. Hist. Rev. 44 (2), 206–222.

Liu, X., Ai, Y., Zhang, F., Lu, S., Zeng, X., Fan, M., 2005. Crop production, nitrogen recovery and water use efficiency in rice–wheat rotation as affected by non-flooded mulching cultivation (NFMC). Nutr. Cycl. Agroecosystems 71, 289–299.

Loyal, A., Pahuja, S.K., Sharma, P., Malik, A., Srivastava, R.K., Mehta, S., 2023. Potential environmental and human health implications of nanomaterials used in sustainable agriculture and soil improvement. Engineered Nanomaterials for Sustainable Agricultural Production, Soil Improvement and Stress Management. Academic Press, pp. 387–412.

Luziatelli, F., Ficca, A.G., Colla, G., Baldassarre Svecova, E., Ruzzi, M., 2019. Foliar application of vegetal-derived bioactive compounds stimulates the growth of beneficial bacteria and enhances microbiome biodiversity in lettuce. Front. Plant Sci. 10, 60.

Lv, W., Qiu, Y., Xie, Z., Wang, X., Wang, Y., Hua, C., 2019. Gravel mulching effects on soil physicochemical properties and microbial community composition in the Loess Plateau, northwestern China. Eur. J. Soil. Biol. 94, 103115.

Martin-Closas, L., Costa, J., Pelacho, A.M., 2017. Agronomic effects of biodegradable films on crop and field environment. Soil degradable bioplastics for a sustainable modern agriculture, pp. 67–104.

Masarirambi, M.T., Mndzebele, M.E., Wahome, P.K., Oseni, T.O., 2013. Effects of white plastic and sawdust mulch on 'Savoy' baby cabbage (Brassica oleracea var. bullata) growth, yield and soil moisture conservation in summer in Swaziland. Am-Eurasian. J. Agric. Environ. Sci. 13 (2), 261–268.

Miceli, A., Moncada, A., Vetrano, F., 2021. Use of microbial biostimulants to increase the salinity tolerance of vegetable transplants. Agronomy 11, 1143.

Moore, J.C., Wszelaki, A.L., 2019. The use of biodegradable mulches in pepper production in the southeastern United States. HortScience 54 (6), 1031–1038.

Nardi, S., Pizzeghello, D., Muscolo, A., Vianello, A., 2002. Physiological effects of humic substances on higher plants. Soil. Biol. Biochem. 34 (11), 1527–1536.

Nithisha, A., Bokado, K., Charitha, K.S., 2022. Mulches: their impact on the crop production.

Nwodo, U.U., Green, E., Okoh, A.I., 2012. Bacterial exopolysaccharides: functionality and prospects. Int. J. Mol. Sci. 13 (11), 14002–14015.

Organic Food Global Market Report, 2023. By Product Type, By Application, By Distribution Channel – Market Size, Trends, And Global Forecast 2023–2032. https://www.thebusinessresearchcompany.com/report/organic-food-global-market-report

Phelan, M., Aherne, A., FitzGerald, R.J., O'Brien, N.,M., 2009. Casein-derived bioactive peptides: Biological effects, industrial uses, safety aspects and regulatory status. Int. Dairy J. 19 (11), 643–654.

Ramya, S.S., Vijayanand, N., Rathinavel, S., 2015. Foliar application of liquid biofertilizer of brown alga Stoechospermum marginatum on growth, biochemical and yield of Solanum melongena. Int. J. Recycl. Org. 4, 167–173.

Rani, K., Sharma, P., Kumar, S., Wati, L., Kumar, R., Gurjar, D.S., et al., 2019. Legumes for sustainable soil and crop management. Sustainable Management of Soil and Environment, pp. 193–215.

Rani, K., Rani, A., Sharma, P., Dahiya, A., Punia, H., Kumar, S., et al., 2022. Legumes for agroecosystem services and sustainability. Advances in Legumes for Sustainable Intensification. Academic Press, pp. 363–380.

Rouphael, Y., Colla, G., 2020a. Biostimulants in agriculture. Front. Plant Sci. 11, 40.

Rouphael, Y., Colla, G., 2020b. Editorial: Biostimulants in agriculture. Front. Plant Sci. 1–7.

Rouphael, Y., Carillo, P., Cristofano, F., Cardarelli, M., Colla, G., 2021. Effects of vegetal-versus animal-derived protein hydrolysate on sweet basil morpho-physiological and metabolic traits. Sci. Hortic. 284, 110123.

Ruzzi, M., Aroca, R., 2015. Plant growth-promoting rhizobacteria act as biostimulants in horticulture. Sci. Hortic. 196, 124–134.

Saia, S., Colla, G., Raimondi, G., Di Stasio, E., Cardarelli, M., Bonini, P., et al., 2019. An endophytic fungi-based biostimulant modulated lettuce yield, physiological and functional quality responses to both moderate and severe water limitation. Sci. Hortic. 256, 108595.

Sangwan, S., Sharma, P., Wati, L., Mehta, S., 2023. Effect of chitosan nanoparticles on growth and physiology of crop plants. Engineered Nanomaterials for Sustainable Agricultural Production, Soil Improvement and Stress Management. Academic Press, pp. 99–123.

Sekara, A., Pokluda, R., Cozzolino, E., del Piano, L., Cuciniello, A., Caruso, G., 2019. Plant growth, yield, and fruit quality of tomato affected by biodegradable and non-degradable mulches. Hort. Sci. 46 (3), 138–145.

Sharma, P., Sharma, M.M.M., Kapoor, D., Rani, K., Singh, D., Barkodia, M., 2020a. Role of microbes for attaining enhanced food crop production. Microbial Biotechnology: Basic Research and Applications. Springer, pp. 55–78.

Sharma, P., Sharma, M.M.M., Patra, A., Vashisth, M., Mehta, S., Singh, B., et al., 2020b. The role of key transcription factors for cold tolerance in plants. Transcription Factors for Abiotic Stress Tolerance in Plants. Academic Press, pp. 123–152.

Sharma, M.M.M., Sharma, P., Kapoor, D., Beniwal, P., Mehta, S., 2021a. Phytomicrobiome community: an Agrarian perspective towards resilient agriculture. Plant Performance Under Environmental Stress: Hormones, Biostimulants and Sustainable Plant Growth Management. Springer, pp. 493–534.

Sharma, P., Pandey, V., Sharma, M.M.M., Patra, A., Singh, B., Mehta, S., et al., 2021b. A review on biosensors and nanosensors application in agroecosystems. Nanoscale Res. Lett. 16, 1–24.

Sharma, P., Sharma, M.M.M., Malik, A., Vashisth, M., Singh, D., Kumar, R., et al., 2021c. Rhizosphere, rhizosphere biology, and rhizospheric engineering. Plant growth-promoting microbes for sustainable biotic and abiotic stress management. Springer International Publishing, Cham, pp. 577–624.

Sharma, P., Meyyazhagan, A., Easwaran, M., Sharma, M.M.M., Mehta, S., Pandey, V., et al., 2022a. Hydrogen sulfide: a new warrior in assisting seed germination during adverse environmental conditions. Plant Growth Regul. 98 (3), 401–420.

Sharma, P., Sangwan, S., Kumari, A., Singh, S., Kaur, H., 2022b. Impact of climate change on soil microorganisms regulating nutrient transformation. Plant Stress Mitigators: Action and Application. Springer Nature Singapore, Singapore, pp. 145–172.

Sharma, P., Sangwan, S., Mehta, S., 2023. Emerging role of phosphate nanoparticles in agriculture practices. Engineered Nanomaterials for Sustainable Agricultural Production, Soil Improvement and Stress Management. Academic Press, pp. 71–97.

Singh, A., Sharma, P., Kumari, A., Kumar, R., Pathak, D.V., 2019. Management of root-knot nematode in different crops using microorganisms. Plant biotic interactions: state of the art, pp. 85–99.

Singh, S., Sangwan, S., Sharma, P., Devi, P., Moond, M., 2021. Nanotechnology for sustainable agriculture: an emerging perspective. J. Nanosci. Nanotechnol. 21 (6), 3453–3465.

Somanathan, H., Sathasivam, R., Sivaram, S., Kumaresan, S.M., Muthuraman, M.S., Park, S.U., 2022. An update on polyethylene and biodegradable plastic mulch films and their impact on the environment. Chemosphere 307, 135839.

Sorrentino, M., Diego, De, Ugena, N., Spíchal, L., Lucini, L., Miras-Moreno, L., et al., 2021. Seed priming with protein hydrolysates improves Arabidopsis growth and stress tolerance to abiotic stresses. Front. Plant Sci. 12, 626301.

Tan, Z., Yi, Y., Wang, H., Zhou, W., Yang, Y., Wang, C., 2016. Physical and degradable properties of mulching films prepared from natural fibers and biodegradable polymers. Appl. Sci. 6 (5), 147.

Thompson, A.A., Samuelson, M.B., Kadoma, I., Soto-Cantu, E., Drijber, R., Wortman, S.E., 2019. Degradation rate of bio-based agricultural mulch is influenced by mulch composition and biostimulant application. J. Polym. Environ. 27, 498–509.

Trouvelot, S., Heloir, M.C., Poinssot, B., Gauthier, A., Paris, F., Guillier, C., et al., 2014. Carbohydrates in plant immunity and plant protection: roles and potential application as foliar sprays. Front. Plant Sci. 592.

Van Oosten, M.J., Pepe, O., De Pascale, S., Silletti, S., Maggio, A., 2017. The role of biostimulants and bioeffectors as alleviators of abiotic stress in crop plants. Chem. Bio. Technol. Agric. 4, 1–12.

Walan, P., Davidsson, S., Johansson, S., Hook, M., 2014. Phosphate rock production and depletion: regional disaggregated modeling and global implications. Resour. Conserv. Recycl. 93, 178–187.

Wang, F.X., Feng, S.Y., Hou, X.Y., Kang, S.Z., Han, J.J., 2009. Potato growth with and without plastic mulch in two typical regions of Northern China. Field Crop. Res. 110 (2), 123–129.

Wilson, H.T., Amirkhani, M., Taylor, A.G., 2018. Evaluation of gelatin as a biostimulant seed treatment to improve plant performance. Front. Plant Sci. 9, 1006.

Yakhin, O.I., Lubyanov, A.A., Yakhin, I.A., Brown, P.H., 2017. Biostimulants in plant science: a global perspective. Front. Plant Sci. 7, 2049.

Yobterik, A.C., Timmer, V.R., 1994. Nitrogen mineralization of agroforestry tree mulches under saline soil conditions. In: Bryan, R.B. (Ed.), Advances in Geoecology; Soil Erosion, Land Degradation and Social Transition, 27. Catena Verlag, pp. 181–194.

Zeng, K., Hwang, H.M., Yu, H., 2002. Effect of dissolved humic substances on the photochemical degradation rate of 1-aminopyrene atrazine. Int. J. Mol. Sci. 3 (10), 1048–1057.

Zhang, X., Wang, K., Ervin, E.H., 2010. Optimizing dosages of seaweed extract-based cytokinins and zeatin riboside for improving creeping bentgrass heat tolerance. Crop. Sci. 50 (1), 316–320.

Zhang, H., Miles, C., Ghimire, S., Benedict, C., Zasada, I., DeVetter, L., 2019. Polyethylene and biodegradable plastic mulches improve growth, yield, and weed management in floricane red raspberry. Sci. Hortic. 250, 371–379.

Zulfiqar, F., Casadesus, A., Brockman, H., Munne-Bosch, S., 2020. An overview of plant-based natural biostimulants for sustainable horticulture with a particular focus on moringa leaf extracts. Plant Sci. 295, 110194.

CHAPTER

13

Plant biochemistry and yield in response to biostimulants

Bushra Ahmad[1], Arshad Jamil[2], Dure Shahwar[1], Aisha Siddique[1] and Umama Syed[1]

[1]Department of Biochemistry, Shaheed Benazir Bhutto Women University, Peshawar, Pakistan
[2]Department of Plant Breeding and Genetics, University of Agriculture, Dera Ismail Khan, Pakistan

13.1 Introduction

Plant biochemistry deals with the overall chemistry of plants, for example, which types of metabolites are produced and how they are regulated and metabolized. The major concern of plant biochemistry is to study the primary and secondary metabolites and regulate plant growth and development. This is a vast field covering all the aspects taking place in plants, from photosynthesis to genetic engineering.

To increase the quality of yields, agricultural production looks for technical alternatives. Due to their favorable effects on plant growth, development, and yield, as well as their safety for humans and environmental friendliness (Caruso et al., 2019a; Caruso et al., 2019b), biostimulants are becoming more common. Biostimulants, independent of their nutrient content, are defined by Du Jardin (2015) as "any substance or microorganism administered to plants with the objective of increasing nutrition efficiency, abiotic stress tolerance, and/or crop quality attributes." Also, the European Biostimulants Industry Council was established to improve regulatory provisions surrounding the registration of biostimulants depending on their specific mode of activity. However, at the moment, their registration is based on legal regulations regarding pesticides and fertilizers, and some of them have a marketing gap (Traon et al., 2014; Du Jardin, 2015; Matyjaszczyk, 2015; Matyjaszczyk, 2019). Biostimulant is a recently coined term used in the field of plant biochemistry. This topic is serving as a hotspot in terms of plant growth and overall yield of crops, along with the mitigation of biotic stress induced in plants due to climate change, as shown in Fig. 13.1. The word biostimulant was basically devised by experts working in

Biostimulants in Plant Protection and Performance
DOI: https://doi.org/10.1016/B978-0-443-15884-1.00005-1

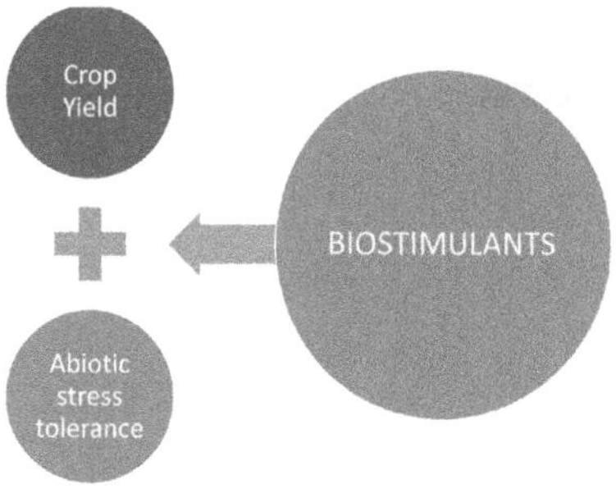

FIGURE 13.1 Role of biostimulants.

the domain of horticulture for relating substances to promote plant growth without using soil nutrients or pesticides. Biostimulants are organic compounds, microbes, or an amalgamation of both that could regulate plant growth behavior through molecular alteration and physiological, biochemical, and anatomical modulations. Their nature is diverse due to the varying composition of bioactive compounds, and they function through various modes of action.

Due to climate change and the anticipated population growth, by 2050, the danger of hunger is likely to increase by 30% (Van Dijk et al., 2021). To increase agricultural output and guarantee food security and safety, a variety of traditional (like fertilization) and cutting-edge bioengineering technologies (like genetically modified crops) have been extensively explored (Bailey-Serres et al., 2019). Plant biostimulants (PBs) are possibly a strategy to minimize climate change-induced stress and reduce reliance on chemical fertilizers, as there is a widespread understanding that emerging synthetic fertilizers pose environmental hazards to local and global ecosystems, as explained in Fig. 13.2 (Hunter et al., 2017; Koli et al., 2019).

By 2050, the European Commission wants biobased alternatives to replace 30% of chemical fertilizers (Hansen, 2018). To maintain crop productivity in the face of low fertilizer levels, PB application is a more environmentally friendly farming practice (Gupta et al., 2020). According to Regulation (EU) 2019/1009, 2019 of the European Commission, PBs fall under the umbrella of fertilizing goods. In a nutshell, it says that PBs are substances that promote plant growth while also enhancing one or more additional functions, such as nutrient usage efficiency, abiotic stress tolerance, crop quality features, and the availability of limited nutrients in the soil or plant rhizosphere, as described in Fig. 13.2. Moreover, Regulation (EU) 2019/1009 divided stimulatory bioactivity into two groups based on whether it originated from microorganisms or not, and this may call for an even more precise classification (Rouphael and Colla, 2020).

Microbial PBs include arbuscular mycorrhizal fungi (AMF) and plant growth-promoting rhizobacteria (PGPR) within the most widely acknowledged subcategories (Rouphael and Colla, 2020). Non-microbial PBs are often divided into six subcategories: silicon, chitosan, humic and fulvic acids, animal and plant protein hydrolysates, phosphites, and seaweed extracts (Si). The plant extract (PE)-based PBs, which include PBs other than seaweed extract (SWE) and have recently attracted a lot of interest, are regarded as a separate class of PBs (Du Jardin, 2015; Bio4Safe, I.S.P, 2021). In the product description of several PBs, the phrase "crop yield enhancement" is frequently used (Ricci

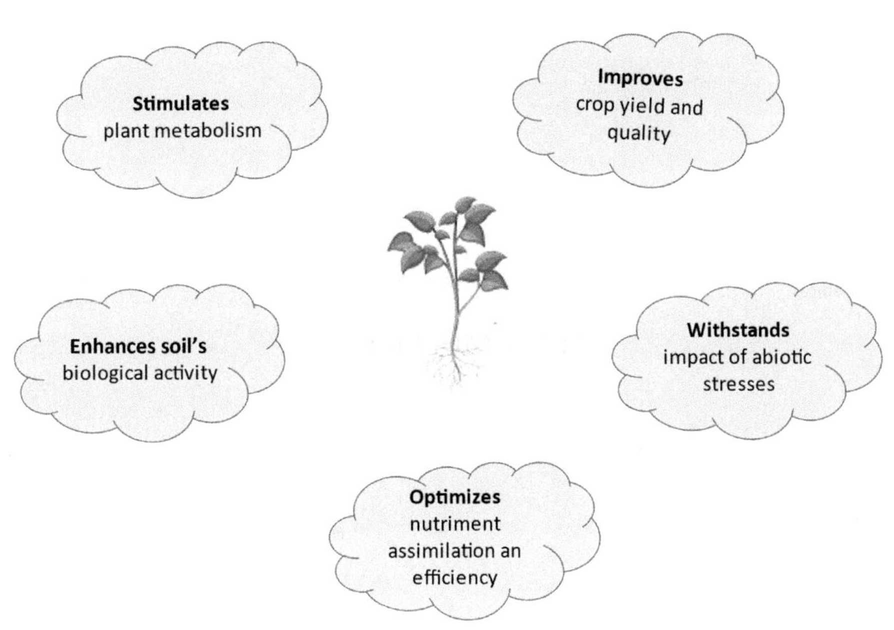

FIGURE 13.2 Effects of biostimulants on plants.

et al., 2019). Given that diverse environmental variables and management techniques have an impact on yield performance (Liliane and Charles, 2020), empirical data depends on varied experimental work.

There are natural and synthetic biostimulants, depending on the source. Both the former and the latter are derived from biological material and are physically and functionally identical to it (Matyjaszczyk, 2018). Preparations based on free amino acids, humic substances (HS), seaweed or fruit extracts, chitin and its derivative, chitosan, or microbial inoculants (free-living bacteria, fungus, and AMF) are included in the category of natural biostimulants, as shown in Fig. 13.3 (Calvo et al., 2014; Colla and Rouphael, 2015; Rouphael et al., 2017). The most significant class of compounds that promote plant growth and development within this class is biostimulants, which have an impact on a plant's metabolic processes, promoting the production or activity of phytohormones, promoting root development, and enhancing nutrient uptake, translocation, and utilization, all of which affect the yield's quality as shown in Fig. 13.4 (Calvo et al., 2014; Battacharyya et al., 2015; Colla and Rouphael, 2015; Du Jardin, 2015; Colla et al., 2017; Rouphael et al., 2017;). Additionally, biostimulants increase plant resistance to abiotic stress factors like drought, frost, salinity, and heavy metal contamination in the environment, which is likely a result of changes in the enzymatic activity of antioxidant compounds and an increase in their synthesis (Basak, 2008; Calvo et al., 2014), including protein hydrolysates and seaweed extracts (Battacharyya et al., 2015; Colla et al., 2017), as shown in Fig. 13.2.

The effects of biostimulants on the production of the bean crop are discussed in a recent study. The goal of the study was to determine how often using the biostimulants Kelpak, Terra Sorb Complex, and Fylloton affected bean production and nutritional qualities. They

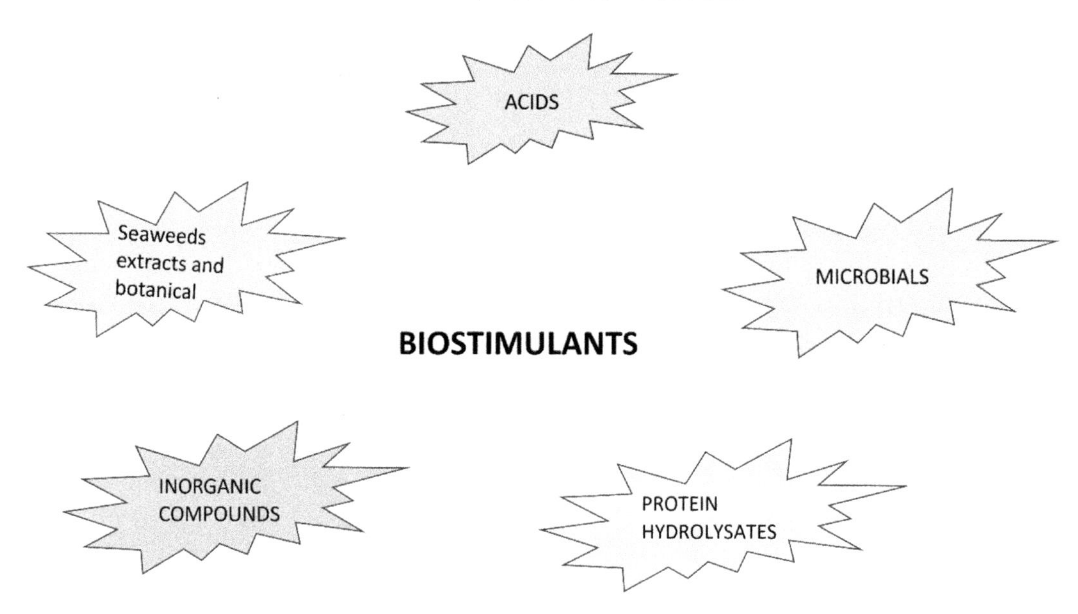

FIGURE 13.3 Major types of biostimulants.

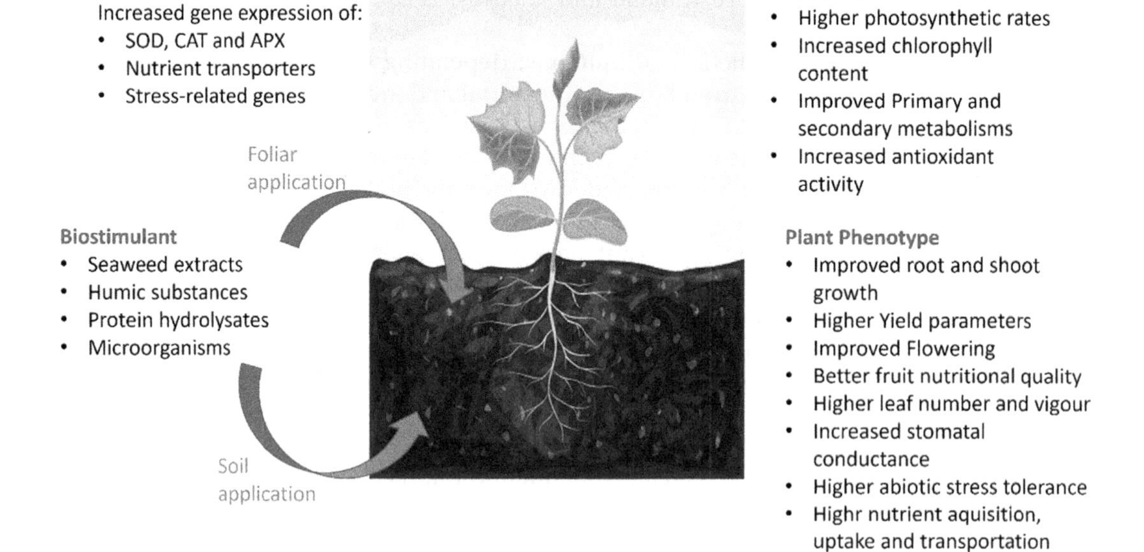

FIGURE 13.4 Effects of biostimulants on cellular biochemical, and physiological parameters.

concluded that the flavonoid content was raised by the use of all biostimulants. The seed yield was increased by biostimulants containing seaweed (Kelpak-Ecklonia maxima extract) or amino-acid extracts (Fylloton-Ascophyllum nodosum extract and amino acids or Terra Sorb Complex-amino acids), which also improved the quality of the seed by raising its protein, polyphenol, and flavonoid content. It was concluded that using Kelpak biostimulanys twice enhances the yeald and quality of crop.

13.2 Major types of plant biostimulants

There is no legal or regulatory definition of PBs anywhere in the world, including in the European Union and the United States, despite recent efforts to clarify the regulatory status of biostimulants. A thorough listing and classification of the compounds and microbes included in the idea are precluded by the current circumstances. Despite this, some major categories that include both substances and microorganisms are broadly acknowledged by scientists, regulators, and stakeholders, as shown in Fig. 13.3 (Du Jardin, 2012; Calvo et al., 2014; Halpern et al., 2015a; Halpern et al., 2015b). Beneficial fungi and bacteria, primarily PGPRs, are examples of microorganisms. They may be endosymbiotic, free-living, or rhizospheric.

13.2.1 N-containing compounds and protein hydrolysates

By chemically and enzymatically hydrolyzing proteins from agricultural by-products, such as crop leftovers and animal wastes like collagen and epithelial tissues, amino-acid and peptide combinations are produced (Du Jardin, 2012; Calvo et al., 2014; Halpern et al., 2015a,b). Single or combined molecules can also be made through chemical synthesis. Betaines, polyamines, and "non-protein amino acids," which are diverse in higher plants but poorly understood in terms of their physiological and ecological activities, are other nitrogenous compounds (Vranova et al., 2011). A unique instance of an amino-acid derivative with well-known anti-stress characteristics is glycine betaine (Chen and Murata, 2011). These compounds have been demonstrated to display multiple functions as biostimulants responsible for plant growth (Du Jardin, 2012; Calvo et al., 2014; Halpern et al., 2015a,b), as summarized in Table 13.1. Various effects on plants include modulation of N uptake and assimilation, regulation of the enzymes involved in N assimilation and their structural genes, and acting on the signaling pathway of N acquisition in roots. They are also responsible for C and N metabolisms through the regulation of TCA cycle enzymes.

Chelating effects are reported for some amino acids (like proline), which may protect plants against heavy metals but also contribute to micronutrient mobility and intake. Antioxidant activity is confirmed by free radical quenching by nitrogenous compounds, for example, glycine, betaine, and proline, which are responsible for the management of environmental stress. Indirect effects of biostimulants on plant growth and nutrition are

TABLE 13.1 Effects of biostimulants on crop productions, physiological functions, to agricultural/horticultural functions, and predictable economic and environmental benefits (Dobbelaere et al., 1999; Huang et al., 2010; Shabala et al., 2012).

Functions	Humic acids	Seaweed extracts	Protein hydrolysate	Glycine betaine	Plant growth-promoting rhizobacteria
Cellular mechanism (i.e., interaction with cellular components and processes)	Activate plasma membrane proton pumping ATPases, promote cell wall loosening and cell elongation in maize roots (*Zea mays*) (Jindo et al., 2012)	*Ascophyllum nodosum* extracts stimulate the expression of genes encoding transporters of micronutrients (e.g., Cu, Fe, Zn) in oilseed rape (*Brassica napus*) (Billard et al., 2014)	Enzymatic hydrolysate from alfalfa (*Medicago sativa*) stimulates phenylalanine ammonia-lyase (PAL) enzyme and gene expression, and the production of flavonoids under salt stress (Ertani et al., 2013)	Protects photosystem II against salt-induced photodamage in quinoa (Shabala et al., 2012), likely via activation of scavengers of reactive oxygen (Chen & Murata, 2011)	*Azospirillum brasilense* releases auxins and activates auxin signaling pathways involved in root morphogenesis in winter wheat (*Triticum aestivum*) (Dobbelaere et al., 1999)
Physiological function (i.e., action on whole-plant processes)	Increased linear growth of roots, root biomass	Increased tissue concentrations and root-to-shoot transport of micronutrients	Protection by flavonoids against UV and oxidative damage (Huang et al., 2010)	Maintenance of leaf photosynthetic activity under salt stress	Increased lateral root density and surface of root hairs
Agricultural/ horticultural function (i.e., output traits relevant for crop performance)	Increased root foraging capacity, enhanced nutrient use efficiency	Improved mineral composition of plant tissues	Increased crop tolerance to abiotic (e.g., salt) stress	Increased crop tolerance to abiotic (e.g., high salinity) stress	Increased root foraging capacity, enhanced nutrient use efficiency
Economic and environmental benefits (i.e., changes in yield, products' quality, ecosystem services)	Higher crop yield, savings of fertilizers, and reduced losses to the environment	Enhanced nutritional value, "biofortification" of plant tissues (increased contents in S, Fe, Zn, Mg, and Cu)	Higher crop yield under stress conditions (e.g., high salinity)	Higher crop yield under stress conditions (e.g., high salinity)	Higher crop yield, savings of fertilizers, and reduced losses to the environment

also considered very significant in agriculture when protein hydrolysates are added to soils. Protein hydrolysates are responsible for enhancing microbial biomass production and activity, soil respiration, and soil fertility. The chelating and complex formation activities of certain amino acids and peptides are important to enhance nutrient availability and acquisition by plant roots.

Several commercial products derived from animal and plant protein hydrolysates have been placed on the market. Overall, keeping in view various biostimulants, they are supposed to contribute significant improvements in yield and quality traits in many crops in agricultural and horticultural crops, as shown in Fig. 13.4 (Calvo et al., 2014). The safety measures regarding hydrolyzed proteins of animal origin were recently investigated and hence concluded to have no genotoxicity, phytotoxicity, or ecotoxicity based on bioassays involving yeasts and plants as test organisms (Corte et al., 2014).

13.2.2 Humic and fulvic acids

Natural components of soil organic matter, HS, are produced by the breakdown of plant, animal, and microbial wastes, as well as the metabolic activity of soil bacteria utilizing these substrates. The original classification of HS into humins, humic acids, and fulvic acids was based on their molecular weights and solubility. HSs are collections of heterogeneous molecules. Moreover, these compounds exhibit intricate dynamics of association and dissociation into supramolecular colloids, which are regulated by plant roots through exudate and proton release. Hence, the interaction between organic matter, microorganisms, and plant roots leads to the formation of humic compounds and their complexes in the soil.

To produce the desired results, every attempt to use humic compounds to promote plant growth and agricultural yield must optimize these interactions. This explains why the use of humic materials—soluble humic and fulvic acid fractions—shows variable, though generally favorable, effects on plant growth. Rose et al. (2014) reported that HS applied to plants resulted in an overall dry weight increase of 22.4% for shoots and 21.6% for roots.

The source of the HS, the environmental circumstances, the receiving plant, the dose, and the application method all contribute to the heterogeneity in HS effects (Fig. 13.4) (Rose et al., 2014). According to Du Jardin (2012), the sources of HS include naturally humified organic matter (such as peat or volcanic soils), composts and vermicomposts, mineral deposits, and composts and vermicomposts (leonardite, an oxidation form of lignite). Moreover, agricultural by-products are amenable to controlled breakdown and oxidation by chemical methods rather than decomposing in soil or through composting, producing "humic-like compounds" that are advocated as a replacement for natural HS (Eyheraguibel et al., 2008). Humic compounds are known for acting on the physical, physicochemical, chemical, and biological characteristics of the soil and are considered to be important contributors to soil fertility.

The majority of HS's biostimulant effects are related to different processes that improve root nourishment. One of them is the higher absorption of macro- and micronutrients as a result of the soil's enhanced cation exchange capacity and the increased availability of phosphorus caused by the polyanionic HS's interference with calcium phosphate precipitation (Fig. 13.4). The stimulation of plasma membrane H + -ATPases, which transforms the free energy supplied by ATP hydrolysis into a trans-membrane electrochemical potential used for the import of nitrate and other nutrients, is another significant contribution of HS to root nutrition. Proton pumping *via* plasma membrane ATPases promotes food uptake in addition to cell wall thinning, cell enlargement, and organ development (Jindo et al., 2012), as shown in Table 13.1.

By providing C substrates, HS appears to improve invertase and respiration activities. Furthermore, hormonal effects are recorded; however, it is frequently unclear whether HS includes functional groups recognized by the reception/signaling complexes of plant hormonal pathways, releases entrapped hormonal substances, or activates hormone-producing microbes (Du Jardin, 2012). Stress prevention is also mentioned in the HS biostimulation activity proposal. The formation of phenolic chemicals depends on phenylpropanoid metabolism, which is also important for secondary metabolism and a variety of stress reactions. In hydroponically grown maize seedlings, it has been demonstrated that high-molecular-mass HS increases the activity of crucial enzymes involved in this metabolism, indicating that HS may be used to modulate stress responses (Schiavon et al., 2010; Olivares et al., 2015). The overall effect of humic acid substances is summarized in Table 13.1.

13.2.3 Seaweed extracts and botanicals

In the agricultural sciences, it is a common practice to use fresh seaweeds as a source of organic matter and fertilizer. In this way, the commercial use of seaweed extracts and purified compounds, including the polysaccharides laminarin, alginates, and carrageenans, and their breakdown products, is supported. Plant growth promotion is also influenced by other constituents like micro- and macronutrients, sterols, N-containing compounds like betaines, and hormones (Khan et al., 2009; Craigie, 2011). Most of these compounds are quite unique to their specific algal sources, which elaborates on the increasing interest of scientists and industry in these taxonomic groups. Most of the algal species belong to the phylum of brown algae, but carrageenans are produced from red seaweeds, which is a different phylogenetic line. In a study, Khan et al. (2009) described more than 20 seaweed products that are in use as plant growth biostimulants. Seaweeds are added to soils and applied to plants (Craigie et al., 2008; Khan et al., 2009; Craigie, 2011). They can be added to soils in the form of hydroponic solutions or as foliar treatments, as shown in Fig. 13.4. In soils, the seaweed polysaccharides are responsible for gel formation, water retention, and aeration of the soil. The polyanionic compounds are described as playing an important role in the fixation and exchange of cations, which is very significant for soil remediation and heavy metal fixation. With the promotion of plant growth-promoting bacteria and pathogen antagonists in suppressive soils, the positive effects *via* the soil microflora are also described. The micro- and macronutrients in plants act as fertilizers in addition to their nutritional roles. Hormonal effects, which are one of the main causes of biostimulation activity in plants, are associated with seed germination, plant establishment, and further growth and development (Table 13.1). Various plant hormones, for example, cytokinins, auxins, gibberellin, and abscisic acid, are reported to be present in seaweed extracts by bioassays and immunological tools (Craigie, 2011). The hormonal effects of extracts of the brown seaweed *Ascophyllum nodosum* are confirmed by the up- and downregulation of hormone biosynthetic genes in the tissues of plants and also by the hormonal contents of the seaweed extracts themselves (Wally et al., 2013a,b). Anti-stress effects are also described, and it was concluded that both protective compounds, like antioxidants and regulators of endogenous stress-responsive genes, within the seaweed extracts, are responsible, as shown in Table 13.1 (Calvo et al., 2014).

The term botanical refers to the moieties derived from plants that are play important roles in pharmaceutical, food, and cosmetic products, as well as in plant self-defense products (Seiber et al., 2014). The role of botanicals as biostimulants is not much studied, which needs attention to confirm their use as biostimulants as well (Ziosi et al., 2012; Ertani et al., 2013).

13.2.4 Biopolymers and chitosan

Chitosan is a deacetylated form of chitin biopolymer that is synthesized naturally and industrially. The physiological effects of chitosan oligomers in plants are due to their polycationic compound capacity to bind a wide range of cellular components, including DNA, cell wall constituents, and cell membrane, but also to attach to exact receptors involved in the activation of defense genes (El Hadrami et al., 2010; Yin et al., 2010; Hadwiger, 2013; Katiyar et al., 2015). Chitin and chitosan, in fact, use distinctive receptors and signaling pathways. Keeping in view the cellular consequences of their binding to more or less specific cell receptors, the accumulation of hydrogen peroxide and Ca^2+ leakage in the cell have been determined, which probable cause various physiological changes involved in the signaling of stress responses (Povero et al., 2011; Ferri et al., 2014). In agriculture, chitosan has been considered very important, keeping in view their potential in plant protection against fungal pathogens, but also on tolerance to abiotic stress (drought, salinity, and cold stress) and on quality traits regarding primary and secondary metabolisms. Several polymers and oligomers of biological origin have been utilized in agriculture as boosters of plant defense. Although a distinction has to be made between biocontrol and biostimulation (e.g., enhancing abiotic stress), signaling pathways may be interlinked, and the effects of both may result from using similar inducers (Gozzo and Faoro, 2013).

13.2.5 Fungi as biostimulants

Fungi interact with plant roots in a variety of ways, from parasitism to mutualistic symbioses (i.e., when both organisms live in close proximity to one another and form connections that are mutually advantageous) (Behie and Bidochka, 2014). Since the beginning of terrestrial plants, plants and fungi have coevolved, and the mutualism-parasitism continuum idea is valuable for describing the wide diversity of connections that have emerged during evolutionary periods (Bonfante and Genre, 2010; Johnson and Graham, 2013). Around 90% of all plant species have symbiotic relationships with mycorrhizal fungi, which constitute a varied collection of taxa. AMF is a common kind of endomycorrhiza related to crop and horticultural plants, where fungal hyphae of Glomeromycota species enter root cortical cells and produce branching structures known as arbuscules (Bonfante and Genre, 2010; Behie and Bidochka, 2014). With the widely acknowledged advantages of the symbioses to nutrition efficiency (for both macronutrients, notably P, and micronutrients), water balance, and plant protection from biotic and abiotic stress, there is growing interest in the use of mycorrhiza to enhance sustainable agriculture (Augé, 2001; Harrier and Watson, 2004; van der Heijden et al., 2004; Hamel and Plenchette, 2007; Siddiqui et al., 2008; Gianinazzi et al., 2010). Current research further suggests that hyphal networks exist, connecting not just fungi and plants but also specific plants within a plant community. Given that there is evidence to suggest that the fungal conduits permit interplant signaling, this might have substantial ecological and agricultural ramifications (Simard et al., 2012; Johnson and Gilbert, 2015;). AMF creates tripartite connections with plants and rhizobacteria as an additional topic of study that is important in real-world field circumstances (Siddiqui et al., 2008). Crop management strategies and plant cultivars should be adjusted to the interaction with microorganisms to profit from mycorrhizal connections (Plenchette et al., 2005; Hamel and Plenchette, 2007; Gianinazzi et al., 2010; Sheng et al., 2011). An intriguing method for tracking and researching microbial communities in the rhizosphere is metagenomics. Inoculating soils and plant propagules is a complementary strategy to these (Sensoy et al., 2007; Sorensen et al., 2008; Candido et al., 2013; Colla et al., 2015a; Colla et al., 2015b; Sarkar et al., 2015). Biostimulants should include fungus-based treatments used on plants to increase crop output, stress tolerance, nutrition efficiency, and product quality. Its biotrophic nature makes it technically challenging to spread AMF on a broad scale (Dalpé and Monreal, 2004), but more importantly, it makes it difficult to study the factors that influence host specificities and population dynamics of mycorrhizal communities in agroecosystems. However, unlike mycorrhizal species, some fungal endophytes, such as *Trichoderma* spp. (Ascomycota) and *Sebacinales* (Basidiomycota), utilizing *Piriformospora indica* as a model organism, may live at least a portion of their life cycle independently of the plant, invade roots, and, as recently demonstrated, transport nutrients to their hosts through poorly known methods (Behie and Bidochka, 2014). Growing interest is being paid to both, plant inoculants that are easier to propagate in vitro and as models for examining the processes of nutrient transfer between fungal endosymbionts and their hosts. For their myco-parasitic biopestcidal and biocontrol (inducer of disease resistance) abilities, several of these fungi, namely Trichoderma spp., have received substantial study and usage. Biotechnological businesses have taken advantage of these fungi as sources of enzymes (Mukherjee et al., 2012; Nicolás et al., 2014). There is strong evidence that a variety of plant responses are also produced, including improved abiotic stress tolerance, nutrient usage effectiveness, and organ development and morphogenesis (Shoresh et al., 2010; Colla et al., 2015a; Colla et al., 2015b). These fungus endophytes can be classified as biostimulants based on the fact that they are now being used in agriculture and have been labeled as biopesticides.

13.2.6 Bacteria as biostimulants

According to Ahmad et al. (2008), bacteria may interact with plants in a variety of ways (Fig. 13.4). Similar to fungi, bacteria can be: (1) mutualistic or parasitic; (2) bacterial habitats stretch from the soil to the inside of cells; (3) partnerships can be temporary or long-lasting; and (4) certain bacteria can even be transferred vertically through seeds. The roles that plants play in bio-geochemical cycles, nutrient availability, improved nutrient use efficiency, disease resistance induction, improved abiotic stress tolerance, and plant growth regulator-mediated morphogenesis are all functions that have an impact on plant life (Table 13.1). Within this taxonomic, functional, and ecological diversity, two main types of biostimulants should be taken into account with regard to agricultural applications: (1) mutualistic endosymbionts of the type Rhizobium, and (2) mutualistic rhizospheric PGPRs ("plant growth-promoting rhizobacteria"). Rhizobium and similar taxa are sold as biofertilizers or microbial inoculants that help plants absorb nutrients. In-depth reviews of the biology and agricultural applications of the Rhizobium-based symbioses may be found in textbooks and scholarly publications. PGPRs are multifunctional and affect all facets of plant life, including nutrition and growth, morphogenesis and development reaction to biotic and abiotic stress, and interactions with other organisms in agroecosystems (Ahmad et al., 2008; Babalola, 2010; Berendsen et al., 2012; Bhattacharyya and Jha, 2012; Gaiero et al., 2013; Philippot et al., 2013; Vacheron et al., 2013; Berg et al., 2014). In terms of how biostimulants are used in agriculture, there are two primary categories. Several of these tasks are often carried out by the same organisms, while others require strain-specific adaptations or the cooperation of other bacterial consortia. Because of their intricacy, the varying reactions of the plant cultivars, and the receiving settings, PGPRs cannot be used in agriculture. Moreover, the formulation of the inoculants is technically challenging, which leads to variable outcomes in practice (Arora et al., 2011; Brahmaprakash and Sahu, 2012). Despite this, the global market for bacterial biostimulants is expanding, and PGPR inoculants are increasingly viewed as a type of plant "probiotic," or effective providers of nutrition and immunity for plants, as explained in Table 13.1 (Berendsen et al., 2012).

13.2.7 Inorganic compounds

The term "beneficial element" refers to a chemical element that fosters plant development and may be necessary for some species but is not necessary for all plants (Pilon-Smits et al., 2009). Al, Co, Na, Se, and Si are the five primary helpful elements, and they are found in soils, plants, and other inorganic salts, as well as insoluble forms such as amorphous silica in graminaceous species. These positive properties can manifest under specific environmental conditions, such as pathogen assault for selenium and osmotic stress for sodium, or they might be constitutive, like the strengthening of cell walls by silica deposits. So, the definition of helpful components must take into account their unique settings where the positive impacts on plant development and stress response may be seen, in addition to their chemical properties. It is possible to suppose that the physiological actions of the beneficial components found in some complex biostimulants, such as extracts of seaweeds, agricultural residues, or animal wastes, are what cause their bioactivity (Fig. 13.4). The scientific literature reports several positive impacts of beneficial components that support plant development, product quality, and abiotic stress tolerance. Cell wall stiffening, osmoregulation, decreased transpiration by crystal deposits, heat regulation *via* radiation reflection, enzyme activity by cofactors (Table 13.1), plant nutrients *via* interactions with other elements during uptake and mobility, antioxidant prevention, interactions with symbiotic organisms, pathogen, and herbivore responses, protection against heavy metal toxicity, and plant hormone synthesis and signaling are some of these processes (Pilon-Smits et al., 2009). Fungicides have been made from inorganic salts of helpful and necessary elements, such as chlorides, phosphates, phosphites, silicates, and carbonates (Deliopoulos et al., 2010). These inorganic substances have an impact on osmotic, pH, redox homeostasis, hormone signaling, and stress response enzymes, while the exact mechanisms of action are still not entirely understood (e.g., peroxidases), as described in Table 13.1. Further consideration should be given to their role as biostimulants of plant development, working on nutrition efficiency and abiotic stress tolerance, which is separate from their fungicidal activity and their role as suppliers of nutrients in fertilizers. Taken together, it has been concluded that biostimulants increase plant resistance in adverse situations such as drought, frost, salinity, heavy metals, and so on.

References

Ahmad, I., Pichtel, J., Hayat, S., 2008. Plant-Bacteria Interactions. Strategies and Techniques to Promote Plant Growth. WILEY-VCH Verlag GmbH and Co., KGaA, , Weinheim, Germany.

Arora, N.K., Khare, E., Maheshwari, D.K., 2011. Plant growth promoting rhizobacteria: constraints in bioformulation, commercialization, and future strategies. In: Maheshwari, D.K. (Ed.), Plant Growth and Health Promoting Bacteria. Springer, Berlin/Heidelberg, Germany, pp. 97–116.

Augé, R.M., 2001. Water relations, drought and vesicular-arbuscular mycorrhizal symbiosis. Mycorrhiza 11, 3–42.

Babalola, O.O., 2010. Beneficial bacteria of agricultural importance. Biotechnol. Lett. 32, 1559–1570.

Bailey-Serres, J., Parker, J.E., Ainsworth, E.A., Oldroyd, G.E., Schroeder, J.I., 2019. Genetic strategies for improving crop yields. Nature 575, 109–118. Available from: https://doi.org/10.1038/s41586-019-1679-0.

Basak, A., 2008. Biostimulators - definitions, classification legislation. In: Gawro'nska, H. (Ed.), . Biostimulators in Modern Agriculture, General Aspects. Editorial House Wie's Jutra, Warszawa, Poland, pp. 7–17.

Battacharyya, D., Babgohari, M.Z., Rathor, P., Prithiviraj, B., 2015. Seaweed extracts as biostimulants in horticulture. Sci. Hortic. (196), 39–48.

Behie, S.W., Bidochka, M.J., 2014. Nutrient transfer in plant-fungal symbioses. Trends Plant Sci. 19, 734–740.

Berendsen, R.L., Pieterse, C.M., Bakker, P.A., 2012. The rhizosphere microbiome and plant health. Trends Plant Sci. 17, 1360–1385.

Berg, G., Grube, M., Schloter, M., Smalla, K., 2014. Unraveling the plant microbiome: looking back and future perspectives. Front. Microbiol. 5, 1–7. Article 148.

Bhattacharyya, P.N., Jha, D.K., 2012. Plant growth-promoting rhizobacteria (PGPR): emergence in agriculture. World J. Microbiol. Biotechnol. 28, 1327–1350.

Bio4Safe, I.S.P., 2021. Biostimulant Database. https://bio4safe.eu/ (accessed March 10, 2022).

Billard, V., Etienne, P., Jannin, L., Garnica, M., Cruz, F., Garcia-Mina, J.M., Yvin, J.-C, Ourry, A., 2014. Two biostimulants derived from algae or humic acid induce similar responses in the mineral content and gene expression of winter oilseed rape (Brassica napus L.). J. Plant Growth Regul. 33, 305–316. Available from: https://doi.org/10.1007/s00344-013-9372-2.

Bonfante, P., Genre, A., 2010. Interactions in mycorrhizal symbiosis. Nat. Commun. 1 (1), 11.

Brahmaprakash, G.P., Sahu, P.K., 2012. Biofertilizers for sustainability. J. Indian. Inst. Sci 92, 37–62.

Calvo, P., Nelson, L., Kloepper, J.W., 2014. Agricultural uses of plant biostimulants. Plant Soil 383, 3–41.

Candido, V., Campanelli, G., Addabbo, T.D., Castronuovo, D., Renco, M., Camele, I., 2013. Growth and yield promoting effect of artificial mycorrhization combined with different fertiliser rates on field-grown tomato. Ital. J. Agron. 8, 168–174.

Caruso, G., De Pascale, S., Cozzolino, E., Giordano, M., El-Nakhel, C., Cuciniello, A., et al., 2019a. Protein hydrolysate or plant extract-based biostimulants enhanced yield and quality performances of greenhouse perennial wall rocket grown in different seasons. Plants 8, 208.

Caruso, G., De Pascale, S., Cozzolino, E., Cuciniello, A., Cenvinzo, V., Bonini, P., et al., 2019b. Yield and nutritional quality of Vesuvian Piennolo tomato PDO as affected by farming system and Biostimulant application. Agronomy 9, 505.

Chen, T.H.H., Murata, N., 2011. Glycine betaine protects plants against abiotic stress: mechanisms and biotechnological applications. Plant Cell Environ. 34, 1–20.

Colla, G., Rouphael, Y., 2015. Biostimulants in horticulture. Sci. Hortic. 196, 1–2.

Colla, G., Nardi, S., Cardarelli, M., Ertani, A., Lucini, L., Canaguier, R., et al., 2015a. Protein hydrolysates as biostimulants in horticulture. Sci. Hortic. 196, 28–38. Available from: https://doi.org/10.1016/j.scienta.2015.08.037.

Colla, G., Rouphael, Y., Di Mattia, E., El-Nakhel, C., Cardarelli, M., 2015b. Co-inoculation of *Glomus intraradices* and *Trichoderma atroviride* acts as abiostimulant to promote growth, yield and nutrient uptake of vegetable crops. J. Sci. Food Agric. 95, 1706–1715.

Colla, G., Hoagland, L., Ruzzi, M., Cardarelli, M., Bonini, P., Canaguier, R., et al., 2017. Biostimulant action of protein hydrolysates: unraveling their effects on plant physiology and microbiome. Front. Plant Sci. 8, 2202.

Corte, L., Dell'Abate, M.T., Magini, A., Migliore, M., Felici, B., Roscini, L., et al., 2014. Assessment of safety and efficiency of nitrogen organic fertilizers from animal-based protein hydrolysates-a laboratory multidisciplinary approach. J. Sci. Food Agric. 94, 235–245.

Craigie, J.S., 2011. Seaweed extract stimuli in plant science and agriculture. J. Appl. Phycol. 23, 371–393.

Craigie, J.S., MacKinnon, S.L., Walter, J.A., 2008. Liquid seaweed extracts identified using 1H NMR profiles. J. Appl. Phycol. 20, 665–671.

Dalpé, Y., Monreal, M., 2004. Arbuscular mycorrhiza inoculum to support sustainable cropping systems. Online. Symposium Proceeding. Crop Management network. Available from: https://doi.org/10.1094/CM-2004-0301-09-RV.

Deliopoulos, T., Kettlewell, P.S., Hare, M.C., 2010. Fungal disease suppression by inorganic salts: a review. Crop Prot. 29, 1059–1075.

Dobbelaere, S., Croonenborghs, A., Thys, A., Broek, A.V., Vanderleyden, J., 1999. Phytostimulatory effect of *Azospirillum brasilense* wild type and mutantstrains altered in IAA production on wheat. Plant Soil 212, 155–164.

Du Jardin, P., 2012. The Science of Plant Biostimulants—A bibliographic analysis. Adhoc Study Report to the European Commission DG ENTR. 2012. http://ec.europa.eu/enterprise/sectors/chemicals/files/fertilizers/final.report.bio.2012en.pdf.

Du Jardin, P., 2015. Plant biostimulants: definition, concept, main categories and regulation. Sci. Hortic. 196, 3–14.

El Hadrami, A., Adam, L.R., El Hadrami, I., Daayf, F., 2010. Chitosan in plant protection. Mar. Drugs 8, 968–987.

Ertani, A., Schiavon, M., Muscolo, A., Nardi, S., 2013. Alfalfa plant-derived biostimulant stimulate short-term growth of salt stressed *Zea mays* L. plants. Plant Soil 364, 145–158.

Eyheraguibel, B., Silvestre, J., Morard, P., 2008. Effects of humic substances derived from organic waste enhancement on the growth and mineral nutrition of maize. Bioresour. Technol. 99, 4206–4212.

Ferri, M., Franceschetti, M., Naldrett, M.J., Saalbach, G., Tassoni, A., 2014. Effects of chitosan on the protein profile of grape cell culture subcellular fractions. Electrophoresis 35, 1685–1692.

Gaiero, J.R., McCall, C.A., Thompson, K.A., Dayu, N.J., Best, A.S., Dunfield, K.E., 2013. Inside the root microbiome: bacterial root endophytes and plant growth promotion. Am. J. Bot. 100, 1738–1750.

Gianinazzi, S., Gollotte, A., Binet, M.-N., van Tuinen, D., Redecker, D., Wipf, D., 2010. Agroecology: the key role of arbuscular mycorrhizas in ecosystem services. Mycorrhiza 20, 519–530.

Gozzo, F., Faoro, F., 2013. Systemic acquired resistance (50 Years after discovery): moving from the lab to the field. J. Agric. Food Chem. 61, 12473–12491.

Gupta, A., Rico-Medina, A., Cano-Delgado, A.I., 2020. The physiology of plant responses to drought. Science 368, 266–269. Available from: https://doi.org/10.1126/science.aaz7614.

Hadwiger, L.A., 2013. Multiple effects of chitosan on plant systems: solid science or hype. Plant Sci. 208, 42–49.

Halpern, M., Bar-Tal, A., Ofek, M., Minz, D., Muller, T., Yermiyahu, U., 2015a. The use of biostimulants for enhancing nutrient uptake. Adv. Agron. 130, 141–174. Available from: https://doi.org/10.1016/bs.agron.2014.10.001.

Halpern, M., Bar-Tal, A., Ofek, M., Minz, D., Muller, T., Yermiyahu, U., 2015b. The use of biostimulants for enhancing nutrient uptake. In: Sparks, D.L. (Ed.), Advances in Agronomy, 129. Elsevier, pp. 141–174.

Hamel, C., Plenchette, C., 2007. Mycorrhizae in Crop Production. The Haworth Press Inc., New York, USA.

Hansen, J., 2018. EU must get serious about promoting the circular economy [WWW Document]. https://www.theparliamentmagazine.eu/articles/partner_article/fertilizers-europe/eu-must-get-serious-aboutpromoting-.

Harrier, L.A., Watson, C.A., 2004. The potential role of arbuscular mycorrhizal (AM)fungi in the bioprotection of plants against soil-borne pathogens in organic and/or other sustainable farming systems. Pest Manag. Sci. 60, 149–157.

Huang, J., Gu, M., Lai, Z., Fan, B., Shi, K., Zhou, H., et al., 2010. Functional analysis of the Arabidopsis PAL gene family in plant growth, development, and response to environmental stress. Plant Physiol. 153, 1526–1538.

Hunter, M.C., Smith, R.G., Schipanski, M.E., Atwood, L.W., Mortensen, D.A., 2017. Agriculture in 2050: recalibrating targets for sustainable intensification. Bioscience 67, 386–391. Available from: https://doi.org/10.1093/biosci/bix010.

Jindo, K., Martim, S.A., Navarro, E.C., Aguiar, N.O., Canellas, L.P., 2012. Root growth promotion by humic acids from composted and non-composted urban organic wastes. Plant Soil 353, 209–220.

Johnson, D., Gilbert, L., 2015. Interplant signalling through hyphal networks. New Phytol. 205, 1448–1453.

Johnson, N.C., Graham, J.H., 2013. The continuum concept remains a useful framework for studying mycorrhizal functioning. Plant Soil 363, 411–419.

Katiyar, D., Hemantaranjan, A., Singh, B., 2015. Chitosan as a promising natural compound to enhance potential physiological responses in plant: a review. Indian. J. Plant Physiol. 20, 1–9.

Khan, W., Rayirath, U.P., Subramanian, S., Jithesh, M.N., Rayorath, P., Hodges, D.M., et al., 2009. Seaweed extracts as biostimulants of plant growth and development. J. Plant Growth Regul. 28, 386–399.

Koli, P., Bhardwaj, N.R., Mahawer, S.K., 2019. Agrochemicals: harmful and beneficial effects of climate changing scenarios. In: Choudhary, K.K., Kumar, A., Singh, A.K. (Eds.), Climate Change and Agricultural Ecosystems. Elsevier, Duxford, pp. 65–94.

Liliane, T.N., Charles, M.S., 2020. Factors affecting yield of crops, Agronomy- Climate Change & Food Security, 9. IntechOpen, pp. 1–16. Available from: https://doi.org/10.5772/intechopen.90672.

Matyjaszczyk, E., 2015. The introduction of biostimulants on the Polish market. The present situation and legal requirements. Przemysl. Chem. 10, 1841–1844.

Matyjaszczyk, E., 2018. Biorationals in integrated pest management strategies. J. Plant Dis. Prot. 125, 523–527.

Matyjaszczyk, E., 2019. Problems of implementing compulsory integrated pest management. Pest. Manag. Sci. 75.

Mukherjee, P.K., Horwitz, B.A., Herrera-estrella, A., Schmoll, M., Kenerley, C.M., 2012. Trichoderma research in the genome era. Annu. Rev. Phytopathol. 51, 105–129.

Nicolás, C., Hermosa, R., Rubio, B., Mukherjee, P.K., Monte, E., 2014. Trichoderma genes in plants for stress tolerance-status and prospects. Plant Sci. 228, 71–78.

Olivares, F.L., Aguiar, N.O., Rosa, R.C.C., Canellas, L.P., 2015. Substrate biofortification in combination with foliar sprays of plant growth promoting bacteria and humic substances boosts production of organic tomatoes. Sci. Hortic. 183, 100–108.

Philippot, L., Raaijmakers, J.M., Lemanceau, P., Putten, W.H.V.D., 2013. Going back to the roots: the microbial ecology of the rhizosphere. Nat. Rev. Microbiol. 11, 789–799.

Pilon-Smits, E.A.H., Quinn, C.F., Tapken, W., Malagoli, M., Schiavon, M., 2009. Physiological functions of beneficial elements. Curr. Opin. Plant Biol. 12, 267–274..

Plenchette, C., Clermont-Dauphin, C., Meynard, J.M., Fortin, J.A., 2005. Managing arbuscular mycorrhizal fungi in cropping systems. Can. J. Plant Sci. 85, 31–40.

Povero, G., Loreti, E., Pucciariello, C., Santaniello, A., Di Tommaso, D., Di Tommaso, G., et al., 2011. Transcript profiling of chitosan-treated Arabidopsis seedlings. J. Plant Res. 124, 619–629.

Ricci, M., Tilbury, L., Daridon, B., Sukalac, K., 2019. General principles to justify plant biostimulant claims. Front. Plant Sci. 10494. Available from: https://doi.org/10.3389/fpls.2019.00494.

Rose, M.T., Patti, A.F., Little, K.R., Brown, A.L., Jackson, W.R., Cavagnaro, T.R., 2014. A meta-analysis and review of plant growth response to humic substances: practical implications for agriculture. Adv. Agron. 124, 37–89. Available from: https://doi.org/10.1016/B978-0-12-800138-7.00002-4.

Rouphael, Y., Colla, G., 2020. Editorial: biostimulants in agriculture. Front. Plant Sci. 1140. Available from: https://doi.org/10.3389/fpls.2020.00040.

Rouphael, Y., Cardarelli, M., Bonini, P., Colla, G., 2017. Synergistic action of a microbial-based biostimulant and a plant derived-protein hydrolysate enhances lettuce tolerance to alkalinity and salinity. Front. Plant Sci. 8, 131.

Sarkar, A., Asaeda, T., Wang, Q., Rashid, M.H., 2015. Arbuscular mycorrhizal influences on growth, nutrient uptake, and use efficiency of *Miscanthus sacchariflorus* growing on nutrient-deficient river bank soil. Flora: Morphol. Distrib. Funct. Ecol. Plants 212, 46–54.

Schiavon, M., Pizzeghello, D., Muscolo, A., Vaccaro, S., Francioso, O., Nardi, S., 2010. High molecular size humic substances enhance phenylpropanoid metabolism in maize (*Zea mays* L.). J. Chem. Ecol. 36, 662–669.

Seiber, J.N., Coats, J., Duke, S.O., Gross, A.D., 2014. Biopesticides: state of the art and culture opportunities. J. Agric. Food Chem. 62, 11613–11619.

Sensoy, S., Demir, S., Turkmen, O., Erdinc, C., Burak, O., 2007. Responses of some different pepper (*Capsicum annuum* L.) genotypes to inoculation with two different arbuscular mycorrhizal fungi. Sci. Hortic. 113, 92–95.

Shabala, L., Mackay, A., Jacobsen, S., Erik, Z., Hou, D., Shabala, S., 2012. Oxidative stress protection and stomatal patterning as components of salinity tolerance mechanism in quinoa (*Chenopodium quinoa*). Physiol. Plant. 146, 26–38.

Sheng, P.-P., Li, M., Liu, R.-J., 2011. Effects of agricultural practices on community structure of arbuscular mycorrhizal fungi in agricultural ecosystem. A review. Chin. J. Appl. Ecol. 22, 1639–1645.

Shoresh, M., Harman, G.E., Mastouri, F., 2010. Induced systemic resistance and plant responses to fungal biocontrol agents. Annu. Rev. Phytopathol. 48, 21–43.

Siddiqui, Z.A., Akhtar, M.S., Futai, K., 2008. Mycorrhizae: Sustainable Agriculture and Forestry. Springer, Berlin/Heidelberg, Germany..

Simard, S.W., Beiler, K.J., Bingham, M.A., Deslippe, J.R., Philip, L.J., Teste, F.P., 2012. Mycorrhizal networks: mechanisms, ecology and modelling. Fungal Biol. Rev. 26, 39–60.

Sorensen, J.N., Larsen, J., Jakobsen, I., 2008. Pre-inoculation with arbuscular mycorrhizal fungi increases early nutrient concentration and growth of field-grown leeks under high productivity conditions. Plant Soil 307, 135–147.

Traon, D., Amat, L., Zotz, F., Du Jardin, P., 2014. A legal framework for plant biostimulants and agronomic fertiliser additives in the EU. Report for the European Commission Enterprise & Industry Directorate—General; publications.

Vacheron, J., Desbrosse, G., Bouffaud, M.-L., Touraine, B., Moënne-Loccoz, Y., Muller, D., et al., 2013. Plant growth-promoting rhizobacteria and root system functioning. Front. Plant Sci. 4, 1–19. Article 356.

van der Heijden, M.G.A., Van Der Streitwolf-engel, R., Riedl, R., Siegrist, S., Neudecker, A., Boller, T., et al., 2004. The mycorrhizal contribution to plant productivity, plant nutrition and soil structure in experimental grassland. New Phytol. 172, 739–752.

Van Dijk, M., Morley, T., Rau, M.L., Saghai, Y., 2021. A meta-analysis of projected global food demand and population at risk of hunger for the period 2010–2050. Nat. Food 2, 494–501. Available from: https://doi.org/10.1038/s43016-021-00322-9.

Vranova, V., Rejsek, K., Skene, K.R., Formanek, P., 2011. Non-protein amino acids: plant, soil and ecosystem interactions. Plant Soil 342, 31–48.

Wally, O.S.D., Critchley, A.T., Hiltz, D., Craigie, J.S., Han, X., Zaharia, L.I., et al., 2013a. Regulation of phytohormone biosynthesis and accumulation in arabidopsis following treatment with commercial extract from the marine macroalga *Ascophyllum nodosum*. J. Plant Growth Regul. 32, 324–339.

Wally, O.S.D., Critchley, A.T., Hiltz, D., Craigie, J.S., Han, X., Zaharia, L.I., et al., 2013b. Erratum to: regulation of phytohormone biosynthesis and accumulation in arabidopsis following treatment with commercial extract from the marine macroalga *Ascophyllum nodosum*. J. Plant Growth Regul. 32, 340–341.

Yin, H., Zhao, X.M., Du, Y.G., 2010. Oligochitosan: a plant diseases vaccine—a review. Carbohydr. Polym. 82, 1–8.

Ziosi, V., Zandoli, R., Di Nardo, A., Biondi, S., Antognoni, F., Calandriello, F., 2012. Biological activity of different botanical extracts as evaluated by means of an array of in vitro and in vivo bioassays. Acta Hortic. 1009, 61–66.

CHAPTER

14

Biostimulants mediated imprints on seed physiology in crop plants

Riya Johnson[1], *Joy M. Joel*[1], *E. Janeeshma*[2] *and Jos T. Puthur*[1]

[1]Plant Physiology and Biochemistry Division, Department of Botany, University of Calicut, Malappuram, Kerala, India [2]Department of Botany, MES KEVEEYAM College, Valanchery, Malappuram, Kerala, India

14.1 Introduction

There is virtually no question about the fact that the effects of climate change are increasing the strain on available food supplies and shooting up the biotic and abiotic pressures on food crops, which is prompting concerns about the planet's ability to feed its inhabitants. According to FAO 2019, a significant proportion of individuals experience moderate food insecurity, including inadequate access to healthy and nutrient-rich food. By 2050, there will be 9.7 billion people on the planet, making it challenging to produce 70% more food crops to feed them (Lorenz & Lal, 2022). Over 100 million people (11% of the population), according to estimates, live in the region and experience moderate to severe food insecurity. The FAO established this indicator based on the Food Insecurity Experience Scale to monitor progress toward achieving zero hunger. The attainment of nutrition and food security may be hampered by this condition. To achieve the global goals of the 2030 Agenda for Sustainable Development, focused interventions on nutrition, rural development, and inequality are required.

The major factors contributing to negative influences on the growth and maturation of crops are abiotic and biotic stresses. Abiotic stresses that influence plant species, such as extreme temperatures, high salinity, drought conditions, and nutrient deficiency, frequently degrade crop quality and restrict plant growth and survival, along with too high or too low irradiation and water stagnation (Khalid et al., 2022). Mild dosages of abiotic and biotic stresses can enhance gene expression levels by alleviating stress through adjustment of the cellular mechanism (Tripathi et al., 2022) and plant tolerance by upregulating various genes and accumulating proteins which may directly or indirectly

Biostimulants in Plant Protection and Performance
DOI: https://doi.org/10.1016/B978-0-443-15884-1.00026-9

involve alleviating the negative consequences of stress through adjustment of the cellular mechanism and plant tolerance (Liu et al., 2023).

A broad category of compounds/substances known as "plant biostimulants" are derived from various organic as well as inorganic compounds, either from microbes or from inorganic materials, to increase nutrition efficiency and enhance abiotic stress tolerance levels (Du, 2015). Plant biostimulants have drawn a lot of research in the past 25 years because they present a potentially new method for controlling physiological processes in plants to promote growth, improve stress tolerance, and boost yield (Kumari et al., 2022). Modern biostimulants may be complex blends made from raw materials with a wide range of origins, such as waste from the paper and food industries (Kumar & Korat, 2022). They have a wide range of biological functions and are regarded as environmentally safe (Hayat et al., 2022). Both the academic community and the seed industry have turned their attention to the quest for novel compounds with the potential to operate as biostimulants. Among these new products, biostimulants could play a crucial role as a seed-treatment tool (Masondo et al., 2018). These factors have led them to find a significant use in contemporary agricultural production (Yakhin et al., 2017). It has become crucial to maintain the ultimate crop output and yield by employing biostimulants to combat stress (Bulgari et al., 2019).

Biostimulants such as plant growth regulators and various extracts of algae, when added to the soil, influence the seed's physiological activities to increase their success rate of germination and thereby strengthen the plant's health. Accordingly, biostimulants can increase the availability of nutrients, the capacity of plants to retain water, enhance antioxidant levels, and improve metabolism with an increase in chlorophyll synthesis. The term "biostimulant" has steadily expanded to include additional compounds and modes of action over the past several years. To break seed dormancy, stimulate seed germination, promote field emergence, accelerate seedling development till maturity, and maintain plant adaptability under ideal and abiotic stress conditions, biostimulants are increasingly being studied by seed researchers (Ronga et al., 2019).

14.2 Seed physiology

Since the advent of agriculture around 10,000 years ago, several seed plants have been tamed and cultivated and have been extremely important in providing food for the human population. The sporophyte, reserve storage compartment, and protective seed coat are the three elements of an ideal seed (Sabelli & Larkins, 2015). Understanding the quality of the seed, maturity, dormancy, germination, and lifespan of the seed requires knowledge of seed physiology. The main goal of an in-depth study in seed physiology should be to fully understand how changes in various environmental factors such as temperature, moisture, oxygen, and light, as well as other factors like germination-inhibiting or stimulating substances along with phytohormones, affect seed germination and seedling emergence (Gupta et al., 2022).

Triticum durum seeds were treated with a variety of endophytic microbes to see if the fungi could help seedlings emerge and thrive. The seeds were exposed to a combination of endophytic microorganisms that included *Rhizoglomus intraradices* BEG72 and *Trichoderma atroviride.*

When compared to untreated wheat seedlings, the microbe-treated seedlings had a substantial impact on the number of leaves, shoots, and root dry biomass of 17-days-old seedlings (Colla et al., 2015). Similarly, the preventive properties of *Bacillus subtilis* (strain 10-4) against drought stress were investigated to boost germination and seedling development. The studies also confirmed the enhancement in plant growth of 6-day-old wheat seedlings by specifically reducing the lipid peroxidation process, proline content, and electrolyte leakage in 21-day-old seedlings (Lastochkina et al., 2020). Various biostimulants affect the germination of seeds in various plant species in different ways. They can be applied to the seeds by seed priming, coating, and conditioning methodologies. The humic substance, when used as a biostimulant at a concentration of 1000 mg/L by the imbibition of *Sesamum indicum* seeds, enhanced the germination index as well as the coefficient of velocity of germination (Souguir & Hannachi, 2017). Knowledge of seed physiology may serve as a foundation for the technical advancement of environment-friendly substitutes for toxic chemical fertilizers, maximizing crop yield, and incorporating green agriculture methodologies, providing a major contribution to a more thorough and holistic understanding.

14.3 Strategies to improve seed quality

To develop stress tolerance, several approaches were occasionally developed. Traditional breeding techniques like selection and hybridization are among them, as are contemporary techniques like mutation and polyploidy breeding and genetic engineering (Ahmar et al., 2020). The huge manpower investment, energy, and other resource needs of conventional breeding methods are their known limitations, which have largely proven unsuccessful because of the obvious complexities of abiotic stress tolerance traits. Although transgenic plant production can withstand different types of stresses, the deliberate introduction of individual, heterologous characteristics into different elite crop lines is genetic engineering's restriction, limiting its ability to quickly and predictably impact stress tolerance (Gust et al., 2010). The incorporation of new traits by transgenes through breeding programs is a relatively difficult task to achieve, as the effects of pleiotropy and gene silencing can affect the continuity of the transmission of the desired traits into superior crop lines (Flowers et al., 1997).

Additionally, the use of transgenics is hampered by current biosafety laws and regulations as well as the expense of these procedures (Jisha et al., 2013). Because of the aforementioned limitations, it is crucial to acquire relatively simpler methodologies and solutions to impart tolerance ability to plants subjected to various stresses that can be readily adopted by farmers.

Therefore, it has become crucial to adopt alternative solutions to impart tolerance to plants against various stresses. The substitutive strategy should be simple, affordable, and readily adopted by the farmers without any complications, while at the same time being effective in mitigating stress (Jisha et al., 2013). Nowadays, different attempts are made to develop plants that can withstand different abiotic stresses. One of the most popular and acceptable methods for improving the plant's ability to withstand stress is seed priming (Murgia et al., 2015). It entails the exposure of plants to an elicitor or situation for the first time to increase their tolerance to stress (Ibrahim, 2016). The priming methodology strengthens several physio-biochemical defense mechanisms and thus empowers the seedlings to withstand a variety of environmental stresses (Ibrahim 2016). Moreover, under stressful circumstances, the priming methodology promotes quick and uniform seed germination along with good crop establishment (Sharma et al., 2014). According to Abdel Latef et al. (2019), among several strategies of biostimulant application methods, seed-priming has been regarded as a dependable strategy for boosting plant survival under harmful abiotic stress conditions.

14.4 Biostimulants

The use of biostimulants in crop cultivation has grown significantly in recent years and has emerged as a viable method to lessen crop loss in harsh environmental circumstances (Calvo et al., 2014). Plant biostimulants are materials that, when given to plants or rhizospheric regions, stimulate natural processes in a way that improves crop quality, tolerance to abiotic stress, nutrient uptake, and efficiency, regardless of the amount of nutrients present in them (Ricci et al., 2019). According to Traon et al. (2014), "A plant biostimulant is any substance or microorganism, applied to plants, seeds, or the root environment with the intention to stimulate natural processes of plants benefiting nutrient use efficiency and tolerance to abiotic stress, regardless of its nutrients content, or any combination of such substances and microorganisms intended for this use." Biostimulants, as defined by Yakhin et al. (2017), are "a formulated product of biological origin that improves plant productivity as a consequence of the novel or emergent properties of the complex constituents and not as a sole consequence due to the presence of known essential plant nutrients, plant growth regulators, or plant protective compounds." Plant biostimulants contain substances and/or microorganisms that, when applied to the plants or the rhizospheric region, stimulate natural processes to enhance and thus improve crop quality, nutritional uptake efficiency, and uplift tolerance levels to abiotic stresses. By using diverse biostimulants containing amino acids, humic acids, vitamins, and minerals, oxidative stress symptoms can be reduced.

Amino acids, hormones, and peptide combinations obtained from plant or animal sources through chemical or enzymatic degradation typically make up biostimulants (Halpern et al., 2015). Newer research is increasingly focusing on the use of biostimulants to encourage the more sensible use of mineral fertilizers, which is crucial for environmental conservation (Le Mire et al., 2016). To encourage natural processes, biostimulants are typically delivered to plants as foliar applications, although some can also be applied directly to the soil (or other growing media) or as a seed treatment. Although a large portion of the processes that biostimulants affect take place inside the plant, some also take place in the soil surrounding the roots of the plant. The benefits of biostimulants cannot be compared to fertilizer impacts because they are applied at such low concentrations despite possibly containing the same mineral elements as fertilizers. "Phytostimulators, metabolic boosters, and plant conditioners are some other names for biostimulants (Yakhin et al., 2017)." Farmers regularly use them to encourage and accelerate growth, the production of phytohormones, the activity of rhizosphere microorganisms, and soil enzymes, as well as other biological processes throughout the growth cycle of different plants (Brescia et al., 2020).

14.5 Effect of different types of biostimulants on seed physiology

Biostimulants can be applied to seeds or seedlings and are created from natural or synthetic materials. They boost the nutrient uptake and abiotic stress tolerance of the seed, which stimulates the seed's natural process and raises the crop's quality. Different authors have recognized various types of biostimulants over time, depending on the source material, manner of action, and other elements (Yakhin et al., 2017). Including humic and fulvic acids, protein hydrolysates (PHs), seaweed extracts, chitosan, inorganic compounds, and beneficial fungi and bacteria, biostimulants are categorized into seven groups (Du Jardin, 2015). The main biostimulants come from a variety of origins and have different physiological properties, and these features are considered when categorizing the biostimulants. The active components found in various types of biomass with potential biostimulant activity belong to a wide variety of molecules, which include phytohormones such as auxins, cytokinins, gibberellins, ethylene, abscisic acids, and brassinosteroids. These phytohormones play a significant role in plant growth and development, including seed germination, and they increase agricultural production and output by modifying plant metabolism in both favorable and unfavorable situations (Bulgari et al., 2019). How these biostimulants alter the physiological processes of seed physiology is still not revealed. Examples of physiological functions are the reduction of reactive oxygen scavenging by antioxidants or increased synthesis of gibberellin transporters, the protection of photosynthetic machinery against photodamage, or the initiation of lateral roots. So based on the active ingredients and mode of action, biostimulants are divided into two types: non-microbial biostimulants and microbial biostimulants (Table 14.1).

14.5.1 Nonmicrobial based biostimulants

14.5.1.1 Seaweed extracts

Nowadays, the use of seaweed extracts as "plant biostimulants" is considerably more widespread. The manufacture of biostimulants has long relied on macroalgae and their extracts. In general, seaweed extracts, even at low concentrations, are capable of generating a variety of physiological plant responses, including promoting plant growth, improving flowering and production, and enhancing product quality, nutrient content, and shelf life. Phytohormones have been found in seaweed extracts and are thought to represent potential bioactive components of this class of biostimulants. Important phytohormones like GA, auxins, and cytokinins can be found in biostimulants like seaweed extracts (Stirk et al., 2020). Algal extracts can contain a variety of carbohydrates in addition to hormones. Algal extracts contain a variety of polysaccharides, including alginate, fucoidan, and betaines, as well as proteins and minerals that support plant growth in addition to hormones (Polat et al., 2021). Applications of various extract types have also been shown to increase plants' resistance to a variety of abiotic stressors, such as salinity, drought, and severe temperatures. Numerous investigations on the chemical makeup of different seaweed extracts have shown that the nutrient content, typically macronutrients such as N, P, and K, is insufficient to cause physiological reactions at the seaweed's usual concentrations. Therefore, it has long been hypothesized that growth-promoting substances and

TABLE 14.1 Different biostimulant, composition and effects on seed germination.

Biostimulants	Major components/types	Impact on seed germination	Key references
Humic substances	Proteins, carbohydrates, aliphatic biopolymers, and lignin	Increased seed germination percentage and growth rate of seedling	Galambos et al. (2020)
Protein hydrolysates (PHs)	2–20 aminoacids (lysine, leucine, and valine) antioxidant, antihypertensive, immunomodulatory, and antimicrobial peptides. Modified compositions are available	Increased biomass and mineral acquisition Increased the biosynthesis of proteins, chlorophylls, and phenols. Elicited auxin (IAA)-like and gibberellin (GA)-like activities	Ertani et al. (2019)
Vermicompost leachate	Humic acid Fulvic acid microbes	Improved the germination rate and the formation of Roots	Gupta et al. (2022)
Microalgae extract	Chlorophylls, carotenoids, fucoxanthin, enzymes, vitamins (E and C), mycosporine-like amino acids, polysaccharides, hormones, and polyphenols	Stimulate seed germination	Tolpeznikaite et al. (2021)
Seaweed extract	Alginic acid, crude protein, mannitol, ash, potassium, and iodine	Application of high concentrations resulted in inhibitory effects At lower concentrations, promoted seed germination and growth	Voko et al. (2022)
Chitosan	Chitin	Enhanced seed germination	Gupta et al. (2022)
Silicon	Inorganic biostimulant	Germination percentage, germination rate and growth of seedlings were increased under both abiotic stress and nonstress conditions	Gupta et al. (2022)
Selenium (Se)	Inorganic biostimulant	At high concentrations it is toxic to the plants and in low concentration it improved seed germination and growth of young plants	Rocha et al. (2022)
Organic acids	Oxalic, tartaric, quinic, malic, malonic, ascorbic, citric, fumaric, succinic, salicylic, and benzoic acid	Reduced the radicle growth	Lynch (1980)
Vitamins	Vitamin B12 and CoQ10	Increase in growth	Rehim et al. (2021)
Melatonin	Phytomelatonin/chemical melatonin	Increased seed germination	Arnao and Hernández-Ruiz. (2019)
Growth regulators	Auxin, abscisic acid, gibberellins, cytokinins, ethylene, polyamines, jasmonic acid, and salicylic acid	Increase in seedling biomass	Dantas et al. (2012)

elicitors play a significant role in mediating the physiological effects of seaweed extracts. It is plausible that seaweed-derived biostimulants enhance plant resistance to environmental stresses by affecting natural stress responses and molecular priming mechanisms. Seaweed-derived biostimulants have been shown to enhance plant nutrient uptake and metabolic activity while also increasing plant resistance to a diverse range of environmental stresses (Moyo et al., 2021) (Fig. 14.1).

14.5.1.2 Amino acids

Biostimulants based on amino acids increase chlorophyll synthesis to increase photosynthesis in plants. Amino-acid and peptide combinations are created by chemically and enzymatically hydrolyzing agricultural wastes, such as crop residues, and animal wastes (such as collagen and epithelial tissues) (du Jardin, 2015; Calvo et al., 2014, Halpern et al., 2015). Chemical synthesis can also produce isolated or mixed compounds. Other nitrogenous substances include betaines, polyamines, and "non-protein amino acids," which are diverse in higher plants but are little known in terms of their physiological and ecological activities. Glycine betaine is a special amino acid derivative with well-known antistress

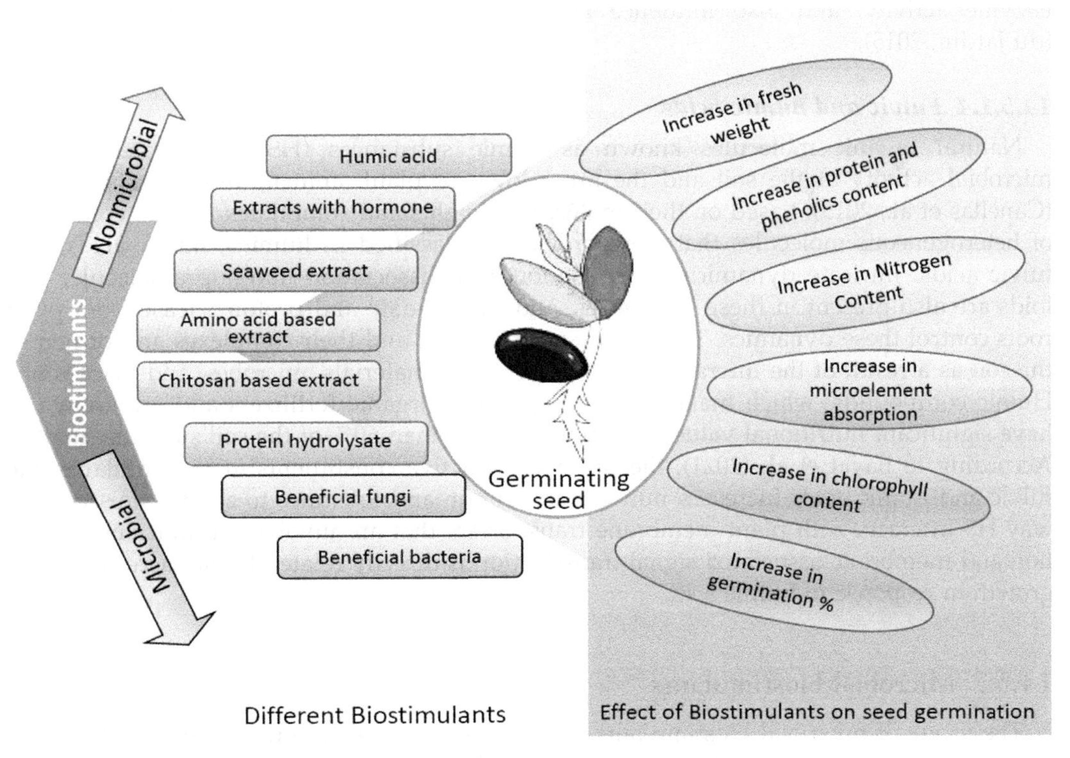

FIGURE 14.1 The impact of biostimulants on seed germination.

properties (Chen and Murata, 2011). Beta-aminobutyric acid, a nonprotein amino acid, has been found to improve Arabidopsis' ability to withstand salt and drought by activating defense pathways (Zulfiqar et al., 2020).

14.5.1.3 Protein hydrolysates

Protein-rich sources that have undergone partial or substantial hydrolysis to create polypeptides, oligopeptides, and amino acids known as PHs are frequently used as biostimulants (Colla et al., 2015). PHs are primarily created by the controlled enzymatic breakdown of complete protein sources by the proper proteolytic enzymes, followed by posthydrolysis processing to separate the required and potent bioactive peptides from the complicated mixture of active and inert peptides (Nasri, 2017). Peptides' bioactivity is primarily determined by the mix and sequence of their amino acids, and they display multifunctional properties based on their structure as well as their hydrophobicity and charge. The distinctions between the idea of the mode of action and the mechanisms of action should be highlighted while talking about PHs. The majority of research has been on the subject of how PHs work, concentrating on their only beneficial impacts on plant productivity, photosynthesis, plant growth, nutrient uptake, and water efficiency (Yakhin et al., 2017). These PHs influence nitrogen absorption and assimilation, control Krebs cycle enzyme activity, and also influence how carbon and nitrogen metabolisms interact (du Jardin, 2015).

14.5.1.4 Fulvic and humic acids

Natural organic molecules known as humic substances (HS) are created by the microbial activity of the soil and the breakdown of plant, animal, and microbial waste (Canellas et al., 2015). Based on their molecular weights and solubilities, HS are collections of heterogeneous molecules that were previously classified as humins, humic acids, and fulvic acids. Intricate dynamics of association and dissociation into supramolecular colloids are also present in these substances, and the exudate and proton release from plant roots control these dynamics. Thus, humic compounds and their complexes are formed in the soil as a result of the interaction between organic materials, microbes, and plant roots. Humic compounds, which make up the majority of organic fertilizers and are known to have significant nutritional values, account for more than 60% of the soil's organic matter. According to Bayat et al. (2021), the presence of active constituents in biostimulants like fulvic and humic acids increases nutrient absorption and tolerance to abiotic stresses. The way HS interacts with plant membrane transporters that are involved in nutrient absorption and membrane-associated signal transduction is directly related to the increased plant growth in response to humic acid.

14.5.2 Microbial biostimulants

The usage of microbial biostimulants, which many farmers are unaware of, is the most recent practice in the agricultural sector. Nonpathogenic bacteria, mycorrhizal fungi, and other soil microorganisms are examples of microbial biostimulants. Microbes from natural or synthetic crops are used to create microbial biostimulants, which can be used on plants or crops (Calvo et al., 2014). This class of biostimulants brings together all the factors that support seed germination, root growth, and chlorophyll production. They boost the crop's quality by enhancing the plant's natural nutrient uptake and tolerance to abiotic stresses. This stimulates the plant's natural processes. Given that many products are made from natural materials or contain microorganisms that synthesize hormones, it is not surprising that the physiological effects of biostimulants appear to produce hormone-like phenotypes (Castiglione et al., 2021). Fungi and bacteria make up the majority of microbial biostimulants, although many other microorganisms are now included.

14.5.2.1 *Fungal biostimulants*

Plants and fungi have coevolved since the origin of terrestrial plants, and the mutualism-parasitism continuum theory helps characterize the vast range of linkages that have formed over the course of evolutionary time. Various symbiotic relationships between fungus and plant roots include parasitism and mutualistic symbioses, in which the two species coexist in close proximity and collaborate for mutual benefit. Plants can be treated with fungus-based biostimulants to improve their nutrient uptake, stress resistance, crop yield, and product quality. There are indications that fungi work together to facilitate interplant communication, which could have significant environmental and agricultural ramifications (Johnson and Graham, 2013). Mycorrhiza is a common type of endomycorrhiza associated with crop and horticultural plants, where fungi of Glomeromycota species produce branched structures known as arbuscules by penetrating root cortical cells (Behie and Bidochka, 2014). Arbuscular mycorrhiza establishes important tripartite relationships between plants and rhizobacteria, leading to nutrition efficiency, water balance, and biotic and abiotic stress protection in plants (Diagne et al., 2020). Recent information further suggests that hyphal networks exist, connecting not only fungi and plants but also specific plants within a plant community. However, other fungal endophytes, distinct from mycorrhizal species, can live at least a portion of their life cycle away from the plant, colonize roots, and transfer nutrients to their hosts using poorly understood mechanisms. These include *Trichoderma* spp. (Ascomycota) and Sebacinales (Basidiomycota), which use *Piriformospora indica* as a model organism (Behie and Bidochka, 2014).

14.5.2.2 *Bacterial biostimulants*

Generally, bacteria interact with plants through the soil and into the interior of cells in the rhizosphere and rhizoplane, and these interactions can be either brief or long-term. Some bacteria can even spread vertically through seeds and affect plant life by participating in bio-geochemical cycles, supplying nutrients, and increasing nutrient uptake (Hamid et al., 2021). Rhizobium-type mutualistic endosymbionts and mutualistic rhizospheric PGPRs, also referred to as "plant growth-promoting rhizobacteria," are two types of bacterial biostimulants based on their functional and ecological variety. Commercially available biofertilizers, or microbial inoculants that facilitate plant nutrient absorption, include Rhizobium and related species. Scientific literature and articles provide in-depth analyses of the biology and agricultural applications of symbioses based on Rhizobium. PGPRs, which serve a variety of activities, have an impact on all aspects of plant life, including nutrition and growth, morphogenesis and development, responses to biotic and abiotic challenges, and interactions with other species in agroecosystems (Basu et al., 2021). While many of these duties are normally carried out by one species, others rely on the interactions of various bacterial strains working in concert.

14.6 Effect of biostimulant application in seeds under abiotic stress

Abiotic stressors can be dealt with by enhancing the circumstances for plant growth and by providing seeds with water, nutrients, and plant growth regulators (PGRs—auxins, cytokinins, gibberellins, strigolactones, and brassinosteroids). Biostimulants have been emphasized as a promoter of productivity optimization by altering physiological processes in plants in addition to these conventional methods. Biostimulants provide a potentially innovative method for controlling and altering physiological processes in seeds to promote growth, reduce stress-related restrictions, and boost yield (Polat et al., 2021). When applied to *Gossypium hirsutum* seeds, the biostimulant based on hormones such as cytokinin, indole butyric acid, and gibberellic acid promoted seedling growth by increasing leaf area and height, increasing fruit size, improving the root system, and increasing seedling emergence. Crop growth, development, and productivity are all better and more balanced as a result of all these factors (Rathore et al., 2009). Biostimulants and their function in sustainable agriculture are still being studied by scientists, investors, regulators, consumers, and other interested parties.

Under low temperatures and high osmotic potential circumstances, the seaweed extract Kelpak, which is widely used in South Africa, dramatically boosted the germination percentage of *Ceratotheca triloba* seeds compared to the control (Masondo et al., 2018). Secondary metabolites in plant extracts may influence seed germination and plant development in a positive or selective manner, depending on the species or cultivar. Barley seed germination was not significantly affected by the aqueous extract of *Calotropis procera*; however, wheat seed germination was increased at a low concentration of 5% (vol/vol) (Radwan et al., 2019). In comparison to the untreated control, pepper seeds treated with various types of chitosan in cold conditions exhibited improved germination percentages, germination rates, and germination energy (Samarah et al., 2016). The performance of plants under stress is known to be stimulated by humic acids. Sofi et al. (2018) undertook a study to look at the impact of humic acid on the germination of *Medicago sativa* seeds under salt stress, among other things. Under normal circumstances, humic acid had little effect on germination parameters; however, in saline environments, humic compounds greatly increased germination values. The germination percentage of *Capsicum frutescens* was not significantly affected by the various humic substance doses; however, the mean germination time and germination velocity index were significantly enhanced (Vieira et al., 2018). According to Bento et al. (2020), there was no significant change in the germination of maize between treatments using humic-like chemicals, HS, and control.

14.7 New modern molecular approaches

The effects of biostimulants on the plant under study can be clearly shown in molecular studies that look at the genome, mRNA, protein, and metabolite levels. By sequencing the entire genome, we may look for modifications or mutations in the regulatory components of genes that may result from stress reactions. *Bacillus licheniformis*, a potential biostimulant for agriculture, has a draft genome sequence, according to Crovadore et al. (2020). This strain can be used as a biostimulant to improve agronomic applications because its genome contains multiple protein-coding genes involved in biocontrol as well as a total of four genes for the auxin production pathway. Similar to this, a previous study used a yellow fluorescent protein tag to analyze root colonization and conduct whole genome sequencing. Similar to this, a recent study used a yellow fluorescent protein tag to analyze root colonization and perform whole genome sequencing on the cucumber to get new insights into the potential of *Bacillus subtilis* as a biocontrol and growth promoter (Samaras et al., 2021). The aforementioned investigations will help us comprehend the function and potential of these microbes as biostimulants in agriculture. The genetic underpinnings of plant adaptation and acclimation will be discovered by gene ontology investigations. We can learn about the expression of genes involved in stress response, adaptation, and critical physiological processes related to the healthy growth of plants by monitoring the activity of key enzymes involved in the synthesis of secondary metabolites or in specific physiological processes, as well as by using transcriptome profiling to identify upregulated and downregulated genes. Epigenomics, which quantifies changes in epigenetics after environmental stress through the study of DNA methylation, posttranslational histone modifications, and noncoding RNAs, can be combined with the study of expression (siRNAs and miRNAs). To develop a durable and transgenerational memory of the stress response by biostimulants, any alteration in epigenetic markers caused by the influence of biostimulants must be investigated (Villagómez-Aranda et al., 2022).

14.8 Omics approach

Biomolecule separation methods, data gathering platforms, chemical libraries, bioinformatics tools, databases, and visualization approaches have become crucial instruments in the quest to discover, measure, and functionally characterize the cellular pool. The exact decoding mode of action of biostimulants on the crop can be made possible by high-throughput omics approaches, which include metabolomics, phenomics, genomics, proteomics, and multiomics methodologies (Fig. 14.2). Sudiro et al. (2022) studied the phenomenon underlying water deficit resistance in tomatoes through foliar sprays of 4-Vita (a potent biostimulant), using phenomics research together with mass spectrometric untargeted metabolomics. Under drought conditions, several coordinative biochemical responses to the biostimulant treatment were identified, along with the alteration of lipids in the thylakoid membrane together with the elevation of xanthin levels, which are primarily involved in the detoxification of ROS, and chlorophyll synthesis, supporting the high resilience of tomatoes to drought stress. Genome editing methods such as CRISPR/Cas9 (Clustered Regulatory Interspaced Short Palindromic Repeats/CRISPR-associated protein 9 system) can be used to edit transcriptomes and to generate stress-resistant crops. The identification of newer proteins along with their characterization is helpful for the development and production of plants and is made possible by proteome analysis, which also considerably aids in understanding the translated proteins engaged in key

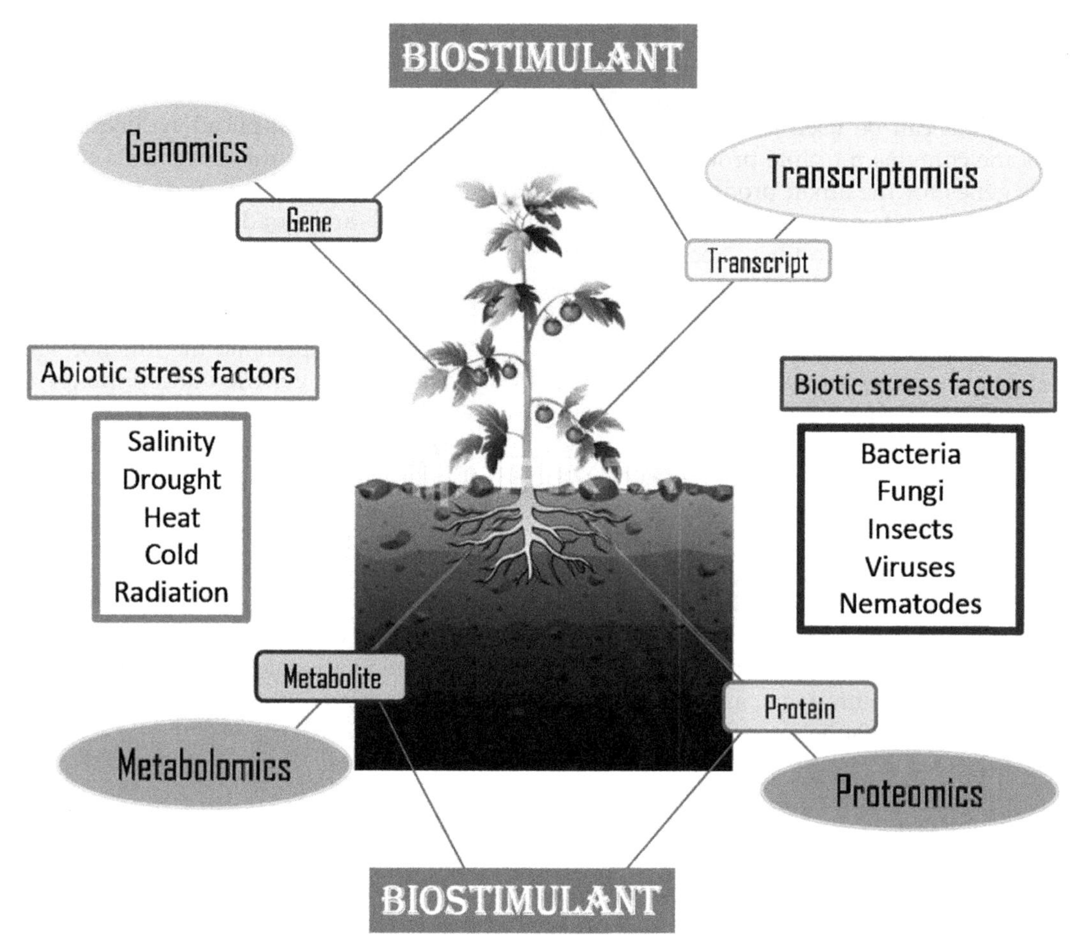

FIGURE 14.2 A general representation of biostimulants induced stress tolerance characterization using omics approach.

physiological processes. Targeted metabolomics has considerably increased our understanding of how biostimulants affect plant growth and stress tolerance by revealing the amounts of numerous major and minor metabolites in treated plants. Vaghela et al. (2022) characterized and carried out metabolomic profiling of *Kappaphycus alvarezii* (red seaweed) and discovered kinetin, dodecanamide, 1-phosphatidyl-1D-myoinositol, sulfa benzamide, and several other compounds that have the potential to promote plant growth promotion, photoprotection, and other functions.

Here, we can gain a clear understanding of the associated metabolic pathways using cutting-edge computational and bioinformatics methods, which will aid in the development/design of novel biostimulants. Various phenotypic features must be evaluated at the cellular, organ, and whole plant levels as part of the laborious process of studying phenotypic changes. The most recent technologies for measuring large-scale phenotypic features include advanced imaging systems, which include magnetic resonance imaging, sensor-based systems, and multispectrum-utilizing imaging systems (Rico-Chávez et al., 2022). Despite the fact that recent research has shown how important the various "omics" methodologies are for understanding how biostimulants work—particularly under transcriptomics, proteomics, and metabolomics—each of the approaches still needs to be revealed.

14.9 Conclusion and future prospects

Robust and healthy seedlings that emerge quickly can make use of the resources at hand, tolerating biotic and abiotic challenges and other unfavorable environmental factors. The effectiveness of any biostimulant depends on several factors, such as the concentration, type, and growing conditions of a plant subjected to a particular biostimulant. In general, the higher concentration of a biostimulant used often inhibits the advantageous properties that a biostimulant can instill at lower concentrations. The large-scale use of biostimulants in agriculture has been linked to better nutrient absorption, higher plant yield, and increased stress tolerance, which can invariably lessen the reliance on chemical fertilizers, which have well-established deleterious repercussions. To study the stimulating properties of any known biostimulant, knowledge of the seed's physiological process during germination and maturation stages is needed.

Moreover, careful application of biostimulants to the field should be followed only after their preceding studies on the effect of the same on the crops subjected to stressed conditions, as the behavior of the biostimulant and its effects might vary, which requires intense research. As the application of different biostimulants in the field is relatively in its infancy, the molecular mechanisms underlying the observed responses of crop plants should be a matter of intense research to decipher the hidden pathways leading to the beneficial qualities of a biostimulant. Research on the exploration of new organisms that are relatively cheaper, economically feasible, and practically effective with the ease to develop the same to be used as a biostimulant could be the subject of intense research. Even though epigenetic, molecular, and omics approaches during biostimulant application are discussed by various researchers, complete characterization of different biostimulants is a prerequisite for an efficient biostimulant conferring tolerance to various stresses on plants.

Acknowledgment

RJ acknowledges the Department of Botany, University of Calicut, and JJM gratefully acknowledge the financial assistance in the form of a Junior Research Fellowship from the University Grants Commission (UGC).

References

Abdel Latef, A.A.H., Mostofa, M.G., Rahman, M.M., Abdel-Farid, I.B., Tran, L.S.P., 2019. Extracts from yeast and carrot roots enhance maize performance under seawater-induced salt stress by altering physio-biochemical characteristics of stressed plants. J. Plant Growth Regul. 38, 966–979.

Ahmar, S., Gill, R.A., Jung, K.H., Faheem, A., Qasim, M.U., Mubeen, M., et al., 2020. Conventional and molecular techniques from simple breeding to speed breeding in crop plants: recent advances and future outlook. Int. J. Mol. Sci. 21 (7), 2590.

Arnao, M.B., Hernández-Ruiz, J., 2019. Melatonin as a chemical substance or as phytomelatonin rich-extracts for use as plant protector and/or biostimulant in accordance with EC legislation. Agronomy 9 (10), 570.

Basu, A., Prasad, P., Das, S.N., Kalam, S., Sayyed, R.Z., Reddy, M.S., et al., 2021. Plant growth promoting rhizobacteria (PGPR) as green bioinoculants: recent developments, constraints, and prospects. Sustainability 13 (3), 1140.

Bayat, H., Shafie, F., Aminifard, M.H., Daghighi, S., 2021. Comparative effects of humic and fulvic acids as biostimulants on growth, antioxidant activity and nutrient content of yarrow (Achillea millefolium L.). Sci. Horticulturae 279, 109912.

Behie, S.W., Bidochka, M.J., 2014. Nutrient transfer in plant–fungal symbioses. Trends Plant. Sci. 19 (11), 734–740.

Bento, L.R., Melo, C.A., Ferreira, O.P., Moreira, A.B., Mounier, S., Piccolo, A., et al., 2020. Humic extracts of hydrochar and Amazonian Dark Earth: molecular characteristics and effects on maize seed germination. Sci. Total. Environ. 708, 135000.

Brescia, F., Marchetti-Deschmann, M., Musetti, R., Perazzolli, M., Pertot, I., Puopolo, G., 2020. The rhizosphere signature on the cell motility, biofilm formation and secondary metabolite production of a plant-associated Lysobacter strain. Microbiological Res. 234, 126424.

Bulgari, R., Franzoni, G., Ferrante, A., 2019. Biostimulants application in horticultural crops under abiotic stress conditions. Agronomy 9 (6), 306.

Calvo, P., Nelson, L., Kloepper, J.W., 2014. Agricultural uses of plant biostimulants. Plant. soil. 383, 3–41.

Canellas, L.P., Olivares, F.L., Aguiar, N.O., Jones, D.L., Nebbioso, A., Mazzei, P., et al., 2015. Humic and fulvic acids as biostimulants in horticulture. Sci. Horticulturae 196, 15–27.

Castiglione, A.M., Mannino, G., Contartese, V., Bertea, C.M., Ertani, A., 2021. Microbial biostimulants as response to modern agriculture needs: composition, role and application of these innovative products. Plants 10 (8), 1533.

Chen, T.H., Murata, N., 2011. Glycinebetaine protects plants against abiotic stress: mechanisms and biotechnological applications. Plant Cell Environ 34 (1), 1–20.

Colla, G., Nardi, S., Cardarelli, M., Ertani, A., Lucini, L., Canaguier, R., et al., 2015. Protein hydrolysates as biostimulants in horticulture. Sci. Horticulturae 196, 28–38.

Crovadore, J., Cochard, B., Grizard, D., Chablais, R., Baillarguet, M., Comby, M., et al., 2020. Draft genome sequence of *Bacillus licheniformis* strain UASWS1606, a plant biostimulant for agriculture. Microbiology Resour. Announcements 9 (37), e00740-20.

Dantas, A.C.V.L., Queiroz, J.M.D.O., Vieira, E.L., Almeida, V.D.O., 2012. Effect of gibberellic acid and the biostimulant Stimulate® on the initial growth of tamarind. Rev. Brasileira de. Fruticultura 34, 8–14.

Diagne, N., Ngom, M., Djighaly, P.I., Fall, D., Hocher, V., Svistoonoff, S., 2020. Roles of arbuscular mycorrhizal fungi on plant growth and performance: importance in biotic and abiotic stressed regulation. Diversity 12 (10), 370.

Du, J., P., 2015. Plant biostimulants: definition, concept, main categories and regulation. Sci. Horticulturae 196, 3–14.

Ertani, A., Nardi, S., Francioso, O., Sanchez-Cortes, S., Foggia, M.D., Schiavon, M., 2019. Effects of two protein hydrolysates obtained from chickpea (*Cicer arietinum* L.) and Spirulina platensis on *Zea mays* (L.) plants. Front. Plant. Sci. 10, 954.

Flowers, T.J., Garcia, A., Koyama, M., Yeo, A.R., 1997. Breeding for salt tolerance in crop plants—the role of molecular biology. Acta Physiologiae Plant. 19, 427–433.

Galambos, N., Compant, S., Moretto, M., Sicher, C., Puopolo, G., Wäckers, F., et al., 2020. Humic acid enhances the growth of tomato promoted by endophytic bacterial strains through the activation of hormone-, growth-, and transcription-related processes. Front. Plant. Sci. 11, 582267.

Gupta, S., Doležal, K., Kulkarni, M.G., Balázs, E., Van Staden, J., 2022. Role of non-microbial biostimulants in regulation of seed germination and seedling establishment. Plant. Growth Regul. 97 (2), 271–313.

Gust, A.A., Brunner, F., Nürnberger, T., 2010. Biotechnological concepts for improving plant innate immunity. Curr. Opin. Biotechnol. 21 (2), 204–210.

Halpern, M., Bar-Tal, A., Ofek, M., Minz, D., Muller, T., Yermiyahu, U., 2015. The use of biostimulants for enhancing nutrient uptake. Adv. Agron. 130, 141–174.

Hamid, B., Zaman, M., Farooq, S., Fatima, S., Sayyed, R.Z., Baba, Z.A., et al., 2021. Bacterial plant biostimulants: a sustainable way towards improving growth, productivity, and health of crops. Sustainability 13 (5), 2856.

Hayat, S., Ahmad, A., Ahmad, H., Hayat, K., Khan, M.A., Runan, T., 2022. Garlic, from medicinal herb to possible plant bioprotectant: a review. Sci. Horticulturae 304, 111296.

Ibrahim, E.A., 2016. Seed priming to alleviate salinity stress in germinating seeds. J. plant. Physiol. 192, 38–46.

Jisha, K.C., Vijayakumari, K., Puthur, J.T., 2013. Seed priming for abiotic stress tolerance: an overview. Acta Physiologiae Plant. 35, 1381–1396.

Johnson, N.C., Graham, J.H., 2013. The continuum concept remains a useful framework for studying mycorrhizal functioning. Plant. Soil. 363, 411–419.

Khalid, M.F., Shafqat, W., Khan, R.I., Abbas, M., Ahmed, T., Ahmad, S., et al., 2022. Citrus responses and tolerance against temperature stress. Citrus Production. CRC Press, pp. 149–155.

Kumar, P., Korat, H., 2022. Mass production methods, markets, and applications of chitosan and chitin oligomer as a biostimulant. Industrial Microbiology Based Entrepreneurship: Making Money from Microbes. Springer Nature Singapore, Singapore, pp. 265–285.

Kumari, M., Swarupa, P., Kesari, K.K., Kumar, A., 2022. Microbial inoculants as plant biostimulants: a review on risk status. Life 13 (1), 12.

Lastochkina, O., Garshina, D., Ivanov, S., Yuldashev, R., Khafizova, R., Allagulova, C., et al., 2020. Seed priming with endophytic *Bacillus subtilis* modulates physiological responses of two different *Triticum aestivum* L. cultivars under drought stress. Plants 9 (12), 1810.

Le Mire, G., Nguyen, M., Fassotte, B., du Jardin, P., Verheggen, F., Delaplace, P., et al., 2016. Implementing biostimulants and biocontrol strategies in the agroecological management of cultivated ecosystems. Biotechnologie, Agronomie, Société et. Environ. .

Liu, X., Pei, Y., Wang, C., Zhu, D., Cheng, F., 2023. Hydrogen sulfide, regulated by VvWRKY30, promotes berry color changes in grapevine cabernet sauvignon. Sci. Horticulturae 309, 111605.

Lorenz, K., Lal, R., 2022. Challenges and opportunities for the global food system. Org. Agriculture Clim. Change 219–232.

Lynch, J.M., 1980. Effects of organic acids on the germination of seeds and growth of seedlings. Plant, Cell & Environ. 3 (4), 255–259.

Masondo, N.A., Kulkarni, M.G., Finnie, J.F., Van Staden, J., 2018. Influence of biostimulants-seed-priming on *Ceratotheca triloba* germination and seedling growth under low temperatures, low osmotic potential and salinity stress. Ecotoxicol. Environ. Saf. 147, 43–48.

Moyo, M., Aremu, A.O., Amoo, S.O., 2021. Potential of seaweed extracts and humate-containing biostimulants in mitigating abiotic stress in plants. Biostimulants for Crops from Seed Germination to Plant Development. Academic Press, pp. 297–332.

Murgia, I., Giacometti, S., Balestrazzi, A., Paparella, S., Pagliano, C., Morandini, P., 2015. Analysis of the transgenerational iron deficiency stress memory in *Arabidopsis thaliana* plants. Front. Plant. Sci. 6, 745.

Nasri, M., 2017. Protein hydrolysates and biopeptides: production, biological activities, and applications in foods and health benefits. A review. Adv. food Nutr. Res. 81, 109–159.

Polat, S., Trif, M., Rusu, A., Šimat, V., Čagalj, M., Alak, G., et al., 2021. Recent advances in industrial applications of seaweeds. Crit. Rev. food Sci. Nutr. 1–30.

Radwan, A.M., Alghamdi, H.A., Kenawy, S.K., 2019. Effect of *Calotropis procera* L. plant extract on seeds germination and the growth of microorganisms. Ann. Agric. Sci. 64 (2), 183–187.

Rathore, S.S., Chaudhary, D.R., Boricha, G.N., Ghosh, A., Bhatt, B.P., Zodape, S.T., et al., 2009. Effect of seaweed extract on the growth, yield and nutrient uptake of soybean (*Glycine max*) under rainfed conditions. South. Afr. J. Botany 75 (2), 351–355.

Rehim, A., A Bashir, M., Raza, Q.U.A., Gallagher, K., Berlyn, G.P., 2021. Yield enhancement of biostimulants, Vitamin B12, and CoQ10 compared to inorganic fertilizer in radish. Agronomy 11 (4), 697.

Ricci, M., Tilbury, L., Daridon, B., Sukalac, K., 2019. General principles to justify plant biostimulant claims. Front. Plant. Sci. 10, 494.

Rico-Chávez, A.K., Franco, J.A., Fernandez-Jaramillo, A.A., Contreras-Medina, L.M., Guevara-González, R.G., Hernandez-Escobedo, Q., 2022. Machine learning for plant stress modeling: a perspective towards hormesis management. Plants 11 (7), 970.

Rocha, L., Silva, E., Pavia, I., Ferreira, H., Matos, C., Osca, J.M., et al., 2022. Seed soaking with sodium selenate as a biofortification approach in bread wheat: effects on germination, seedling emergence, biomass and responses to water deficit. Agronomy 12 (8), 1975.

Ronga, D., Biazzi, E., Parati, K., Carminati, D., Carminati, E., Tava, A., 2019. Microalgal biostimulants and biofertilisers in crop productions. Agronomy 9 (4), 192.

Sabelli, P.A., Larkins, B.A., 2015. New insights into how seeds are made. Front. Plant. Sci. 6, 196.

Samarah, N.H., Wang, H., Welbaum, G.E., 2016. Pepper (*Capsicum annuum*) seed germination and vigour following nanochitin, chitosan or hydropriming treatments. Seed Sci. Technol. 44 (3), 609–623.

Samaras, A., Nikolaidis, M., Antequera-Gómez, M.L., Cámara-Almirón, J., Romero, D., Moschakis, T., et al., 2021. Whole genome sequencing and root colonization studies reveal novel insights in the biocontrol potential and growth promotion by *Bacillus subtilis* MBI 600 on cucumber. Front. Microbiology 11, 600393.

Sharma, A.D., Rathore, S.V.S., Srinivasan, K., Tyagi, R.K., 2014. Comparison of various seed priming methods for seed germination, seedling vigour and fruit yield in okra (*Abelmoschus esculentus* L. Moench). Sci. Horticulturae 165, 75–81.

Sofi, A., Ebrahimi, M., Shirmohammadi, E., 2018. Effect of humic acid on germination, growth, and photosynthetic pigments of *Medicago sativa* L. under salt stress. Ecopersia 6 (1), 21–30.

Souguir, M., & Hannachi, C., 2017. Response of sesame seedlings to different concentrations of humic acids or calcium nitrate at germination and early growth.

Sudiro, C., Guglielmi, F., Hochart, M., Senizza, B., Zhang, L., Lucini, L., et al., 2022. A phenomics and metabolomics investigation on the modulation of drought stress by a biostimulant plant extract in tomato (*Solanum lycopersicum*). Agronomy 12 (4), 764.

Tolpeznikaite, E., Bartkevics, V., Ruzauskas, M., Pilkaityte, R., Viskelis, P., Urbonaviciene, D., et al., 2021. Characterization of macro-and microalgae extracts bioactive compounds and micro-and macroelements transition from algae to extract. Foods 10 (9), 2226.

Traon, D., Amat, L., Zotz, F., & du Jardin, P., 2014. A Legal Framework for Plant Biostimulants and Agronomic Fertiliser Additives in the EU-Report to the European Commission, DG Enterprise & Industry.

Tripathi, D., Singh, M., Pandey-Rai, S., 2022. Crosstalk of nanoparticles and phytohormones regulate plant growth and metabolism under abiotic and biotic stress. Plant. Stress. 100107.

Vaghela, P., Das, A.K., Trivedi, K., Anand, K.V., Shinde, P., Ghosh, A., 2022. Characterization and metabolomics profiling of Kappaphycus alvarezii seaweed extract. Algal Res. 66, 102774.

Vieira, J.H., d S Silva, L.K., de Oliveira, L.C., do Carmo, J.B., Rosa, L.M.T., Botero, W.G., 2018. Evaluation of germination of chilli pepper using humic substances and humic acids. IOSR J. Environ. Sci. Toxicol. Food Technol. 12, 33–39.

Villagómez-Aranda, A.L., Feregrino-Pérez, A.A., García-Ortega, L.F., González-Chavira, M.M., Torres-Pacheco, I., Guevara-González, R.G., 2022. Activating stress memory: Eustressors as potential tools for plant breeding. Plant. Cell Rep. 41 (7), 1481–1498.

Voko, M.P., Kulkarni, M.G., Ngoroyemoto, N., Gupta, S., Finnie, J.F., Van Staden, J., 2022. Vermicompost leachate, seaweed extract and smoke-water alleviate drought stress in cowpea by influencing phytochemicals, compatible solutes and photosynthetic pigments. Plant. Growth Regul. 97 (2), 327–342.

Yakhin, O.I., Lubyanov, A.A., Yakhin, I.A., Brown, P.H., 2017. Biostimulants in plant science: a global perspective. Front. Plant. Sci. 7, 2049.

Zulfiqar, F., Akram, N.A., Ashraf, M., 2020. Osmoprotection in plants under abiotic stresses: new insights into a classical phenomenon. Planta 251, 1–17.

Stirk, W.A., Rengasamy, K.R., Kulkarni, M.G., van Staden, J., 2020. Plant biostimulants from seaweed: An overview. The Chemical Biology of Plant Biostimulants pp.31-55.

CHAPTER

15

Role of biostimulant in adventitious rooting via stimulation of phytohormones

Arshdeep Kaur[1], *Manik Devgan*[1], *Radhika Sharma*[2], *Antul Kumar*[3], *Anuj Choudhary*[3], *Ravi Pratap Singh*[4], *Dadireddy Madhusudan Reddy*[5], *Ajaykumar Venkatapuram*[5,6], *Sahil Mehta*[7] and *Azamal Husen*[8,9,10]

[1]Department of Plant Breeding and Genetics, Punjab Agricultural University, Ludhiana, Punjab, India [2]Department of Soil Science, Punjab Agricultural University, Ludhiana, Punjab, India [3]Department of Botany, Punjab Agriculture University, Ludhiana, Punjab, India [4]Department of Ocean Studies and Marine Biology, Pondicherry University, Port Blair, Andaman and Nicobar Islands, India [5]Department of Microbiology, Palamuru University, Mahbubnagar, Telangana, India [6]Crop Improvement Group, International Center for Genetic Engineering and Biotechnology, Delhi, New Delhi, India [7]Department of Botany, Hansraj College, University of Delhi, New Delhi, India [8]Department of Biotechnology, Smt. S. S. Patel Nootan Science & Commerce College, Sankalchand Patel University, Visnagar, Gujarat, India [9]Department of Biotechnology, Graphic Era (Deemed to be University), Dehradun, Uttarakhand, India [10]Wolaita Sodo University, Wolaita, Ethiopia

15.1 Introduction

Due to their elite ability for adaptation, plants have remarkably developed capacities for regrowth and reproduction. Amid eukaryotes, plants are one of the only few higher species with clonal reproductive capabilities. The growth of adventitious roots (ARs) is one all-too-important factor responsible for clonal propagation in plants (Husen 2002,

Biostimulants in Plant Protection and Performance
DOI: https://doi.org/10.1016/B978-0-443-15884-1.00013-0

2008, 2013; Husen et al., 2017; Pant et al., 2022a; Pant et al., 2022b). ARs are postembryonically developed aerial organs evolved from de-differentiated cells, which are actually non-root tissues. ARs show tremendous plasticity, particularly to conduct specialization tasks ranging from improving water acquisition (epiphytes), bracing plant stands (pillar roots), food storage (sweet potato), exploring soil nutrients (crown roots in monocots), etc (Mhimdi and Pérez–Pérez, 2020). Consequently, ARs can be seen as plants' ultimate response to a wide range of developmental and environmental challenges (Kumar et al., 2022; Husen 2022; Arya et al., 2022a; Pant and Husen, 2022c; Pant and Husen, 2022d).

Consensus-based in-depth research in several species indicates that active phytohormone control is either based on exogenous treatment or as a result of externally induced stress signals, including water logging, injury, abiotic stress, and nutritional deficiencies (Druege et al., 2019; Husen and Zhang, 2023). The development of ARs in plants proves their finely tuned coordinated responses against external stimuli, guided through advanced molecular mechanisms (Lakehal and Bellini, 2019; Arya et al., 2022b). However, there is a lack of comprehensive knowledge regarding the regulation and operation of plant hormone homeostasis as well as the complex signaling network during AR organogenesis. Certain barriers restricting our knowledge include the lack of a basic AR model, complex hormonal activities, limited availability of natural phytohormones, and the complexity of the techniques involved in studying AR developmental activities.

The manipulation of these phytohormones using biostimulants offers some hope. Biostimulants are any substances, commercial products, or microorganisms that, when used sparingly in plants, influence several metabolic processes, including respiration, photosynthesis, nucleic acid synthesis, and ion uptake, thereby encouraging their growth and development (Abbas, 2013). Retrieving plant-based biostimulants is a promising method for improving the value, sustainability, and environmental safety of the agricultural production chain (Ogunsanya et al., 2022). Hormone-like activities of biostimulants imply that they act on hormone synthesis and degradation, hormone signaling and response, hormone conjugation, or hormone sequestration and movement, but they do not necessarily mean that they contain plant growth regulators. There is not much experimental evidence to distinguish between these potential levels of action. Tools have recently been created that use molecular reporters to differentiate between changes in plant hormone steady-state concentrations in the cell and changes in their signaling pathways (Waadt et al., 2015). These methods might offer fresh perspectives on the dynamics of hormone actions at the tissue level in plants that have received biostimulant treatment. Through this focused chapter, we try to resolve the conundrum of intricate mechanisms and roles phytohormones play in AR development and cite strategies on how to control these via biostimulants so that we can exploit these systems to promote agriculture.

15.2 Adventitious root development

The principles underpinning AR creation, which signals a shift of differentiated somatic cells into a new developmental pathway, are extremely fascinating. Unraveling interesting processes that underlie the capacity of plants to regenerate helps bring forth new vistas for the sustainable and effective use of plant genetic resources (Fig. 15.1). Unlike main and

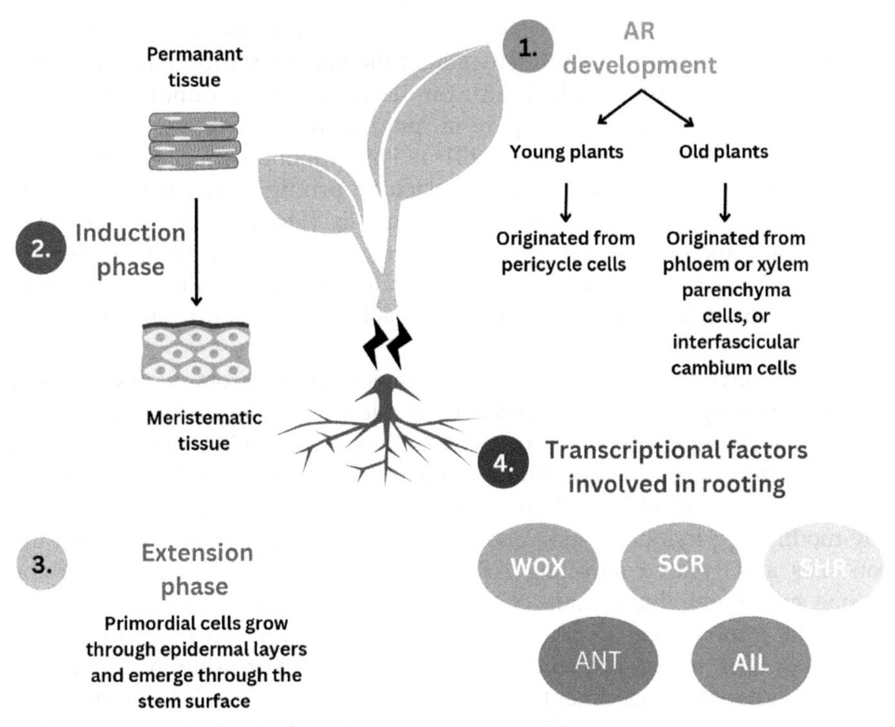

FIGURE 15.1 Overview of adventitious root development in plants.

lateral roots, which develop from the pericycle, the evolution of ARs in plants is quite different, as the latter initiates with dedifferentiation, i.e., loss of function in primary somatic cells, followed by induction and redifferentiation (Massoumi, 2016). The dedifferentiation step is sometimes considered an early phase or initial phase of induction involving an anatomical lag phase during which reprogramming occurs. Multiple species in the plant kingdom are already root-competent, i.e., directly following root-inducing signals. However, for others, this stage is inevitable to acquire root competence via dedifferentiation (Ikeuchi et al., 2016). The origin of ARs is modified by a variation in the lineage (age/type) as seen in hypogeal plants, in which ARs in young plants originate from pericycle cells, whereas older hypocotyls, nonhypocotyl stems, and petioles of detached leaves originate in other tissues near the vascular tissues, such as phloem or xylem parenchyma cells or interfascicular cambium cells (Druege et al., 2016). Finally, after codification and primordia formation, the meristematic cells of primordia prolifically divide to differentiate into root layers: epidermis, cortex, endodermis, vasculature, meristem, and root cap (Guan et al., 2015). The induction phase is followed by a redifferentiation or extension phase in which primordial cells grow through epidermal layers and emerge through the stem surface (Guan et al., 2019).

As stated, lineage variation affects dedifferentiation; moreover, a positive association between cells bearing ARs and vascular tissues has been deciphered (Gonin et al., 2019). In *Arabidopsis* sp. (dicot), de-rooted hypocotyls have shown emergence from vascular tissues (Verstraeten et al., 2014), whereas in petiolar tissues, xylem- or pericycle-like cells show massive division to form microcallus followed by redifferentiation for cell specification (Bustillo–Avendaño et al., 2018). Similarly, vascular bundle cells peripherally confined to ground meristem cells play a pivotal role in crown root formation in monocots, particularly rice (Wang et al., 2016). This "phenomenon of AR emergence" is vastly influenced by phytochemical and molecular prodigies in plants, particularly by plant hormonal balance. They frequently engage in intricate interactions that deliver the reliable spatial and temporal inputs necessary for AR development. Before delving into the roles and mechanisms of various phytohormones manipulating AR formation in plants, we need to understand the molecular backgrounds controlling their expression in plants.

Genetic studies have revealed a hierarchy of transcription factors necessary for AR initiation, and newer findings have underscored their additional role in rooting regulation. ARs are predominant features in several species, thereby ratifying constitutive regulation (Xu et al., 2015); however, both constitutive and induced regulation have been observed in transcription factors controlling AR formation in plants. Transcription factors eventually operate by modifying phytohormones, and several genetic studies have revealed their active expression sites and genes controlling AR development. A WUSCHEL-related homeobox (WOX) is one such highlighted group of transcription factors observed lavishly across plant species, controlling more than 15 protein products participating in root formation in plants (Hao et al., 2019). Gene expression studies about WOX overexpression in *Arabidopsis* sp. have revealed promoted root development (Li et al., 2020). Similarly, GRAS-family transcription factors, including SCARECROW (SCR) and SHORT ROOT (SHR) (Horstman et al., 2014), play an active role in inducing rooting in cutting as well as wood maturation in plants. Rigal et al. (2012) observed a highly synergistic effect of the AINTEGUMENTA LIKE1 Homeotic Transcription Factor (PtAIL1) in inducing ARs in *Populus trichocarpia* via overexpression analysis. Similarly, ectopic expression of the *Brassica rapa* AINTEGUMENTA gene (BrANT-1) in *Arabidopsis* showed upregulation in organ development (Ding et al., 2018). Multiple transcription factors, genes, and QTLs controlling AR development in plants have been formalized in Table 15.1. Genetic constitution plays a very prominent role in providing accurate phytohormone balance. However, the ultimate masters behind the "phenomenon of AR formation" are undeniably phytohormones.

15.3 Phytohormones action in adventitious root formation

Phytohormones coordinate and direct every stage of AR, from the initial process of cell reprogramming to emergence and expansion, in addition to numerous other internal and external cues. AR formation is a result of an integrated activity monitored explicitly by phytohormone balance (Pacurar et al., 2014) (Fig. 15.2). Regardless of the cell type or specification from which AR arises, environmental cues and phytohormones work together to direct every stage of AR development. However, there is a lack of comprehensive knowledge regarding the regulation and operation of plant hormone homeostasis as well as the complex signaling network during AR formation.

TABLE 15.1 Molecular insight of different phytohormones in adventitious root under various plant species.

Plant species	Gene family	Gene-Expression	AR model	Phyto-hormone affected	References
Oryza sativa	*CRL6* (crown rootless6)	Downregulation	Crown root	Auxin	Wang et al. (2016)
Arabidopsis	*PLETHORA1 (PLT1)*	Upregulation	Hypocotyl derived AR	Auxin	Xiong et al. (2020)
Castanea sativa	*CsSCL1* (SCARECROW-like transcript)	Upregulation	Cuttings	Auxin	Vielba et al. (2011)
Juglans nigra	*ARF6* and *ARF8*	Upregulation	Cuttings	Auxin	Stevens et al. (2018)
Arabidopsis	Yucca (*YUC* genes)	Upregulation	Cuttings	Auxin	Chen et al. (2016)
Pisum sativum	*Max1* (more auxiliary growth1)	Downregulation	Cuttings	Strigolactones	Rasmussen et al. (2012)
Brassica rapa	AINTEGUMENTA gene (BrANT-1)	Upregulation	Leaf	Auxin	Ding et al. (2018)
Arabidopsis	*AINTEGUMENTA*-LIKE (*AIL*)	Downregulation	Cuttings	Ethylene	Horstman et al. (2014)
Petunia	*COI1 (CORONATINE INSENSITIVE1)*	Downregulation	Cuttings	Jasmonic acid	Wasternack and Song (2017)
Arabidopsis	*ASA1 (ANTHTHRANILATE SYNTHASE a1)*	Upregulation	Cuttings	Auxin	Fattorini et al. (2017)
Arabidopsis	*WIND* (Wound Induced Dedifferentiation)	Upregulation	Cuttings	Cytokinin	Bustillo–Avendaño et al. (2018)
Arabidopsis	*CYTOKININ OXIDASE (CKX1)*	Downregulation	Cuttings	Cytokinin	Dob et al. (2021)
Avicennia marina	*IAA19*	–	Pneumatophores	Ethylene, ABA	Hao et al. (2021)
Oryza sativa	*NEVER-RIPE (NR)*	Upregulation	Crown roots	Ethylene	Steffens and Rasmussen (2016)
Zea mays	*RTCS* gene	Upregulation	Seminal roots	Auxin	Taramino et al. (2007)
Strawberry	*RTY*	Upregulation	Fibrous roots	Auxin	Li et al. (2021)
Saccharum officinarum	*DEGs* (Differentially expressed genes)	Upregulation	Nodes	Auxin	Li et al. (2020)

15.3.1 Auxin "THE WIZARD"

A highly precise balance of phytohormones forms the basis of the successful AR emergence phenomenon. Behind this intricate and complex set of reactions is

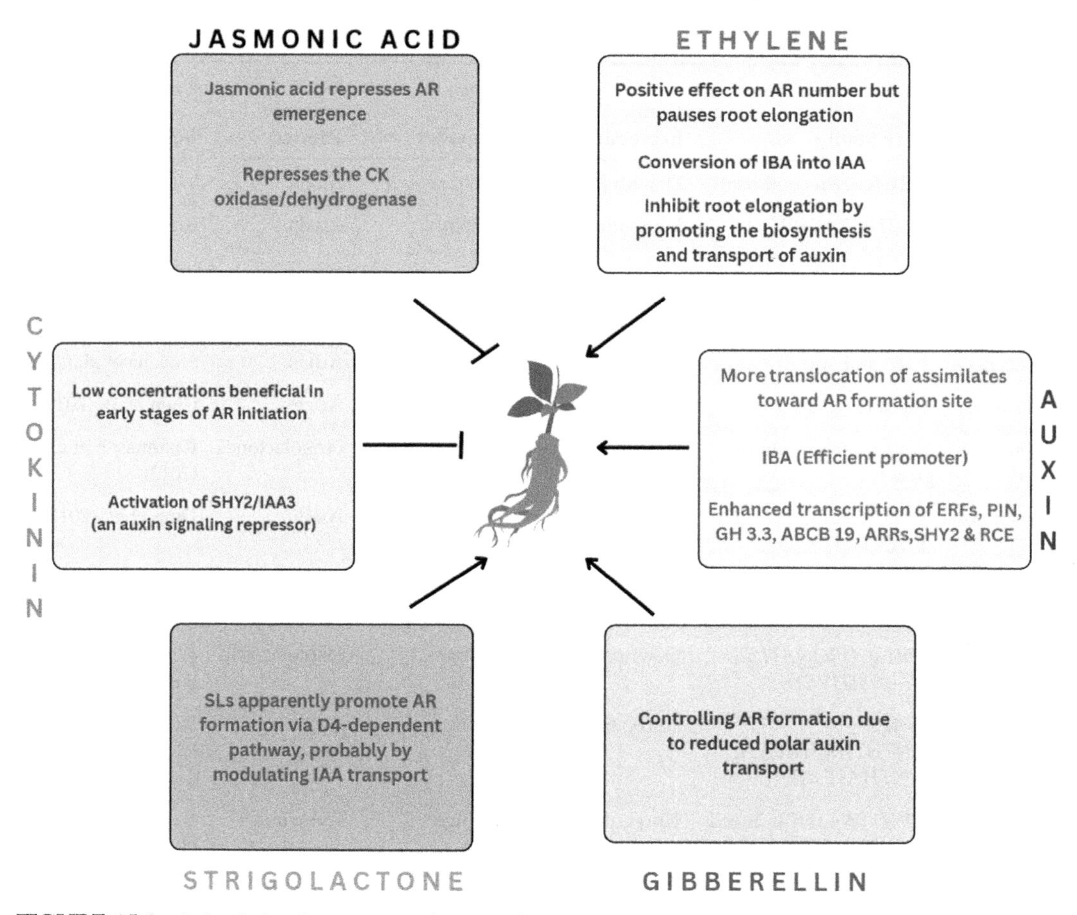

FIGURE 15.2 Role of phytohormones in shaping adventitious root development.

"AUXIN: the wizard." Throughout the plant kingdom, its exogenous treatment consistently results in the de novo formation of roots (Lavenus et al., 2013). Auxin is involved in each aspect of AR formation in angiosperms, starting from cell cycle acquisition to meristem initiation, emergence, and elongation (Arya and Husen, 2022). The particular stage of AR root formation in plants is a varied lineage of auxin treatment, i.e., the simultaneous product of auxin concentration and period of treatment (Abarca, 2021). Low concentrations of auxin are prevalent during root initiation and the early induction phase, followed by a rise from a 24–48 hour period of induction (Jing et al., 2020). Various plant models suggest that auxin application increases the translocation of assimilates from the leaves and the abundance of sugar at the site of root development (Agulló–Antón et al., 2011). This reallocation of metabolic resources can be interpreted as a combined effect of auxin accumulation at the base of excision.

It has been widely accepted that various auxin types encourage the development of AR in various species (Bellini et al., 2014). Indole 3-acetic acid (IAA), the most prevalent natural auxin, regulates numerous plant developmental processes, including AR formation. However, in routine commercial grafting and cutting programs, IBA (Indole 3-butyric acid) is the most commonly used auxin instead of IAA, which is an efficient promoter of AR in many species (Da Costa et al., 2013). The next crucial stage of AR initiation in hypocotyl or stem cuttings of several species is polar auxin transport (Sukumar et al., 2013). In most plant species, three classes of carriers and transporters present at the plasma membrane contribute as auxin influx and efflux carriers that are transcriptionally and posttranscriptionally regulated and mediate cell-to-cell auxin transport (Armengot et al., 2016). The plant-specific PIN family of efflux carriers and the ATP-binding cassette (ABC) superfamily of transporters, primarily the B-type (ABCB/multidrug resistance [MDR]/phosphoglycoprotein [PGP]), are the two main categories that exhibit auxin-efflux activity. The auxin influx symporters are encoded by the AUXIN1/LIKE-AUX1 (AUX/LAX) gene family (Titapiwatanakun et al., 2009; Adamowski and Friml, 2015). Each class of carrier is coded to the influx and efflux movement of a particular auxin homolog. For instance, AUX1 acts as an influx carrier for IAA and not for IBA. Similarly, ABCB19, ABCB1, PIN2, and PIN7 act as efflux carriers for IAA, and PDR appears to be efflux IBA but not IAA (Strader and Bartel, 2009).

This signaling pathway and control of auxin were previously considered to be transcriptionally regulated, but recent epigenetic studies have suggested otherwise. Genes controlling auxin concentration and signaling pathways are termed auxin-responsive genes, which control auxin-responsive factors (ARFs). The majority of the ARFs control IAA in vitro synthesis in plants and, therefore, control auxin levels in particular tissues (Lakehal et al., 2019). ARF6 and ARF8 have been found to have positive regulation of auxin, whereas ARF17 has negative regulation (Ibáñez et al., 2019). ARF6 and ARF8 induce GRETCHEN HAGEN3 (GH3.3), GH3.5, and GH3.6 enzymes that encode acyl acid amino synthases, which alter hormonal balance by conjugating with jasmonic acids (JAs) and therefore modifying ARs (Gutierrez et al., 2012).

Some prominent mutant phenotypes for auxin-controlling genes have been identified, including auxin-overproducing superroot2 (*sur2*) mutants (Pacurar et al., 2014a). Superroot2−1 mutant (*sur2−1* mutant) in *Arabidopsis* has been found to overproduce auxin, leading to excessive adventitious rooting in cuttings. Some particular sites have been discovered recently, including COP9 SINGALOSOME SUBUNIT4, AUXIN RESPONSE1, SHORT HYPOCOTYL2 (SHY2), and RUB-CONJUGATING ENZIME1 (RCE1) (Pacurar et al., 2017). Auxin efflux carriers like ABCB19 and PIN1 that are involved in auxin accumulation at the base of the stem are major contributors to AR formation, but when loss-of-function mutant genes *abcb19−1* and *pin1−1* are present, drastic effects on AR number in mutant phenotypes have been observed (Sukumar et al., 2013). AR formation studies have also revealed the active involvement of some major signaling molecules, including INDOLE-3-ACETIC ACID INDUCIBLE28 (IAA28), CRANE (IAA18), WOODEN LEG, ARABIDOPSIS RESPONSE REGULATORS1 (ARR1), ARR10, and ARR12 molecules in petiolar tissues as revealed by gain-of-function mutants (*crane-2* and *solitary root-1* [*slr-1*]) (Bustillo−Avendaño et al., 2018).

Extensive characterization studies have revealed the active role of auxin in AR formation and its major influencing genetic components. This solidifies the fact that modulation of auxin homeostasis directly affects AR formation. No doubt, auxin is the master regulator of virtually all aspects of AR formation in several plants, not only individually but through exerting its myriad effects in coordination with other biostimulants.

15.3.2 Ethylene "THE BITTERSWEET HORMONE"

As the description projects, the role of ethylene in AR formation is subtle and complex. In conjugation with auxin, ethylene acts as both antagonistic and synergistic by regulating its signals and transport (Stepanova and Alonso, 2009). Ethylene is a stress-related phytohormone that is released by plants in response to unfavorable environmental stimuli such as flooding, maturity, and diseases. 1-aminocyclopropane-1-carboxylic acid, an ethylene precursor, seems to have a strong effect on AR formation in plants. In an interesting predicament, it has been observed that ethylene has a positive effect on AR number but pauses root elongation (Druege et al., 2016). Moreover, the effect of ethylene on AR formation is also dependent on the presence of auxin. Elevated ethylene concentrations may indirectly inhibit root elongation by promoting the biosynthesis and transport of auxin into the elongation zone (Muday et al., 2012). Genetically, ethylene is controlled by multiple responsive factors known as ethylene responsive factors (ERFs) or APETALA2. While ERF proteins primarily respond as a result of external stimuli, AP2 TFs in the AP2/ERF superfamily primarily regulate developmental programs (Licausi et al., 2013). Several AP2/ERF have pronounced effects on the development of shoots in different species; for instance, ectopic expression of WOUND INDUCED DEDIFFERENTIATION 1 (WIND1) has been observed to control dedifferentiation in *Arabidopsis*. The control is so magnificent that it does not require the exogenous application of auxins or cytokinins (Iwase et al., 2011). Ethylene also manipulates auxin homology by fastening the conversion of IBA into IAA and therefore promoting AR formation (Veloccia et al., 2016). Numerous findings prove the bittersweet role of ethylene in controlling AR development in plants. Although there is no direct control of ethylene in AR development per se, it is still important for auxin activity and, thus, sound photo-hormonal machinery.

15.3.3 Jasmonic acid and Strigolactones "THE MODULATORS"

Over the last decade, several new classes of hormones have been identified that, either alone or in combination with other phytohormones, silently modulate the development of AR in plants. JA, or jasmonate, is an oxylipin-derived class of biomodulators shown to play a key regulatory role in AR development (Wasternack and Song 2017). The rooting action of JA has been observed in cases of injury or site of wounds in response to protecting plants from pathogenic attack, where it is rapidly and transiently induced. Molecular studies have confirmed JA represses AR emergence through transcriptional regulation of genes controlling the *APETALA2/ETHYLENE RESPONSE FACTOR115* (*ERF115*) transcription factor (Lakehal et al., 2020; Dob et al., 2021), showing the repressive action of JA on gene *CKX1*, which encodes the cytokinin-degrading enzyme *CK oxidase/dehydrogenase*, ultimately inhibiting AR initiation. JA and cytokinins simultaneously accumulate at the site of AR, respectively, after the former reaches its peak (Keuchi et al., 2017). Moreover, JA and cytokinins additively inhibit AR initiation through the overexpression of the *RAP2.6L* transcription factor gene. Recent research has demonstrated that jasmonate is a significant regulator of AR formation, most likely as an inhibitor of ARI, though its actual role is a bit more complex and conservative.

Just like JA, strigolactones have been recently recognized as a class of hormones that participate interactively in the phenomenon of AR development. First characterized by Rasmussen et al. (2012), in *Arabidopsis* and peas, these phytohormones were observed for their inhibitory effect on adventitious rooting. Strigolactones are carotenoid-derived phytohormones abundantly found in plant roots (Waters et al., 2017). Recent studies have demonstrated its pivotal role in crown root elongation in rice, where genotype dwarf10 (d10), a strigolactone-deficient rice mutant, showed a lower number of ARs as compared to wild-type (Sun et al., 2015). The biochemically responsible factor for strigolactone is auxin inhibition by modulation in its transport pathway; however, this is species-dependent. Recent studies have hypothesized that strigolactones perform independently of other phytohormones such as IAA, ethylene, and cytokinins.

15.3.4 Cytokinins "THE ANTI-HERO"

Cytokinins are an auxin-antagonistic group of phytohormones that are known for their suppressiveaction against AR root formation in several species (Kitomi et al., 2011; Arya et al., 2022b). Instead of direct involvement in AR formation, cytokinins extract a hefty toll on the process by modulating auxin activity in plants. Besides suppressing of auxin content, cytokinins are involved in many plant processes, including cell division and shoot and root morphogenesis. Low concentrations of cytokinins are very beneficial for the early stages of AR initiation in multiple species (Ricci et al., 2008). Downregulation of auxin activity due to cytokinin concentration is an integrated involvement of the latter in auxin transport assemblies, particularly PIN transporters (Šimášková et al., 2015). Cytokinin response factors transcriptionally modify cis-regulatory elements present in PIN-FORMED genes, thereby altering their expression, eventually leading to auxin distribution mutants. In addition to this, cytokinins are also responsible for the activation of SHY2/IAA3 (SHY2), an auxin signaling repressor. Cytokinin-controlling genes (CK oxidase/dehydrogenase) have a prominent effect on the frequency of AR formation in plants. For instance, the overexpression of CK oxidase/dehydrogenase genes in *Arabidopsis* led to decreased in vitro levels of cytokinins, followed by a positive response in AR activity (Wang et al., 2014). These results imply the prominent role of cytokinins in AR formation in plants and their impact on other phytohormones.

15.3.5 The EXIGENT classes of hormones

In addition to the previously mentioned phytohormones, many plant compounds have been identified and demonstrated to play a role in symbiotic interactions, either cohesively or abrasively, but their impact on root development remains unknown. Other classes of phytohormones modifying AR root formation in plants include gibberellins (GA), salicylic acid, abscisic acid, brassinosteroids, etc. Most of these biostimulants modulate AR initiation by inhibiting or promoting auxin signaling pathways. GA is one such class of inhibitors that seem to be involved in controlling AR formation due to reduced polar auxin transport. High levels of GA through ectopic expression harm arsin, tobacco, rice, etc (Niu et al., 2013). In another study, increased application of GA4 in poplar resulted in a decreased number of ARs as compared to wild-type (Mauriat et al., 2014). Unlike GA, brassinosteroids have a synergistic effect with auxins, and they appear to promote lateral root development through cohesive reactions with auxin (Bellini et al., 2014). Plant growth regulators are part and parcel of "the AR emergence phenomenon." These stimulants are responsible for cell specification, which ultimately determines cell fate.

15.4 The nonhormonal alliance for a flawless AR establishment

AR formation can be compared to a particle-by-particle building process just like building furniture, where auxin is a "mallet" and other phytohormones are the "iron nails." Auxin plays a central role in AR formation (Pacurar et al., 2014). Any injury or cut to a plant triggers a hormone-regulated cascade of responses mediated mainly via auxin canalization toward the target cells, together with other growth regulators like ethylene, cytokinin, gibberellic acids, abscisic acid, strigolactones, brassinosteroids, salicylic acid, JA, polyamides $(PA)_n$, and NO_n, as well as environmental factors like light, temperature, or nutrition. Besides these, hormonal signaling and function implementation are mediated by altered sugar metabolism, cell cycle modulation (microtubule assembly), and cell wall reorganization (Druege et al., 2016). Previous sections highlighted some important phytohormones coacting with auxins for AR development, but one cannot ignore the importance of nonhormonal controls, which act as silent modulators during the "AR paradigm." The ultimate product of hormones is cell division and elongation; microtubule (MT)-related transcripts play an important role in the modification of cell walls (Landrein and Hamant, 2013). Null mutant studies have shown that downregulation in MT-associated protein (KATANIN) leads to a reduction in the auxin-induced primordial formation of AR (Abu–Abied et al., 2015; Abu–Abied et al., 2018), also observed a significant control of XI myosin protein in control of polar auxin transport and cell division during AR formation. Mutant analysis reveals the comprehensive role of cyclin-dependent kinases as well as cyclins in inducing explant ARs in *Arabidopsis* sp. (Bustillo–Avendaño et al., 2018).

Sugar metabolism is another exigent aspect of nonhormonal control of AR development; its supply is a critical metabolic bottleneck that is necessary for the production of energy and metabolic activity that generates distinct sugar and amino acid profiles, paying dividends at different stages of AR growth (Agulló-Antón et al., 2014). Sucrose is an important modulator, affecting the expression of multiple genes at specific phases. Several agonistic functions of sucrose have been found in auxin signaling (Ljung et al. (2015) and Barbier et al. (2015) demonstrated the effect of sucrose in downregulating strigolactone transduction in *Rosa hybrida*). Similarly, the effect of visible light wavelength has been observed to be correlated with AR formation in plants. In a study, far-red light significantly increased *ARF6* and *ARF8* expression levels during ARI in *Eucalyptus globulus* compared to white light treatment (Ruedell et al., 2015). Numerous different phytohormones, like strigolactones, have also been shown to interact with the visible region in this manner

(Rasmussen et al., 2012). As far as nutritional modulators are concerned, studies have revealed that certain inorganic nutrients, particularly nitrate, are active auxin influx facilitators, and low nitrate levels consequently repress root development (Krouk et al., 2010). Drüge (2019) observed that a high C/N ratio in pelargonium significantly impacted AR formation. In contrast to nitrate, phosphorus is a limiting nutrient in AR formation, and this limitation is believed to trigger adaptive responses in plants via AR formation (Bellini et al., 2014). These studies justify the role of genotype, the specific system, the AR-generating tissue, and the environment in influencing the specific interactions and bottlenecks along with phytohormones.

15.5 Application of natural biostimulants for phytohormonal control of adventitious roots

Now that we are aware of the mechanisms by which various phytohormones contribute to the development of AR, how can we connect this knowledge to real-world experience and introduce the idea of biostimulants to create a link that will help to clarify this intricate mechanism of AR formation? We can answer this question one at a time. The source and type of the biostimulant to be used must first be understood, then its impact on AR and a specific phytohormone, and finally the practical application of that biostimulant in horticulture. Source-based classification of biostimulants was proposed by Patrick du Jardin (Du Jardin, 2012), which sorted biostimulants into eight classes as presented in Table 15.2. Out of these eight categories, three biostimulants—protein hydrolysates, seaweed extracts, and humic substances (Yakhin et al., 2017)—share some common features with phytohormones. They are active at a low dose and often cause growth inhibition when applied at high concentrations or doses. For instance, Trevisan et al. (2017) highlighted that the application of protein hydrolysate (APR) modified the response of maize (B73) seedlings to abiotic stress, which was confirmed with mRNA analysis of transcriptome profiles. Active involvement of brassinosteroids was seen. Similarly, in another study, UHPLC/QTOF-Ms metabolomics revealed a complex biochemical response to the biostimulant treatment, in which brassinosteroids and their interactions with other hormones appeared to be crucial in promoting root growth in melon (Lucini et al., 2018). In another study, NMR-based bioactivity characterization revealed that when Belgian endive extracts were applied as an aqueous solution to *Arabidopsis* as biostimulants, major root promotion activity was seen (Ogunsanya et al., 2022).

Recent studies have proved that the application of biostimulants to vegetative propagules like cutting seedlings increases the formation of ARs and root branching. Wise et al. (2020) observed that the application of biostimulant root nectar (kelp and aloe vera) and willow bark extract increased AR development in chrysanthemum and lavender cuttings. Similarly, Kim et al. (2019) illustrated the active involvement of vegetal-derived biostimulants in inducing ARs on basil and tomato cuttings via brassinosteroids. Seaweed biostimulant is an entirely new concept in phytohormone manipulation of ARs. In a joint study (India-Europe), the effect of seaweed biostimulant "Kelpak SL" on onion bulbs

TABLE 15.2 Different categories of substances act as biostimulants.

Biostimulant category	Source	Property	Effect	Reference
Humic substances (HS)	Soil organic matter	• Supra-colloidal • Heterogeneous	• Increased nutrient uptake • Stress protection	da Silva et al. (2021)
Protein hydrolysates and N-containing compounds	Agro-industry	• Single and Mixed compounds • Nonprotein amino acids	• Increased N-assimilation • Stress protection against heavy metals • Hormonal activity	Colla et al. (2014)
Seaweed extracts	Seaweeds Brown algae	• Polysaccharide • Laminarin • carrageenans	• Affecting hormonal gene expression • Antistress effects	Du Jardin (2015)
Chitosan and chitin derivatives	Chitosan is a deacetylated form of chitin	• Poly- and oligomers • Variable sizes	• Enhance cellular responses • Modify signaling pathways	Katiyar et al. (2015)
Antitranspirants	Agro-industry	• Kaolin • Polyacrylamides	Increased stress tolerance	Rouphael and Colla (2020)
Inorganics	Inorganic salts	• Al • Co • Ni • Se • Si	• Antistress • Osmoregulation • Increased enzyme activity	Rouphael and Colla (2020)
Beneficial micro-organisms	Mycorrhizal fungi Rhizobacteria	–	• Increased nutrient uptake • Enzyme activity • Antistress	Du and Jardin (2012)
Complex organic materials	Plant roots Shoots Barks	• Amino acids • Organic acids	• Increased plant productivity • Antistress • Increased tolerance	Ogunsanya et al. (2022)

was practically examined, and results revealed the effects to be highly synergistic (Szczepanek et al., 2017). In another research, a seaweed biostimulant called "Goteo" was found to promote rooting on semihardwood cuttings of *Lantana camara* and Grandiflora (Loconsole et al., 2022). Similarly, "Eckol," a biostimulant from brown seaweed Ecklonia maxima, positively increased the number of seminal roots in maize on application (Rengasamy et al. 2015).

Such discoveries have encouraged new researchers to actively participate in this field of study and investigate the active use of biostimulants for use in commercial agriculture. AR formation is indispensable for vegetative asexual propagation. Therefore, identification and characterization of the factors regulating AR formation and development are essential for understanding and potentially manipulating AR formation. As it sounds, AR development serves as an enormous task in plants, but this serves as the foundation for clonal multiplication, a technology that can be utilized for the breeding and production of horticultural plants. The elucidation of the agricultural function and action mechanisms of phytohormones will enable the creation of a second generation of biostimulants with functionally designed benefits and supplementary mechanisms.

15.6 Future perspective and conclusion

In this chapter, several findings have been discussed regarding the hormonal regulation of various cellular events in AR formation and have brought new insight into the involvement of metabolic and nutrient homeostasis in the whole process. Precise knowledge regarding the control of AR formation provides insight into the fascinating processes underlying the regeneration ability of plants and opens up new perspectives for sustainable and efficient utilization of plant genetic resources. The above-discussed facts are invaluable for scientists trying to manipulate and exfoliate the prodigies of AR formation in plants. An important area of research can be explored toward the identification of the involvement of new genes, phytohormones, the interaction of hormones with other signaling molecules, and more precisely with miRNAs.

The established and growing molecular approaches using various plant species led to the recognition of diverse and conserved regulatory loops indulged in the initial phases of AR development. The downstream events involved in the endogenous level of phytohormone homeostasis strengthen the activation of root stem cells to locally control signaling molecules such as reactive oxygen species. The discussed applications of biostimulants in adventitous root initiation and establishment widen our prospective about the enormous possibilities these present in this field of agriculture and how to harness their potential effectively.

References

Abarca, D., 2021. Identifying molecular checkpoints for adventitious root induction: are we ready to fill the gaps? Front. Plant. Sci. 12, 341.

Abbas, S.M., 2013. The influence of biostimulants on the growth and on the biochemical composition of Viciafaba CV. Giza 3 beans. Rom. Biotechnol. Lett. 18, 8061–8068.

Abu–Abied, M., Belausov, E., Hagay, S., Peremyslov, V., Dolja, V., Sadot, E., 2018. Myosin XI–K is involved in root organogenesis, polar auxin transport, and cell division. J. Exp. Bot. 69, 2869–2881.

Abu–Abied, M., Rogovoy, O., Mordehaev, I., Grumberg, M., Elbaum, R., Wasteneys, G.O., et al., 2015. Dissecting the contribution of microtubule behaviour in adventitious root induction. J. Exp. Bot. 66, 2813–2824.

Adamowski, M., Friml, J., 2015. PIN–dependent auxin transport: action, regulation, and evolution. Plant. Cell 27, 20–32.

Agulló-Antón, M.Á., Ferrández-Ayela, A., Fernández-García, N., Nicolás, C., Albacete, A., Pérez-Alfocea, F., et al., 2014. Early steps of adventitious rooting: morphology, hormonal profiling and carbohydrate turnover in carnation stem cuttings. Physiol. Plant. 150, 446–462.

Agulló–Antón, M.Á., Sánchez–Bravo, J., Acosta, M., Druege, U., 2011. Auxins or sugars: what makes the difference in the adventitious rooting of stored carnation cuttings? J. Plant. Growth Regul. 30, 100–113.

Armengot, L., Marquès–Bueno, M.M., Jaillais, Y., 2016. Regulation of polar auxin transport by protein and lipid kinases. J. Exp. Bot. 67, 4015–4037.

Arya, A., Gola, D., Tyagi, P.K., Husen, A., 2022a. Molecular control of adventitious root formation. In: Husen, A. (Ed.), Environmental, Physiological and Chemical Controls of Adventitious Rooting in Cuttings. Elsevier Inc, Cambridge, MA, USA, pp. 25–46.

Arya, A., Sharma, V., Tyagi, P.K., Gola, D., Husen, A., 2022b. Role of cytokinins in adventitious root formation. In: Husen, A. (Ed.), Environmental, Physiological and Chemical Controls of Adventitious Rooting in Cuttings. Elsevier Inc, Cambridge, MA, USA, pp. 239–249.

Arya, A., Husen, A., 2022. Role of various auxins in adventitious root formation. In: Husen, A. (Ed.), Environmental, Physiological and Chemical Controls of Adventitious Rooting in Cuttings. Elsevier Inc, Cambridge, MA, USA, pp. 213–238.

Barbier, F., Péron, T., Lecerf, M., Perez–Garcia, M.D., Barrière, Q., Rolčík, J., et al., 2015. Sucrose is an early modulator of the key hormonal mechanisms controlling bud outgrowth in *Rosa hybrida*. J. Exp. Bot. 66, 2569–2582.

Bellini, C., Pacurar, D.I., Perrone, I., 2014. Adventitious roots and lateral roots: similarities and differences. Annu. Rev. Plant. Biol. 65, 639–666.

Bustillo–Avendaño, E., Ibáñez, S., Sanz, O., Sousa Barros, J.A., Gude, I., Perianez–Rodriguez, J., et al., 2018. Regulation of hormonal control, cell reprogramming, and patterning during de novo root organogenesis. Plant. Physiol. 176, 1709–1727.

Chen, L., Tong, J., Xiao, L., Ruan, Y., Liu, J., Zeng, M., et al., 2016. YUCCA–mediated auxin biogenesis is required for cell fate transition occurring during de novo root organogenesis in *Arabidopsis*. J. Exp. Bot. 67, 4273–4284.

Colla, G., et al., 2014. Biostimulant action of a plant-derived protein hydrolysate produced through enzymatic hydrolysis. Front. Plant. Sci. 5.

Da Costa, C.T., de Almeida, M.R., Ruedell, C.M., Schwambach, J., Maraschin, F.S., Fett–Neto, A.G., 2013. When stress and development go hand in hand: main hormonal controls of adventitious rooting in cuttings. Front. Plant. Sci. 4, 133.

da Silva, A.P.S., et al., 2021. Attenuations of bacterial spot disease *Xanthomonas euvesicatoria* on tomato plants treated with biostimulants. Chem. Biol. Technol. Agric. 8 (42). Available from: https://doi.org/10.1186/s40538-021-00240-9.

Ding, Q., Cui, B., Li, J., Li, H., Zhang, Y., Lv, X., et al., 2018. Ectopic expression of a *Brassica rapa* AINTEGUMENTA gene (BrANT–1) increases organ size and stomatal density in *Arabidopsis*. Sci. Rep. 8, 1–13.

Dob, A., Lakehal, A., Novak, O., Bellini, C., 2021. Jasmonate inhibits adventitious root initiation through repression of CKX1 and activation of RAP2. 6L transcription factor in *Arabidopsis*. J. Exp. Bot. 72, 7107–7118.

Drüge, U., 2019. Physiological and molecular control of adventitious root formation in ornamental cuttings. Hannover: Gottfried Wilhelm Leibniz Universität, Habil.-Schr., 2018. Available from: https://doi.org/10.15488/7659.

Druege, U., Franken, P., Hajirezaei, M.R., 2016. Plant hormone homeostasis, signaling, and function during adventitious root formation in cuttings. Front. Plant. Sci. 7, 381.

Druege, U., Hilo, A., Pérez–Pérez, J.M., Klopotek, Y., Acosta, M., Shahinnia, F., et al., 2019. Molecular and physiological control of adventitious rooting in cuttings: phytohormone action meets resource allocation. Ann. Bot 123, 929–949.

Du Jardin, P., 2012. The Science of Plant Biostimulants – A Bibliographic Analysis. Contract 30–CE0455515/00–96, Ad hoc Study on Bio–stimulants Products.

Du Jardin, P., 2015. Plant biostimulants: Definition, concept, main categories and regulation. Sci. Hortic. 196, 3–14. Available from: https://doi.org/10.1016/j.scienta.2015.09.021.

Fattorini, L., Veloccia, A., Della Rovere, F., D'Angeli, S., Falasca, G., Altamura, M.M., 2017. Indole–3–butyric acid promotes adventitious rooting in *Arabidopsis thaliana* thin cell layers by conversion into indole–3–acetic acid and stimulation of anthranilate synthase activity. BMC Plant. Biol. 17, 121.

Gonin, M., Bergougnoux, V., Nguyen, T.D., Gantet, P., Champion, A., 2019. What makes adventitious roots? Plants 8, 240.

Guan, L., Murphy, A.S., Peer, W.A., Gan, L., Li, Y., Cheng, Z.M., 2015. Physiological and molecular regulation of adventitious root formation. Crit. Rev. Plant. Sci. 34, 506–521.

Guan, L., Tayengwa, R., Cheng, Z.M., Peer, W.A., Murphy, A.S., Zhao, M., 2019. Auxin regulates adventitious root formation in tomato cuttings. BMC Plant. Biol. 19, 1–16.

Gutierrez, L., Mongelard, G., Floková, K., Pacurar, D., Novák, I., Staswick, O., et al., 2012. Auxin controls *Arabidopsis* adventitious root initiation by regulating jasmonic acid homeostasis. Plant. Cell Rep. 24, 2515–2527.

Hao, Q., Zhang, L., Yang, Y., Shan, Z., Zhou, X.A., 2019. Genome–wide analysis of the WOX gene family and function exploration of GmWOX18 in soybean. Plants 8, 215.

Hao, S., Su, W., Li, Q.Q., 2021. Adaptive roots of mangrove *Avicennia marina*: structure and gene expressions analyses of pneumatophores. Sci. Total. Environ. 757, 143994.

Horstman, A., Willemsen, V., Boutilier, K., Heidstra, R., 2014. AINTEGUMENTA–LIKE proteins: hubs in a plethora of networks. Trends Plant. Sci. 19, 146–157.

Husen, A., 2002. Physiological Effects of Phytohormones and Mineral Nutrients on Adventitious Root formation and Clonal Propagation of *Tectona grandis Linn*. f. PhD thesis, Forest Research Institute, Dehra Dun, India.

Husen, A., 2008. Clonal propagation of *Dalbergia sissoo* Roxb. and associated metabolic changes during adventitious root primordium development. N. For. 36, 13–27.

Husen, A., 2013. Clonal multiplication of teak (*Tectona grandis*) by using moderately hard stem cuttings: effect of genotypes (FG1 and FG11 clones) and IBA treatment. Adv. Forestry Lett. 2, 14–19.

Husen, A., Iqbal, M., Siddiqui, S.N., Sohrab, S.S., Meshresa, G., 2017. Effect of indole-3-butyric acid on clonal propagation of mulberry (*Morus alba* L.) stem cuttings: rooting and associated biochemical changes. Proc. Natl Acad. Sci., India Sect. B: Biol. Sciences 87, 161–166.

Husen, A., 2022. Environmental, Physiological and Chemical Controls of Adventitious Rooting in Cuttings. Elsevier Inc, Cambridge, MA, USA.

Husen, A., Zhang, W., 2023. Hormonal Cross-Talk, Plant Defense and Development. Elsevier Inc, Cambridge, MA, USA.

Ibáñez, S., Ruiz–Cano, H., Fernández, M.Á., Sánchez–García, A.B., Villanova, J., Micol, J.L., et al., 2019. A network–guided genetic approach to identify novel regulators of adventitious root formation in *Arabidopsis thaliana*. Front. Plant. Sci. 10, 461.

Ikeuchi, M., Ogawa, Y., Iwase, A., Sugimoto, K., 2016. Plant regeneration: cellular origins and molecular mechanisms. Development 143, 1442–1451.

Iwase, A., Mitsuda, N., Koyama, T., Hiratsu, K., Kojima, M., Arai, T., et al., 2011. The AP2/ERF transcription factor WIND1 controls cell dedifferentiation in *Arabidopsis*. Curr. Biol. 21, 508–514.

Jing, T., Ardiansyah, R., Xu, Q., Xing, Q., Müller–Xing, R., 2020. Reprogramming of cell fate during root regeneration by transcriptional and epigenetic networks. Front. Plant. Sci. 11, 317.

Katiyar, Deepmala, et al., 2015. Chitosan as a promising natural compound to enhance potential physiological responses in plant: a review. Indian journal of plant physiology 20, 1–9. Available from: https://doi.org/10.1007/s40502-015-0139-6.

Keuchi, M., Iwase, A., Rymen, B., Lambolez, A., Kojima, M., Takebayashi, Y., et al., 2017. Wounding triggers callus formation via dynamic hormonal and transcriptional changes. Plant. Physiol. 175, 1158–1174.

Kim, H.J., Ku, K.M., Choi, S., Cardarelli, M., 2019. Vegetal–derived biostimulant enhances adventitious rooting in cuttings of basil, tomato, and chrysanthemum via brassino steroid–mediated processes. Agronomy 9, 74.

Kitomi, Y., Ito, H., Hobo, T., Aya, K., Kitano, H., Inukai, Y., 2011. The auxin responsive AP2/ERF transcription factor CROWN ROOTLESS5 is involved in crown root initiation in rice through the induction of OsRR1, a type-A response regulator of cytokinin signaling. Plant. J. 67, 472–484.

Krouk, G., Lacombe, B., Bielach, A., Perrine–Walker, F., Malinska, K., Mounier, E., et al., 2010. Nitrate–regulated auxin transport by NRT1.1 defines a mechanism for nutrient sensing in plants. Dev. Cell 18, 927–937.

Kumar, A., Choudhary, A., Kaur, H., Sangeetha, K., Mehta, S., Husen, A., 2022. Physiological and environmental control of adventitious root formation in cuttings. In: Husen, A. (Ed.), Environmental, Physiological and Chemical Controls of Adventitious Rooting in Cuttings. Elsevier Inc, Cambridge, MA, USA, pp. 1–24.

Lakehal, A., et al., 2019. Molecular framework for TIR1/AFB-Aux/IAA-dependent auxin sensing controlling adventitious rooting in Arabidopsis. Mol. Plant. Available from: https://doi.org/10.1016/j.molp.2019.09.001.

Lakehal, A., Bellini, C., 2019. Control of adventitious root formation: insights into synergistic and antagonistic hormonal interactions. Physiol. Plant. 165, 90–100.

Lakehal, A., Dob, A., Rahneshan, Z., Novák, O., Escamez, S., Alallaq, S., et al., 2020. Ethylene response factor 115 integrates jasmonate and cytokinin signaling machineries to repress adventitious rooting in *Arabidopsis*. N. Phytol. 228, 1611–1626.

Landrein, B., Hamant, O., 2013. How mechanical stress controls microtubule behavior and morphogenesis in plants: history, experiments and revisited theories. Plant. J. 75, 324–338.

Lavenus, J., Goh, T., Roberts, I., Guyomarc'h, S., Lucas, M., De Smet, I., et al., 2013. Lateral root development in *Arabidopsis*: fifty shades of auxin. Trends Plant. Sci. 18, 450–458.

Li, A., Lakshmanan, P., He, W., Tan, H., Liu, L., Liu, H., et al., 2020. Transcriptome profiling provides molecular insights into auxin–induced adventitious root formation in sugarcane (*Saccharum* spp. interspecific hybrids) Microshoots. Plants 9, 931.

Li, M., Yang, Y., Raza, A., Yin, S., Wang, H., Zhang, Y., et al., 2021. Heterologous expression of *Arabidopsis thaliana* rty gene in strawberry (*Fragaria* × *ananassa* Duch.) improves drought tolerance. BMC Plant. Biol. 21, 1–20.

Li, Z., Liu, D., Xia, Y., Li, Z., Jing, D., Du, J., et al., 2020. Identification of the WUSCHEL–Related Homeobox (WOX) gene family, and interaction and functional analysis of TaWOX9 and TaWUS in wheat. Inter. J. Mol. Sci. 21, 1581.

Licausi, F., Ohme-Takagi, M., Perata, P., 2013. APETALA 2/Ethylene responsive factor (AP 2/ERF) transcription factors: mediators of stress responses and developmental programs. N. Phytol. 199, 639–649.

Ljung, K., Nemhauser, J.L., Perata, P., 2015. New mechanistic links between sugar and hormone signalling networks. Curr. Opin. Plant. Biol. 25, 130–137.

Loconsole, D., Cristiano, G., De Lucia, B., 2022. Improving aerial and root quality traits of two landscaping shrubs stem cuttings by applying a commercial brown seaweed extract. Sci. Hortic. 8, 806.

Lucini, L., Rouphael, Y., Cardarelli, M., Bonini, P., Baffi, C., Colla, G., 2018. A vegetal biopolymer–based biostimulant promoted root growth in melon while triggering brassinosteroids and stress–related compounds. Front. Plant. Sci. 9, 472.

Massoumi, M., 2016. Adventitious root formation in Arabidopsis: underlying mechanisms and applications, 192 pages. PhD thesis, Wageningen University, Wageningen, NL.

Mauriat, M., Petterle, A., Bellini, C., Moritz, T., 2014. Gibberellins inhibit adventitious rooting in hybrid aspen and *Arabidopsis* by affecting auxin transport. Plant. J. 78, 372–384.

Mhimdi, M., Pérez–Pérez, J.M., 2020. Understanding of adventitious root formation: what can we learn from comparative genetics? Front. Plant. Sci. 11, 582020.

Muday, G.K., Rahman, A., Binder, B.M., 2012. Auxin and ethylene: collaborators or competitors? Trends Plant. Sci. 17, 181–195.

Niu, S., Li, Z., Yuan, H., Fang, P., Chen, X., Li, W., 2013. Proper gibberellin localization in vascular tissue is required to regulate adventitious root development in tobacco. J. Exp. Bot. 64, 3411–3424.

Ogunsanya, H.Y., Motti, P., Li, J., Trinh, H.K., Xu, L., Bernaert, N., et al., 2022. Belgian endive–derived biostimulants promote shoot and root growth in vitro. Sci. Rep. 12, 1–13.

Pacurar, D.I., Pacurar, M.L., Lakehal, A., Pacurar, A.M., Ranjan, A., Bellini, C., 2017. The *Arabidopsis* Cop9 signalosome subunit 4 (CNS4) is involved in adventitious root formation. Sci. Rep. 7, 628.

Pacurar, D.I., Pacurar, M.L., Bussell, J.D., Schwambach, J., Pop, T.I., Kowalczyk, M., et al., 2014a. Identification of new adventitious rooting mutants amongst suppressors of the *Arabidopsis thaliana* superroot2 mutation. J. Exp. Bot. 65, 1605–1618.

Pacurar, D.I., Perrone, I., Bellini, C., 2014. Auxin is a central player in the hormone cross–talks that control adventitious rooting. Physiol. Plant. 151, 83–96.

Pant, M., Bhandari, A., Husen, A., 2022a. Adventitious root formation and clonal propagation of forest-based tree species. In: Husen, A. (Ed.), Environmental, Physiological and Chemical Controls of Adventitious Rooting in Cuttings. Elsevier Inc, Cambridge, MA, USA, pp. 471–490.

Pant, M., Gautam, A., Chaudhary, S., Singh, A., Husen, A., 2022b. Adventitious root formation in ornamental and horticultural plants. In: Husen, A. (Ed.), Environmental, Physiological and Chemical Controls of Adventitious Rooting in Cuttings. Elsevier Inc, Cambridge, MA, USA, pp. 455–469.

Pant, M., Husen, A., 2022c. Micropropagation in mature trees by manipulation of phase change, stress and culture environment. In: Husen, A. (Ed.), Environmental, Physiological and Chemical Controls of Adventitious Rooting in Cuttings. Elsevier Inc, Cambridge, MA, USA, pp. 421–437. , ISBN: 9780323906364.

Pant, M., Husen, A., 2022d. *In vitro* micrografting to induce juvenility and improvement of rooting. In: Husen, A. (Ed.), Environmental, Physiological and Chemical Controls of Adventitious Rooting in Cuttings. Elsevier Inc, Cambridge, MA, USA, pp. 439–453.

Rasmussen, A., Mason, M.G., De Cuyper, C., Brewer, P.B., Herold, S., Agusti, J., et al., 2012. Strigolactones suppress adventitious rooting in *Arabidopsis* and pea. Plant. Physiol. 158, 1976–1987.

Rengasamy, K.R., Kulkarni, M.G., Stirk, W.A., Van Staden, J., 2015. Eckol—a new plant growth stimulant from the brown seaweed *Ecklonia maxima*. J. Appl. Phycol. 27, 581–587.

Ricci, A., Rolli, E., Dramis, L., Diaz–Sala, C., 2008. N,N –bis–(2,3–methylenedioxyphenyl)urea and N,N– bis–(3,4–methylenedioxyphenyl)urea enhance adventitious rooting in Pinusradiata and affect expression of genes induced during adventitious rooting in the presence of exogenous auxin. Plant. Sci. 175, 356–363.

Rigal, A., et al., 2012. The AINTEGUMENTA LIKE1 homeotic transcription factor PtAIL1 controls the formation of adventitious root primordia in poplar. Plant Physiol. 160 (4), 1996–2006. Available from: https://doi.org/10.1104/pp.112.204453.

Rouphael, Y., Colla, G., et al., 2020. Editorial: Biostimulants in gariculture. Front. Plant. Sci. 11. Available from: https://doi.org/10.3389/fpls.2020.00040.

Ruedell, C.M., de Almeida, M.R., Fett–Neto, A.G., 2015. Concerted transcription of auxin and carbohydrate homeostasis–related genes underlies improved adventitious rooting of microcuttings derived from far–red treated *Eucalyptus globulus* Labill mother plants. Plant. Physiol. Biochem. 97, 11–19.

Šimášková, M., O'Brien, J.A., Khan, M., Van Noorden, G., Ötvös, K., Vieten, A., et al., 2015. Cytokinin response factors regulate PIN–FORMED auxin transporters. Nat. Commun. 6, 1–11.

Steffens, B., Rasmussen, A., 2016. The physiology of adventitious roots. Plant. Physiol. 170, 603–617.

Stepanova, A.N., Alonso, J.M., 2009. Ethylene signaling and response: where different regulatory modules meet. Curr. Opin. Plant. Biol. 12, 548–555.

Stevens, M.E., Woeste, K.E., Pijut, P.M., 2018. Localized gene expression changes during adventitious root formation in black walnut (Juglansnigra L.). Tree Physiol. 38, 877–894.

Strader, L.C., Bartel, B., 2009. The *Arabidopsis* pleiotropic drug resistance8/ABCG36 ATP binding cassette transporter modulates sensitivity to the auxin precursor indole–3–butyric acid. Plant. Cell 21, 1992–2007.

Sukumar, P., Maloney, G.S., Muday, G.K., 2013. Localized induction of the ATP–binding cassette B19 auxin transporter enhances adventitious root formation in *Arabidopsis*. Plant. Physiol. 162, 1392. –40.

Sun, H., Tao, J., Hou, M., Huang, S., Chen, S., Liang, Z., et al., 2015. A strigolactone signal is required for adventitious root formation in rice. Ann. Bot. 115, 1155–1162.

Szczepanek, M., Wszelaczyńska, E., Pobereżny, J., Ochmian, I., 2017. Response of onion (*Allium cepa* L.) to the method of seaweed biostimulant application. Acta Sci. Pol. Hortorum. Cultus 16, 113–122.

Taramino, G., Sauer, M., Stauffer Jr, J.L., Multani, D., Niu, X., Sakai, H., et al., 2007. The maize, *Zea mays* L. *RTCS* gene encodes a LOB domain protein that is a key regulator of embryonic seminal and post–embryonic shoot–borne root initiation. Plant. J. 50, 649–659.

Titapiwatanakun, B., Blakeslee, J.J., Bandyopadhyay, A., Yang, H., Mravec, J., Sauer, M., et al., 2009. ABCB19/PGP19 stabilises PIN1 in membrane microdomains in *Arabidopsis*. Plant. J. 57, 27–44.

Trevisan, S., Manoli, A., Ravazzolo, L., Franceschi, C., Quaggiotti, S., 2017. mRNA–sequencing analysis reveals transcriptional changes in root of maize seedlings treated with two increasing concentrations of a new biostimulant. J. Agric. Food Chem. 65, 9956–9969.

Veloccia, A., Fattorini, L., Rovere, F., Sofo, A., Angeli, S., Betti, C., et al., 2016. Ethylene and auxin interaction in the control of adventitious rooting in *Arabidopsis thaliana*. J. Exp. Bot. 67, 6445–6458.

Verstraeten, I., Schotte, S., Geelen, D., 2014. Hypocotyl adventitious root organogenesis differs from lateral root development. Front. Plant. Sci. 5, 1–13.

Vielba, J.M., Diaz–Sala, C., Ferro, E., Rico, S., Lamprecht, M., Abarca, D., et al., 2011. CsSCL1 is differentially regulated upon maturation in chestnut microshoots and is specifically expressed in rooting–competent cells. Tree Physiol. 3, 1152–1160.

Waadt, R., Hsu, P.K., Schroeder, J.I., 2015. Abscisic acid and other plant hormones: methods to visualize distribution and signaling. BioEssays 37, 1338–1349.

Wang, Y., Liu, H., Xin, Q., 2014. Genome–wide analysis and identification of cytokinin oxidase/dehydrogenase (CKX) gene family in foxtail millet (*Setariaitalica*). Crop. J. 2, 244–254.

Wang, Y., Wang, D., Gan, T., Liu, L., Long, W., Wang, Y., et al., 2016. CRL6, a member of the CHD protein family, is required for crown root development in rice. Plant. Physiol. Biochem. 105, 185–194.

Wasternack, C., Song, S., 2017. Jasmonates: biosynthesis, metabolism, signaling by proteins activating repressing transcription. J. Exp. Bot. 68, 1303–1321.

Waters, M.T., Gutjahr, C., Bennett, T., Nelson, D.C., 2017. Strigolactone signaling and evolution. Annu. Rev. Plant. Biol. 68, 291–322.

Wise, K., Gill, H., Selby–Pham, J., 2020. Willow bark extract and the biostimulant complex Root Nectar® increase propagation efficiency in chrysanthemum and lavender cuttings. Sci. Hortic. 263, 109108.

Xiong, F., Zhang, B.K., Liu, H.H., Wei, G., Wu, J.H., Wu, Y.N., et al., 2020. Transcriptional regulation of *PLETHORA1* in the root meristem through an importin and its two antagonistic cargos. Plant. Cell 32, 3812–3824.

Xu, M., Xie, W., Huang, M., 2015. Two WUSCHEL–related HOMEOBOX genes, PeWOX11a and PeWOX11b, are involved in adventitious root formation of poplar. Physiol. Plant. 155, 446–456.

Yakhin, O.I., Lubyanov, A.A., Yakhin, I.A., Brown, P.H., 2017. Biostimulants in plant science: a global perspective. Front. Plant. Sci. 7, 2049.

CHAPTER

16

Arbuscular mycorrhizal fungi inoculation in the modulation of plant yield and bioactive compounds

Weria Weisany

Department of Agriculture and Food Science, Science and Research Branch, Islamic Azad University, Tehran, Iran

16.1 Introduction

Sustainable management of agricultural ecosystems relies heavily on mycorrhizal symbioses. Arbuscules are distinctive structures found within the cortical cells of many plant roots and some mycothalli that have been colonized by AM fungus, hence the name "arbuscular." These structures have been used as diagnostic indicators for AM symbioses alongside intracellular and intercellular storage vesicles. In order to boost the development and productivity of host plants, arbuscular mycorrhizal fungus (AMF) aids in water and nutrient absorption and triggers physiological changes (Gupta and Janardhanan, 1991). Despite its lack of universality, the enhanced biomass allocation to leaves in AM plants is taken as a significant indicator of favorable AM colonization.

The ability of the AM fungus to enhance the plant's phosphorus uptake is well established. AM has also been found to improve nitrogen acquisition (Govindarajulu et al., 2005; Hawkins et al., 2000), but this is more contentious. By establishing a vast belowground hyphal network from colonized roots into the soil environment, AMF significantly contributes to soil fertility and plant nutrition (Giovannetti and Avio, 2002; Smith and Read, 2008). This is accomplished by increasing the uptake and translocation of mineral nutrients, primarily P, N, S, K, Ca, Fe, Cu, and Zn, from the soil to host plants.

Carotenoids, essential oils, antioxidants, and flavors are just some of the types of bioactive compounds commonly added to food products in an effort to improve their flavor, aroma, texture, and overall appeal, as well as to increase their nutritional and health value. The inoculation of AMF has been suggested as a means of optimizing the synthesis of

Biostimulants in Plant Protection and Performance
DOI: https://doi.org/10.1016/B978-0-443-15884-1.00002-6

these biomolecules (Silva et al., 2014; Hazzoumi et al., 2015; Oliveira et al., 2015a,b). Plants benefit from symbiotic relationships with AMF (Selvaraj et al., 2009; Karagiannidis et al., 2011) because they boost the production of secondary compounds that raise the value of the plant's phytomass and, perhaps, its medicinal potential (Oliveira et al., 2013; Zitter-Eglseer et al., 2015).

Secondary metabolites, such as phenolics in roots and aerial parts, and essential oils of host plants, can be influenced by symbiotic AMF (Devi and Reddy, 2002; Copetta et al., 2006; Rojas-Andrade et al., 2003; Toussaint et al., 2007). The condition of mycorrhizal association and its variation in medicinal and aromatic plants is of special interest, and so is the increasing number of pharmaceutical chemicals available for study. The pattern of secondary plant chemicals is just one of many chemical and biological factors that are altered in plants during the establishment of AM symbiosis.

Plants that have been colonized by AMF have been reported to accumulate a wide variety of secondary metabolites, including phytoalexins (Yao et al., 2003), phenolic compounds (Rojas-Andrade et al., 2003), cyclohexanone derivatives and apocarotenoids (Fester et al., 2002; Vierheilig et al., 2000a, b), flavonoids (Larose et al., 2002), triterpenoids (Akiyama and Hayashi, 2002), and glucosinolates (Vierheilig et al., 2000c). Several reports have found that AM fungus is frequently found in close proximity to aromatic and therapeutic plants like dill (*Anethum graveolens*) (Weisany et al., 2015), pelargonium (*Pelargonium peltatum*) (Perner et al., 2007), basil (*Ocimum basilicum*) (Khalediyan et al., 2021), lavender (*Lavandula angustifolia*) (Tsuro et al., 2001), herb known as oregano (*Origanum onites*) (Khaosaad et al., 2006). fennel (*Foeniculum vulgare*) (Kapoor et al., 2004), black cumin (*Nigella sativa* L.) (Darakeh et al., 2022; Weisany et al., 2017), the herb mint (*Mentha requienii*) (Freitas et al., 2004a,b; Cabello et al., 2005), the herb coriander (*Coriandrum sativum*) (Weisany et al., 2017), sage (*Salvia officinalis*) (Nell et al., 2009), and fenugreek (*Trigonella foenum graecum* L.) (Weisany et al., 2017). Several mechanisms have been proposed to explain how mycorrhizal symbiosis causes an increase in the production of compounds involved in secondary plant metabolism. These include an upregulation of gene expression (Gupta et al., 2002; Mandal et al., 2015), the activation of metabolic routes (Lohse et al., 2005), and changes in the activity of key enzymes (Zhang et al., 2013). Only one study has been undertaken in Brazil; Silva et al. (2014) reported a rise in the foliar content of gallic acid in mycorrhizal *Libidibia ferrea*, suggesting a relationship between secondary metabolite synthesis and infected plants in the field.

16.2 Bioactive compounds

Secondary metabolites, often known as bioactive compounds, are present in low concentrations throughout a wide range of organisms. Most of these chemicals are insoluble in water, or hydrophobic. Terpenes and terpenoids, alkaloids, phenolic compounds, and saponins make up the bulk of the bioactive substances found in plants. The chemical structures of the most important bioactive chemicals used in active packaging in the last few years are shown in Fig. 16.1. Several different classes of compounds are represented among the bioactive molecules; polysaccharides and triterpenes are particularly common examples. An abundance of bioactive compounds and, by extension, therapeutic effects

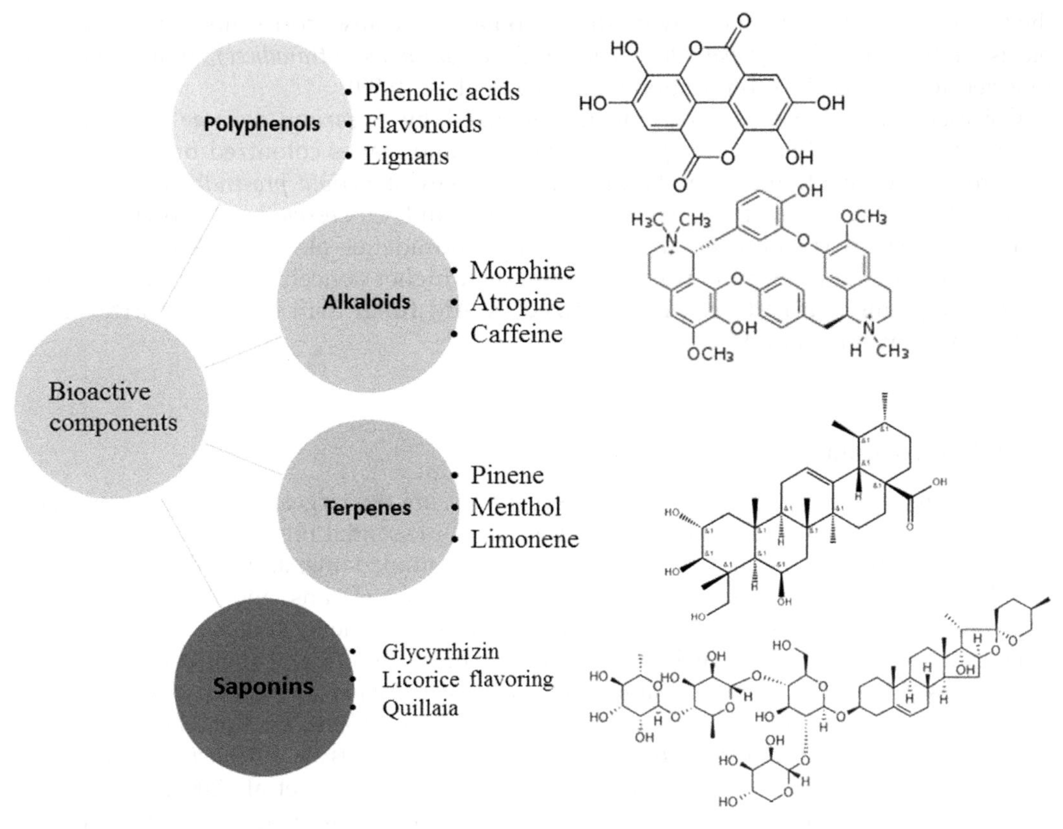

FIGURE 16.1 Chemical structures of most common groups of plant-derived bioactive components.

may be present in a single species. Most notably, Ganoderma lucidum, which contains over 120 unique triterpenes in addition to polysaccharides, proteins, and other bioactive substances, is the best example of this type of plant (Zhou and Gao, 2002).

16.3 Common bioactive compounds

16.3.1 Alkaloids

Plants create alkaloids, which are nitrogen-containing compounds, in response to predators, pathogens, or other stresses (Jan et al., 2021). Analgesia, local anesthetic, heart stimulation, respiratory stimulation and relaxation, vasoconstriction, muscular relaxation, anticancer activity, and blood pressure modulation are only some of the many pharmacological actions they display (Table 16.1) (Hussein and El-Anssary, 2018). Numerous researches have confirmed a beneficial role of AM fungus in the accumulation of alkaloids, beginning with the initial finding by Wei and Wang (1989) that AMF colonization can increase the overall amount of scopolamine and hyoscyamine in the plant. Leaves from the *Castanospermum australe* tree were shown to have a positive correlation with AMF symbiosis (a mixture of *Gigaspora margarita* and *Rhizophagus intraradices*), which has been reported to decrease HIV replication (Abu-Zeyad et al., 1999).

Colchicine in *Gloriosa superba* tubers colonized by *Fusarium mosseae* growing in a greenhouse, trigonelline in *Prosopis laevigata* roots and leaves colonized by *G. rosea* in an in vitro environment, and scopolamine in the leaves of *Eclipta prostrata* colonized by a combination of *C. etunicatum*, *F. claroideoglomus*, and *C. Fossey* in a growth chamber were synthesized (Pandey et al., 2014; Rojas-Andrade et al., 2003; Vo et al., 2019). The alkaloid castanospermine was found in much higher concentrations in the leaves of *C. australe* plants that had been mycorrhized by AM fungi than in nonmycorrhizal controls (Abu-Zeyad et al., 1999).

16.3.2 Terpenoids

The primary elements of essential oils, terpenoids, are the largest and most diverse class of secondary metabolites (Cox-Georgian et al., 2019). Essential oils are frequently used as perfumes and fragrances, as well as antimicrobials, antioxidants, and medicines. They are a volatile lipophilic mixture of secondary metabolites (Deans and Waterman, 1993). Multiple studies have demonstrated that AMF affects the chemical composition of essential oils extracted from medicinal and aromatic plants (Table 16.1). Essential oil production has increased in AMF-colonized plants such as *Trachyspermum ammi*, *S. ofcinalis*, *Inula ensifolia*, *Satureja macrostema*, *Thymus vulgaris*, *Plectranthus amboinicus*, *Corianderum sativum*, *Origanum vulgare* and *Origanum onites*, *Thymus daenensis*, *Artemisia umbelliformis*, *Atractylodes lancea*, and *F. vulgare* (Kapoor et al., 2002a; Kapoor et al., 2004; Zubek et al., 2010; Binet et al., 2011; Karagiannidis et al., 2012; Carreón-Abud et al., 2015; Rydlová et al., 2016; Liang et al., 2018; Sete da Cruz et al., 2019; Merlin et al., 2020; Arpanahi et al., 2020; Machiani et al., 2021). The terpenoid artemisinin in *Artemisia annua* leaves was found to be significantly increased when either *Glomus macrocarpum* or *G. fasciculatum* was available (Kapoor et al., 2007).

There were increased quantities of artemisinin, a sesquiterpene lactone chemical found in *Artemisia annua* and widely known for its effects on cancer (Chaudhary et al., 2008; Krishna et al., 2008; Huang et al., 2011; Domokos et al., 2018), in the leaves of plants colonized by *F. mosseae G. macrocarpum*, or a mix of *G. macrocarpum* and *R. epigaea*, *Diversispora fasciculatus*, and *R. irregularis* is grown in fields or in pots. Greenhouse-grown *Coleus forskohlii* roots injected with *G. bagyarajii* had much higher concentrations of the diterpene forskolin, which is used to treat cardiovascular disease, glaucoma, asthma, and even some malignancies (Kavitha et al., 2010; Sailo and Bagyaraj 2005). Tubers of *Coleus forskohlii* linked to *R. fasciculatus* growing in organic field settings have a higher forskolin concentration, as reported by Singh et al. (2013). Researchers have also looked at how AMF symbiosis affects tissue-cultured medicinal plants. Micropropagated *Origanum vulgare* subsp. hirtum showed increased concentrations of carvacrol after being exposed to the aphid-feeding fungus *Septoglomus viscosum* (Sharif-Rad et al., 2018; Morone-Fortunato and Avato, 2008).

TABLE 16.1 Some studies on the arbuscular mycorrhizal fungi associated with some medicinal and aromatic plants.

Plant species	Plant organ	Secondary compounds	AMF	Change in Secondary compounds	Reference
Leucas aspera	Shoots and leaves	Alkaloids	*Funneliformis mosseae*	Increased	Tejavathi and Jayashree (2011)
Anethum graveolens	Shoots and leaves	Limonene, phellandrene, terpinen-4-ol, cryptone and carvone	*Funneliformis mosseae*	Increased	Weisany et al. (2015)
Trachyspermum ammi	Fruits	Thymol	*Rhizophagus fasciculatus*	Increased	Kapoor et al. (2002a)
Anethum graveolens	Seed	Phellandrene, limonene, dill-ether and carvone	*Funneliformis mosseae*	Increased	Weisany et al. (2016a)
Trigonella foenum graecum	Shoots and leaves	Dill_apiole, Linalool	*Funneliformis mosseae*	Increased	Weisany et al. (2017)
Coriandrum sativum, Nigella sativa	Shoots and leaves	Dill_apiole, Carvone	*Funneliformis mosseae*	Increased	Weisany et al. (2017)
Ocimum basilicum	Shoots and leaves	Linalol, methyl chavicol, trans-geraniol, camphor, and limonene	*Glomus intraradices and G. mosseae*	Increased	Khalediyan et al. (2021)
Satureja hortensis	Shoots and leaves	Carvacrol, thymol, p-cymene, α-terpinene, and γ-terpinene	*Glomus intraradices and G. mosseae*	Increased	Khalediyan et al. (2021)
Nigella sativa	Shoots and leaves	Dillapiole, Carvone	*Glomus mosseae*	Increased	Darakeh et al. (2022)
Gymnema sylvestre	Shoots and leaves	Gymnemic acid	*Funneliformis mosseae* and *Rhizophagus fascuculatus*	Increased	Zimare et al. (2013)
Cynara cardunculus	—	Phenolics	*Rhizophagus intraradices*	Not significant	Colonna et al. (2016)
Euphorbia hirta	Leaves	Phenols, terpenoids, flavonoids, and alkaloids,	*Funneliformis mosseae*	Increased	Tejavathi and Jayashree (2011)
Libidibia ferrea	Leaves	Total lavonoids	*Claroideoglomus etunicatum*	Increased	Silvia et al. (2014)
Libidibia ferrea	Shoots and leaves	Flavonoids	*Acaulospora Longula* and *Claroideoglomus etunicatum*	Increased	Dos Santos et al. (2017)

(Continued)

TABLE 16.1 (Continued)

Plant species	Plant organ	Secondary compounds	AMF	Change in Secondary compounds	Reference
Curcuma longa	Rhizomes	Curcumin	*Glomus, Gigaspora,*	Increased	Dutta and Neog (2016)
Viola tricolor	Shoots and leaves	Cafeic acid	*Rhizophagus irregularis*	Increased	Zubek et al. (2015)
Withania somnifera	Root	Withaferin-A	*Rhizophagus irregularis*	Increased	Johny et al. (2021)
Thymus vulgaris	Shoots and leaves	Thymol, p-cymene, and γ-terpinene	*Funneliformis mosseae*	Increased	Machiani et al. (2021)
Salvia miltiorrhiza	Root	Phenolic acids	*Acaulospora laevis*	Increased	Wu et al. (2021)

16.3.3 Phenolics

A large family of substances known as phenolics (Cosme et al., 2020) includes molecules including tannins, curcuminoids, phenolic acids, stilbenes, coumarins, lignans, flavonoids, and quinones. These substances all share one or more phenolic groups (Hussein and El-Anssary, 2018). When mycorrhizal fungus is present, it has been discovered that the amount of phenol in pharmaceutical plants increases (Table 16.1). For example, formononetin (which has antibacterial, antioxidant, antilipidemic, antidiabetic, anticancer, and neuroprotective effects; Vishnuvathan et al., 2016; Volpin et al., 1994) was more abundant in *Medicago sativa* grown with *R. intraradices*. In greenhouse-grown *Curcuma longa*, AMF species from the genera *Glomus/Rhizophagus, Gigaspora,* and *Acaulospora* sp. boosted curcumin production by about 26%. The antiinflammatory, antioxidant, anticarcinogenic, antiseptic, antiplasmodial, astringent, digestive, and diuretic properties of curcumin are only a few of its many advantageous properties (Dutta and Neog 2016). Total tannin concentration increased by 40% when *L. ferrea* was inoculated with *Acaulospora longula* in the field (Santos et al., 2020). This substance may be utilized to treat tonsillitis, pharyngitis, hemorrhoids, and skin eruptions (Britannica, 2021). Furthermore, it has been demonstrated that p-hydroxybenzoic acid and rutin concentrations are up in *R. irregularis*-colonized *Viola tricolor* and that cichoric acid concentrations are elevated in *R. intraradices*-colonized *Echinacea purpurea* (Zubek et al., 2015; Araim et al., 2009). When *Gigaspora albida* and three different isolates of *R. intraradices*. colonized *L. ferrea*, exposure to *R. intraradices* dramatically increased the overall amount of flavonoids in *L. ferrea* and gallic acid in *Valeriana jatamansi* (Silvia et al., 2014; Jugran et al., 2015).

16.3.4 Saponins

Triterpenoidal saponins, or steroidal saponins, are triterpenoids attached to a mono- or oligosaccharide chain (Hussein and El-Anssary, 2018). The pharmacological effects of some of these compounds, including their anticancer, sedative, expectorant, analgesic, and antiinflammatory characteristics, have been documented (Hussein and El-Anssary, 2018). Saponin synthesis in medicinal plants has been shown to be increased by AMF (Table 16.1). The triterpenoid saponin glycyrrhizic acid, used to treat bronchitis, gastritis, and jaundice, improved in concentration in *Glycyrrhiza glabra* (liquorice) plants cultivated in sand in a greenhouse by 0.38–1.07-fold after 4 months and by 1.34–1.43-fold after 30 months (Pastorino et al., 2018). A similar increase in glycyrrhizic acid concentration was reported by Johny et al. (2021) for *G. glabra* inoculated with *C. etunicatum* and grown in a greenhouse.

16.4 Effect of AMF on growth and secondary metabolite production

As plants develop and mature, they produce a wide variety of secondary metabolites that serve a variety of purposes. Initial studies (Copetta et al., 2006; Gupta et al., 2002; Toussaint et al., 2007) on the effects of AM on culinary and fragrant herbs like basil and fennel focused on the buildup of secondary chemicals in the plant's aerial portions. Since the ground-breaking study of Wei and Wang (1991), which demonstrated the favorable effect of AMF inoculation of *Datura stramonium* and *Schizonepeta tenuifolia* on the formation of active chemical compounds, other studies have been conducted. Reports regarding these AM-treated plants show that, while the essential oil's quality is generally improved, the quantity of oil produced is also impacted, and, in some cases, the makeup of the oil's individual constituents is very slightly altered (Copetta et al., 2006; Toussaint et al., 2007; Weisany et al., 2015, 2016a, 2016b). Both the *Ocimum basilicum* and the *Satureja hortensis* arbuscular mycorrhizae and rhizobacteria inoculation considerably boost oil content and yield compared to nonmycorrhizal plants, as shown by Khalediyan et al. (2021). An increase of 89% in the essential oil and menthol concentrations of *M. arvensis* plants was also noted by Freitas et al. (2004a,b). Biomass increase and bioactive chemical production were the primary areas of research for the vast majority of medicinal plants exposed to AMF (Table 16.1).

Mycorrhizal infected *O. basilicum* had a greater increase in glandular hair abundance and essential oil yield, as reported by Copetta et al. (2006). Evidence from studies of *Coriandrum*, *Anethum*, and *F. vulgare* (Kapoor et al., 2002a, b; 2004) show that root colonization by AM fungi improves the quality of essential oil by modifying its components. However, essential oil levels in plants treated with P are unaffected, suggesting that the increase in essential oil concentration in mycorrhizal oregano plants is not due to improved P nutrition but rather directly depends on association with the AM fungus. Two oregano (*Origanum* sp.) genotypes associated with *G. mosseae* have significant increases in essential oil concentration (Khaosaad et al., 2006). For several medicinal plants, including *S. miltiorrhiza*, *Dioscorea spp.*, *Gymnema sylvestre*, *Chlorophytum borivilianum*, *S. macrostema*, *G. uralensis*, *L. ferrea*, and *O. basilicum*, the concentration of secondary metabolites has been shown to correlate directly with the biomass of AMF-colonized plants (Zolfaghari et al., 2013; Silvia et al., 2014; Carreón-Abud et al., 2015; Chen et al., 2017; Yang et al., 2017; Khalediyan, et al., 2021). On the other hand, the yield of *Cynara cardunculus* colonized by *R. intraradices* and *F. mosseae* has increased significantly, but the phenolic content has decreased. (Colonna et al., 2016). Studies with *Hypericum perforatum* found no increase in shoot biomass after inoculation with *R. intraradices* or a mixture of *F. mosseae, F. constrictum, F. geosporum,* and *R. intraradices*, and studies with *V. ofcinalis* found a negative effect on rhizome growth after inoculation with *R. intraradices* or a mixture of six AMF species (*F. mosseae, R. intraradices, G. cladoideum, R. microaggregatum, F. caledonium,* and *C. etunicatum*) (Zubek et al., 2012; Nell et al., 2010).

16.5 How the symbiotic relationship between AMFs and medicinal plants facilitates secondary metabolism

The reasons for the AM's impacts on secondary metabolites, as reviewed by Toussaint et al. (2007), are still unknown, despite theories linking this effect to the plants' improved phosphorus status or altered hormonal balance. According to Zeng et al. (2013) review, it is assumed that the increased levels of numerous types of secondary metabolites (such as flavonoids and phenolics) in AMF-colonized plants are the result of various defense response mechanisms. Increased levels of terpenoids in the carotenoid pathway, flavonoids, phenolic compounds, and certain alkaloids (such as hyoscyamine and scopolamine) in the phenylpropanoid pathway are frequently found in plants that have been invaded by AMF (Kaur and Suseela, 2020).

These pathways serve a variety of purposes throughout the plant-AMF symbiosis, including signaling, stress tolerance, nutrient uptake, and resistance to biotic and abiotic stresses. The exact mechanism by which AMF causes shifts in phytochemical concentrations in plant tissues remains obscure (Toussaint et al., 2007). The processes through which AMF regulates plant production of terpenoids, phenolic compounds, and alkaloids have been the subject of a great deal of research. Both the methyleritrophosphate (MEP) and the mevalonic acid (MVA) pathways utilize isoprene units to produce terpenoids (Zhi et al., 2007). The malonic acid pathway and the shikimic acid pathway both lead to the formation of phenylpropanoids, which are then used in the synthesis of phenolic compounds (such as phenols, flavonoids, proanthocyanidins, and tannins) (Oksana et al., 2012). The shikimic acid pathway synthesizes most alkaloids from a variety of biological substrates (most amino acids) like tyrosine and tryptophane (Facchini, 2001). Secondary metabolite production in AMF-colonized plants is higher because of several shared nutritional and nonnutritional factors (Kapoor et al., 2017; Sharma et al., 2017; Dos Santos et al., 2017).

Initially, the boost in food sources was attributed to the improved nutrient uptake by AMF-colonized plants (Oliveira et al., 2015a,b; Ezzati et al., 2021; Weisany et al., 2021). Some examples of well-established roles for phosphorus include the MVA (acetyl-CoA, ATP, and NADPH) and MEP (glyceraldehyde phosphate and pyruvate) routes in the biosynthesis of terpenoids (Kapoor et al., 2017). Higher concentrations of high-energy pyrophosphate molecules like isopentenyl pyrophosphate (IPP) and dimethylallyl pyrophosphate (DMAPP) are responsible for phosphorus's beneficial effects on terpenoid biosynthesis (Kapoor et al., 2002a, b, 2004; Zubek et al., 2010).

Plants treated with P showed no change in essential oil concentration, but two *Origanum* sp. genotypes infected with *F. mosseae* showed a significant rise in essential oil concentration (Khaosaad et al., 2006). This indicates that the interaction with the fungus may be directly responsible for the elevated content of essential oils in AMF-colonized *Origanum* sp. Another study by Zubek et al. (2012) found that when AMFs colonized *Hypericum perforatum*, the plant's hypericin and pseudohypericin concentrations increased, likely due to better plant P and/or N intake and the increased availability of the fungi. The greater production of these metabolites in plants is also explained by the increased development of AMF-colonized plants as a result of the improved uptake of nutrients and water.

It has been well established that the AMF symbiosis causes *O. basilicum* to have larger shoots in terms of biomass, length, and nodes (Rasouli-Sadaghiani et al., 2010; Copetta et al., 2006; Khaosaad et al., 2008; Copetta et al., 2006). With increased leaf biomass, the total amount of photosynthates (such as ATP, carbon substrate, glyceraldehyde-3-phosphate, pyruvate, phosphoenolpyruvate, or erythrose-4-phosphate) required for the production of terpenoids, phenolics, and alkaloids rises (Cao et al., 2008; Dave et al., 2011; Hofmeyer et al., 2010; Zubek et al., 2010). Changes in the levels of phytohormones in AMF-colonized plants may represent their increased production as nonnutritional factors (Zubek et al., 2012; Mandal et al., 2013, 2015). In fact, studies have shown that the AMF symbiosis modifies the levels of phytohormones in plants, including jasmonic acid (JA), gibberellic acid (GA3), and cytokinins (Hause et al., 2002; Shaul-Keinan et al., 2002). Phytohormones may play a part in plants' secondary metabolism, according to data (An et al., 2011; Maes et al., 2008; Maes and Goossens, 2010). Sesquiterpenoid biosynthesis gene expression in *Artemisia annua* is known to be coordinated by JA (Maes et al., 2011). There is a favorable correlation between the increased content of terpenoids in plant leaves and the phytohormonal modifications of GA3, BAP (6-benzylaminopurine), and JA (Maes et al., 2011). Plants produce and store terpenoids in glandular trichomes, which are epidermal secretory structures (Covello et al., 2007). Several plants (including Mentha x piperita, *Phaseolus lunatus*, and *Lavendula angustifolia*) have been found to have an increased content of terpenoids (essential oils) and an increased density of glandular trichomes (Bartram et al., 2006; Behnam et al., 2006).

It is therefore generally accepted that the effect of mycorrhization on increased trichome density also accounts for a higher concentration of terpenoids (Copetta et al., 2006; Kapoor et al., 2007; Morone-Fortunato and Avato 2008). Changes in secondary metabolite levels in AMF-plants may also be brought on by signaling between the host plant and the fungus (Rojas-Andrade et al., 2003; Xie et al., 2018). With the presence of *F. mosseae* on *Trifolium repens*, it has been demonstrated the production of signaling molecules like nitric oxide, salicylic acid (SA), and hydrogen peroxide has increased. These molecules all play a role in the activation of crucial enzymes in phenolic biosynthesis, including l-phenylalanine ammonia-lyase (PAL) and chalcone synthase (CHS).

AMF may also promote the expression of genes encoding enzymes involved in the production of these substances in mycorrhizal plants (Battini et al., 2016; Mandal et al., 2015, Xie et al., 2018). It is likely that the altered terpenoid profile in AMF-colonized tomato plants can be attributed to the induction of the terpene synthase (TPS) family genes TPS31, TPS32, and TPS33 (Zouari et al., 2014). The amount of artemisinin in Artemisia annua leaves that had been inoculated with R. intraradices was reported to have increased (Mandal et al., 2015).

Increased JA levels were shown to be associated with increased expression of genes involved in critical biosynthetic pathways. One such gene is the allene oxidase synthase gene, which encodes a key enzyme in JA production. Furthermore, AMF may improve the biosynthesis of these compounds by inducing metabolic biosynthetic pathways (Zimare et al., 2013; Dos Santos et al., 2017) and/or by inducing important synthase enzymes (Mandal et al., 2013; Shrivastava et al., 2015; Sharma et al., 2017; Dos Santos et al., 2017). The transcript levels of 1-deoxy-D-xylulose 5-phosphate synthase (DXS) and 1-deoxy-D-xylulose 5-phosphate reductoisomerase (DXR) in wheat roots have been shown to increase after mycorrhizal colonization (R. intraradices) (Walter et al., 2000).

Two enzymes are necessary for the synthesis of isoprenoids: DXS, which catalyzes the MEP route's initial step, and DXR, which is found immediately after DXS in the MEP pathway (Walter et al., 2000). In a different study, Walter et al. (2002) discovered that the AMF (a mix of *F. mosseae* and *R. intraradices*) colonization of *Medicago truncatula* roots significantly raised DXS2 mRNA levels, which were linked to the accumulation of carotenoids and apocarotenoids. Additionally, plant defense responses to AMF invasion may cause shifts in the production of these secondary metabolites (Zubek et al., 2012, 2015; Mechri et al., 2015; Torres et al., 2015).

Many reports have shown that adding exogenous jasmonate to cell cultures of Boragenaceous plants (such as *Lithospermum erythrorhizon*, *Alkanna tinctoria*, and *Arnebia euchroma*) results in higher levels of alkanin/shikonin and its derivatives (A/S) (Gaisser and Heide 1996; Urbanek et al., 1996). Root endogenous jasmonate levels have been shown to increase after AMF colonization in several different plant species, including *Hordeum vulgare*, *Cucumis sativus*, *Medicago truncatula*, and *Glycine max* (Hause et al., 2002; Vierheilig and Piche, 2002; Stumpe et al., 2005; Meixner et al., 2005).

Hormonal regulators known as jasmonates (which include jasmonic acid and its analogs) play a role in plant responses to abiotic and biotic stimuli and in plant growth and development (Wasternack 2007). It has been established that the plant produces more endogenous jasmonate in response to injury or infection. There has been no definitive research into how AMF might influence A/S production in these Boragenaceous medicinal plants. These results, however, point to AMF as a plausible component boosting A/S synthesis in mycorrhizal Boraginaceae plants by regulating jasmonate. Essential oil components, such as phenolic compounds and terpenoids, are thought to be a defense mechanism against fungal colonization. It has been hypothesized that mycorrhizal plants may enhance their synthesis of fungicide-containing compounds in response to the presence of AM fungus (Copetta et al., 2006).

16.6 Conclusion

The accumulation of secondary compounds, antioxidant photosynthesis, and mineral nutrition in the aerial sections of mycorrhizal plants has received only a small amount of study. In order to provide the framework for more study in this area, we have sought to provide a thorough overview of the published research on AMF and secondary metabolites of medicinal plants in this study. The presence of AM fungi has been shown to improve plant growth, photosynthetic activity, and phosphorus content, as well as alter the concentration of plant metabolites. Photosynthetic rates in mycorrhizal plants may be higher than those in nonmycorrhizal plants. As a result, N, P, and K absorption were also dramatically boosted by AM inoculation. In conclusion, more work is needed to learn about the role of AM symbiosis and the utilization of mycorrhizal technology in the organic production of herbal products.

References

Abu-Zeyad, R., Khan, A., Khoo, C., 1999. Occurrence of arbuscular mycorrhiza in *Castanospermum austral* A. Cunn. & *C. Fraser* and effects on growth and production of castanospermine. Mycorrhiza 9, 111–117.

Akiyama, K., Hayashi, H., 2002. Arbuscular mycorrhizal fungus promoted accumulation of two new triterpenoids in cucumber roots. Biosci. Biotechnol. Biochem. 66, 762–769.

An, L., Zhou, Z., Yan, A., Gan, Y., 2011. Progress on trichome development regulated by phytohormone signaling. Plant. Signal. Behav. 6 (12), 1959–1962.

Araim, G., Saleem, A., Arnason, J.T., Charest, C., 2009. Root colonization by an arbuscular mycorrhizal (AM) fungus increases growth and secondary metabolism of purple conefower, *Echinacea purpurea* (L.) Moench. J. Agric. Food Chem. 57 (6), 2255–2258.

Arpanahi, A.A., Feizian, M., Mehdipourian, G., Khojasteh, D.N., 2020. Arbuscular mycorrhizal fungi inoculation improve essential oil and physiological parameters and nutritional values of *Thymus daenensis* Celak and *Thymus vulgaris* L. under normal and drought stress conditions. Eur. J. Soil. Biol. 100, 103217.

Bartram, S., Jux, A., Gleixner, G., Boland, W., 2006. Dynamic pathway allocation in early terpenoid biosynthesis of stress-induced lima bean leaves. Phytochemistry 67 (15), 1661–1672.

Battini, F., Bernardi, R., Turrini, A., Agnolucci, M., Giovannetti, M., 2016. *Rhizophagus intraradices* or its associated bacteria affect gene expression of key enzymes involved in the rosmarinic acid biosynthetic pathway of basil. Mycorrhiza 26, 699–707.

Behnam, S., Farzaneh, M., Ahmadzadeh, M., Tehrani, A.S., 2006. Composition and antifungal activity of essential oils of *Mentha piperita* and *Lavendula angustifolia* on post-harvest phytopathogens. Commun. Agric. Appl. Biol. Sci. 71 (3 Pt B), 1321–1326.

Binet, M.N., van Tuinen, D., Deprêtre, N., Koszela, N., Chambon, C., Gianinazzi, S., 2011. Arbuscular mycorrhizal fungi associated with *Artemisia umbelliformis* Lam, an endangered aromatic species in Southern French Alps, influence plant P and essential oil contents. Mycorrhiza 21, 523–535.

Britannica, 2021. The editors of encyclopaedia. Tannin. Encyclopedia Britannica. Available from: https://www.britannica.com/science/tannin, Accessed 9 May 2022.

Cabello, M., Irrazabal, G., Bucsinszky, A.M., Saparrat, M., Schalamuk, S., 2005. Effect of an arbuscular mycorrhizal fungus, *Glomus mosseae,* and a rock-phosphate-solubilizing fungus, *Penicillium thomii,*on Mentha piperita growth in a soilless medium. J. Basic. Microbiol. 45, 182–189.

Cao, B., Dang, Q.L., Yu, X., Zhang, S., 2008. Effects of (CO2) and nitrogen on morphological and biomass traits of white birch (*Betula papyrifera*) seedlings. For. Ecol. Manag. 254 (2), 217–224.

Carreón-Abud, Y., Torres-Martínez, R., Farfán-Soto, B., HernándezGarcía, A., Ríos-Chávez, P., BelloGonzález, M. Á., et al., 2015. Arbuscular mycorrhizal symbiosis increases the content of volatile terpenes and plant performance in *Satureja macrostema* (Benth.) Briq. Bol. Latinoam. Caribe Plant. Med. Aromat. 14 (4), 273–279.

Chaudhary, V.L., Kapoor, R., Bhatnagar, A.K., 2008. Effectiveness of two arbuscular mycorrhizal fungi on concentrations of essential oil and artemisinin in three accessions of *Artemisia annua*. Appl. Soil. Ecol. 40, 174–181.

Chen, M.L., Yang, G., Sheng, Y., Li, P.Y., Qiu, H.Y., Zhou, X.T., et al., 2017. *Glomus mosseae* inoculation improves the root system architecture, photosynthetic efficiency and flavonoids accumulation of liquor ice under nutrient stress. Front. Plant. Sci. 8, 931.

Colonna, E., Rouphael, Y., De Pascale, S., Barbieri, G., 2016. Effects of mycorrhiza and plant growth promoting rhizobacteria on yield and quality of artichoke. Acta Hortic. 1147, 43–50.

Copetta, A., Lingua, G., Berta, G., 2006. Effects of three AM fungi on growth, distribu-tion of glandular hairs, and essential oil production in *Ocimum basilicum* L. Var. Genovese. Mycorrhiza 16, 485–494.

Cosme, P., Rodríguez, A.B., Espino, J., Garrido, M., 2020. Plant phenolics: bioavailability as a key determinant of their potential health promoting applications. Antioxidants 9, 1263.

Covello, P.S., Teoh, K.H., Polichuk, D.R., Reed, D.W., Nowak, G., 2007. Functional genomics and the biosynthesis of artemisinin. Phytochemistry 68 (14), 1864–1871.

Cox-Georgian, D., Ramadoss, N., Dona, C., Basu, C., 2019. Therapeutic and medicinal uses of terpenes. J. Med. Plant. Res. 333–359.

Darakeh, S.A.S.S., Weisany, W., Abdul-Razzak Tahir, N., Schenk, P.M., 2022. Physiological and biochemical responses of black cumin to vermicompost and plant biostimulants: arbuscular mycorrhizal and plant growth-promoting rhizobacteria. Ind. Crop. Prod. 188, 115557.

Dave, A., Hernández, M.L., He, Z., Andriotis, V.M., Vaistij, F.E., Larson, T.R., Graham, I.A., 2011. 12-Oxo-phytodienoic acid accumulation during seed development represses seed germination in Arabidopsis. Plant Cell 23, 583–599.

Deans, S.G., Waterman, P.G., 1993. Biological activity of volatile oils. In: Waterman, P.G. (Ed.), Volatile Oil Crops: Their Biology, Biochemistry and Pro-duction (Hay RKM). Longman Scientific and Technical, Essex, England, pp. 97–111.

Devi, M.C., Reddy, M.N., 2002. Phenolic acid metabolism of groundnut (*Arachis hypogaea* L.) plants inoculated with VAM fungus and Rhizobium. Plant. Growth Regul. 37, 151–156.

Domokos, E., Jakab-Farkas, L., Darkó, B., Bíró-Janka, B., Mara, G., Albert, C., et al., 2018. Increase in *Artemisia annua* plant biomass artemisinin content and guaiacol peroxidase activity using the arbuscular mycorrhizal fungus *Rhizophagus irregularis*.. Front. Plant. Sci. 9478.

Dos Santos, E.L., Alves da Silva, F., Barbosa da Silva, F.S., 2017. Arbuscular mycorrhizal fungi increase the phenolic compounds concentration in the bark of the stem of *Libidibia Ferrea* in field conditions. Open. Microbiol. J. 11, 283–291.

Dutta, S.C., Neog, B., 2016. Accumulation of secondary metabolites in response to antioxidant activity of turmeric rhizomes co-inoculated with native arbuscular mycorrhizal fungi and plant growth promoting rhizobacteria. Sci. Hortic. 204, 179–184.

Ezzati, Z., Weisany, W., Abdul- razzak Tahir, N., 2021. Arbuscular mycorrhizal fungi species improve the fatty acids profile and nutrients status of soybean cultivars grown under drought stress. J. Appl. Microbiol. 132, 2177–2188.

Facchini, P.J., 2001. Alkaloid biosynthesis in plants: biochemistry, cell biology, molecular regulation, and metabolic engineering applications. Annu. Rev. Plant.. Physiol. Plant Mol. Biol. 52, 29–66.

Fester, T., Hause, B., Schmidt, D., Halfmann, K., Schmidt, J., Wray, V., et al., 2002. Occurrence and localization of apocarotenoids in arbuscular mycorrhizal plant roots. Plant. Cell Physiol. 43, 256–265.

Freitas, M.S., Martins, M.A., Carvalho, A.J., Carneiro, R.F., 2004a. Crescimento e produção de fenóis totais em carqueja (*Baccharis trimera* (Less.) DC.) na presença e na ausência de adubação mineral. Rev. Bras. Pl. Med. 6, 30–34.

Freitas, M.S.M., Martins, M.A., Curcino Vieira, I.J., 2004b. Yield and quality of essential oils of *Mentha arvensis* in response to inoculation with arbuscular mycorrhizal fungi. Pesqui. Agropecu. Bras. 39, 887–894.

Gaisser, S., Heide, L., 1996. Inhibition and regulation of shikonin biosynthesis in suspension cultures of Lithospermum. Phytochemistry 41 (4), 1065–1072.

Giovannetti, M., Avio, L., 2002. Biotechnology of arbuscular mycorrhizas. In: Khachatourians, G.G., Arora Dilip, K. (Eds.), Applied Mycology and Biotechnology, Volume 2, Agriculture and Food Production. Elsevier, Amsterdam, pp. 275–310.

Govindarajulu, M., Pfeffer, P.E., Jin, H.R., Abubaker, J., Douds, D.D., Allen, J.W., et al., 2005. Nitrogen transfer in the arbuscular mycorrhizal symbiosis. Nature 435, 819–823.

Gupta, M.L., Janardhanan, K.K., 1991. Mycorrhizal association of *Glomus aggregatum* with palmarosa enhances growth and biomass. Plant. Soil. 131, 261–263.

Gupta, M.L., Prasad, A., Ram, M., Kumar, S., 2002. Effect of the vesicular–arbuscular mycorrhizal (VAM) fungus *Glomus fasciculatum* on the essential oil yield related characters and nutrient acquisition in the crops of different cultivars of menthol mint (*Mentha arvensis*) under field conditions. Bioresour. Technol. 81, 77–79.

Hause, B., Maier, W., Miersch, O., Kramell, R., Strack, D., 2002. Induction of jasmonate biosynthesis in arbuscular mycorrhizal barley roots. Plant. Physiol. 130, 1213–1220.

Hawkins, H.J., Johansen, A., George, E., 2000. Uptake and transport of organic and inorganic nitrogen by arbuscular mycorrhizal fungi. Plant. Soil. 226, 275–285.

Hazzoumi, Z., Moustakime, Y., Elharchli, E.H., Joutei, K.A., 2015. Effect of arbuscular mycorrhizal fungi (AMF) and water stress on growth, phenolic compounds, glandular hairs, and yield of essential oil in basil (*Ocimum gratissimum* L.). Chem. Biol. Technol. Agric. 2, 2–11.

Hofmeyer, P.V., Seymour, R.S., Kenefc, L.S., 2010. Production ecology of *Thuja occidentalis*. Can. J. Res. 40 (6), 1155–1164.

Huang, J.H., Tan, J.F., Jie, H.K., Zeng, R.S., 2011. Effects of inoculating arbuscular mycorrhizal fungi on *Artemisia annua* growth and its officinal components. J. Appl. Ecol. 22 (6), 1443–1449.

Hussein, R.A., El-Anssary, A.A., 2018. Plants secondary metabolites: the key drivers of the pharmacological actions of medicinal plants. In: Builders, P. (Ed.), Herb. Med. Intech Open, Chapter 2.

Jan, R., Asaf, S., Numan, M., Lubna, K.K.M., 2021. Plant secondary metabolite biosynthesis and transcriptional regulation in response to biotic and abiotic stress conditions. Agronomy 11, 968.

Johny, L., Cahill, D.M., Adholeya, A., 2021. AMF enhance secondary metabolite production in ashwagandha, licorice, and marigold in a fungi-host specific manner. Rhizosphere 17.

Jugran, A.K., Bahukhandi, A., Dhyani, P., Bhatt, I.D., Rawal, R.S., Nandi, S.K., et al., 2015. The effect of inoculation with mycorrhiza: AM on growth, phenolics, tannins, phenolic composition and antioxidant activity in *Valeriana jatamansi* Jones. J. Soil. Sci. Plant. Nutr. 15 (4), 1036–1049.

Kapoor, R., Anand, G., Gupta, P., Mandal, S., 2017. Insight into the mechanisms of enhanced production of valuable terpenoids by arbuscular mycorrhiza. Phytochem. Rev. 16.

Kapoor, R., Chaudhary, V., Bhatnagar, A.K., 2007. Effects of arbuscular mycorrhiza and phosphorus application on artemisinin concentration in *Artemisia annua* L. Mycorrhiza 17 (7), 581–587.

Kapoor, R., Giri, B., Mukerji, K.G., 2004. Improved growth and essential oil yield and quality in *Foeniculum vulgare* mill on mycorrhizal inoculation supplemented with P-fertilizer. Bioresour. Technol. 93 (3), 307–311.

Kapoor, R., Giri, B., Mukerij, K.G., 2002a. *Glomus macrocarpus*: a potential bioinoculant to improve essential oil quality and concentration in Dill (*Anethum graveolens* L.) and Carum (*Trachyspermum ammi* (Linn.) Sprague). World J. Microbiol. Biotechnol. 18, 459–463.

Kapoor, R., Giri, B., Mukerji, K.G., 2002b. Mycorrhization of coriander (*Coriandrum sativum* L.) to enhance the concentration and quality of essential oil. J. Sci. Food Agric. 82, 339–342.

Karagiannidis, N., Thomidis, T., Panou-Filotheou, E., Karagiannidou, C., 2012. Response of three mint and two oregano species to *Glomus etunicatum* inoculation. AJCS 6 (1), 164–169.

Karagiannidis, N., Thomidis, T., Lazari, D., Panou-filotheou, E., Karagiannidou, C., 2011. Effect of three Greek arbuscular mycorrhizal fungi in improving the growth, nutrient concentration, and production of essential oils of oregano and mint plants. Sci. Hortic. (Amsterdam). 129, 329–334.

Kaur, S., Suseela, V., 2020. Unraveling arbuscular mycorrhiza-induced changes in plant primary and secondary metabolome. Metabolites 10 (8), 335.

Kavitha, C., Rajamani, K., Vadivel, E., 2010. Coleus forskohlii: a comprehensive review on morphology, phytochemistry and pharmacological aspects. J. Med. Plant. Res. 4 (4), 278–285.

Khalediyan, N., Weisany, W., Schenk, P.M., 2021. Arbuscular mycorrhizae and rhizobacteria improve growth, nutritional status and essential oil production in *Ocimum basilicum* and *Satureja hortensis*. Ind. Crop. Prod. 160, 113163.

Khaosaad, T., Krenn, L., Medjakovic, S., Ranner, A., Lossl, A., Nell, M., et al., 2008. Effect of mycorrhization on the isofavone content and the phytoestrogen activity of red clover. J. Plant. Physiol. 165, 1161–1167.

Khaosaad, T., Vierheilig, H., Nell, M., Zitterl-Eglseer, K., Novak, J., 2006. Arbuscular mycorrhiza alters the concentration of essential oils in oregano (*Origanum* sp., Lamiaceae). Mycorrhiza 16, 443–446.

Krishna, S., Bustamante, L., Haynes, R.K., Staines, H.M., 2008. Artemisinins: their growing importance in medicine. Trends Pharmacol. Sci. 29 (10), 520–527.

Larose, G., Chenevert, R., Moutoglis, P., Gagne, S., Piché, Y., Vierheilig, H., 2002. Flavonoid levels in roots of *Medicago sativa* are modulated by the developmental stage of the symbiosis and the root colonizing arbuscular mycorrhizal fungus. J. Plant. Physiol. 159, 1329–1339.

Liang, X.F., Tang, M.J., Lu, L.X., Zhao, X.Y., Dai, C.C., 2018. Effects of three arbuscular mycorrhizal fungi (AMF) species on the growth, physiology, and major components of essential oil of *Atractylodes Lancea*. Chin J. Plant. Ecol. 37 (6), 1871–1879.

Lohse, S., Schliemann, W., Ammer, C., Kopka, J., Strack, D., Fester, T., 2005. Organization and metabolism of plastids and mitochondria in arbuscular mycorrhizal roots of *Medicago truncatula*.. Plant. Physiol. 139 (1), 329–340.

Machiani, M.A., Javanmard, A., Morshedloo, M.R., Aghaee, A., Maggi, F., 2021. *Funneliformis mosseae* inoculation under water deficit stress improves the yield and phytochemical characteristics of thyme in intercropping with soybean. Sci. Rep. 11, 15279.

Maes, L., Goossens, A., 2010. Hormone-mediated promotion of trichome initiation in plants is conserved but utilizes species and trichomespecifc regulatory mechanisms. Plant. Signal. Behav. 5 (2), 205–207.

Maes, M., Kubera, M., Leunis, J.C., 2008. The gut-brain barrier in major depression: Intestinal mucosal dysfunction with an increased translocation of LPS from gram negative enterobacteria (leaky gut) plays a role in the inflammatory pathophysiology of depression. Neuro Endocrinol. Lett. 29, 117–124.

Maes, L., Van Nieuwerburgh, F.C., Zhang, Y., Reed, D.W., Pollier, J., Vande Casteele, S.R., et al., 2011. Dissection of the phytohormonal regulation of trichome formation and biosynthesis of the antimalarial compound artemisinin in *Artemisia annua* plants. N. Phytol. 189 (1), 176–189.

Mandal, S., Evelin, H., Giri, B., Singh, V.P., Kapoor, R., 2013. Arbuscular mycorrhiza enhances the production of stevioside and rebaudioside-A in *Stevia rebaudiana* via nutritional and non-nutritional mechanisms. Appl. Soil. Ecol. 72, 187–194.

Mandal, S., Evelin, H., Giri, B., Singh, V.P., Kapoor, R., 2015. Enhanced production of steviol glycosides in mycorrhizal plants: a concerted effect of arbuscular mycorrhizal symbiosis on transcription of biosynthetic genes. Plant. Physiol. Biochem. 89, 100–106.

Mechri, B., Tekaya, M., Cheheb, H., Attia, F., Hammami, M., 2015. Accumulation of flavonoids and phenolic compounds in olive tree roots in response to mycorrhizal colonization: a possible mechanism for regulation of defense molecules. J. Plant. Physiol. 185, 40–43.

Meixner, C., Ludwig-Müller, J., Miersch, O., Gresshof, P., Staehelin, C., Vierheilig, H., 2005. Lack of mycorrhizal autoregulation and phytohormonal changes in the supernodulating soybean mutant nts1007. Planta 222, 709–715.

Merlin, E., Melato, E., Emerson, L.B.L., Ezilda, J., Arquimedes, G.J., Rayane, M.S.C., et al., 2020. Inoculation of arbuscular mycorrhizal fungi and phosphorus addition increase coarse mint (*Plectranthus amboinicus* Lour.) plant growth and essential oil content. Rhizosphere 15, 100217.

Morone-Fortunato, I., Avato, P., 2008. Plant development and synthesis of essential oils in micropropagated and mycorrhiza inoculated plants of *Origanum vulgare* L. ssp. hirtum (Link) Ietswaart. Plant. Cell Tiss. Organ. Cult. 93, 139–149.

Nell, M., Wawrosch, C., Steinkellner, S., Vierheilig, H., Kopp, B., Lössl, A., et al., 2010. Root colonization by symbiotic arbuscular mycorrhizal fungi increases sesquiterpenic acid concentrations *in Valeriana ofcinalis* L. Planta Med. 76 (4), 393–398.

Nell, M., Vötsch, M., Vierheilig, H., Steinkellner, S., Zitterl-Eglseer, K., Franz, C., et al., 2009. Effect of phosphorus uptake on growth and secondary metabolites of garden sage (*Salvia officinalis* L.). J. Sci. Food Agric. 89, 1090–1096.

Oksana, S., Marian, B., Mahendra, R., Bo, S.H., 2012. Plant phenolic compounds for food, pharmaceutical and cosmetics production. J. Med. Plant. Res. 6, 2526–2539.

Oliveira, M.S., Campos, M.A., Albuquerque, U.P., Silva, F.S., 2013. Arbuscular mycorrhizal fungi (AMF) affects biomolecules content in *Myracrodruon urundeuva* seedlings. Ind. Crop. Prod. 50, 244–247.

Oliveira, M.S., Campos, M.A.S., Silva, F.S.B., 2015a. Arbuscular mycorrhizal fungi and vermicompost to maximize the production of foliar biomolecules in *Passifora alata* Curtis seedlings. J. Sci. Food Agric. 95, 522–528.

Oliveira, P.T., Alves, G.D., Silva, F.A., Silva, F.S., 2015b. Foliar bioactive compounds in *Amburana cearensis* (Allemao) A. C. Smith seedlings: increase of biosynthesis using mycorrhizal technology. J. Med. Plants Res. 9, 712–718.

Pandey, D.K., Malik, T., Dey, A., Singh, J., Banik, R.M., 2014. Improved growth and colchicine concentration in *Gloriosa superba* on mycorrhizal inoculation supplemented with phosphorus-fertilizer. Afr. J. Tradit. Complement. Altern. Med. 11 (2), 439–446.

Pastorino, G., Cornara, L., Soares, S., Rodrigues, F., Oliveira, M.B.P.P., 2018. Liquorice (*Glycyrrhiza glabra*): a phytochemical and pharmacological review. Phytother. Res. 32 (12), 2323–2339.

Perner, H., Schwarz, P., Bruns, C., Maeder, P., George, E., 2007. Effect of arbuscular mycorrhizal colonization and two levels of compost supply on nutrient uptake and flowering of pelargonium plants. Mycorrhiza. 17, 469–474.

Rasouli-Sadaghiani, M., Hassani, A., Barin, M., Danesh, Y.R., Sefdkon, F., 2010. Effects of arbuscular mycorrhizal (AM) fungi on growth, essential oil production and nutrients uptake in basil. J. Med. Plants Res. 4 (21), 2222–2228.

Rojas-Andrade, R., Cerda-García-Rojas, C., Frías Hernández, J., Dendooven, L., Olalde-Portugal, V., RamosValdivia, A., 2003. Changes in the concentration of trigonelline in a semi-arid leguminous plant

(*Prosopis laevigata*) induced by an arbuscular mycorrhizal fungus during the presymbiotic phase. Mycorrhiza 13 (1), 49–52.

Rydlová, J., Jelínková, M., Dušek, K., Dušková, E., Vosátka, M., Püschel, D., 2016. Arbuscular mycorrhiza differentially affects synthesis of essential oils in coriander and dill. Mycorrhiza 26 (2), 123–131.

Sailo, G.S., Bagyaraj, D.J., 2005. Influence of different AM-fungi on the growth, nutrition and forskolin content of *Coleus forskohlii*. Mycol. Res. 109 (7), 795–798.

Santos, E.L., Ferreira, M.R.A., Soares, L.A.L., Sampaio, E.V.S.B., Silva, W.A.V., Silva, F.A., 2020. *Acaulospora longula* increases the content of phenolic compounds and antioxidant activity in fruits of *Libidibia ferrea*. Open. Microbiol. J. 14 (1), 132–139.

Selvaraj, T., Nisha, M.C., Rajeshkumar, S., 2009. Effect of indigenous arbuscular mycorrhizal fungi on some growth parameters and phytochemical constituents of *Pogostemon patchouli* Pellet. J. Sci. Technol. 3, 222–234.

Sete da Cruz, R.M., Dragunski, D.C., Gonçalves, A.C., Alberton, O., Hulse, S., Sete da Cruz, G.L., 2019. Inoculation with arbuscular mycorrhizal fungi alters content and composition of essential oil of sage (*Salvia ofcinalis*) under different phosphorous levels. Aust. J. Crop. Sci. 13 (10), 1617–1624.

Sharif-Rad, M., Varoni, E.M., Iriti, M., Martorell, M., Setzer, W.N., Del Mar, C.M., et al., 2018. Carvacrol and human health: a comprehensive review. Phytother. Res. 32 (9), 1675–1687.

Sharma, E., Anand, G., Kapoor, R., 2017. Terpenoids in plant and arbuscular mycorrhiza-reinforced defense against herbivorous insects. Ann. Bot. 119, 791–801.

Shaul-Keinan, O., Gadkar, V., Ginzberg, I., Grunzweig, J.M., Chet, I., Elad, Y., et al., 2002. Hormone concentrations in tobacco roots change during arbuscular mycorrhizal colonization with *Glomus intraradices*. N. Phytol. 154 (2), 501–507.

Shrivastava, G., Ownley, B.H., Augé, R.M., Toler, H., Dee, M., Vu, A., et al., 2015. Colonization by arbuscular mycorrhizal and endophytic fungi enhanced terpene production in tomato plants and their defense against a herbivorous insect. Symbiosis 65 (2), 65–74.

Silva, F.A., Ferreira, M.R., Soares, L.A., Sampaio, E.V., Silva, F.S., Maia, L.C., 2014. Arbuscular mycorrhizal fungi increase gallic acid production in leaves of field grown *Libidibia ferrea* (Mart. ex Tul.) L. P. Queiroz. J. Med. Plants Res. 8, 1110–1115.

Silvia, F.A., Silva, F.S.B., Maia, L.C., 2014. Biotechnical application of arbuscular mycorrhizal fungi used in the production of foliar biomolecules in ironwood seedlings (*Libidibia ferrea* (Mart. ex Tul.) L.P. Queiroz var. ferrea). J. Med. Plant. Res. 8 (20), 814–819.

Singh, R., Soni, S.K., Kalra, A., 2013. Synergy between *Glomus fasciculatum* and a beneficial Pseudomonas in reducing root diseases and improving yield and forskolin content in *Coleus forskohlii* Briq. under organic field conditions. Mycorrhiza 23 (1), 35–44.

Smith, S.E., Read, D., 2008. Mycorrhizal Symbiosis. Academic Press, London, UK.

Stumpe, M., Carsjens, J.G., Stenzel, I., Gobel, C., Lang, I., Pawlowski, K., et al., 2005. Lipid metabolism in arbuscular mycorrhizal roots of *Medicago truncatula*. Phytochemistry 66, 781–791.

Tejavathi, D.H., Jayashree, D., 2011. Effect of AM fungal association on the growth performance of selected medicinal herbs. Indian J. Appl. Res. 3 (7), 12–15.

Torres, N., Goicoechea, N., Antolín, M.C., 2015. Antioxidant properties of leaves from different accessions of grapevine (*Vitis vinifera* L.) cv. Tempranillo after applying bioticand/or environmental modulator factors. Ind. Crop. Prod. 76, 77–85.

Toussaint, J.P., Smith, F.A., Smith, S.E., 2007. Arbuscular mycorrhizal fungi can induce the production of photochemicals in sweet basil irrespective of phosphorus nutrition. Mycorrhiza 17, 291–297.

Tsuro, M., Inoue, M., Kameoka, H., 2001. Variation in essential oil components in regenerated lavender (*Lavandula vera* DC) plants. Sci. Hortic. 88, 309–317.

Urbanek, H., Katarzyna Bergier, K., Marian Saniewski, M., Patykowski, J., 1996. Effect of jasmonates and exogenous polysaccharides on production of alkannin pigments in suspension cultures of *Alkanna tinctoria*. Plant. Cell Rep. 15 (8), 637–641.

Vierheilig, H., Piche, Y., 2002. Signalling in arbuscular mycorrhiza: facts and hypothesis. In: Buslig, B.S., Manthey, J.A. (Eds.), Flavonoids in Cell Function. Advances in Experimental Medicine and Biology, vol. 505. Springer, Boston, MA, pp. 23–39.

Vierheilig, H., Bennett, R., Kiddle, G., Kaldorf, M., Ludwig-Müller, J., 2000c. Differences in glucosinolate patterns and arbuscular mycorrhizal status of glucosinolate-containing plant species. N. Phytol. 146, 343–352.

Vierheilig, H., Gagnon, H., Strack, D., Maier, W., 2000b. Accumulation of cyclohexenone derivatives in barley, wheat and maize roots in response to inoculation with different arbuscular mycorrhizal fungi. Mycorrhiza 9, 291–293.

Vierheilig, H., Maier, W., Wyss, U., Samson, J., Strack, D., Piché, Y., 2000a. Cyclohexenon e derivative and phosphate-levels in split-root systems and their role in the systemic suppression of mycorrhization in precolonized barley plants. J. Plant. Physiol. 157, 593–599.

Vishnuvathan, V.J., Lakshmi, K.S., Srividya, A.R., 2016. Medicinal uses of formononetin—a review. J. Ethnobiol. Ethnomedicine 126, 1197–1209.

Vo, A.T., Haddidi, I., Daood, H., Mayer, Z., Posta, K., 2019. Impact of arbuscular mycorrhizal inoculation and growth substrate on biomass and content of polyphenols in *Eclipta prostrata*. Hortic. Sci. 54 (11), 1976–1983.

Volpin, H., Elkind, Y., Okon, Y., Kapulnik, Y., 1994. A vesicular arbuscular mycorrhizal fungus (*Glomus intraradix*) induces a defense response in alfalfa roots. Plant. Physiol. 104 (2), 683–689.

Walter, M.H., Fester, T., Strack, D., 2000. Arbuscular mycorrhizal fungi induce the non-mevalonate methylerythritol phosphate pathway of isoprenoid biosynthesis correlated with accumulation of the 'yellow pigment' and other apocarotenoids. Plant. J. 21, 571–578.

Walter, M.H., Hans, J., Strack, D., 2002. Two distantly related genes encoding 1-deoxy-d-xylulose 5-phosphate synthases: differential regulation in shoots and apocarotenoid-accumulating mycorrhizal roots. Plant. J. 31, 243–254.

Wasternack, C., 2007. Jasmonates: an update on biosynthesis, signal transduction and action in plant stress response, growth and development. Ann. Bot. 100 (4), 681–697.

Wei, G.T., Wang, H.G., 1989. Effects of VA mycorrhizal fungi on growth, nutrient uptake and effective compounds in Chinese medicinal herb *Datura stramonium* L. Sci. Agr. Sin. 22 (5), 56–61.

Wei, G.T., Wang, H.G., 1991. Effect of vesicular-arbuscular mycorrhizal fungi on growth, nutrient uptake and synthesis of volatile oil in *Schizonepeta tenuifolia*. Chin.. J. Chin Mater. Med. 16 (3), 139–142.

Weisany, W., Abdul- razzak Tahir, N., Schenk, P.M., 2021. Coriander/soybean intercropping and mycorrhizae application lead to overyielding and changes in essential oil profiles. Eur. J. Agron. 126, 126283.

Weisany, W., Raei, Y., Ghasemi-Golezani, K., 2016b. *Funneliformis mosseae* alters seed essential oil content and composition of dill in intercropping with common bean. Ind. Crop. Prod. 79, 29–38.

Weisany, W., Raei, Y., Pertot, I., 2015. Changes in the essential oil yield and composition of dill (*Anethum graveolens* L.) as response to arbuscular mycorrhiza colonization and cropping system. Ind. Crop. Prod. 77, 295–306.

Weisany, W., Raei, Y., Salmasi, S.Z., Sohrabi, Y., Ghassemi-Golezani, K., 2016a. Arbuscular mycorrhizal fungi induced changes in rhizosphere, essential oil and mineral nutrients uptake in dill/common bean intercropping system. Ann. Appl. Biol. 169 (3), 384–397.

Weisany, W., Sohrabi, Y., Siosemardeh, A., Ghassemi-Golezani, K., 2017. *Funneliformis mosseae* fungi changed essential oil composition in *Trigonella foenum graecum* L., *Coriandrum sativum* L. and *Nigella sativa* L. J. Essent. Oil Res. 29 (3), 276–287.

Wu, Y.H., Wang, H., Liu, M., Li, B., Chen, X., Ma, Y.T., et al., 2021. Effects of native arbuscular mycorrhizae isolated on root biomass and secondary metabolites of *Salvia miltiorrhiza* Bge. Front. Plant. Sci. 12.

Xie, W., Hao, Z., Zhou, X., Jiang, X., Xu, L., Wu, S., et al., 2018. Arbuscular mycorrhiza facilitates the accumulation of glycyrrhizin and liquiritin in *Glycyrrhiza uralensis* under drought stress. Mycorrhiza 28, 285–300.

Yang, Y., Ou, X.H., Yang, G., Xia, Y.S., Chen, M.L., Guo, L.P., et al., 2017. Arbuscular mycorrhizal fungi regulate the growth and phyto-active compound of *Salvia miltiorrhiza* seedlings. Appl. Sci. 7 (1), 68.

Yao, M.K., Desilets, H., Charles, M.T., Boulanger, R., Tweddell, R.J., 2003. Effect of mycorrhization on the accumulation of rishitin and solavetivone in potato plantlets challenged with *Rhizoctonia solani*. Mycorrhiza 13, 333–336.

Zeng, Y., Guo, L.P., Chen, B.D., Hao, Z.P., Wang, J.Y., Huang, L.Q., et al., 2013. Arbuscular mycorrhizal symbiosis and active ingredients of medicinal plants: current research status and prospectives. Mycorrhiza 23 (4), 253–265.

Zhang, R.Q., Zhu, H.H., Zhao, H.Q., Yao, Q., 2013. Arbuscular mycorrhizal fungal inoculation increases phenolic synthesis in clover roots via hydrogen peroxide, salicylic acid and nitric oxide signaling pathways. J. Plant. Physiol. 170 (1), 74–79.

Zhi, L.Y., Chuan, C.D., Lian, Q.C., 2007. Regulation and accumulation of secondary metabolites in plant-fungus symbiotic system. Afr. J. Biotechnol. 6, 1266–1271.

Zhou, S., Gao, Y., 2002. The immunomodulating effects of *Ganoderma lucidum* (Curt.:Fr.) P.Karst (LingZhi, Reishi Mushroom) (Aphylloromycetidae). Int. J. Med. Mushrooms 4, 1–11.

Zimare, S.B., Borde, M.Y., Jite, P.K., Malpathak, N.P., 2013. Effect of AM fungi (Gf, Gm) on biomass and gymnemic acid content of Gymnema sylvestre (Retz.) R. Br. ex Sm. Proc. Natl. Acad. Sci., India, Sect. B Biol. Sci. 83, 439–445.

Zitter-Eglseer, K., Nell, M., Lamien-Meda, A., Steinkellner, S., Wawrosh, C., Kopp, B., et al., 2015. Effects of root colonization by symbiotic arbuscular mycorrhizal fungi on the yield of pharmacologically active compounds in *Angelica archangelica* L. Acta Physiol. Plant. 37, 2–11.

Zolfaghari, M., Nazeri, V., Sefdkon, F., Rejali, F., 2013. Effect of arbuscular mycorrhizal fungi on plant growth and essential oil content and composition of *Ocimum basilicum* L. Iran. J. Plant. Physiol. 3, 643–650.

Zouari, I., Salvioli, A., Chialva, M., Novero, M., Miozzi, L., Tenore, G.C., et al., 2014. From root to fruit: RNA-Seq analysis shows that arbuscular mycorrhizal symbiosis may affect tomato fruit metabolism. BMC Genomics 15, 2215.

Zubek, S., Mielcarek, S., Turnau, K., 2012. Hypericin and pseudo hypericin concentrations of a valuable medicinal plant *Hypericum perforatum* L. are enhanced by arbuscular mycorrhizal fungi. Mycorrhiza 22, 149–156.

Zubek, S., Rola, K., Szewczyk, A., Majewska, M.L., Katarzyna, T., 2015. Enhanced concentrations of elements and secondary metabolites in *Viola tricolor* L. induced by arbuscular mycorrhizal fungi. Plant. Soil. 390, 129–142.

Zubek, S., Stojakowska, A., Anielska, T., Turnau, K., 2010. Arbuscular mycorrhizal fungi alter thymol derivative contents of *Inula ensifolia* L. Mycorrhiza 20 (7), 497–504.

CHAPTER

17

Significance of endophytes in plant growth and performance

Hui Yee Chong and Wendy Ying Ying Liu

School of Applied Sciences, Faculty of Integrated Life Sciences, Quest International University, Ipoh, Perak, Malaysia

17.1 Introduction

The ever-increasing global population has augmented the demand for global agriculture production and food security (Van Dijk et al., 2021). Hence, producers have resorted to the massive use of agrochemicals, including synthetic chemical fertilizers and pesticides, in order to increase agricultural productivity. However, the extensive use of such chemicals in agriculture has resulted in detrimental effects on the environment and human health, including biodiversity loss, a decrease in soil organic carbon (C) content and beneficial soil microbiota, greenhouse gas emissions, and a surge in numerous human diseases (Bisht and Singh Chauhan, 2021; Sivojiene et al., 2021; Menegat et al., 2022).

However, the widespread use of such chemicals in agriculture has had a negative impact on the environment and human health, including biodiversity loss, a decrease in beneficial soil microbial flora and organic C in soil, greenhouse gas emissions, eutrophication, and an increase in the occurrence of numerous human diseases. As a result, alternative ways for promoting sustainable agriculture production that are less harmful to the ecosystem are being researched. Optimizing the utilization of competent plant-beneficial microbes to improve crop growth, vigor, and nutrient usage efficiency while minimizing negative environmental effects could be a potential approach (Lopes et al., 2021).

Beneficial plant microbial associations have been widely explored in order to boost sustainable plant growth and performance to harvest food, biofuels, and bioactive metabolites (Basit et al., 2021; Das et al., 2022). Free-living, symbiotic, and endophytic microorganisms are among the few microorganisms shown to promote plant growth and performance viz. direct and/or indirect mechanisms (Eid et al., 2021). Endophytes are microorganisms that survive for at least part of their lives within the intracellular and intercellular sections of

Biostimulants in Plant Protection and Performance
DOI: https://doi.org/10.1016/B978-0-443-15884-1.00014-2

plants without causing damage to their hosts. De Bary (1866) coined the term "endophyte" to distinguish these microorganisms from epiphytic organisms that live on the surface of plants.

Endophytes form symbiotic relationships with their host plants, providing direct and indirect protection against a variety of stresses and pathogens in exchange for nutrients and protection from their hosts (Card et al., 2016; Grabka et al., 2022). They develop strategies for living and surviving alongside their hosts during the association period (Mengistu, 2020). As a result, they may influence plant growth and development, plant community diversity, and ecosystem functioning (Ortíz-Castro et al., 2009; Chen et al., 2019). Endophytic strains may directly promote plant growth via improvement in nutrient acquisition (i.e., nitrogen (N), phosphorus (P), and iron (Fe)), as well as, modulation of plant hormone levels. Furthermore, they may indirectly promote plant development by aiding their host plants in alleviating abiotic stresses and reducing inhibitory agents such as harmful pathogens in soil by producing siderophores, hydrogen cyanide (HCN), antibiotics, and other compounds (Afzal et al., 2019; Eid et al., 2021).

Nonetheless, there are instances when an endophyte can be beneficial or pathogenic to the plant it colonizes. While many endophytes do not exert negative effects on various plant species, some may be pathogenic when tested on other plant species, which may be exacerbated by environmental factors, host genotypes, and biotic interactions (Fadiji and Babalola, 2020). For instance, inoculation of fluorescent *Pseudomonas* spp. (that were typically beneficial to plants) onto *Rumohra adiantiformis* resulted in fern distortion syndrome under specific conditions (Kloepper et al., 2013). As a result, rather than their functions, endophytes are defined hereafter based on their colonization niche.

There have been no reports of plant species that are not associated with endophytes colonizing their tissues to date, indicating the abundance and critical roles of endophytes in association with their host plants. Despite this, plant growth-enhancing rhizobacteria have garnered significantly more attention than endophytes as a viable alternative to the usage of agrochemicals in promoting plant growth and performance. Thus, this review focuses on endophytic distribution, colonization behavior, mechanisms of action in plant growth promotion, and future prospects in the agriculture industry.

17.2 Endophytic distribution and colonization behavior

Endophytes have the ability to enter and colonize any plant part, including the seed embryo, via vertical seeding, horizontal transmission from plant or soil to plant, or both (Frank et al., 2017; Verma et al., 2021). The majority of fungal endophytes are transmitted vertically through seeds, whereas endophytic bacteria are more prone to horizontal transmission (Frank et al., 2017; Gundel et al., 2020). Some endophytes are able to colonize various plant species while some are more restricted and can only colonize specific plant species (Baron and Rigobelo, 2022; Byregowda et al., 2022).

Endophytes can be divided into three major groups based on their approach to colonizing their host plants. Obligate endophytes are unable to reproduce outside of plant tissues and are typically transferred vertically via seeds rather than horizontally through the soil. On the other hand, facultative endophytes have a biphasic lifestyle in which they alternate between living in the environment for a period of time and colonizing the inner issues of plants. Finally, passive endophytes are microorganisms that colonize plant tissues unintentionally as a result of conditions such as wounds and cracks on root hairs (Fadiji and Babalola, 2020).

Endophytic colonization of host plants is dependent on a number of factors that govern the entire colonization process, which necessitates complex communication between the host and endophytes. Host plant genotype, plant tissue type, native soil composition, endophytic strain type, and various biotic and abiotic environmental factors are all factors that influence the success of endophytic colonization (Mengistu, 2020). Endophytes typically begin colonization at the roots by recognizing specific molecules or compounds found in root exudates before potentially migrating to above-ground plant tissues for systematic colonization (Pinski et al., 2019). Nonetheless, these endophytes could enter the plant endosphere through aerial parts like stems, flowers, and leaves (Frank et al., 2017).

For root entry by bacterial endophytes, the preliminary endophytic colonization process begins when an early communication is initiated between the plant root exudates and associated rhizospheric microorganisms. The root exudates contain important organic compounds (i.e., organic acids, proteins, and amino acids) that aid in the recruitment of microorganisms in the rhizosphere. Subsequently, they attach themselves to the plant surface as they swim toward the roots and attach to the root surface (Hori and Matsumoto, 2010). The secretion of extracellular polymeric substances (EPS) by the bacterial endophytes aids in their attachment to the root surface (Meneses et al., 2011). This is critical during the early stages of endophyte colonization of a plant. Meneses et al. (2011) discovered that EPS is essential for the endophytic bacterium *Gluconacetobacter diazotrophicus* Pal5 to attach to and colonize rice root surfaces. Furthermore, certain structures, including cell surface polysaccharides, fimbriae and flagella, are prone to play important roles in endophytic attachment to the plant surface. It was reported that bacterial lipopolysaccharides were needed for endophytic *Herbaspirillum seropedicae* to attach and colonize the maize plants (Balsanelli et al., 2010).

After establishing their presence on the epiphytic surfaces, bacterial endophytes can enter the host plant via root openings where root hairs and lateral roots emerge, or via wounds or pores in plant shoots (Hardoim et al., 2015). Some endophytes can alter plant cell walls by secreting cellulolytic enzymes (e.g., cellulase, endoglucanase, pectinase, and xylanase), allowing bacteria to enter and spread within plant tissues (Compant et al., 2005). Furthermore, studies have revealed that some bacteria colonize plant root cortexes and vascular tissues via root apexes and hairs (Prieto et al., 2011; Rangel de Souza et al., 2016).

Bacterial endophytes typically inhabit intercellular spaces in various parts of a plant due to the abundance of nutrients (e.g., amino acids and carbohydrates) in those areas (Elbeltagy et al., 2001; Hardoim et al., 2015). They are typically found first in root hairs and then in the cortex of the roots. Compant et al. (2005) discovered that lower stems had greater colonization of endophytic bacteria than stems closer to the shoot apex. Furthermore, cellulolytic enzymes produced by the cells may aid in them diverging to aerial plant parts such as stems and leaves (Balsanelli et al., 2010; Santi et al., 2013).

Meanwhile, for fungal endophytes, the germinative tube of the endophyte approaches the root closely, causing the root to lose apical dominance. The hyphal penetration apparatus is then formed and enters the root cortex with the infected hyphae, kicking off the colonization process. As the fungus approaches the inner cortex, the hyphae penetrate the plant cell wall and colonize nearby tissues such as the xylem and phloem.

17.3 Mechanisms of action of endophytes in plant growth promotion

Endophytes have various effects on plant activities, and their presence in plant tissues generally benefits their host plants by actively and/or passively promoting plant growth (Fadiji and Babalola, 2020). Some of the important direct mechanisms of promotion of plant growth comprise nutrient acquisition and phytohormone modulation, while conferring tolerance to biotic and abiotic stresses is an indirect mechanism through which endophytes can promote growth (Eid et al., 2021; Liu and Poobathy, 2021). Tables 17.1–17.4 show examples of endophytic mechanisms of action in promoting plant growth and performance.

17.3.1 Facilitation of nutrient acquisition

Many beneficial endophytes have been found to boost plant growth and performance by aiding their host plants in acquiring nutrients, such as N, P, potassium (K), Fe, and others (Tables 17.1 and 17.2).

17.3.1.1 Nitrogen uptake

Nitrogen is an essential macronutrient for the growth and development of plants, and it is found in proteins, nucleic acids, chlorophyll, and other substances (Krapp, 2015). Despite the abundance of N_2 in the atmosphere, plants are unable to utilize this form of N and they typically utilize N via the form of nitrate (NO_3^-) and ammonium (NH_4^+) (Xu et al., 2012; Zhu et al., 2021). As N is a main limiting factor to plant growth, the utilization of microorganisms that are capable of facilitating N uptake by their host plants can address this issue and increase the soil N budget, thus reducing the reliance on chemical N fertilizers.

Some endophytic bacteria can fix atmospheric N_2 into ammonia, via the nitrogenase enzyme, which their host plants could utilize for their growth. These N_2-fixing endophytic bacteria belong to various phyla with Proteobacteria accounting for the vast majority (82%), followed by Firmicutes (9%), Actinobacteria (6%), and Bacteroidetes (3%) (Rana et al., 2023). For example, endophytic *Curtobacterium* sp. A02 isolated from cassava roots in China had nitrogenase activity of 95.81 nmol/mL/h and its re-inoculation onto cassava significantly increased the biomass of various plant parts by 10.3% − 17.6% in comparison to the non-inoculated control (Zhang et al., 2022). Meanwhile, some bacterial endophytes are able to produce ammonia and subsequently facilitate their host plants' N uptake. For instance, Borah et al. (2019) stated that 66 out of 129 bacterial isolates sampled from the root tissues of *Camilla sinensis* were able to produce ammonia with *Brevibacterium* sp. strain S91 showing the highest ammonia production.

Fungal endophytes, on the other hand, can facilitate N acquisition in their host plants by (1) altering N uptake and distribution and (2) translocating N derived from insects to

TABLE 17.1 Examples of direct mechanisms of action of bacterial endophytes in promoting plant growth and performance.

Mechanisms of action	Phylum	Bacterial endophyte	Host plant/organ	Remarks	Reference
Facilitation of nutrient acquisition					
Ammonia production	Firmicutes	*Bacillus* spp.	*Cicer arietinum* (roots)	–	Brígido et al. (2019)
		Brevibacillus spp. Af.13, and Af.14	*Achillea fragrantissima* (leaves)	Increase dry tissue weight and P concentration in shoots.	Alkahtani et al. (2020)
	Proteobacteria	*Enterobacter* spp. and *Pseudomonas* spp.	*Cicer arietinum* (roots)	–	Brígido et al. (2019)
Nitrogen fixation	Proteobacteria	*Burkholderia vietnamiensis* WPB	*Poa pratensis*	Increase N content and dry weight in 50 days.	Xin et al. (2009)
		Klebsiella variicola DX120E	ROC22 and GT21 sugarcane (roots)	Increase dry weight, N, P and K content of roots, stems and leaves.	Wei et al. (2014)
		Rahnella strain WP5	*Lycopersicon lycopericum* cv "Glacier"	Increase root and shoot biomass.	Khan et al. (2012)
Phosphate solubilization	Firmicutes	*Bacillus subtilis*	Micropropagated strawberry	Increase root number, length and dry weight and increase in leaf number, petiole length and dry weight.	Dias et al. (2009)
		Bacillus megaterium and *Bacillus amyloliquefasciens*	*Triticum aestivum*	–	Verma et al. (2015)
	Proteobacteria	*Pantoea* sp.	*Arachis hypogaea*	Increase aerial dry weight.	Taurian et al. (2010)
		Pantoea dispersa	*Manihot esculenta* Crantz	Promote P uptake due to secretion of salicylic and benzene acetic acids	Chen et al. (2014)
		Sphingopyxis sp.	Micropropagated strawberry	Increase root number, length and dry weight and increase in leaf number, petiole length and dry weight.	Dias et al. (2009)
Potassium solubilization	Proteobacteria	*Burkholderia* sp. strain FDN2–1	*Zea mays* (root)	Increase root length and dry weight of corn seedling.	Baghel et al. (2020)
		Pseudomonas strains (K03, Y04, and N05)	*Nicotiana tobacum* (seed)	Increase the root enzyme activity and root growth.	Li et al. (2018)

(*Continued*)

TABLE 17.1 (Continued)

Mechanisms of action	Phylum	Bacterial endophyte	Host plant/organ	Remarks	Reference
Siderophore production	Actinobacteria	*Micrococcus lutues*	Ginseng (stem)	Positive for plant growth promoting traits such as IAA production.	Vendan et al. (2010)
	Bacillota	*Lysinibacillus fusiformis*			
	Proteobacteria	*Bacillus megaterium* and *Bacillus cereus*			
		Bacillus sp. B19, *Bacillus* sp. P12 and *Bacillus amyloliquefaciens* B14	*Phaseolus vulgaris*	Increase crop germination potential, root length and stem length.	Sabaté et al. (2018)
		Methylobacterium mesophilicum	*Citrus* sp.	Release hydroxamate-type siderophores.	Lacava et al. (2008)
Phytohormone modulation					
Abscisic acid	Proteobacteria	*Azospirillum lipoferum*	*Zea mays*	Neutralize the reduced relative water content during drought stress.	Cohen et al. (2009)
Auxin	Firmicutes	*Bacillus subtilis, Bacillus flexus, Bacillus cereus, Bacillus megaterium* and *Bacillus endophyticus*	*Capsicum annuum* cv Jalapeno	Increase stem diameter, root volume, fresh biomass of roots and total fresh biomass.	Peña-Yam et al. (2016)
		Brevibacillus agri (PB5)	*Phaseolus vulgaris* (root)	Increase the vegetative growth of common bean plants.	Ismail et al. (2021)
	Proteobacteria	*Burkholderia kururiensis*	*Oryza sativa*	Increase number of lateral roots resulted in higher yield.	Mattos et al. (2008)
Cytokinin	Firmicutes	*Bacillus subtilis*	*Lactuca sativa*	Increase in plant shoot and root weight.	Arkhipova et al. (2005)
Ethylene	Proteobacteria	*Pseudomonas brassicacearum* SVB6R1	Sorghum, cucumber and tomato	Increase plant biomass and reduce salinity stress symptoms.	Gamalero et al. (2020)
Gibberellin	Firmicutes	*Bacillus amyloliquefaciens* RWL-1	*Oryza sativa* (seed)	Improve host-plant physiology and regulate endogenous phytohormones.	Shahzad et al. (2016)
	Proteobacteria	*Azospirillum lipoferum*	*Zea mays*	Neutralized the reduced relative water content during drought stress.	Cohen et al. (2009)

TABLE 17.2 Examples of indirect mechanisms of action of bacterial endophytes in promoting plant growth and performance.

Mechanisms of action	Phylum	Bacterial endophyte	Host plant/ organ	Remarks	Reference
Stress tolerance					
Low temperature	Proteobacteria	*Burkholderia phytofirmans* strain PsJN	*Vitis vinifera*	Increase in plant biomass, starch, proline, and phenolics.	Ait Barka et al. (2006)
Drought tolerance	Firmicutes	*Bacillus pumilus*	*Glycyrrhiza uralensis* Fisch.	Modification of antioxidant accumulation and increase in total biomass.	Xie et al. (2019)
	Proteobacteria	*Burkholderia phytofirmans* strain PsJN *and Enterobacter* sp. FD17	*Zea mays*	Increase leaf relative water content and better growth and physiological status of seedlings under drought stress.	Naveed et al. (2014)
		Gluconacetobacter diazotrophicus PAL5	Sugarcane cv. SP70−1143	Activation of drought-responsive markers and hormone pathways (ABA, ethylene).	Vargas et al. (2014)
Heavy metal	Proteobacteria	*Pseudomonas stutzeri* A1501	*Oryza sativa*	Secretion of ACC deaminase that can enhance the plant growth.	Han et al. (2015)
Salinity tolerance	Proteobacteria	*Pseudomonas fluorescens* YsS6, *Pseudomonas migulae* 8R6	*Solanum lycopersicum*	Increase fresh and dry biomass.	Ali et al. (2014)
		Pseudomonas pseudoalcaligenes	*Oryza sativa*	Increase concentration of glycine betaine-like quaternary compounds and increase shoot biomass.	Jha et al. (2011)
Disease control					
Fungi	Firmicutes	*Bacillus* sp.	–	Produce lipopeptides which hydrolyze the fungal hyphal membranes.	Ongena and Jacques (2008)
Competition for space and nutrients					
Antibiosis	Firmicutes	*Bacillus* sp.	*Bacopa monnieri*	Production of surfactin, iturin and fengycin which are toxic to pathogens and control the symbiotic relationship.	Jasim et al. (2016)
Induced Systemic Resistance	Firmicutes	*Bacillus velezensis* YC7010	*Arabidopsis*	Induce systemic resistance against green peach aphid by accumulation of hydrogen peroxide.	Rashid et al. (2017)

TABLE 17.3 Examples of direct mechanisms of action of fungal endophytes in promoting plant growth and performance.

Mechanisms of action	Phylum	Fungal endophyte	Host plant/ organ	Remarks	Reference
Facilitation of nutrient acquisition					
Ammonia production	Ascomycota	*Penicillium chrysogenum* Pc_25, *Alternaria alternata* Aa_27 and sterile hyphae Sh_26	*Asclepias sinaica*	Increase root growth.	Fouda et al. (2015)
		Penicillium chrysogenum and *Penicillium crustosum*	*Teucrium polium*	Increase root length and fresh root weight.	Hassan (2017)
	Ascomycota and Basidiomycota	*Agaricus bisporous* (Basidiomycetes) and *Mycolepto discus* (Ascomycota)	*Vanda cristata*	Higher number of roots and root length (without tryptophan induction) and higher number of shoots (with tryptophan).	Chand et al. (2020)
Nitrogen fixation	Ascomycota	*Phomopsis liquidambaris*	*Arachis hypogaea*	Improve nodulation by nodulation-related strains (*Bradyrhizobium* sp.).	Xie et al. (2019)
		Colletotrichum tropicale, *Pestalotiopsis* sp. and *Colletotrichum theobromicola*	*Theobroma cacao* (leave)	Increase N uptake and total biomass.	Christian et al. (2019)
	Oomycota	*Phytophthora palmivora*			
Phosphate solubilization	Ascomycota	*Fusarium verticillioides* RK01 and *Humicola* sp. KNU01	*Glycine max* (roots)	Protect soybean plants against oxidative damage	Radhakrishnan et al. (2015)
		Penicillium crustosum EP-2, *Penicillium chrysogenum* EP-3, and *Aspergillus flavus* EP-14	*Ephedra pachyclada*	Increase root length and vegetative growth.	Khalil et al. (2021)
Siderophore production	Ascomycota	*Penicillium chrysogenum* (CAL1), *Aspergillus sydowii* (CAR12) and *Aspergillus terreus*	*Cymbidium aloifolium*	Inhibit virulent plant pathogen (*Ralstonia solanacearum*).	Chowdappa et al. (2020)
Phytohormone production					
Abscisic acid	Ascomycota	*Nigropsora* sp. TC2–054	*Fragaria virginiana*	Exhibit antimycobacterial activity against *Mycobacterium tuberculosis* H37Ra.	Clark et al. (2013)

Auxin	Ascomycota	*Penicillium chrysogenum* Pc_25, *Alternaria alternata* Aa_27 and sterile hyphae Sh_26	*Asclepias sinaica*	Increase root growth.	Fouda et al. (2015)
		Penicillium commune EP-5, *Aspergillus flavus* EP-14, and *Aspergillus niger* EP-15	*Ephedra pachyclada*	Increase root length and P content in shoot.	Khalil et al. (2021)
		Penicillium chrysogenum and *Penicillium crustosum*	*Teucrium polium*	Increase root length and fresh root weight.	Hassan (2017)
Cytokinin	Basidiomycota	*Serendipita indica*	*Poncirus trifoliata*	Increase shoot biomass, root length, and total chlorophyll content.	Liu et al. (2022)
	Glomeromycota	*Funneliformis mosseae*			
Ethylene	Ascomycota	*Fusarium solani*	*Solanum lycopersicum*	Protect tomato plant against the root pathogen *Fusarium oxysporum* f.sp. *radicis-lycopersici*.	Kavroulakis et al. (2007)
Gibberellin	Ascomycota	*Phoma glomerata* LWL2 and *Penicillium* sp. LWL3	*Waito-C* and Dongjin-beyo rice	Increase shoot length, chlorophyll content, shoot fresh and dry weight.	Waqas et al. (2012)
	Basidiomycota	*Porostereum spadiceum* AGH786	*Glycine max*	Increase fresh and dry biomass and chlorophyll content.	Hamayun et al. (2017)

TABLE 17.4 Examples of indirect mechanisms of action of fungal endophytes in promoting plant growth and performance.

Mechanisms of action	Phylum	Fungal endophyte	Host plant/organ	Remarks	Reference
Stress tolerance					
Drought tolerance	Ascomycota	*Ampelomyces* sp. and *Penicillium* sp.	*Solanum lycopersicum*	Increase average shoot and root dry weight yield and tolerance to salt and H_2O_2.	Morsy et al. (2020)
		Phoma spp.	*Pinus tabulaeformis* (root)	Secrete extracellular abscisic acid which triggered the protective mechanisms	Zhou et al. (2021)
Heat tolerance	Ascomycota	*Penicilium funiculosum* (A2) and *Endomelanconiopsis endophytica* (X5)	*Oryza sativa*	Increase shoot and root growth.	Arpitha et al. (2022)
	Basidiomycota	*Ceriporia lacerate* (A7)			
Heavy metal	Ascomycota	*Phialocephala fortinii* and *Rhizoscyphus* sp.	*Clethra barbinervis*	Increase K uptake, number of leaves, height, root length, and root fresh weight.	Yamaji et al. (2016)
	Glomeromycota	*Rhizodermea veluwensis*			
Salinity tolerance	Ascomycota	*Microsphaeropsis arundinis*	*Triticum aestivum*	Increase shoot length and yield.	Manjunatha et al. (2022)
		Stemphylium lycopersici	*Zea mays*	Higher fresh and dry biomass, chlorophyll contents, number of flowers and buds.	Ali et al. (2022)
Competition for space and nutrients					
Antibiosis	Ascomycota	*Epicoccum* sp. and *Pleosporales* sp.	*Echinopsis chiloensis* and *Baccharis linearis*	Inhibit the growth of *Botrytis cinerea*.	Castro et al. (2022)
Induced Systemic Resistance	Ascomycota	*Fusarium solani*	*Solanum lycopersicum*	Protect against tomato foliar pathogen *Septoria lycopersici* (by ethylene).	Kavroulakis et al. (2007)

the plants. *Colletotrichum tropicale*, a foliar endophytic fungus, significantly increased ^{15}N uptake in *Theobroma cacao* plants compared to the control plants, indicating the endophyte's ability to modify N uptake and distribution in its host plant (Christian et al., 2019). Meanwhile, Behie et al. (2012) reported that, after 28 days of inoculation, endophytic *Metarhizium robertsii* was able to translocate N derived from an insect pathogen, waxmoth larvae, and contribute 12% and 48% of N content to haricot bean and switch grass, respectively. Additionally, it was demonstrated that in the tripartite interaction, in exchange for insect-derived N, haricot bean delivered photosynthate to the fungus (Behie et al., 2017).

17.3.1.2 Phosphate solubilization

Following N, P is a vital nutrient for overall plant growth and performance. Because of its structural, functional, and metabolic properties, it is frequently a constraint for plant growth. Regardless of large soil P pools in organic and inorganic forms, the availability of P is strongly influenced by biogeochemical cycling reactions (Matos et al., 2017). As most soil P is insoluble and plants can only access soluble P, it is inaccessible to plants (Ikhajiagbe et al., 2020). Though chemical P fertilizer provides soluble P, it is quickly immobilized after application (Bindraban et al., 2020). Phosphorus solubilizing microorganisms (PSMs) have been widely studied as an alternative to increase P availability to plants through solubilization and mineralization of complex P compounds (Zhu et al., 2012). In comparison to non-endophytic PSMs, endophytic PSMs are thought to be more competitive as they are able to invade their host plant cells without triggering their defense pathways (Mehta et al., 2019).

Depending on the type of P sources available in the soil, PSMs utilize different mechanisms viz. inorganic and organic P solubilization. For inorganic P solubilization, it is recognized that insoluble P is mainly solubilized via the secretion of various organic acids (OA), including gluconic, acetic, citric, and oxalic acid (Rawat et al., 2021). The type of OA secreted is highly dependent on the form of insoluble P while the amount of OA relies on the type of microbial strain (Marciano Marra et al., 2012; Alori et al., 2017). The OA secreted will result in the chelation of mineral ions or lowered pH level, and this will subsequently acidify the microbial cells and their surroundings, causing P-ion to be released from P-mineral (Trivedi and Sa, 2008). For example, Chen et al. (2014) reported that the endophytic *Pantoea dispersa*, isolated from cassava roots, was able to solubilize phosphate via the production of OA such as benzeacetic acid, succinate, oxalic acid, salicylic acid, and citric acid.

On the other hand, organic P mineralization is facilitated by various enzymes including phosphatases and phytases released by PSMs that catalyze the hydrolysis of phosphoric esters. For instance, endophytic *Penicillium* sp. and *Aspergillus* sp. strains isolated from roots of *Taxus wallichiana* were reported to solubilize phosphate via the phosphate substrate utilization and phosphate and phytase enzyme production (Adhikari and Pandey, 2019).

17.3.1.3 Potassium solubilization

Plants require K which is a key macronutrient to ensure their growth and development whereby it has a crucial role in various processes such as photosynthesis, protein synthesis, enzyme activation, osmotic balance, and phloem transport (Hasanuzzaman et al.,

2018). As plants can only acquire K from soil, and majority of K exists as insoluble rocks and silicate minerals, the amount of soluble K that plants can utilize is usually low (Singh and Sindhu, 2013). Potassium-solubilizing endophytes can aid their host plants in acquiring K via OA production which can dissolve K from various insoluble minerals including muscovite, illite, glauconite, micas, and orthoclases (Jyoti et al., 2016). Baghel et al. (2020) reported that eight out of twenty-four bacterial endophytic isolates sampled from corn root were able to solubilize potassium from glauconite and muscovite. Subsequent seed inoculation with *Burkholderia* sp. FDN-2–1 in soil modified with glauconite showed significantly increased potassium concentration after 90 days of growth.

17.3.1.4 Siderophore production

Because iron is a component of redox enzymes, which aid in oxygen transport, cellular proliferation, chlorophyll production, and nucleic acid synthesis, it is an essential trace element required for plant growth (Tripathi et al., 2018). Though Fe is the fourth most abundant element, plants and microorganisms are unable to readily assimilate it as it exists mainly as Fe^{3+} and, under aerobic conditions, it forms oxides and hydroxides that are insoluble (Schröder et al., 2003). Plants typically employ two strategies in Fe sequestration and the strategy used is highly dependent on plant type under Fe-limiting conditions. First, the acidification of the rhizosphere takes place via proton extrusion and reduction of chelated Fe^{3+} by ferric chelate reductase in monocots and dicots, thus resulting in successive Fe^{2+} uptake into the root cells. Secondly, monocotyledonous plants, particularly grass, secrete phytosiderophores, which are dependent on mugineic acid, to solubilize and transport Fe into root cells (Guerinot, 2010; Altomare and Tringovska, 2011; Radzki et al., 2013; Martín-Barranco et al., 2021). Nonetheless, both techniques are inadequate to produce adequate Fe for plant use in Fe-deficient circumstances. Thus, using microorganisms capable of producing siderophores serves as an alternative method of acquiring Fe under Fe-limiting conditions (Zuo and Zhang, 2011; Radzki et al., 2013).

Siderophores are classified into three types: catecholates, carboxylates, and hydroxamates, with most of them being water-soluble (Ahmed and Holmström, 2014). As low-molecular-mass iron chelators, siderophores bind to Fe^{3+} ions to form Fe-siderophore complexes, which reduce Fe^{3+} to Fe^{2+} in the cytosol, and this Fe^{2+} is later released into the cell, making Fe available to the microorganism (Noinaj et al., 2010). As such, siderophore production by endophytes can aid their host plants to thrive in low Fe soils. For example, bacterial endophyte *Streptomyces* sp. GMKU 3100, which is capable of producing siderophore, was shown to be able to significantly improve the growth parameters of rice and mung bean under both Fe-sufficient and Fe-deficient conditions (Rungin et al., 2012).

In addition to Fe scavenging, siderophores produced by endophytes can also form complexes with metals found in the environment, assisting their host plants in dealing with high heavy metal concentrations in the soil (Ahmed and Holmström, 2014). For example, *Bacillus thuringiensis* GDB-1 sampled from the root interior of *Pinus sylvestris* was reported to be able to eliminate heavy metals (i.e., lead and zinc) and significantly increased heavy metal accumulation in *Alnus firma* seedlings (Babu et al., 2013). Also, endophytes can also outcompete phytopathogens by producing siderophores which can inhibit their growth via binding to Fe^{3+} irons in the rhizosphere, depriving the pathogens of iron.

Verma et al. (2011) reported that endophytic *Streptomyces* sp. strains promoted the growth of *Azadirachta indica* while suppressing *Alternaria alternata* which causes tomato early blight disease.

17.3.2 Phytohormone production and modulation

Phytohormones can facilitate, impede or modify plant growth and development, especially in response to environmental cues. Plants often strive to modify the levels of endogenous phytohormones when subjected to environmental stresses (such as salinity, drought, high temperature, and low pH) in order to offset the negative consequences of the stresses (Dilworth et al., 2017). Endophytes can produce phytohormones in vitro and modulate phytohormone levels to achieve equilibrium, assisting in the regulation of plant responses to environmental stresses. Some of the common phytohormones produced and/or modulated by endophytes include auxin, gibberellic acid (GA), cytokinin, and ethylene (Chaudhary et al., 2022; Egamberdieva et al., 2017).

Indole acetic acid (IAA), a form of auxin, promotes cell division and differentiation; improves the germination of seeds; initiates and accelerates root development; regulates stress responses in plants, influences photosynthesis and produces metabolites (Fu et al., 2015; Müller and Munné-Bosch, 2021). Plants and microorganisms can synthesize this phytohormone via four and three major pathways, respectively. The pathways for microorganisms are indole-3-acetamide (IAM), indole-3-pyruvic acid (IPA), and indole-3-acetonitrile (IAOx) (Duca et al., 2014). The type of pathway used by the microorganism to synthesize IAA, whether inside or close to the plants, is determined by genetic and/or environmental factors, and can ascertain whether the microorganism is beneficial or harmful to the plants (Spaepen et al., 2007). Most beneficial endophytes use the IPA pathway while the phytopathogens produce IAA via the IAM pathway (Hardoim et al., 2008). Since most plants are sensitive to the amount of IAA available in their tissues at any given time, a beneficial IAA-producing endophyte must provide the accurate dose of IAA while accounting for endogenous IAA produced by the plants; otherwise, steep IAA concentrations can be harmful to the plants (Ali et al., 2017). Endophytic bacteria *Sphingomonas* sp. LK11 isolated from the leaves of *Tephorsia apollinea* was shown to be able to produce IAA and GA and also significantly boost the growth parameters of inoculated tomato plants in comparison to uninoculated plants (Khan et al., 2014).

Gibberellic acids are diterpenoid acids that regulate various plant growth and developmental processes such as germination of seeds, elongation of stems, flowering, formation of fruits, and senescence (Achard and Genschik, 2009). The GA levels produced by plants are much lower when the plants are subjected to environmental stresses such as salinity (Hafeez et al., 2021; Ma et al., 2022). As a result, the ability of endophytes to produce GAs is extremely beneficial, as it has been reported that bioactive GA availability is a critical factor in the negative effects of stresses on plants (Leitão and Enguita, 2016). In comparison to fungal endophytes, the mechanism of GA production by bacterial endophytes is less understood, with reports primarily describing their ability to produce this phytohormone (Bottini et al., 2004; Leitão and Enguita, 2016). The mevalonic acid pathway is used by fungal endophytes to produce GA, with the main end products being GA_1 and GA_3 (Khan et al., 2015).

Promotion of growth was observed in cucumber plants inoculated with the gibberellin-producing fungal endophyte, *Phoma* sp. GAH7 that contains more active GAs (GA_1, GA_3 and GA_4) compared to inactive GAs (GA_9, GA_{15}, GA_{19} and GA_{20}) (Hamayun et al., 2010).

Ethylene can stimulate the initiation of roots, inhibit elongation of roots, accelerate ripening of fruits, reduce wilting, promote the germination of seeds, and activate the synthesis of other phytohormones (Růzicka et al., 2007; Schaller, 2012; Ahammed et al., 2020;). However, at high concentrations in stressed plants, ethylene can negatively affect plant growth and even result in cell death (Ali and Kim, 2018). Some endophytes can regulate the levels of plant ethylene by the means of 1-aminoacyclopropane-1-carboxylate (ACC), an immediate precursor in ethylene production via ACC deaminase. ACC is hydrolyzed into α-ketobutyrate and NH3, which are used as C and N sources by plants, lowering ethylene concentrations (Singh et al., 2015). Endophytic wild-type *Pseudomonas* spp. with ACC deaminase, for example, significantly improved the growth parameters in tomato plants (i.e., fresh and dry weight and chlorophyll content) when inoculated, compared to controls and plants treated with ACC deaminase deficient mutants under both saline and non-saline conditions (Ali et al., 2014).

17.3.3 Biocontrol of plant diseases

Endophytes have the capacity to mitigate or prevent the adverse effects of phytopathogens and thus have a significant potential for use as biocontrol agents (BCAs) in agriculture to promote plant growth and performance (Xia et al., 2022). Many endophytes with biocontrol abilities use a combination of mechanisms, including induced systemic resistance, competition, parasitism, antibiosis, and signal interference via quorum sensing (Chaudhary et al., 2022). For most mechanisms, it is essential to assess whether the mechanisms used by the endophyte are indeed induced during host colonization by endophytes when determining the mechanisms employed by the endophyte. Many studies have used in vitro experiments rather than *in planta* experiments or both, to investigate the mechanisms of BCAs, which involve a three-way interaction between the host plant, endophyte, and pathogen. As a result, these findings may be unable to provide much meaningful information and may produce misleading conclusions for some of the mechanisms. For instance, Knudsen et al. (1997) reported that *Clonostachys rosea* did not inhibit the growth of *Fusarium culmorum* in vitro but it was demonstrated to be effective in controlling *Fusarium* seedling blight in barley.

17.3.3.1 Induced systemic resistance

Host plants typically recognize microbial components as foreign and subsequently prepare themselves for potential pathogens via induced defense responses (Kaur et al., 2022). Plants recognize infections and endophytes in the same way, but their responses to both are distinct (Xu et al., 2015). Induced resistance is a crucial mechanism employed by endophytes in suppressing diseases, particularly considering the favorable relationship between plants and endophytes (Latz et al., 2018). Endophytes can prime their host plants to produce faster and more intense defense responses in the face of attacks from pathogen at a small physiological cost to the host plant (Liu et al., 2017). Cell wall components, protein compounds, lipids, volatile organic compounds (VOCs), and other chemicals involved in hormonal responses are examples of endophytic components that might elicit plant defense mechanisms (Lugtenberg et al., 2016; Kaddes et al., 2019). The microbial components are identified as microbe-associated molecular patterns (MAMPs) or pathogen-associated molecular patterns (PAMPs), and they are responsible for stimulating MAMP/PAMP-triggered immunity (Newman et al., 2013). For example, components of the fungal cell wall (i.e., β-glucans and chitin) are MAMPs that are identified by the pattern recognition receptors (PPRs), triggering immunity to boost the plant defense system against pathogens (Fesel and Zuccaro, 2016).

There are two types of systemic induced plant resistance: induced systemic resistance (ISR) and systemic acquired resistance (SAR). Plant hormones, ethylene, and jasmonic acid (JA) have crucial roles in ISR while salicylic acid (SA) is vital in SAR (Xia et al., 2022). The local resistance is induced by SAR as a response to pathogens and preexisting pathogen infections by triggering hypersensitive reactions through SA and associated pathogenesis-related (PR) proteins to the infected and neighboring plant regions, thereby defending the plant against biotrophic pathogens. SAR provides long-term protection against a wide variety of microorganisms (Vlot et al., 2017). In the ISR mechanism, however, JA and associated PR proteins are triggered to infected parts, leading to plant defense against necrotrophic pathogens. This mechanism does not directly kill or inhibit the pathogen but rather strengthens plant physical barriers to reduce the plant's susceptibility (Pieterse et al., 2014).

In order to determine whether an endophytic strain indeed induces resistance against a specific pathogen, two criteria must be fulfilled: (1) the induced responses should control the targeted plant pathogens and subsequent observed expressions should be related to hindering infection and; (2) the elimination process of pathogens should be correlated with Koch's postulates (Adeleke et al., 2022). For instance, endophytic *Trichoderma citrinoviride* HT-1 sampled from *Rheum palmatum* was reported to inhibit *Fusarium oxysporum*, the causative agent of root rot (Chen et al., 2022). Inhibition of *F. oxysporum* was demonstrated through (1) significant inhibition of the pathogen in dual culture (71.85%); (2) suppression of the growth of mycelia of *F. oxysporum* by endophytic fermentation metabolites; (3) suppression of root rot of *R. palmatum* (with low disease index); and (4) significant increase in induced defense-associated enzymes and expression of defense-response-associated genes.

17.3.3.2 Antibiosis

Endophytes can produce a plethora of molecules that benefit their hosts, in addition to stimulating the plant to produce molecules used for defense against pathogens and herbivores (Mengistu, 2020; Grabka et al., 2022). Endophytes can use bioactive compounds such as alkaloids, flavonoids, peptides, polyketones, phenols, steroids, quinones, terpenoids, and VOCs to directly combat potential plant pathogens via antibiosis (Lugtenberg et al., 2016). These bioactive compounds have been shown to possess multifaceted functions including antibacterial, antifungal, antiviral, and insect action properties (Latz et al., 2018). When endophytes colonize their host plants, the endophytes and/or plants produce metabolites that restrict the growth of potential pathogens (Kusari et al., 2012). Under certain circumstances, both host plant and endophyte may share segments of a particular pathway with each contributing significantly to the production of metabolites, or either endophyte or host may stimulate the metabolism of the other. Although there have been numerous reports on the ability of endophytes to produce metabolites and inhibit pathogens via endophytic cultures or crude extracts, the studies were frequently conducted in vitro, providing only evidence that the metabolites produced by endophytes can restrict pathogens but not on their efficacy to do so *in planta* (Latz et al., 2018).

Alkaloid is a well-known example of a bioactive compound produced by fungal endophytes that is toxic to certain pathogens when accumulated. Endophytic fungus *Epichloë* of perennial ryegrass and tall fescue, for example, is commercially used in pastures to confer resistance to phytopathogens, insects, and herbivores (Vassiliadis et al., 2023). Furthermore, many endophytes isolated from medicinal plants were found to have broad-spectrum biocontrol activities against a variety of pathogens. For instance, Chareprasert et al. (2006) found that 37 bacterial endophytes isolated from rain and teak trees could produce compounds that effectively repressed the growth of *Staphylococcus aureus*, *Bacillus subtilis* and *Escherichia coli* in vitro.

17.3.3.3 Competition

Endophytes use microbial competition as one of the important mechanisms to prevent phytopathogens from colonizing their host plants by mitigating the microbiomes of pathogens (Muthu Narayanan et al., 2022). They can quickly colonize various tissues of their host plants, whether locally or systematically, and scavenge for any available nutrients (Eid et al., 2021). As a result, they are able to occupy a niche that pathogens could have infiltrated (Muthu Narayanan et al., 2022). Perea-Molina et al. (2022) demonstrated that endophytic *Bacillus velezensis* IBUN 2755 could reduce the negative effects of pathogenic *Burkholderia glumae* in rice. The endophytic strain was able to reduce the population of *B. glumae* in both the root and shoot, as well as disease symptoms (e.g., presence of empty grains). It was proposed that *B. velezensis* acted through niche competition because both strains can use alanine, proline, glucuronic acid, arabinose, and manose as their sole C source, indicating a niche overlap.

In biological control, competitive exclusion is more likely to happen in conjunction with other processes than alone (Adeleke et al., 2022). Lahlali and Hijri (2010) reported that six fungal endophytic strains sampled from Norway maple were able to suppress *Rhizoctonia solani*, a pathogen that affects potato plants, both in vitro and in planta, via one or a combination of mechanisms (inclusive of microbial competition). For example, endophytic *Phomospsis* sp. was shown, through confocal microscopy, to its biocontrol capability via nutrient and space competition.

17.3.3.4 Parasitism

Fungal parasitism (mycoparasitism), a biocontrol mechanism used by endophytic fungi, occurs when one fungus relies solely on another fungus for nutrients. This parasitism can occur via biotrophism, in which the parasite obtains nutrients from living host cells, or neotrophism, in which the parasite lives on the host's dead cells (Kim and Vujanovic, 2016). The occurrence of mycoparasitism can take place directly or indirectly, reliant on whether the parasite produces compounds that releases nutrient from the host at close proximity or at a distance (Latz et al., 2018). To allow for the release and acquisition of nutrients, the parasite produces a variety of compounds, including antibiotics, toxins, and cell wall degrading enzymes (Viterbo and Horwitz, 2010). Putri et al. (2022) demonstrated that endophytic *Trichoderma asperellum*, *Curvularia chiangmaiensis* and *Fusarium solani* strains were able to inhibit *Pyricularia oryzae*, pathogen that causes rice blast diseases. These strains produced hydrolytic enzymes (i.e., chitinase and cellulase) that are able to lyse phytopathogens by damaging and degrading their cell walls.

It is frequently arduous to establish the actual mechanisms that are involved in the observed control due to the prospective overlap between antibiosis and fungal parasitism (Jeffries, 1995). Nonetheless, this further re-iterates that an endophyte may utilize more than one mechanism for biocontrol at the same time (Adeleke et al., 2022). Most studies involve in vitro experiments to verify the occurrence of mycoparasitism via culturing in Petri plates and microscopy to determine the interactions (Latz et al., 2018). Endophytes may make close communication, infiltrate, and coil around the prey hyphae, disrupting the prey hyphae (Adeleke et al., 2022). It is difficult to verify the transfer of nutrients through *in planta* experiments and observation of close association between two fungi is not sufficient to confirm the presence of mycoparasitism (Latz et al., 2018).

17.3.4 Protection against abiotic stresses

With global climate change looming, plants are facing challenging conditions for their growth and the abiotic stresses negatively affect both plant morphology and physiology (Okoro Gideon et al., 2019). Endophytes have been reported to assist their host plant in alleviating and adapting to these stresses via various mechanisms (Afzal et al., 2019; Kamran et al., 2022). When posed with drought, heat, and/or salinity stresses, endophytes are capable of inducing changes in the lipid composition of cell membranes, preventing electrolyte leakage (Baron and Rigobelo, 2022). Endophytes have also been shown to confer tolerance to reactive oxidative species and reduce lipid peroxidation during oxidative stress, though the mechanism is unknown (González-Teuber et al., 2022). Thus, endophytes have the potential to be utilized to mitigate abiotic stresses as they can affect the physiological responses of their host plants to stress. As a result of their ability to influence plant physiological responses to stress, endophytes have the potential to be used to mitigate abiotic stresses, boost resilience, and improve crop productivity (Kamran et al., 2022). For example, endophytic *Bacillus* sp. strains sampled from *Atriplex halimus* and *Tamarix aphylla* from Algeria (which are haplotypes) were able to alleviate salinity stress in tomato and wheat by showing 50%–58% improvement in dry weight in comparison to the control (Belaouni et al., 2022). The genome of *Bacillus* sp. BH32 was mined and revealed the genes involved in stress tolerance, plant growth promotion, production of antioxidants, and secondary metabolites.

17.4 Potential of endophytes in sustainable agriculture

Endophytic bacteria and fungi have enormous potential as environmentally friendly resources, such as biofertilisers, biostimulants, and/or biocontrol agents, that can be used in boosting agro-based production to enable sustainable crop intensification in the face of increased global population and climate change, thereby promoting food security.

With their multifaceted mechanisms of action, the application of endophytes in the fields to promote agricultural crop production could be advantageous. It has been proposed that endophytes have more advantages and potential for successful field application than other plant growth-promoting microorganisms because they colonize the inner tissues of plants, reducing their exposure to compete with natural microbiota found in the rhizosphere and establishing their populations (Card et al., 2016). Furthermore, formulations and delivery techniques for seed-transmitted endophytes are not required, providing an additional benefit for commercialization. Endophytes do, however, have their own set of challenges and limitations that must be addressed before they can be used successfully in the fields.

To date, only a small percentage of known plant species (1%–2%) have been studied for their relationship with endophytes, with the majority of them being land plants (Khare et al., 2018; Strobel, 2018; Strobel and Daisy, 2003). There are still gaps in our understanding of endophytes and their associated genes, despite the fact that various endophytes and their associated genes have now been identified, providing much-needed insights into their behaviors and mechanisms of action. Thus, it is critical to study the communities and functionalities of endophytes using advanced biotechnological tools (e.g., genomics, proteomics, transcriptomics, and metabolomics) in order to fully exploit their potential to improve plant growth and performance.

In addition, while it has been widely reported that various endophytic strains possess various functional traits that may promote plant growth via direct and/or indirect mechanisms, the majority of studies have been conducted in vitro and/or *in planta* under controlled conditions. However, due to the influence of plant-microbe interactions and diverse environmental scenarios, some endophytes may be rendered ineffective in providing benefits to their host plants when applied in the fields. Endophytes' activities, including their biocontrol ability, can be negatively influenced by changes in humidity, pH, and temperature. Endophytes may also interact with the plant's existing microbiome, influencing the responses of its growth promotion mechanisms (Busby et al., 2016). The variation in responses may be related to the defense mechanism of their respective host plant, influencing the success of endophytic colonization.

It is also important to consider the safety of endophytic applications to humans, animals, and/or other plant species because some endophytic strains may have negative effects as they could be opportunistic pathogens. For example, as mentioned earlier, *Epichloë* spp. which typically colonizes grasses could produce alkaloids that are toxic to several invertebrates and vertebrates at varying concentrations. On the other hand, some fungal endophytes are also capable of producing mycotoxins that are harmful to human health (Chitnis et al., 2020). As such, it is important to avoid selecting any known pathogens to be potentially applied in the fields.

Furthermore, the formulation of endophyte-derived bioproducts should include complementary consortia of various endophytic microbial strains with multiple functional traits to increase their efficacy when applied in the field, where they face ever-changing and unpredictable environmental constraints. Moreover, bioprospecting endophytes from plants from diverse and extreme ecological niches will increase the likelihood of discovering novel endophytes with superior traits that could be beneficial (Omomowo and Babalola, 2019). Lastly, besides assessing the functionality of endophytes, proper formulation (i.e., addition of additives and sticking materials), production methods (i.e., determination of suitable carrier material) and marketing remain some of the limitations that may limit the use of these microbes for agricultural practices.

17.5 Conclusion

Much attention has been recently focused on endophytes as a potential sustainable alternative to promote plant growth as they possess various mechanisms and functions as demonstrated here. Many endophytes from various plant species have been shown to play important roles in facilitating their host plants' nutrient uptake, modulating phytohormone levels, acting as biocontrol agents against pathogens and pests, and inducing their host plants' tolerance to abiotic stresses. Nonetheless, future research should focus in depth on plant-endophyte interactions in order to determine the best way to make them effective for sustainable agricultural production. Also, as various endophytes are identified through culture-dependent techniques, studies on culture-independent techniques via genomics and metagenomics could shed more light on detecting new species and functions.

References

Achard, P., Genschik, P., 2009. Releasing the brakes of plant growth: how GAs shutdown DELLA proteins. JXB 60 (4), 1085–1092.

Adeleke, B.S., Ayilara, M.S., Akinola, S.A., Babalola, O.O., 2022. Biocontrol mechanisms of endophytic fungi. Egypt. J. Biol. Pest. Control. 32 (1), 46.

Adhikari, P., Pandey, A., 2019. Phosphate solubilization potential of endophytic fungi isolated from *Taxus wallichiana* Zucc. roots. Rhizosphere 9, 2–9.

Afzal, I., Shinwari, Z.K., Sikandar, S., Shahzad, S., 2019. Plant beneficial endophytic bacteria: mechanisms, diversity, host range and genetic determinants. Microbiol. Res. 221, 36–49.

Ahammed, G.J., Gantait, S., Mitra, M., Yang, Y., Li, X., 2020. Role of ethylene crosstalk in seed germination and early seedling development: a review. IJPPB 151, 124–131.

Ahmed, E., Holmström, S.J.M., 2014. Siderophores in environmental research: roles and applications. Microb. Biotechnol. 7 (3), 196–208.

Ait Barka, E., Nowak, J., Clément, C., 2006. Enhancement of chilling resistance of inoculated grapevine plantlets with a plant growth-promoting rhizobacterium, *Burkholderia phytofirmans* strain PsJN. Appl. Environ. Microbiol. 72 (11), 7246–7252.

Ali, R., Gul, H., Rauf, M., Arif, M., Hamayun, M., Husna, et al., 2022. Growth-promoting endophytic fungus (*Stemphylium lycopersici*) ameliorates salt stress tolerance in maize by balancing ionic and metabolic status. Front. Plant Sci. 13, 890565.

Ali, S., Kim, W.C., 2018. Plant growth promotion under water: decrease of waterlogging-induced ACC and ethylene levels by ACC deaminase-producing bacteria. Front. Microbiol. 9, 1096.

Ali, S., Charles, T.C., Glick, B.R., 2014. Amelioration of high salinity stress damage by plant growth-promoting bacterial endophytes that contain ACC deaminase. Plant Physiol. Biochem. 80, 160–167.

Ali, S., Charles, T.C., Glick, B.R., 2017. Endophytic phytohormones and their role in plant growth promotion. Functional Importance of the Plant Microbiome. pp. 89–105.

Alkahtani, M.D.F., Fouda, A., Attia, K.A., Al-Otaibi, F., Eid, A.M., Ewais, E.E.-D., et al., 2020. Isolation and characterization of plant growth promoting endophytic bacteria from desert plants and their application as bioinoculants for sustainable agriculture. Agronomy 10 (9), 1325.

Alori, E.T., Glick, B.R., Babalola, O.O., 2017. Microbial phosphorus solubilization and its potential for use in sustainable agriculture. Front. Microbiol. 8, 971.

Altomare, C., Tringovska, I., 2011. Genetics, biofuels and local farming systems. Sustainable Agriculture Reviews. Springer.

Arkhipova, T.N., Veselov, S.U., Melentiev, A.I., Martynenko, E.V., Kudoyarova, G.R., 2005. Ability of bacterium *Bacillus subtilis* to produce cytokinins and to influence the growth and endogenous hormone content of lettuce plants. Plant Soil 272 (1), 201–209.

Arpitha, P., N, E., S, S.K., & H, L.R. 2022. Fungal endophytes of himalayan cold desert induces heat tolerance in rice (*Oryza sativa L.*). Research Square Platform LLC.

Babu, A.G., Kim, J.-D., Oh, B.-T., 2013. Enhancement of heavy metal phytoremediation by Alnus firma with endophytic *Bacillus thuringiensis* GDB-1. J. Hazard. Mater. 250, 477–483.

Baghel, V., Thakur, J.K., Yadav, S.S., Manna, M.C., Mandal, A., Shirale, A.O., et al., 2020. Phosphorus and potassium solubilization from rock minerals by endophytic *Burkholderia* sp. strain FDN2-1 in soil and shift in diversity of bacterial endophytes of corn root tissue with crop growth stage. Geomicrobiol. J. 37 (6), 550–563.

Balsanelli, E., Serrato, R.V., De Baura, V.A., Sassaki, G., Yates, M.G., Rigo, L.U., et al., 2010. *Herbaspirillum seropedicae rfbB* and *rfbC* genes are required for maize colonization. Environ. Microbiol. 12 (8), 2233–2244.

Baron, N.C., Rigobelo, E.C., 2022. Endophytic fungi: a tool for plant growth promotion and sustainable agriculture. J. Mycol. 13 (1), 39–55.

Basit, A., Shah, S., Ullah, I., Ullah, I., & Mohamed, H. 2021. Microbial bioactive compounds produced by endophytes (bacteria and fungi) and their uses in plant health. In (pp. 285–318).

Behie, S.W., Moreira, C.C., Sementchoukova, I., Barelli, L., Zelisko, P.M., Bidochka, M.J., 2017. Carbon translocation from a plant to an insect-pathogenic endophytic fungus. Nat. Commun. 8 (1), 14245.

Behie, S., Zelisko, P., Bidochka, M., 2012. Endophytic insect-parasitic fungi translocate nitrogen directly from insects to plants. Science 336 (6088), 1576–1577.

Belaouni, H.A., Compant, S., Antonielli, L., Nikolic, B., Zitouni, A., Sessitsch, A., 2022. In-depth genome analysis of *Bacillus* sp. BH32, a salt stress-tolerant endophyte obtained from a halophyte in a semiarid region. Appl. Microbiol. Biotechnol. 106 (8), 3113–3137.

Bindraban, P.S., Dimkpa, C.O., Pandey, R., 2020. Exploring phosphorus fertilizers and fertilization strategies for improved human and environmental health. Biol. Fertil. Soils 56 (3), 299–317.

Bisht, N., Singh Chauhan, P., 2021. Excessive and disproportionate use of chemicals cause Soil contamination and nutritional stress. IntechOpen .

Borah, A., Das, R., Mazumdar, R., Thakur, D., 2019. Culturable endophytic bacteria of *Camellia* species endowed with plant growth promoting characteristics. J. Appl. Microbiol. 127 (3), 825–844.

Bottini, R., Cassán, F., Piccoli, P., 2004. Gibberellin production by bacteria and its involvement in plant growth promotion and yield increase. Appl. Microbiol. Biotechnol. 65, 497–503.

Brígido, C., Singh, S., Menéndez, E., Tavares, M., Glick, B., Félix, M., et al., 2019. Diversity and functionality of culturable endophytic bacterial communities in chickpea plants. Plants 8 (2), 42.

Busby, P.E., Ridout, M., Newcombe, G., 2016. Fungal endophytes: modifiers of plant disease. Plant Mol. Biol. 90, 645–655.

Byregowda, R., Prasad, S.R., Oelmüller, R., Nataraja, K.N., Prasanna Kumar, M.K., 2022. Is endophytic colonization of host plants a method of alleviating drought stress? Int. J. Mol. Sci. 23 (16), 9194.

Card, S., Johnson, L., Teasdale, S., Caradus, J., 2016. Deciphering endophyte behaviour: the link between endophyte biology and efficacious biological control agents. FEMS 92 (8).

Castro, P., Parada, R., Corrial, C., Mendoza, L., Cotoras, M., 2022. Endophytic fungi isolated from *Baccharis linearis* and *Echinopsis chiloensis* with antifungal Activity against *Botrytis cinerea*. J. Fungus 8 (2), 197.

Chand, K., Shah, S., Sharma, J., Paudel, M.R., Pant, B., 2020. Isolation, characterization, and plant growth-promoting activities of endophytic fungi from a wild orchid *Vanda cristata*. Plant Signal. Behav. 15 (5), 1744294.

Chareprasert, S., Piapukiew, J., Thienhirun, S., Whalley, A.J., Sihanonth, P., 2006. Endophytic fungi of teak leaves *Tectona grandis* L. and rain tree leaves *Samanea saman* Merr. World J. Microbiol. Biotechnol. 22, 481–486.

Chaudhary, P., Agri, U., Chaudhary, A., Kumar, A., Kumar, G., 2022. Endophytes and their potential in biotic stress management and crop production. Front. Microbiol. 13, 933017.

Chen, C., Chen, H.Y.H., Chen, X., Huang, Z., 2019. Meta-analysis shows positive effects of plant diversity on microbial biomass and respiration. Nat. Commun. 10 (1).

Chen, D., Hou, Q., Fan, B., Zhang, H., Jia, L., Sun, K., 2022. Biocontrol potential of endophytic *Trichoderma citrinoviride* HT-1 against root rot of *Rheum palmatum* through both antagonistic effects and induced systemic resistance. World J. Microbiol. Biotechnol. 38 (5), 88.

Chen, Y., Fan, J.-B., Du, L., Xu, H., Zhang, Q.-H., He, Y.-Q., 2014. The application of phosphate solubilizing endophyte *Pantoea dispersa* triggers the microbial community in red acidic soil. Appl. Soil. Ecol. 84, 235–244.

Chitnis, V.R., Suryanarayanan, T.S., Nataraja, K.N., Prasad, S.R., Oelmüller, R., Shaanker, R.U., 2020. Fungal endophyte-mediated crop improvement: the way ahead. Front. Plant Sci. 11, 561007.

Chowdappa, S., Jagannath, S., Konappa, N., Udayashankar, A.C., Jogaiah, S., 2020. Detection and characterization of antibacterial siderophores secreted by endophytic fungi from *Cymbidium aloifolium*. Biomolecules 10 (10), 1412.

Christian, N., Herre, E.A., Clay, K., 2019. Foliar endophytic fungi alter patterns of nitrogen uptake and distribution in *Theobroma cacao*. N. Phytol. 222 (3), 1573–1583.

Clark, T.N., Ellsworth, K., Li, H., Johnson, J.A., Gray, C.A., 2013. Isolation of the plant hormone (+)-abscisic acid as an antimycobacterial constituent of the medicinal plant endophyte *Nigrospora* sp. Nat. Prod. Commun. 8 (12), 1934578X1300801203.

Cohen, A.C., Travaglia, C.N., Bottini, R., Piccoli, P.N., 2009. Participation of abscisic acid and gibberellins produced by endophytic *Azospirillum* in the alleviation of drought effects in maize. Botany 87 (5), 455–462.

Compant, S., Reiter, B., Sessitsch, A., Nowak, J., Clément, C., Ait Barka, E.D., 2005. Endophytic colonization of *Vitis vinifera* L. by plant growth-promoting bacterium *Burkholderia* sp. strain PsJN. AEM 71 (4), 1685–1693.

Das, P.P., Singh, K.R.B., Nagpure, G., Mansoori, A., Singh, R.P., Ghazi, I.A., et al., 2022. Plant-soil-microbes: a tripartite interaction for nutrient acquisition and better plant growth for sustainable agricultural practices. Environ. Res. 214, 113821.

De Bary, A. 1866. *Morphologie und physiologie der pilze, flechten und myxomyceten* (Vol. 1). W. Engelmann.

Dias, A.C.F., Costa, F.E.C., Andreote, F.D., Lacava, P.T., Teixeira, M.A., Assumpção, L.C., et al., 2009. Isolation of micropropagated strawberry endophytic bacteria and assessment of their potential for plant growth promotion. World J. Microbiol. Biotechnol. 25 (2), 189–195.

Dilworth, L.L., Riley, C.K., Stennett, D.K., 2017. Chapter 5 - Plant constituents: carbohydrates, oils, resins, balsams, and plant hormones. In: Badal, S., Delgoda, R. (Eds.), Pharmacognosy. Academic Press, pp. 61–80.

Duca, D., Lorv, J., Patten, C., Rose, D., Glick, B., 2014. Indole-3-acetic acid in plant-microbe interactions. Anton. Leeuw. Int. J. G. 106.

Egamberdieva, D., Wirth, S.J., Alqarawi, A.A., Abd_Allah, E.F., Hashem, A., 2017. Phytohormones and beneficial microbes: essential components for plants to balance stress and fitness [review]. Front. Microbiol. 8.

Eid, A.M., Fouda, A., Abdel-Rahman, M.A., Salem, S.S., Elsaied, A., Oelmüller, R., et al., 2021. Harnessing bacterial endophytes for promotion of plant growth and biotechnological applications: an overview. Plants 10 (5), 935.

Elbeltagy, A., Nishioka, K., Sato, T., Suzuki, H., Ye, B., Hamada, T., et al., 2001. Endophytic colonization and in planta nitrogen fixation by a *Herbaspirillum* sp. isolated from wild rice species. AEM 67 (11), 5285–5293.

Fadiji, A.E., Babalola, O.O., 2020. Exploring the potentialities of beneficial endophytes for improved plant growth. Saudi J. Biol. Sci. 27 (12), 3622–3633.

Fesel, P.H., Zuccaro, A., 2016. β-glucan: crucial component of the fungal cell wall and elusive MAMP in plants. Fungal Genet. Biol. 90, 53–60.

Fouda, A.H., Hassan, S.E.-D., Eid, A.M., Ewais, E.E.-D., 2015. Biotechnological applications of fungal endophytes associated with medicinal plant *Asclepias sinaica* (Bioss.). AOAS 60 (1), 95–104.

Frank, A., Saldierna Guzmán, J., Shay, J., 2017. Transmission of bacterial endophytes. Microorganisms 5 (4), 70.

Fu, S.-F., Wei, J.-Y., Chen, H.-W., Liu, Y.-Y., Lu, H.-Y., Chou, J.-Y., 2015. Indole-3-acetic acid: a widespread physiological code in interactions of fungi with other organisms. Plant Signal. Behav. 10 (8), e1048052.

Gamalero, E., Favale, N., Bona, E., Novello, G., Cesaro, P., Massa, N., et al., 2020. Screening of bacterial endophytes able to promote plant growth and increase salinity tolerance. Appl. Sci. 10 (17), 5767.

González-Teuber, M., Contreras, R.A., Zúñiga, G.E., Barrera, D., Bascuñán-Godoy, L., 2022. Synergistic association with root endophytic fungi improves morpho-physiological and biochemical responses of *Chenopodium quinoa* to salt stress [original research]. Front. Ecol. Evol. 9.

Grabka, R., D'Entremont, T.W., Adams, S.J., Walker, A.K., Tanney, J.B., Abbasi, P.A., et al., 2022. Fungal endophytes and their role in agricultural plant protection against pests and pathogens. Plants 11 (3), 384.

Guerinot, M., 2010. Cell biology of betals and nutrients, Plant Cell Monographs.

Gundel, P.E., Sun, P., Charlton, N.D., Young, C.A., Miller, T.E.X., Rudgers, J.A., 2020. Simulated folivory increases vertical transmission of fungal endophytes that deter herbivores and alter tolerance to herbivory in *Poa autumnalis*. Ann. Bot. 125 (6), 981–991.

Hafeez, M.B., Raza, A., Zahra, N., Shaukat, K., Akram, M.Z., Iqbal, S., et al., 2021. Chapter 22 - Gene regulation in halophytes in conferring salt tolerance. In: Hasanuzzaman, M., Prasad, M.N.V. (Eds.), Handbook of Bioremediation. Academic Press, pp. 341–370.

Hamayun, M., Hussain, A., Khan, S.A., Kim, H.-Y., Khan, A.L., Waqas, M., et al., 2017. Gibberellins producing pndophytic fungus *Porostereum spadiceum* AGH786 rescues growth of salt affected soybean [original research]. Front. Microbiol. 8.

Hamayun, M., Khan, S.A., Khan, A.L., Tang, D.-S., Hussain, J., Ahmad, B., et al., 2010. Growth promotion of cucumber by pure cultures of gibberellin-producing *Phoma* sp. GAH7. World J. Microbiol. Biotechnol. 26, 889–894.

Han, Y., Wang, R., Yang, Z., Zhan, Y., Ma, Y., Ping, S., et al., 2015. 1-Aminocyclopropane-1-carboxylate deaminase from *Pseudomonas stutzeri* A1501 facilitates the growth of rice in the presence of salt or heavy metals. J. Microbiol. Biotechnol. 25 (7), 1119–1128.

Hardoim, P.R., van Overbeek, L.S., van Elsas, J.D., 2008. Properties of bacterial endophytes and their proposed role in plant growth. Trends Microbiol. 16 (10), 463–471.

Hardoim, P.R., Van Overbeek, L.S., Berg, G., Pirttilä, A.M., Compant, S., Campisano, et al., 2015. The hidden world within plants: ecological and evolutionary considerations for defining functioning of microbial endophytes. MMBR 79 (3), 293–320.

Hasanuzzaman, M., Bhuyan, M., Nahar, K., Hossain, M., Mahmud, J., Hossen, M., et al., 2018. Potassium: a vital regulator of plant responses and tolerance to abiotic stresses. Agronomy 8 (3), 31.

Hassan, S.E., 2017. Plant growth-promoting activities for bacterial and fungal endophytes isolated from medicinal plant of *Teucrium polium* L. J. Adv. Res. 8 (6), 687–695.

Hori, K., Matsumoto, S., 2010. Bacterial adhesion: from mechanism to control. Biochem. Eng. J. 48 (3), 424–434.

Ikhajiagbe, B., Anoliefo, G.O., Olise, O.F., Rackelmann, F., Sommer, M., Adekunle, I.J., 2020. Major phosphorus in soils is unavailable, yet critical for plant development. Not. Sci. Biol. 12 (3), 500–535.

Ismail, M.A., Amin, M.A., Eid, A.M., Hassan, S.E.-D., Mahgoub, H.A.M., Lashin, I., et al., 2021. Comparative study between exogenously applied plant growth hormones versus metabolites of microbial endophytes as plant growth-promoting for *Phaseolus vulgaris* L. Cells 10 (5), 1059.

Jasim, B., Sreelakshmi, K.S., Mathew, J., Radhakrishnan, E.K., 2016. Surfactin, iturin, and fengycin biosynthesis by endophytic *Bacillus* sp. from *Bacopa monnieri*. Microb. Ecol. 72 (1), 106–119.

Jeffries, P., 1995. Biology and ecology of mycoparasitism. Can. J. Bot. 73 (S1), 1284–1290.

Jha, Y., Subramanian, R.B., Patel, S., 2011. Combination of endophytic and rhizospheric plant growth promoting rhizobacteria in *Oryza sativa* shows higher accumulation of osmoprotectant against saline stress. Acta Physiol. Plant. 33 (3), 797–802.

Jyoti, R., Sanwal, P., & Saxena, J. 2016. Potassium and its role in sustainable agriculture. In (pp. 235–253).

Kaddes, A., Fauconnier, M.-L., Sassi, K., Nasraoui, B., Jijakli, M.-H., 2019. Endophytic fungal volatile compounds as solution for sustainable agriculture. MOLEFW 24 (6), 1065.

Kamran, M., Imran, Q.M., Ahmed, M.B., Falak, N., Khatoon, A., Yun, B.-W., 2022. Endophyte-mediated stress tolerance in plants: a sustainable strategy to enhance resilience and assist crop improvement. Cells 11 (20), 3292.

Kaur, S., Samota, M.K., Choudhary, M., Choudhary, M., Pandey, A.K., Sharma, A., et al., 2022. How do plants defend themselves against pathogens-biochemical mechanisms and genetic interventions. Physiol. Mol. Biol. Plants 28 (2), 485–504.

Kavroulakis, N., Ntougias, S., Zervakis, G.I., Ehaliotis, C., Haralampidis, K., Papadopoulou, K.K., 2007. Role of ethylene in the protection of tomato plants against soil-borne fungal pathogens conferred by an endophytic *Fusarium solani* strain. JXB 58 (14), 3853–3864.

Khalil, A.M.A., Hassan, S.E., Alsharif, S.M., Eid, A.M., Ewais, E.E., Azab, E., et al., 2021. Isolation and characterization of fungal endophytes isolated from medicinal plant *Ephedra pachyclada* as plant growth-promoting. Biomolecules 11 (2).

Khan, A.L., Hussain, J., Al-Harrasi, A., Al-Rawahi, A., Lee, I.J., 2015. Endophytic fungi: resource for gibberellins and crop abiotic stress resistance. Crit. Rev. Biotechnol. 35 (1), 62–74.

Khan, A.L., Waqas, M., Kang, S.-M., Al-Harrasi, A., Hussain, J., Al-Rawahi, A., et al., 2014. Bacterial endophyte *Sphingomonas* sp. LK11 produces gibberellins and IAA and promotes tomato plant growth. J. Microbiol. 52, 689–695.

Khan, Z., Guelich, G., Phan, H., Redman, R., Doty, S., 2012. Bacterial and yeast endophytes from poplar and willow promote growth in crop plants and grasses. ISRN Agron. 2012, 1–11.

Khare, E., Mishra, J., Arora, N.K., 2018. Multifaceted interactions between endophytes and plant: developments and prospects. Front. Microbiol. 9, 2732.

Kim, S.H., Vujanovic, V., 2016. Relationship between mycoparasites lifestyles and biocontrol behaviors against *Fusarium* spp. and mycotoxins production. Appl. Microbiol. Biotechnol. 100 (12), 5257–5272.

Kloepper, J.W., McInroy, J.A., Liu, K., Hu, C.-H., 2013. Symptoms of fern distortion syndrome resulting from inoculation with opportunistic endophytic fluorescent *Pseudomonas* spp. PLoS One 8 (3), e58531.

Knudsen, I., Hockenhull, J., Jensen, D.F., Gerhardson, B., Hökeberg, M., Tahvonen, R., et al., 1997. Selection of biological control agents for controlling soil and seed-borne diseases in the field. Eur. J. Plant Pathol. 103, 775–784.

Krapp, A., 2015. Plant nitrogen assimilation and its regulation: a complex puzzle with missing pieces. Curr. Opin. Plant Biol. 25, 115–122.

Kusari, S., Hertweck, C., Spiteller, M., 2012. Chemical ecology of endophytic fungi: origins of secondary metabolites. Chem. Biol. 19 (7), 792–798.

Lacava, P.T., Silva-Stenico, M.E., Araújo, W.L., Simionato, A.V.C., Carrilho, E., Tsai, S.M., et al., 2008. Detection of siderophores in endophytic bacteria *Methylobacterium* spp. associated with *Xylella fastidiosa* subsp. pauca. Pesqui. Agropecu. Bras. 43 (4), 521–528.

Lahlali, R., Hijri, M., 2010. Screening, identification and evaluation of potential biocontrol fungal endophytes against *Rhizoctonia solani* AG3 on potato plants. FEMS 311 (2), 152–159.

Latz, M.A.C., Jensen, B., Collinge, D.B., Jørgensen, H.J.L., 2018. Endophytic fungi as biocontrol agents: elucidating mechanisms in disease suppression. Plant Ecol. Divers. 11 (5–6), 555–567.

Leitão, A.L., Enguita, F.J., 2016. Gibberellins in Penicillium strains: challenges for endophyte-plant host interactions under salinity stress. Microbiol. Res. 183, 8–18.

Li, J., Zheng, B., Hu, R., Liu, Y., Jing, Y., Xiao, Y., et al., 2018. Pseudomonas species isolated from tobacco seed promote root growth and reduce lead contents in *Nicotiana tobacum* K326. Can. J. Microbiol. 65 (3), 214–223.

Liu, H., Carvalhais, L.C., Crawford, M., Singh, E., Dennis, P.G., Pieterse, C.M.J., et al., 2017. Inner plant values: diversity, colonization and benefits from endophytic bacteria [review]. Front. Microbiol. 8.

Liu, R., Yang, L., Zou, Y., Wu, Q., 2022. Root-associated endophytic fungi modulate endogenous auxin and cytokinin levels to improve plant biomass and root morphology of trifoliate orange. Hortic. Plant. J. .

Liu, W.Y.Y., Poobathy, R., 2021. Biofertilizer utilization in forestry. Biofertilizers: Study Impact. pp. 1–37.

Lopes, M.J.d S., Dias-Filho, M.B., Gurgel, E.S.C., 2021. Successful plant growth-promoting microbes: inoculation methods and abiotic factors [review]. Front. Sustain. Food Syst. 5.

Lugtenberg, B.J., Caradus, J.R., Johnson, L.J., 2016. Fungal endophytes for sustainable crop production. FEMS 92 (12).

Ma, L., Liu, X., Lv, W., Yang, Y., 2022. Molecular mechanisms of plant responses to salt stress. Front. Plant Sci. 13, 934877.

Manjunatha, N., Li, H., Sivasithamparam, K., Jones, M.G.K., Edwards, I., Wylie, S.J., et al., 2022. Fungal endophytes from salt-adapted plants confer salt tolerance and promote growth in wheat (*Triticum aestivum* L.) at early seedling stage. Microbiol. SGM 168 (8).

Marciano Marra, L., Fonsêca Sousa Soares, C.R., de Oliveira, S.M., Avelar Ferreira, P.A., Lima Soares, B., de Fráguas Carvalho, R., et al., 2012. Biological nitrogen fixation and phosphate solubilization by bacteria isolated from tropical soils. Plant Soil. 357, 289–307.

Martín-Barranco, A., Thomine, S., Vert, G., Zelazny, E., 2021. A quick journey into the diversity of iron uptake strategies in photosynthetic organisms. Plant Signal. Behav. 16 (11), 1975088.

Matos, A.D., Gomes, I.C., Nietsche, S., Xavier, A.A., Gomes, W.S., DOS SANTOS, J.A., et al., 2017. Phosphate solubilization by endophytic bacteria isolated from banana trees. An. Acad. Bras. Cienc. 89, 2945–2954.

Mattos, K.A., Pádua, V.L.M., Romeiro, A., Hallack, L.F., Neves, B.C., Ulisses, T.M.U., et al., 2008. Endophytic colonization of rice (*Oryza sativa* L.) by the diazotrophic bacterium *Burkholderia kururiensis* and its ability to enhance plant growth. An. Acad. Bras. Ciênc. 80 (3), 477–493.

Mehta, P., Sharma, R., Putatunda, C., Walia, A., 2019. Endophytic fungi: role in phosphate solubilization. Advances in Endophytic Fungal Research: Present Status and Future Challenges. pp. 183–209.

Menegat, S., Ledo, A., Tirado, R., 2022. Greenhouse gas emissions from global production and use of nitrogen synthetic fertilisers in agriculture. Sci. Rep. 12 (1).

Meneses, C.H., Rouws, L.F., Simões-Araújo, J.L., Vidal, M.S., Baldani, J.I., 2011. Exopolysaccharide production is required for biofilm formation and plant colonization by the nitrogen-fixing endophyte *Gluconacetobacter diazotrophicus*. MPMI 24 (12), 1448–1458.

Mengistu, A.A., 2020. Endophytes: colonization, behaviour, and their role in defense mechanism. Int. J. Microbiol. 2020, 6927219.

Morsy, M., Cleckler, B., Armuelles-Millican, H., 2020. Fungal endophytes promote tomato growth and enhance drought and salt tolerance. Plants 9 (7), 877.

Müller, M., Munné-Bosch, S., 2021. Hormonal impact on photosynthesis and photoprotection in plants. Plant Physiol. 185 (4), 1500–1522.

Muthu Narayanan, M., Ahmad, N., Shivanand, P., Metali, F., 2022. The role of endophytes in combating fungal- and bacterial-induced stress in plants. MOLEFW 27 (19), 6549.

Naveed, M., Mitter, B., Reichenauer, T.G., Wieczorek, K., Sessitsch, A., 2014. Increased drought stress resilience of maize through endophytic colonization by Burkholderia phytofirmans PsJN and Enterobacter sp. FD17. EEB 97, 30–39.

Newman, M.-A., Sundelin, T., Nielsen, J., Erbs, G., 2013. MAMP (microbe-associated molecular pattern) triggered immunity in plants [Review]. Front. Plant Sci. 4.

Noinaj, N., Guillier, M., Barnard, T., Buchanan, S., 2010. TonB-Dependent Transporters: regulation, structure, and function. Annu. Rev. Microbiol. 64, 43–60.

Okoro Gideon, O., Onu Ogbonnaya, B., Ngasoh Felix, G., Namessan, N., 2019. The effect of climate change on abiotic plant stress: a review. In: Alexandre Bosco de, O. (Ed.), Abiotic and Biotic Stress in Plants. IntechOpen (pp. Ch. 5).

Omomowo, O.I., Babalola, O.O., 2019. Bacterial and fungal endophytes: tiny giants with immense beneficial potential for plant growth and sustainable agricultural productivity. Microorganisms 7 (11), 481.

Ongena, M., Jacques, P., 2008. Bacillus lipopeptides: Versatile weapons for plant disease biocontrol. Trends Microbiol. 16 (3), 115–125.

Ortíz-Castro, R., Contreras-Cornejo, H.A., Macías-Rodríguez, L., López-Bucio, J., 2009. The role of microbial signals in plant growth and development. Plant Signal. Behav. 4 (8), 701–712.

Peña-Yam, L.P., Ruíz-Sánchez, E., Barboza-Corona, J.E., Reyes-Ramírez, A., 2016. Isolation of mexican *Bacillus* species and their effects in promoting growth of chili pepper (*Capsicum annuum* L. cv Jalapeño). Indian J. Microbiol. 56 (3), 375–378.

Perea-Molina, P.A., Pedraza-Herrera, L.A., Beauregard, P.B., Uribe-Vélez, D., 2022. A biocontrol *Bacillus velezensis* strain decreases pathogen *Burkholderia glumae* population and occupies a similar niche in rice plants. Biol. Control 176, 105067.

Pieterse, C.M.J., Zamioudis, C., Berendsen, R.L., Weller, D.M., Wees, S.C.M.V., Bakker, P.A.H.M., 2014. Induced systemic resistance by beneficial microbes. Annu. Rev. Phytopathol. 52 (1), 347–375.

Pinski, A., Betekhtin, A., Hupert-Kocurek, K., Mur, L.A.J., Hasterok, R., 2019. Defining the genetic basis of plant–endophytic bacteria interactions. Int. J. Mol. Sci. 20 (8), 1947.

Prieto, P., Schilirò, E., Maldonado-González, M.M., Valderrama, R., Barroso-Albarracín, J.B., Mercado-Blanco, J., 2011. Root hairs play a key role in the endophytic colonization of olive roots by *Pseudomonas* spp. with biocontrol activity. Microb. Ecol. 62, 435–445.

Putri, N.D., Sulistyowati, L., Aini, L.Q., Muhibuddin, A., Trianti, I., 2022. Screening of endophytic fungi as potential antagonistic agents of *Pyricularia oryzae* and evaluation of their ability in producing hydrolytic enzymes. Biodiversitas 23 (2).

Radhakrishnan, R., Khan, A.L., Kang, S.M., Lee, I.-J., 2015. A comparative study of phosphate solubilization and the host plant growth promotion ability of *Fusarium verticillioides* RK01 and *Humicola* sp. KNU01 under salt stress. Ann. Microbiol. 65 (1), 585–593.

Radzki, W., Gutierrez Mañero, F.J., Algar, E., Lucas García, J.A., García-Villaraco, A., Ramos Solano, B., 2013. Bacterial siderophores efficiently provide iron to iron-starved tomato plants in hydroponics culture. Anton. Leeuw. 104 (3), 321–330.

Rana, K.L., Kour, D., Kaur, T., Negi, R., Devi, R., Yadav, N., et al., 2023. Endophytic nitrogen-fixing bacteria: untapped treasurer for agricultural sustainability. J. Appl. Biol. Biotechnol. 11 (2), 75–93.

Rangel de Souza, A., De Souza, S., De Oliveira, M., Ferraz, T., Figueiredo, F., Da Silva, N., et al., 2016. Endophytic colonization of *Arabidopsis thaliana* by *Gluconacetobacter diazotrophicus* and its effect on plant growth promotion, plant physiology, and activation of plant defense. Plant Soil 399, 257–270.

Rashid, M.H., Khan, A., Hossain, M.T., Chung, Y.R., 2017. Induction of systemic resistance against aphids by endophytic *Bacillus velezensis* YC7010 via expressing phytoalexin deficient4 in Arabidopsis. Front. Plant Sci. 8, 211.

Rawat, P., Das, S., Shankhdhar, D., Shankhdhar, S.C., 2021. Phosphate-solubilizing microorganisms: mechanism and their role in phosphate solubilization and uptake. J. Soil Sci. Plant Nutr. 21 (1), 49–68.

Rungin, S., Indananda, C., Suttiviriya, P., Kruasuwan, W., Jaemsaeng, R., Thamchaipenet, A., 2012. Plant growth enhancing effects by a siderophore-producing endophytic streptomycete isolated from a Thai jasmine rice plant (*Oryza sativa* L. cv. KDML105). Anton. Leeuw. 102, 463–472.

Růzicka, K., Ljung, K., Vanneste, S., Podhorská, R., Beeckman, T., Friml, J., et al., 2007. Ethylene regulates root growth through effects on auxin biosynthesis and transport-dependent auxin distribution. Plant Cell 19 (7), 2197–2212.

Sabaté, D.C., Brandan, C.P., Petroselli, G., Erra-Balsells, R., Audisio, M.C., 2018. Biocontrol of *Sclerotinia sclerotiorum* (Lib.) de Bary on common bean by native lipopeptide-producer Bacillus strains. Microbiol. Res. 211, 21–30.

Santi, C., Bogusz, D., Franche, C., 2013. Biological nitrogen fixation in non-legume plants. Ann. Bot. 111 (5), 743–767.

Schaller, G.E., 2012. Ethylene and the regulation of plant development. BMC Biol. 10 (1)9.

Schröder, I., Johnson, E., de Vries, S., 2003. Microbial ferric iron reductases. FEMS 27 (2–3), 427–447.

Shahzad, R., Waqas, M., Khan, A.L., Asaf, S., Khan, M.A., Kang, S.M., et al., 2016. Seed-borne endophytic *Bacillus amyloliquefaciens* RWL-1 produces gibberellins and regulates endogenous phytohormones of *Oryza sativa*. Plant. Physiol. Biochem. 106, 236–243.

Singh, P., Sindhu, S., 2013. Potassium solubilization by rhizosphere bacteria: influence of nutritional and environmental conditions. Res. J. Microbiol. 3.

Singh, R.P., Shelke, G.M., Kumar, A., Jha, P.N., 2015. Biochemistry and genetics of ACC deaminase: a weapon to "stress ethylene" produced in plants. Front. Microbiol. 6, 937.

Sivojiene, D., Kacergius, A., Baksiene, E., Maseviciene, A., Zickiene, L., 2021. The influence of organic fertilizers on the abundance of soil microorganism communities, agrochemical indicators, and yield in east lithuanian light soils. Plants 10 (12), 2648.

Spaepen, S., Vanderleyden, J., Remans, R., 2007. Indole-3-acetic acid in microbial and microorganism-plant signaling. FEMS 31 (4), 425–448.

Strobel, G., Daisy, B., 2003. Bioprospecting for microbial endophytes and their natural products. Microbiol. Mol. Biol. Rev. 67 (4), 491–502.

Strobel, G., 2018. The emergence of endophytic microbes and their biological promise. J. Fungi 4 (2), 57.

Taurian, T., Anzuay, M.S., Angelini, J.G., Tonelli, M.L., Ludueña, L., Pena, D., et al., 2010. Phosphate-solubilizing peanut associated bacteria: screening for plant growth-promoting activities. Plant Soil 329 (1), 421–431.

Tripathi, D.K., Singh, S., Gaur, S., Singh, S., Yadav, V., Liu, S., et al., 2018. Acquisition and homeostasis of iron in higher plants and their probable role in abiotic stress tolerance [mini review]. Front. Environ. Sci. 5.

Trivedi, P., Sa, T., 2008. *Pseudomonas corrugata* (NRRL B-30409) mutants increased phosphate solubilization, organic acid production, and plant growth at lower temperatures. Curr. Microbiol. 56, 140–144.

Van Dijk, M., Morley, T., Rau, M.L., Saghai, Y., 2021. A meta-analysis of projected global food demand and population at risk of hunger for the period 2010–2050. Nat. Food 2 (7), 494–501.

Vargas, L., Santa Brígida, A.B., Mota Filho, J.P., De Carvalho, T.G., Rojas, C.A., Vaneechoutte, D., et al., 2014. Drought tolerance conferred to sugarcane by association with gluconacetobacter diazotrophicus: a transcriptomic view of hormone pathways. PLoS One 9 (12), e114744.

Vassiliadis, S., Reddy, P., Hemsworth, J., Spangenberg, G.C., Guthridge, K.M., Rochfort, S.J., 2023. Quantitation and distribution of epichloë-derived alkaloids in perennial ryegrass tissues. Metabolites 13 (2), 205.

Vendan, R.T., Yu, Y.J., Lee, S.H., Rhee, Y.H., 2010. Diversity of endophytic bacteria in ginseng and their potential for plant growth promotion. J. Microbiol. 48 (5), 559–565.

Verma, P., Yadav, A.N., Khannam, K.S., Panjiar, N., Kumar, S., Saxena, A.K., et al., 2015. Assessment of genetic diversity and plant growth promoting attributes of psychrotolerant bacteria allied with wheat (*Triticum aestivum*) from the Northern hills zone of India. Ann. Microbiol. 65 (4), 1885–1899.

Verma, S.K., Sahu, P.K., Kumar, K., Pal, G., Gond, S.K., Kharwar, R.N., et al., 2021. Endophyte roles in nutrient acquisition, root system architecture development and oxidative stress tolerance. J. Appl. Microbiol. 131 (5), 2161–2177.

Verma, V., Singh, S., Prakash, S., 2011. Bio-control and plant growth promotion potential of siderophore producing endophytic *Streptomyces* from *Azadirachta indica* A. Juss. J. Basic. Microbiol. 51 (5), 550–556.

Viterbo, A., Horwitz, B.A., 2010. Mycoparasitism. Cellular and Molecular Biology of Filamentous Fungi. pp. 676–693.

Vlot, A.C., Pabst, E., Riedlmeier, M., 2017. Systemic Signalling in Plant Defence. eLS, pp. 1–9.

Waqas, M., Khan, A.L., Kamran, M., Hamayun, M., Kang, S.-M., Kim, Y.-H., et al., 2012. Endophytic fungi produce gibberellins and indoleacetic acid and promotes host-plant growth during stress. MOLEFW 17 (9), 10754–10773.

Wei, C.-Y., Lin, L., Luo, L.-J., Xing, Y.-X., Hu, C.-J., Yang, L.-T., et al., 2014. Endophytic nitrogen-fixing *Klebsiella variicola* strain DX120E promotes sugarcane growth. Biol. Fertil. Soils 50 (4), 657–666.

Xia, Y., Liu, J., Chen, C., Mo, X., Tan, Q., He, Y., et al., 2022. The Multifunctions and future prospects of endophytes and their metabolites in plant disease management. Microorganisms 10 (5), 1072.

Xie, Z., Chu, Y., Zhang, W., Lang, D., Zhang, X., 2019. Bacillus pumilus alleviates drought stress and increases metabolite accumulation in *Glycyrrhiza uralensis* Fisch. EEB 158, 99–106.

Xin, G., Zhang, G., Kang, J.W., Staley, J.T., Doty, S.L., 2009. A diazotrophic, indole-3-acetic acid-producing endophyte from wild cottonwood. Biol. Fertil. Soils 45 (6), 669–674.

Xu, G., Fan, X., Miller, A.J., 2012. Plant nitrogen assimilation and use efficiency. Annu. Rev. Plant Biol. 63 (1), 153–182.

Xu, X.-H., Wang, C., Li, S.-X., Su, Z.-Z., Zhou, H.-N., Mao, L.-J., et al., 2015. Friend or foe: differential responses of rice to invasion by mutualistic or pathogenic fungi revealed by RNAseq and metabolite profiling. Sci. Rep. 5 (1), 1–14.

Yamaji, K., Watanabe, Y., Masuya, H., Shigeto, A., Yui, H., Haruma, T., 2016. Root fungal endophytes enhance heavy-metal stress tolerance of *Clethra barbinervis* growing naturally at mining sites via growth enhancement, promotion of nutrient uptake and decrease of heavy-metal concentration. PLoS One 11 (12), e0169089.

Zhang, X., Tong, J., Dong, M., Akhtar, K., He, B., 2022. Isolation, identification and characterization of nitrogen fixing endophytic bacteria and their effects on cassava production. PeerJ 10, e12677.

Zhou, X.R., Dai, L., Xu, G.F., Wang, H.S., 2021. A strain of *Phoma* species improves drought tolerance of Pinus tabulaeformis. Sci. Rep. 11 (1), 7637.

Zhu, H.-J., Sun, L.-F., Zhang, Y.-F., Zhang, X.-L., Qiao, J.-J., 2012. Conversion of spent mushroom substrate to biofertilizer using a stress-tolerant phosphate-solubilizing *Pichia farinose* FL7. Bioresour. Technol. 111, 410–416.

Zhu, Y., Qi, B., Hao, Y., Liu, H., Sun, G., Chen, R., et al., 2021. Appropriate $NH_4{}^+/NO_3{}^-$ ratio triggers plant growth and nutrient uptake of flowering chinese cabbage by optimizing the ph value of nutrient solution. Front. Plant Sci. 12, 656144.

Zuo, Y., Zhang, F., 2011. Soil and crop management strategies to prevent iron deficiency in crops. Plant Soil 339, 83–95.

CHAPTER

18

Role of endophytes on plant protection and resilience

Wiwiek Harsonowati[1], *Hafiz Muhammad Ahmad*[2], *Dyah Manohara*[1], *Sri Widyaningsih*[1], *Saira Ishaq*[3], *Sri Widawati*[4], *Suliasih*[4] *and Deciyanto Soetopo*[1]

[1]Research Center for Horticultural and Estate Crops, National Research and Innovation Agency (BRIN), Cibinong, Indonesia [2]Department of Bioinformatics and Biotechnology, Government College University, Faisalabad, Pakistan [3]Department of Food Sciences and Technology, University of Poonch, Rawalakot, Pakistan [4]Research Center for Applied Microbiology, National Research and Innovation Agency (BRIN), Cibinong, Indonesia

18.1 Introduction

A phenomenon referred to as a "cry for help" occurs when a plant is subjected to stressful conditions, whether those conditions are biotic or abiotic in nature. This allows the plant to recruit groups of microbes to alleviate specific potentially harmful effects. Endophytic microbes (EMOs), including bacteria and fungi, are plant-beneficial endosymbionts, living intra- and intercellularly of host tissues and promisingly attained much more applicability for multi-dynamic and sustainable agriculture systems (Ahmad et al., 2022a). The mutualistic interactions by colonizing plant tissues both intercellularly and/or intracellularly is a well-versed component of their lifestyle and most of the modern research clearly shows that the survival and health of plants are very much dependent upon EMOs (Potshangbam et al., 2017). Several agrochemicals, such as fertilizers, pesticides, and insecticides, are used to improve plant health and yields; however, these chemical residues cause substantial problems for agricultural products' quality, human health, and the environment.

Endophytes can directly benefit host plants by increasing nutrient absorption and altering growth and stress-related phytohormones (Lin et al., 2020). Endophytes can indirectly

Biostimulants in Plant Protection and Performance
DOI: https://doi.org/10.1016/B978-0-443-15884-1.00011-7

improve plant health by targeting pests and diseases using antibiotics, lytic enzymes, nutrient limitation, and other methods (Khan et al., 2019), see Fig. 18.1.

EMOs are becoming more popular as a result of the diverse bio-molecule production of potential implications. It is well-known that bio-products synthesized by EMOs have superiority over synthetic origin (Parrino et al., 2019). These EMOs symbiotically live in association with host plants without damaging their tissues (Aroca et al., 2013). The evaluation of the interactive and prospective roles of EMOs and plants is an effective way to meet the future challenges of important crop production and protection for food security.

In this framework, plant-EMO interactions are gaining attention for various endophytic activities to raise awareness about the use of EMOs in organic farming. Additionally, the growing interest in products of EMOs is the synthesis of metal-based nanoparticles, indicating a forceful action against biofilm (Parrino et al., 2019; Wang et al., 2019). These metal-based nanoparticles with medicinal applications have been reported to accelerate defense responses against pathogen attacks by causing their death on the biofilm surface and then moving to its inner surface, directly interacting with the internal pathogen (Parrino et al., 2019). These biofilms are recognized to be intricately linked to the increasing defense against the pathogen, posing a risk to health by the discovery of new biomolecules capable of dispersing pre-synthesized biofilms. One eco-friendly way to combat pathogen communication in biofilms is through the use of quorum sensing (QS) inhibitors, which are produced by EMOs (Mookherjee et al., 2018).

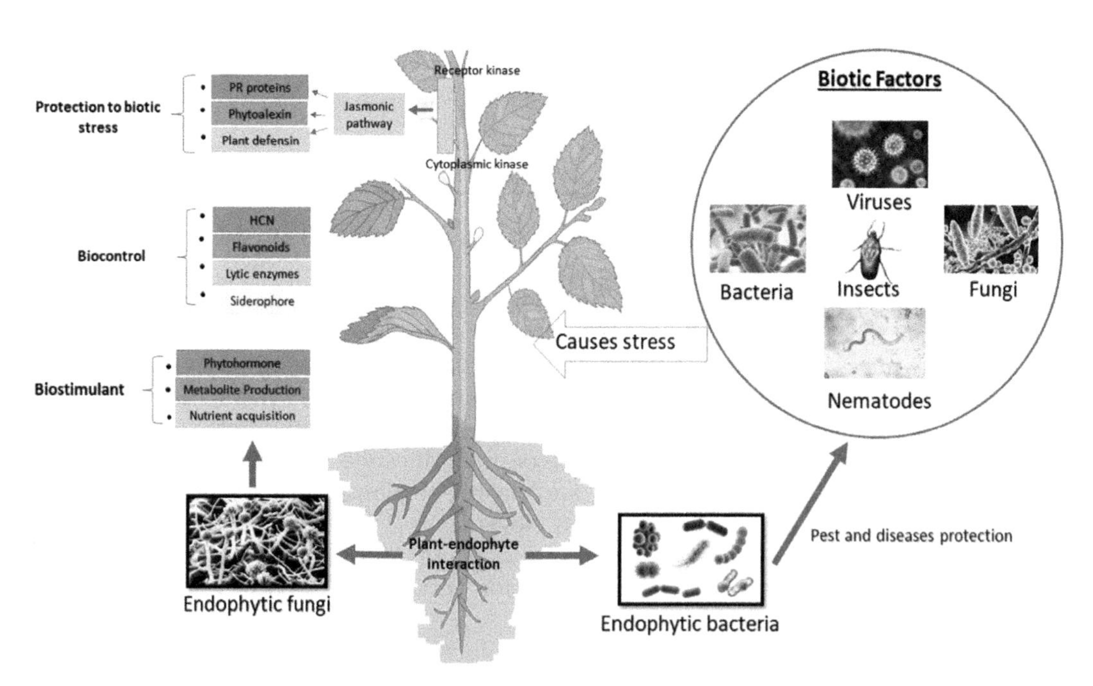

FIGURE 18.1 Fungal and bacterial endophytes substantially improve plant health by targeting biotic (pests and diseases) stresses.

The proteins or enzymes made by EMOs are more stable than those made by plants or animals, which is yet another important feature of these organisms. Consequently, the enzyme of microbial origin has attained much more applicability due to their potential uses like bio-energy, bio-bleaching, bio-fuel production, bio-conversion, etc. Most of the recent drift toward isolation and identification of enzymes from the harshest habitats like amylases, cellulases, laccases, pectinases, and xylanases highlighted their different properties for industrial applicability and biotechnological interest intended for subsequent improvement (Golgeri et al., 2022). Bacterial endophytes, in particular, have the tendency to biosynthesize specific antibiotic compounds that are effective against a wide range of pathogens that have a deterrent effect on plant growth (Jiao et al., 2021).

18.2 Role of endophytes in modulation the crop development

18.2.1 Bacterial endophytes

Microbes most likely endophytic bacteria that reside and move freely in the rhizosphere and inside the plants play a broad-spectrum role in several mechanisms benefiting the plants (Ahmad et al., 2022a). Endophytes present a diverse and mysterious world in both the internal and external of plants. To date, bacterial endophytes have been discovered in all plant species identified and reported. Microbial endophytes establish a network within their host plants via metabolic connections, allowing them to produce signal molecules (Marchin et al., 2020). Furthermore, genetic recombination between EMOs and host plants helps to mimic the plant's physio-chemical functioning and generates similar dynamic signals and molecules (Badri et al., 2013). EMOs produce phytochemicals related to the host plants independently in culture conditions. there are 54 (fifty-four) major bioactive compounds of agricultural importance. The role of endophyte genome mining and "multi-Omics" techniques in agricultural sustainability (Dwibedi et al., 2022).

Root tissues were hampered by a diverse range of endophytic bacteria when compared to leaves and other organs (Gouda et al., 2018). A close and significant association and exchanges among various species of endophytic bacteria and fungi and in different tissues were also found in many economically important plant species. The specific tissue type has been found to play a more viable and dominant role in community structuring modeling than cultivars (Gouda et al., 2018). Endophytic bacteria can both, directly and indirectly, modulate stress and growth-related phytohormone activity by enhancing the uptake of essential nutrients and by targeting pathogenic organisms. The specificity diversity and host range of these novel bacterial endophytes provide a safe and viable use of bio-inoculants to promote the growth potential of the plant kingdom (Afzal et al., 2019).

Bacterial endophytes are reported to be present massively in plant tissues like floral structures making the floral microbiome (Massoni et al., 2021). The origin of bacterial endophytes of floral organs is mostly enthralling since these can be propagated via seed as stated in theories rather than displaying vertical transmissions within host plants. In addition to these seeds, endophytes are gaining popularity among scientists because they are a reliable example of host-microbe co-evolution, as well as a potent strategy for producing crop seeds harboring plant-growth-promoting microbes (Abdelfattah et al., 2022).

Many agronomists and plant scientists are becoming interested in the diverse potential of EMOs to alleviate plant stresses through pathogen-biocontrol (Compant et al., 2021; Rolli et al., 2021). Currently, the most desired anti-biofilm compounds can prevent or inhibit microbial adhesions to restrict the formation of biofilms and disperse the already established biofilm by disrupting extra- and intra-cellular connections for biofilm formation (anti-Quorum sensing). The identification and classification of a larger collection of endophytic species are establishing an important source for the sustainability of agriculture and food production with minimum input in marginal areas by increasing crop yield in exploiting marginal lands and increasing the yield of crops in climatic change systems (Riva et al., 2022). Much more research still needs to be accomplished to resolve these issues before the consistent application of EMOs in the improvement of plant performance (Vaccaro et al., 2022).

By promoting growth and antioxidant activity, the proportion of bacterial endophytes has a significant impact on the health and fitness of seedlings. (Gerna et al., 2020). Their role in maintaining plant vigor at the seedling stage of growth was demonstrated by distinct bacterial colonization in seed embryos and endosperms (Kuźniar et al., 2020). Whereas, insufficiency in potency and resilience of seed was reported in the case of reduction in the population of endophytes (Rodríguez et al., 2020). This knowledge is critical to understanding the characteristic role of seed-inhibiting bacterial communities in improving nutrient utilization for both seed and seedling robustness.

In the case of plant flowers, an interchange of EMOs among plants and various insect species revealed the effects of pollinators presenting distinct forms related to the pollinations using various vectors (Chandra et al., 2019). The intrinsic ability of bacterial endophytes in enhancing plant growth and development, bioremediation, biotic and abiotic stress tolerance, and as bio-fertilizers justify their multi-faceted benefits in agriculture sustainability (Mellidou and Karamanoli, 2022).

18.2.2 Fungal endophytes

Endophytic fungi have been studied extensively in recent decades to better understand how they interact with their hosts, the types of relationships they establish, and the potential effects of this interaction. The plants benefit from endophytic fungi via indirect or direct mechanisms. The indirect beneficial mechanism comprises biotic and abiotic confrontation (Woo et al., 2020). The biodiversity of fungal endophytes is extensive and has been reported from nearly all plants/crops worldwide (Tiwari et al., 2020). According to Rodriguez et al. (2009), fungal endophytes are classified into four groups based on the parts they colonize and their transmittance: (1) Class I are Clavicipitaceous fungi (reside in grasses with vertical transmittance), (2) Class- II-IV are non-Clavicipitaceous fungi (Ascomycota and Basidiomycota), wherein Class-II endophytes inhabits roots, shoots, and rhizomes, and transmit via seeds or rhizomes (i.e., both horizontally and vertically), (3) Class-III inhabits leaves/shoots (transmitted horizontally), and (4) class-IV are typical Ascomycetes, forming conidia and resides in roots (transmitted horizontally) of the plants. Class, I endophytes are free-living symbiotic species such as *Epichloe festucae*, and *Neotyphodium tembladerae* which are associated with grasses, rushes, and sedges. Their associations help plants in nematode resistance, anti-fungal compound production, and abiotic stress mitigation, including drought and metal tolerance (Ripa et al., 2019; Tufail et al., 2021). Class II endophytes including *Mycophycia ascophylli*, *Phoma sp.*, and *Arthrobotrys sp.* increase plant root, and shoot biomass and confer tolerance against disease, drought, desiccation heat, and salinity. Class III contains species such as *Ustilago maydis*, *Phyllosticta sp.*, and *Colletotrichum sp.*, which are known as hyper-diverse fungal endophytes mostly inhabiting above-ground tissues and are not habitat-specific. They promote interaction among various microorganisms' populations (Subramaniam et al., 2020). Class IV endophytes are root-associated sterile Ascomycetous fungi forming melanized structures prevalent in high-stress environments worldwide for example, *Chloridium paucisporum*, *Phialocephala dimorphosphora*, and *Cladophialophora chaetospira* (Chaudhary et al., 2022; Harsonowati et al., 2020; Madhaiyan et al., 2013).

Dark septate endophytes (DSE) are fungal endophytes of the class IV. DSE are conidial or sterile septate fungal endophytes, usually isolated from healthy plants, that form melanized structures including inter- and intracellular hyphae and microsclerotia in the roots (Rodriguez et al., 2009). Several studies reported beneficial associations between DSE and plants including nutrient uptake (Della Monica et al., 2015; WU et al., 2021; Yakti et al., 2018), growth-promoting (Harsonowati et al., 2020; WU et al., 2021), tolerance to abiotic stress, such as drought (He et al., 2019; Li et al., 2018; Liu and Wei, 2021), salinity (Gonzalez Mateu et al., 2020; Hou et al., 2021), and pathogens (Harsonowati et al., 2020; Marian et al., 2022; Tellenbach et al., 2013). To date, other species of DSE have been characterized; however, their roles in the interactions with their hosts are unpredictable potential traits (Harsonowati et al., 2022), thereby, positioning DSE as potential new tools for crop production and achieving sustainable agriculture.

Fungal endophytes may also play significant roles in the host plant through phytoremediation, Phyto-immobilization, and phytotransformation (Yasin et al., 2018), see Fig. 18.2. The direct beneficial mechanism encompasses nitrogen fixation, antimicrobial metabolite production, and phosphate solubilization (Msimbira and Smith, 2020). They may biodegrade biomass and recycle them in the environment, enhancing nitrogen availability, and zinc and phosphorus uptake for the host resulting in phytoimmobilization (Hartman and Tringe, 2019; Shah et al., 2021). Phytoimmobilization eventually enhancements the plants to tolerate abiotic stresses via immobilizing osmolytes and also stabilizes membrane ion conductivity in stress conditions (Hartman and Tringe, 2019). On the other hand, phytotransformation is the procedure of neutralizing pollutants in soil, for example, insecticides. Endophytic fungi such as *Exophiala* sp., *Neotyphodium* sp., *and Piriformospora* sp., related to many crop species, are reported to remediate from metal pollutants in soil, eluding phytotoxicity (Qessaoui et al., 2019). However, bioremediation utilizing fungal endophytes needs emphasis study in the near future integrating endophyte nanosensors to divulge its impending in ecological management (Rajesh and Ravishankar Rai, 2014). Overall, most of the fungal endophytes (*Ascomycota* and *Basidiomycota*) reported in crops produce secondary metabolites (Marchin et al., 2020). These endophytes that deliberate host plants resistance which may be natural like phytohormones or any nutrients or alkaloids, terpenes, quinines, benzopyranones, chinones, phenolic acids, terpenoids, steroids, flavonoids, hydrocarbons, etc., in response to stressed condition (Chouhan et al., 2022); depending upon fungal endophyte and host species (Chaudhary et al., 2022). Therefore, it is essential to gain comprehensive information on their host range, colonization, specificity, transmission, and host-endophyte interaction.

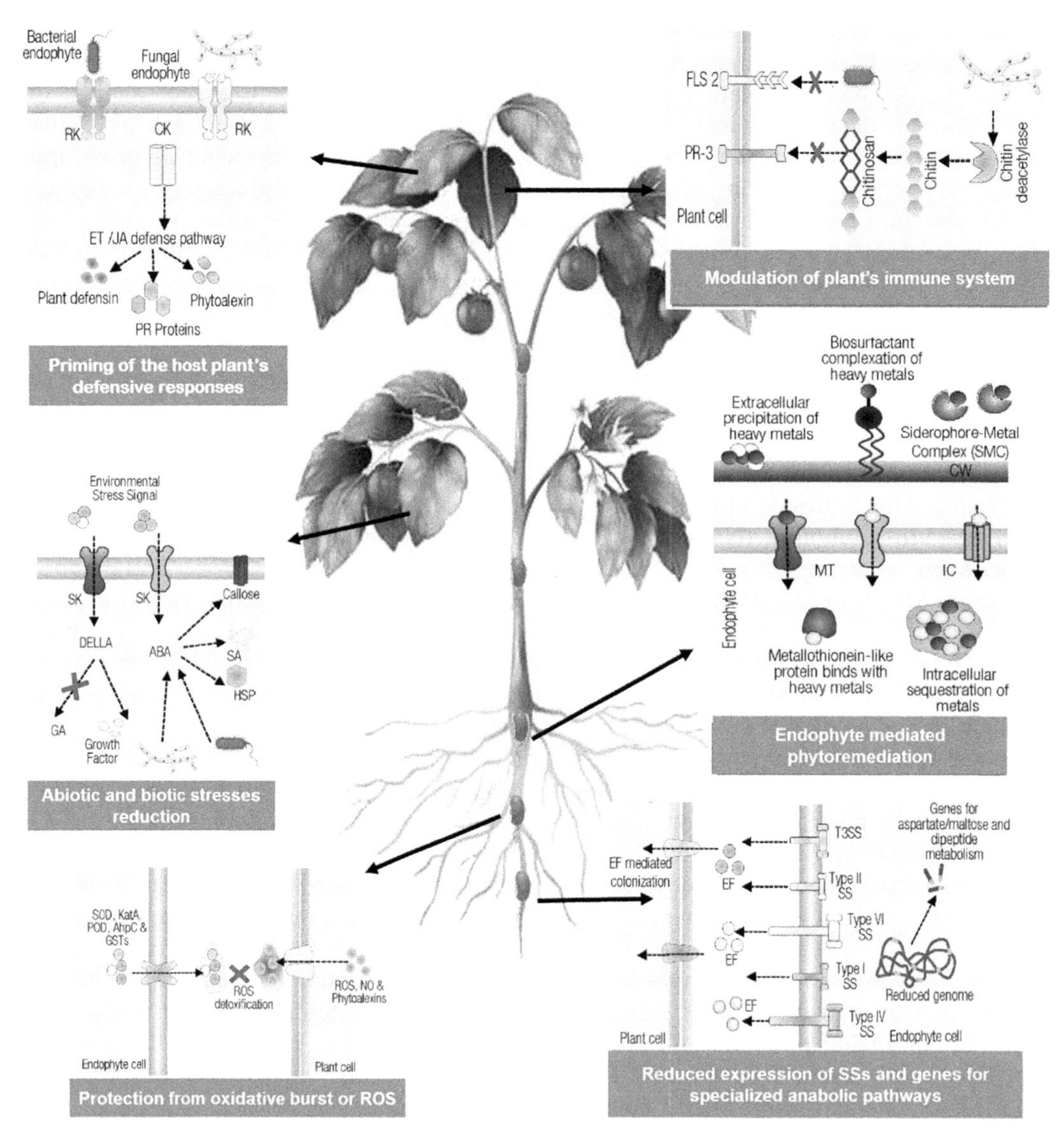

FIGURE 18.2 Plant-EMOs interaction promotes plant health and resilience against abiotic and biotic stresses. Endophytes modulate the host plant's defensive responses to phytopathogens and positively induce stress-responsive gene expression.

18.3 Role of endophytes against abiotic stress

Plants are unable to achieve their full genetic potential for growth and production because of a variety of environmental conditions such as salinity, nutritional fluctuations, and drought (Ahmad et al., 2022b; Ilyas et al., 2020). Abiotic stressors severely affect plant growth; hence, a plant needs a robust system or mechanism for survival. Numerous microbes have been shown to regularly consume nutrients from various plants, with some of their interactions being beneficial to the host and others being harmful. various EMO species that create a network with the host and are closely associated with it include bacteria, fungi, and actinomycetes (Tyagi et al., 2022). These EMOs are also helped because they can exist in plant tissues without doing much harm or receiving anything in return other than a place to call habitat and refuge from unfavorable conditions like a shortage of carbon sources (Ripa et al., 2019). Endophytic bacteria help organisms adapt to low temperatures more effectively by reducing cellular damage, increasing photosynthetic activity, and gathering molecules associated with cold stress such as starch, proline, and phenolic substances (Ritonga and Chen, 2020). Bacterial endophytes such as *Enterobacter cloacae*, *Bacillus subtilis*, *Pseudomonas*, *Arthrobacter*, *Pantoea*, and *Brachybacterium* are examples that may persist in the environment with 5%–15% NaCl salts and enhance plant growth factors (Tyagi et al., 2022).

Drought, which can arise because of either a limited water supply or a high rate of transpiration, is one of the most obvious abiotic stresses (Ahmad et al., 2020). It may have negative effects such as delayed photosynthesis, decreased germination rates, and even membrane degradation (Ahmad et al., 2018; Iqbal et al., 2022). Some plants with endophytes consume less water and generate more biomass than plants without endophytes (Abbasi et al., 2020). A plant hormone called abscisic acid (ABA) acts as a signaling molecule in the absence of water (Zahid et al., 2023). The endophyte either causes the host to produce more ABA or prevents ABA from being broken down (Ahmad et al., 2022b; Sabagh et al., 2021).

Soil salinization is the process by which water-soluble salts accumulate in the soil to the point that they obstruct plant growth and lower soil fertility (Sandhya and Ali, 2015). Endophytic microorganisms increase endurance by regulating the metabolism and signaling activity of phytohormones such as pipevine, gibberellic acid, and (indole acetic acid) IAA (Samaddar et al., 2019). The host's ability to produce biomass is reduced because of the alterations in metabolism (Zahid et al., 2023). Plants can signal changes in metabolism by triggering the generation of ROS, which includes the radicals H_2O_2, O_2, and OH. It has also been noted that more endophytes minimize the damage from oxidative species (Gowtham et al., 2020; Verma et al., 2021).

Temperature is a type of abiotic stress that can have a significant impact on plants' health and results in decreased biomass production. Extreme temperatures can have a negative impact on plant growth since not all organisms can withstand vast temperature variations. Low temperatures disrupt metabolism whereas high temperatures can cause the denaturation of biological proteins (Munir et al., 2022). Endophytic bacteria improve plant defense mechanisms by producing phytohormones and other metabolites to reduce the negative impacts of heat stress (Shaffique et al., 2022).

Flooding is another abiotic stress that is caused by heavy rain or poor soil drainage that leads to reduced crop production. Endophytic bacteria play a role in improving the seedling establishment and crop yield under submergence situations in rice (Ahumada et al., 2022).

The production of phytohormones (ethylene, ABA, and IAA), ROS, and even phospholipids is triggered by the osmotic stress in the presence of endophytes (Zahid et al., 2023), as shown in Fig. 18.2. By increasing the plant's cellular osmolarity, *Piriformospora indica* maintains turgor pressure and activates osmo-protective compounds (Akbar et al., 2019; Tyagi et al., 2022). Environmental problems with a long history include heavy metal concentration, agrochemical pollutants, and other elements (Ijaz et al., 2022). To ensure environmental safety, their negative effects are seriously concerning (Kumar et al., 2021; Munir et al., 2022).

Abiotic and biotic stress signals both positively induce stress-responsive gene expression, preinvasion defense, and callose deposition. ABA, on the other hand, has a negative impact on signals that cause systemic acquired resistance. The endophyte significantly modulates stress by downregulating ABA. Gibberellins produced by plants or endophytes reduce the inhibitory effects of DELLA proteins on plant growth signals. Cell surface receptor kinases (RK) detect the phytopathogen early, and subsequent cytoplasmic kinases (CK) mediate intracellular responses and activate the ethylene/jasmonic acid transduction pathway. Endophytes produce enzymes such as superoxide dismutases (SOD), catalases (CatA), peroxidases (POD), alkyl hydroperoxide reductases (AhpC), and glutathione-S-transferases (GSTs) to neutralize reactive oxygen species (ROS) produced by the plant, see Fig. 18.2.

18.4 Role of endophytes against biotic stresses

Agriculture plays an important role in our lives because not only provides us with food but also provides us resources, sources of income, medicine, and nourishment (Javed et al., 2016; Sobia et al., 2014). Food security is directly related to crop productivity, but the production of crops is reduced due to environmental stresses (Barnawal et al., 2017). There are two types of stresses that plants challenge, such as biotic stresses including nematodes, bacteria, fungi, viruses, insects, arachnids, weeds, and herbivores (Mahto et al., 2022; Ranjith et al., 2022) the other abiotic stresses such as temperature, salinity, chilling, droughts, and flood (Mellidou and Karamanoli, 2022).

Biotic stressors are responsible for the cause of plant death due to the depletion of their host's nutrients. It may contribute to pre- and postharvest losses. It can also affect plant health, crop yield, natural habitats, and ecosystem nutrient cycling. These factors directly deprive economic decisions as well as practical development (Pandey et al., 2017). Various antioxidant enzymes such as chitinase, polyphenol oxidase, phenylalanine ammonia lyase, and lipoxygenase peroxidase as described in Fig. 18.2, may decrease the adverse effect of biotic stresses (Isah, 2019). Suberization, hypersensitive response, crosslinking of phenolics, lignifications, and phytoalexin production are related to peroxidase enzymes (Resende et al., 2019). Fungal endophytes modulate plant immunity by producing lytic enzymes such as chitin deacetylases, which deacetylate chitosan oligomers and thus prevent them from being recognized by chitin-specific receptors (PR-3) of plants that recognize chitin oligomers. Endophytes perceive flagellin (FLS 2) differently than phytopathogens (Noorifar et al., 2021).

Endophytes are effective bacterial sources required to be sightseen for their application in the agriculture sector. The facultative group of endophytes contains the most beneficial growth-promoting species that freely habitat in soil but in favorable conditions colonize crop plants (Lastochkina et al., 2020). Colonization of endophytes is helpful in regulating different plant growth factors, based on the metabolic capacity of the host and microenvironment; they biosynthesize several compounds that arise growth-promoting deeds comparable to rhizospheric microbes (Chaudhary et al., 2022). They sustain stable symbiosis by stashing numerous bioactive compounds subsidizing plant growth and colonization (Gouda et al., 2018). The characteristics related to endophytes include the bioremediation, synthesis of secondary metabolites against phytopathogens-induced systemic resistance, and production of extracellular enzymes (Asghari et al., 2020; Gerna et al., 2020; Lastochkina et al., 2020).

Endophytes produce several compounds that aid plants in impeding pathogens by identifying pathogen-associated structures (Larran et al., 2016). Numerous metabolites such as hormones, antibiotics, and volatiles effectually control the expression of genes correlated to the response to stress and improve plant growth through persuaded resistance (Zahid et al., 2023). Various endophytes are notorious for producing peroxidase enzymes, which play a significant role in the transfiguration of hydrogen peroxide into water (Shah et al., 2020).

Host plants can be protected with the help of endophytic bacteria which will be helpful to colonize different parts and tissues of the plant earlier than the pathogenic microorganisms (Nuangmek et al., 2021). The colonization magnitude of the endophytic bacteria in the host's tissues makes them apt for adaptation in a definite locality (Gámez-Arcas et al., 2022). They were found to be involved in the root colonization of the bacterial endophytes containing type IV pili (Lu et al., 2018), strain-dependent mechanism of the O-antigen chain of bacterial lipopolysaccharides and vitamin B1, NADH dehydrogenases from the bacteria having a high growth rate (Cellini et al., 2021; Gámez-Arcas et al., 2022). Plant endophytes protect the crop as listed in Table 18.1 from pathogens such as stem borer, fruit borer, and root rot and incorporate resistance against these diseases (Qayyum et al., 2015; Renuka et al., 2016).

18.5 Role of endophytes in the alteration of plant secondary metabolites

Endophytes are dependable hosts' symbionts and significantly contribute to the development and growth of hosts as well as the biosynthesis of secondary metabolites (Rong et al., 2022; Sun et al., 2022a, 2022b; Wen et al., 2022). Organisms have evolved two powerful competing strategies to survive. One strategy is to produce allelochemicals that prevent the growth of their competitors and remove any harmful side effects they may have left behind. The other is to produce allelochemicals that help symbiotic hosts or symbionts develop relationships with their producers (Alam et al., 2021). Because of these symbiotic interactions, both partners can survive and reproduce even in the most adverse conditions.

The majority of these allelochemicals are secondary metabolites, a class of typically low-molecular-weight and remarkably diverse molecules that were previously thought to play no role in the growth, development, and reproduction of the organisms that produce them (Alam et al., 2021). Endophytes have been shown to produce low-molecular-weight volatile organic compounds (VOCs) that cause a range of physiological changes in plants, aid in their growth and development, and help them to defend themselves from pest or pathogen attacks or to send warning signals intra- or inter-specifically during such attacks (Munir et al., 2022). *Bipolaris sacchari* and *Ceratocystis paradoxa*, two fungal phytopathogens,

TABLE 18.1 Response of endophytes against biotic and abiotic stresses.

Crop	Endophytes species	Abiotic/biotic controlled	Reference
Zea mays	*Bacillus amyloliquefacien*	Herbivore	Ranjith et al. (2022)
Zea mays	*Gluconacetobacter diazotrophicus*	Drought tolerance	Tufail et al. (2021)
Zea mays	*Cladosporium oxysporum and Rigidoporus vinctus*	Maize stem borer	Renuka et al. (2016)
Triticum aestivum	*Bacillus subtilis*	Salinity tolerance	Hadj Brahim et al. (2022)
Triticum aestivum	*Salicornia brachiata*	Fusarium head blight	Hadj Brahim et al. (2022)
Triticum aestivum	*Trichoderma afroharzianum*	Salt stress	Oljira et al. (2020)
Triticum aestivum	*Trichoderma hamatum, Penicillium*	Tan spot disease	Larran et al. (2016)
Olea europaea	*Aureobasidium pullulans and Sarocladium summerbellii*	Olive anthracnose	Sdiri et al. (2022)
Lolium perenne	*Epichloë festucae*	Drought and pathogen stress	Li et al. (2020)
Glycine max	*Bradyrhizobium japonicum*	Symbiotic association	Egamberdieva et al. (2020)
Glycine max	*Bacillus firmus*	Salt tolerance	El-Esawi et al. (2018)
Glycine max	*Bacillus subtilis*	Salinity stress	Yasmin et al. (2020)
Helianthus annuus	*Bacillus aerophilus*	Copper accumulation	Kumar et al. (2021)
Leymus chinensis	*Mycorrhizal Fungi*	Antioxidant Enzyme Activities	Li et al. (2019)
Capsicum annum	*Pseudomonas vancouverensis*	Salinity stress	Samaddar et al. (2019)
Linum usitatissimum	*Azospirillum brasilense, Pseudomonas geniculata*	Salinity stress	
Sesbania rostrata	*Azorhizobium caulinodans*	Salt stress tolerance	Liu et al. (2022)
Cucumis melo	*Trichoderma phayaoense*	Fusarium equiseti	Nuangmek et al. (2021)
Gossipium hersutum	*Beauveria bassiana*	Cotton leafworm	Sánchez-Rodríguez et al. (2018)

were prevented from proliferating by VOCs (Narayanan and Glick, 2022). The development of fungal phytopathogens was impeded by VOCs. The term "secondary" has been either associated with these substances' "inessentiality," however, it has been demonstrated that these natural metabolites do indirectly serve significant roles in organism growth and development (Alam et al., 2021).

Endophytes or the host plant secrete substances in response to the presence of multiple microbial species within a single plant to prevent the spread of potentially harmful bacteria (Zhou et al., 2016). Chemical metabolites are continuously produced as a result of the relationship between the plant and its endophyte (Shameer and Prasad, 2018). In response to any change in environmental conditions, it can be argued that both the plant and the endophyte are responsible for producing these metabolites. Researchers have made it possible to produce these substances on an industrial scale by identifying these endophytes and developing optimal production conditions.

As a result of the bioprocessing of fungal endophytes, numerous physiologically active compounds, including statins, terpenes, alkaloids, phenolics, steroids, and medications with anticancer, antitumor, antioxidant, hypercholesterolemia, and antidiabetic properties, are produced (Jochum et al., 2019). The secondary metabolites produced by these endophytes are more valuable than the other chemical products because they can be produced on an industrial scale by modifying the organisms. Endophytic bacteria also produce alkaloids, terpenes, non-ribosomal peptides, polyketides, and a few other metabolites (Narayanan and Glick, 2022).

Terpene cyclase acts on the mevalonate pathway to produce isoprene-containing terpenes, whereas the shikimic acid and mevalonate pathways produce alkaloid biosynthesis from aromatic amino acids and dimethylallyl pyrophosphates (Kuzuyama, 2017). Studies indicate that endophytes produce secondary metabolites that are either identical to or comparable to those of their hosts (Ludwig-Müller, 2015). Some of these bioactive substances produced by EMOs and host interaction include natural insecticides azadirachtin and the anticancer medications podophyllotoxin and camptothecin (Rochín-Hernández et al., 2022). Numerous strategies for the concurrent production of biological molecules have been identified. However, even for the same substance, the biosynthetic pathways in plants and their endophytes are fundamentally different in some cases (Rochín-Hernández et al., 2022).

Numerous factors have been reported to have an impact on the methods used to extract endophyte metabolites, including the sampling season, weather conditions at the time, and location (Alam et al., 2021). It has been linked to the development of microbes that can integrate genetic material from higher plants, resulting in improved host adaption and the ability to perform a variety of functions such as protecting themselves from pests, diseases, and other animals. Several endophytes can produce polysaccharides, lipopolysaccharides, and glycoproteins causing plant defense systems and a rise in plant production of secondary metabolites that successfully lessen the physical attack of pathogens (Gouda et al., 2016). However, there is insufficient knowledge regarding how endophytes persist in their host plant when there is a significant increase in the production of secondary metabolites. The majority of endophytes have been shown to produce secondary metabolites, some of which have been shown to have antibacterial and antifungal properties and hinder the growth of phytopathogenic microorganisms (Gámez-Arcas et al., 2022; Mellidou and Karamanoli, 2022). Many different kinds of research are still being done today to find endophyte metabolites that might be useful for industry. The ability of several bioactive substances to control a range of phytopathogens has been studied (Gouda et al., 2018). Additionally, several endophyte metabolites, including flavonoids, peptides, quinones, alkaloids, phenols, steroids, terpenoids, and polyketides, have been discovered to have antibacterial properties (Narayanan and Glick, 2022).

Bioactive metabolites have a prospective future because they are in high demand in the industries of medicine, agriculture, biodegradation, and bioremediation. Endophyte-based nanoparticles are expected to have a significant impact on future medication development (Ghosh and Das, 2022). The development and application of HPCE, HPLC-Ms, and other technologies in recent years have facilitated the rapid identification of active metabolites of plant endophytes.

18.6 Conclusion and prospects

Endophytes, whether bacterial or fungal, have a wide range of unpredictable potential and applications in agriculture due to their cost-effectiveness, environmental friendliness, and promotion of sustainability and agricultural productivity, as science and technology progressed. Chemical pesticides are increasingly being used, which is reflected in the harm they cause not only to crops but also to the environment and people's health. Previous attempts to improve resistance to abiotic stressors through genetic engineering have met with limited success. For this reason, it is important to explore alternative strategies, and endophytes can help by inhibiting the host without harming it, protecting it from pests and diseases, and increasing its tolerance to biotic and abiotic challenges. Endophytes perform various positive impacts on plant health and incorporate resistances against various stresses. Additionally, EMOs are also used for the production of important by-products or bio-active metabolites such as siderophore, allelochemicals, and enzymes without disturbing the environmental balance. Plant biologists have been focusing on the role of these EMOs in plants over the last few decades, but their role has yet to be fully explored.

References

Abbasi, S., Sadeghi, A., Safaie, N., 2020. *Streptomyces* alleviate drought stress in tomato plants and modulate the expression of transcription factors ERF1 and WRKY70 genes. Sci. Hortic. 265, 109206. Available from: https://doi.org/10.1016/j.scienta.2020.109206.

Abdelfattah, A., Tack, A.J.M., Wasserman, B., Liu, J., Berg, G., Norelli, J., et al., 2022. Evidence for host–microbiome co-evolution in apple. New Phytol. 234, 2088–2100. Available from: https://doi.org/10.1111/nph.17820.

Afzal, I., Shinwari, Z.K., Sikandar, S., Shahzad, S., 2019. Plant beneficial endophytic bacteria: Mechanisms, diversity, host range and genetic determinants. Microbiol. Res. 221, 36–49. Available from: https://doi.org/10.1016/j.micres.2019.02.001.

Ahmad, H.M., Fiaz, S., Hafeez, S., Zahra, S., Shah, A.N., Gul, B., et al., 2022a. Plant growth-promoting rhizobacteria eliminate the effect of drought stress in plants: a review. Front. Plant Sci. 13. Available from: https://doi.org/10.3389/fpls.2022.875774.

Ahmad, H.M., Mahmood-ur-Rahman, Azeem, F., Shaheen, T., Irshad, M.A., 2020. Genome-wide analysis of long-chain Acyl-CoA synthetase (LACS) genes in sunflower (*Helianthus annuus*) suggests their role in drought stress. Int. J. Agric. Biol. 24, 863–870. Available from: https://doi.org/10.17957/IJAB/15.1510.

Ahmad, H.M., Mahmood-Ur-Rahman, Azeem, F., Tahir, N., Iqbal, Mu.S., 2018. QTL mapping for crop improvement against abiotic stresses in cereals. J. Anim. Plant Sci. 28, 1558–1573.

Ahmad, H.M., Wang, X., Ijaz, M., Mahmood-Ur-Rahman, Oranab, S., Ali, M.A., et al., 2022b. Molecular aspects of microRNAs and phytohormonal signaling in response to drought stress: a review. Curr. Issues Mol. Biol. 44, 3695–3710. Available from: https://doi.org/10.3390/cimb44080253.

Ahumada, G.D., Gómez-Álvarez, E.M., Dell'Acqua, M., Bertani, I., Venturi, V., Perata, P., et al., 2022. Bacterial endophytes contribute to rice seedling establishment under submergence. Front. Plant Sci. 13. Available from: https://doi.org/10.3389/fpls.2022.908349.

Akbar, M., Aslam, N., Khalil, T., Akhtar, S., Siddiqi, E.H., Iqbal, M.S., 2019. Effects of seed priming with plant growth-promoting rhizobacteria on wheat yield and soil properties under contrasting soils. J. Plant Nutr. 42, 2080–2091. Available from: https://doi.org/10.1080/01904167.2019.1655041.

Alam, B., Lǐ, J., Gě, Q., Khan, M.A., Gōng, J., Mehmood, S., et al., 2021. Endophytic fungi: from symbiosis to secondary metabolite communications or vice versa. Front. Plant Sci. 12. Available from: https://doi.org/10.3389/fpls.2021.791033.

Aroca, R., Ruiz-Lozano, J.M., Zamarreño, Á.M., Paz, J.A., García-Mina, J.M., Pozo, M.J., et al., 2013. Arbuscular mycorrhizal symbiosis influences strigolactone production under salinity and alleviates salt stress in lettuce plants. J. Plant Physiol. 170, 47–55. Available from: https://doi.org/10.1016/j.jplph.2012.08.020.

Asghari, B., Khademian, R., Sedaghati, B., 2020. Plant growth promoting rhizobacteria (PGPR) confer drought resistance and stimulate biosynthesis of secondary metabolites in pennyroyal (*Mentha pulegium* L.) under water shortage condition. Sci. Hortic. 263, 109132. Available from: https://doi.org/10.1016/j.scienta.2019.109132.

Badri, D.V., Chaparro, J.M., Zhang, R., Shen, Q., Vivanco, J.M., 2013. Application of natural blends of phytochemicals derived from the root exudates of arabidopsis to the soil reveal that phenolic-related compounds predominantly modulate the soil microbiome. J. Biol. Chem. 288, 4502–4512. Available from: https://doi.org/10.1074/jbc.M112.433300.

Barnawal, D., Bharti, N., Pandey, S.S., Pandey, A., Chanotiya, C.S., Kalra, A., 2017. Plant growth-promoting rhizobacteria enhance wheat salt and drought stress tolerance by altering endogenous phytohormone levels and TaCTR1 / TaDREB2 expression. Physiol. Plant. 161, 502–514. Available from: https://doi.org/10.1111/ppl.12614.

Cellini, A., Spinelli, F., Donati, I., Ryu, C.-M., Kloepper, J.W., 2021. Bacterial volatile compound-based tools for crop management and quality. Trends Plant Sci. 26, 968–983. Available from: https://doi.org/10.1016/j.tplants.2021.05.006.

Chandra, D., Srivastava, R., Gupta, V.V.S.R., Franco, C.M.M., Sharma, A.K., 2019. Evaluation of ACC-deaminase-producing rhizobacteria to alleviate water-stress impacts in wheat (*Triticum aestivum* L.) plants. Can. J. Microbiol. 65, 387–403. Available from: https://doi.org/10.1139/cjm-2018-0636.

Chaudhary, P., Agri, U., Chaudhary, A., Kumar, A., Kumar, G., 2022. Endophytes and their potential in biotic stress management and crop production. Front. Microbiol. 13. Available from: https://doi.org/10.3389/fmicb.2022.933017.

Chouhan, S., Agrawal, L., Prakash, A., 2022. Amelioration in traditional farming system by exploring the different plant growth-promoting attributes of endophytes for sustainable agriculture. Arch. Microbiol. 204, 151. Available from: https://doi.org/10.1007/s00203-021-02637-4.

Compant, S., Cambon, M.C., Vacher, C., Mitter, B., Samad, A., Sessitsch, A., 2021. The plant endosphere world – bacterial life within plants. Environ. Microbiol. 23, 1812–1829. Available from: https://doi.org/10.1111/1462-2920.15240.

Della Monica, I.F., Saparrat, M.C.N., Godeas, A.M., Scervino, J.M., 2015. The co-existence between DSE and AMF symbionts affects plant P pools through P mineralization and solubilization processes. Fungal Ecol. 17, 10–17. Available from: https://doi.org/10.1016/j.funeco.2015.04.004.

Dwibedi, V., Rath, S.K., Joshi, M., Kaur, R., Kaur, Gurleen, Singh, D., et al., 2022. Microbial endophytes: application towards sustainable agriculture and food security. Appl. Microbiol. Biotechnol. 106, 5359–5384. Available from: https://doi.org/10.1007/s00253-022-12078-8.

Egamberdieva, D., Ma, H., Alimov, J., Reckling, M., Wirth, S., Bellingrath-Kimura, S.D., 2020. Response of soybean to hydrochar-based rhizobium inoculation in loamy sandy soil. Microorganisms 8, 1674. Available from: https://doi.org/10.3390/microorganisms8111674.

El-Esawi, M.A., Alaraidh, I.A., Alsahli, A.A., Alamri, S.A., Ali, H.M., Alayafi, A.A., 2018. *Bacillus firmus* (SW5) augments salt tolerance in soybean (*Glycine max* L.) by modulating root system architecture, antioxidant defense systems and stress-responsive genes expression. Plant Physiol. Biochem. 132, 375–384. Available from: https://doi.org/10.1016/j.plaphy.2018.09.026.

Gámez-Arcas, S., Baroja-Fernández, E., García-Gómez, P., Muñoz, F.J., Almagro, G., Bahaji, A., et al., 2022. Action mechanisms of small microbial volatile compounds in plants. J. Exp. Bot. 73, 498–510. Available from: https://doi.org/10.1093/jxb/erab463.

Gerna, D., Roach, T., Mitter, B., Stöggl, W., Kranner, I., 2020. Hydrogen peroxide metabolism in interkingdom interaction between bacteria and wheat seeds and seedlings. Mol. Plant Microbe Interact. 33, 336–348. Available from: https://doi.org/10.1094/MPMI-09-19-0248-R.

Ghosh, S., Das, S., 2022. Impact of climate change on microbial endophytes: novel nanoscale cell factories. Microbiome Under Changing Climate. Elsevier, pp. 161–185. Available from: https://doi.org/10.1016/B978-0-323-90571-8.00007-9.

Golgeri, M., Mulla, D.B., Bagewadi, S.I., Tyagi, Z.K., Hu, S., Sharma, A., et al., 2022. A systematic review on potential microbial carbohydrases: current and future perspectives. Crit. Rev. Food Sci. Nutr. 1–18. Available from: https://doi.org/10.1080/10408398.2022.2106545.

Gonzalez Mateu, M., Baldwin, A.H., Maul, J.E., Yarwood, S.A., 2020. Dark septate endophyte improves salt tolerance of native and invasive lineages of Ph*ragmites australis*. ISME J. 14, 1943–1954. Available from: https://doi.org/10.1038/s41396-020-0654-y.

Gouda, S., Das, G., Sen, S.K., Shin, H.-S., Patra, J.K., 2016. Endophytes: a treasure house of bioactive compounds of medicinal importance. Front. Microbiol. 7. Available from: https://doi.org/10.3389/fmicb.2016.01538.

Gouda, S., Kerry, R.G., Das, G., Paramithiotis, S., Shin, H.-S., Patra, J.K., 2018. Revitalization of plant growth promoting rhizobacteria for sustainable development in agriculture. Microbiol. Res. 206, 131–140. Available from: https://doi.org/10.1016/j.micres.2017.08.016.

Gowtham, H., S., B.S., M., M., N., S., Prasad, M., Aiyaz, M., et al., 2020. Induction of drought tolerance in tomato upon the application of ACC deaminase producing plant growth promoting rhizobacterium *Bacillus subtilis* Rhizo SF 48. Microbiol. Res. 234, 126422. Available from: https://doi.org/10.1016/j.micres.2020.126422.

Hadj Brahim, A., Ben Ali, Manel, Daoud, L., Jlidi, M., Akremi, I., Hmani, H., et al., 2022. Biopriming of durum wheat seeds with endophytic diazotrophic bacteria enhances tolerance to fusarium head blight and salinity. Microorganisms 10, 970. Available from: https://doi.org/10.3390/microorganisms10050970.

Harsonowati, W., Marian, M., Surono, M., Narisawa, K., 2020. The effectiveness of a dark septate endophytic fungus, *Cladophialophora chaetospira* SK51, to mitigate strawberry fusarium wilt disease and with growth promotion activities. Front. Microbiol. 11, 1–11. Available from: https://doi.org/10.3389/fmicb.2020.00585.

Harsonowati, W., Masrukhin, Narisawa, K., 2022. Prospecting the unpredicted potential traits of *Cladophialophora chaetospira* SK51 to alter photoperiodic flowering in strawberry, a perennial SD plant. Sci. Hortic. 295, 110835. Available from: https://doi.org/10.1016/j.scienta.2021.110835.

Hartman, K., Tringe, S.G., 2019. Interactions between plants and soil shaping the root microbiome under abiotic stress. Biochem. J. 476, 2705–2724. Available from: https://doi.org/10.1042/BCJ20180615.

He, C., Wang, W., Hou, J., 2019. Plant growth and soil microbial impacts of enhancing licorice with inoculating dark septate endophytes under drought stress. Front. Microbiol. 10, 1–16. Available from: https://doi.org/10.3389/fmicb.2019.02277.

Hou, L., Li, X., He, X., Zuo, Y., Zhang, D., Zhao, L., 2021. Effect of dark septate endophytes on plant performance of *Artemisia ordosica* and associated soil microbial functional group abundance under salt stress. Appl. Soil Ecol. 165, 103998. Available from: https://doi.org/10.1016/j.apsoil.2021.103998.

Ijaz, M., Ansari, M.-R., Alafari, H.A., Iqbal, M., Alshaya, D.S., Fiaz, S., et al., 2022. Citric acid assisted phytoextraction of nickle from soil helps to tolerate oxidative stress and expression profile of NRAMP genes in sunflower at different growth stages. Front. Plant Sci. 13. Available from: https://doi.org/10.3389/fpls.2022.1072671.

Ilyas, M., Khan, S.A., Awan, S.I., Rehman, S., Ahmed, W., Khan, M.R., et al., 2020. Preponderant of dominant gene action in maize revealed by generation mean analysis under natural and drought stress conditions. Sarhad J. Agric. 35. Available from: https://doi.org/10.17582/journal.sja/2020/36.1.198.209.

Iqbal, S., Wang, X., Mubeen, I., Kamran, M., Kanwal, I., Díaz, G.A., et al., 2022. Phytohormones trigger drought tolerance in crop plants: outlook and future perspectives. Front. Plant Sci. 12. Available from: https://doi.org/10.3389/fpls.2021.799318.

Isah, T., 2019. Stress and defense responses in plant secondary metabolites production. Biol. Res. 52, 39. Available from: https://doi.org/10.1186/s40659-019-0246-3.

Javed, I., Awan, S., Ahmad, H., Rao, A., 2016. Assesment of genetic diversity in wheat synthetic double haploids for yield and drought related traits through factor and cluster analyses. Plant Gene Trait . Available from: https://doi.org/10.5376/pgt.2016.07.0003.

Jiao, R., Cai, Y., He, Pengfei, Munir, S., Li, X., Wu, Y., et al., 2021. *Bacillus amyloliquefaciens* YN201732 produces lipopeptides with promising biocontrol activity against fungal pathogen *Erysiphe cichoracearum*. Front. Cell. Infect. Microbiol. 11. Available from: https://doi.org/10.3389/fcimb.2021.598999.

Jochum, M.D., McWilliams, K.L., Borrego, E.J., Kolomiets, M.V., Niu, G., Pierson, E.A., et al., 2019. Bioprospecting plant growth-promoting rhizobacteria that mitigate drought stress in grasses. Front. Microbiol. 10. Available from: https://doi.org/10.3389/fmicb.2019.02106.

Khan, M.A., Asaf, S., Khan, A.L., Adhikari, A., Jan, R., Ali, S., et al., 2019. Halotolerant rhizobacterial strains mitigate the adverse effects of nacl stress in soybean seedlings. BioMed. Res. Int. 2019, 1–15. Available from: https://doi.org/10.1155/2019/9530963.

Kumar, A., Tripti, Maleva, M., Bruno, L.B., Rajkumar, M., 2021. Synergistic effect of ACC deaminase producing *Pseudomonas* sp. TR15a and siderophore producing *Bacillus aerophilus* TR15c for enhanced growth and copper accumulation in *Helianthus annuus* L. Chemosphere 276, 130038. Available from: https://doi.org/10.1016/j.chemosphere.2021.130038.

Kuźniar, A., Włodarczyk, K., Grządziel, J., Goraj, W., Gałązka, A., Wolińska, A., 2020. Culture-independent analysis of an endophytic core microbiome in two species of wheat: *Triticum aestivum* L. (cv. 'Hondia') and the first report of microbiota in *Triticum spelta* L. (cv. 'Rokosz'). Syst. Appl. Microbiol. 43, 126025. Available from: https://doi.org/10.1016/j.syapm.2019.126025.

Kuzuyama, T., 2017. Biosynthetic studies on terpenoids produced by *Streptomyces*. J. Antibiot. 70, 811–818. Available from: https://doi.org/10.1038/ja.2017.12.

Larran, S., Simón, M.R., Moreno, M.V., Siurana, M.P.S., Perelló, A., 2016. Endophytes from wheat as biocontrol agents against tan spot disease. Biol. Control 92, 17–23. Available from: https://doi.org/10.1016/j.biocontrol.2015.09.002.

Lastochkina, O., Baymiev, A., Shayahmetova, A., Garshina, D., Koryakov, I., Shpirnaya, I., et al., 2020. Effects of endophytic *Bacillus Subtilis* and salicylic acid on postharvest diseases (*Phytophthora infestans*, *Fusarium oxysporum*) development in stored potato tubers. Plants 9, 76. Available from: https://doi.org/10.3390/plants9010076.

Li, F., Duan, T., Li, Y., 2020. Effects of the fungal endophyte epichloë festucae var. lolii on growth and physiological responses of perennial ryegrass cv. fairway to combined drought and pathogen stresses. Microorganisms 8, 1917. Available from: https://doi.org/10.3390/microorganisms8121917.

Li, J., Meng, B., Chai, H., Yang, X., Song, W., Li, S., et al., 2019. Arbuscular mycorrhizal fungi alleviate drought stress in C3 (*Leymus chinensis*) and C4 (*Hemarthria altissima*) grasses via altering antioxidant enzyme activities and photosynthesis. Front. Plant Sci. 10. Available from: https://doi.org/10.3389/fpls.2019.00499.

Li, X., He, X., Hou, L., Ren, Y., Wang, S., Su, F., 2018. Dark septate endophytes isolated from a xerophyte plant promote the growth of *Ammopiptanthus mongolicus* under drought condition. Sci. Rep. 8, 26–28. Available from: https://doi.org/10.1038/s41598-018-26183-0.

Lin, Y., Watts, D.B., Kloepper, J.W., Feng, Y., Torbert, H.A., 2020. Influence of plant growth-promoting rhizobacteria on corn growth under drought stress. Commun. Soil Sci. Plant Anal. 51, 250–264. Available from: https://doi.org/10.1080/00103624.2019.1705329.

Liu, Y., Liu, X., Dong, X., Yan, J., Xie, Z., Luo, Y., 2022. The effect of *Azorhizobium caulinodans* ORS571 and γ-aminobutyric acid on salt tolerance of Sesbania rostrata. Front. Plant Sci. 13. Available from: https://doi.org/10.3389/fpls.2022.926850.

Liu, Y., Wei, X., 2021. Dark septate endophyte improves the drought-stress resistance of *Ormosia hosiei* seedlings by altering leaf morphology and photosynthetic characteristics. Plant Ecol. 222, 761–771. Available from: https://doi.org/10.1007/s11258-021-01135-3.

Lu, X., Liu, S.-F., Yue, L., Zhao, X., Zhang, Y.-B., Xie, Z.-K., et al., 2018. Epsc involved in the encoding of exopolysaccharides produced by *Bacillus amyloliquefaciens* FZB42 act to boost the drought tolerance of *Arabidopsis thaliana*. Int. J. Mol. Sci. 19, 3795. Available from: https://doi.org/10.3390/ijms19123795.

Ludwig-Müller, J., 2015. Plants and endophytes: equal partners in secondary metabolite production? Biotechnol. Lett. 37, 1325–1334. Available from: https://doi.org/10.1007/s10529-015-1814-4.

Madhaiyan, M., Peng, N., Te, N.S., Hsin I, C., Lin, C., Lin, F., et al., 2013. Improvement of plant growth and seed yield in *Jatropha curcas* by a novel nitrogen-fixing root associated *Enterobacter* species. Biotechnol. Biofuels 6, 140. Available from: https://doi.org/10.1186/1754-6834-6-140.

Mahto, R.K., Ambika, Singh, C., Chandana, B.S., Singh, R.K., Verma, S., et al., 2022. Chickpea biofortification for cytokinin dehydrogenase via genome editing to enhance abiotic-biotic stress tolerance and food security. Front. Genet. 13. Available from: https://doi.org/10.3389/fgene.2022.900324.

Marchin, R.M., Ossola, A., Leishman, M.R., Ellsworth, D.S., 2020. A simple method for simulating drought effects on plants. Front. Plant Sci. 10. Available from: https://doi.org/10.3389/fpls.2019.01715.

Marian, M., Takashima, Y., Harsonowati, W., Murota, H., Narisawa, K., 2022. Biocontrol of pythium root rot on lisianthus using a new dark septate endophytic fungus Hyalo*scypha variabilis* J1PC1. Eur. J. Plant Pathol. 163, 97–112. Available from: https://doi.org/10.1007/s10658-022-02459-0.

Massoni, J., Bortfeld-Miller, M., Widmer, A., Vorholt, J.A., 2021. Capacity of soil bacteria to reach the phyllosphere and convergence of floral communities despite soil microbiota variation. Proc. Natl. Acad. Sci. 118. Available from: https://doi.org/10.1073/pnas.2100150118.

Mellidou, I., Karamanoli, K., 2022. Unlocking PGPR-mediated abiotic stress tolerance: what lies beneath. Front. Sustain. Food Syst. 6. Available from: https://doi.org/10.3389/fsufs.2022.832896.

Mookherjee, A., Singh, S., Maiti, M.K., 2018. Quorum sensing inhibitors: can endophytes be prospective sources? Arch. Microbiol. 200, 355–369. Available from: https://doi.org/10.1007/s00203-017-1437-3.

Msimbira, L.A., Smith, D.L., 2020. The roles of plant growth promoting microbes in enhancing plant tolerance to acidity and alkalinity stresses. Front. Sustain. Food Syst. 4. Available from: https://doi.org/10.3389/fsufs.2020.00106.

Munir, N., Hanif, M., Abideen, Z., Sohail, M., El-Keblawy, A., Radicetti, E., et al., 2022. Mechanisms and strategies of plant microbiome interactions to mitigate abiotic stresses. Agronomy 12, 2069. Available from: https://doi.org/10.3390/agronomy12092069.

Narayanan, Z., Glick, B.R., 2022. Secondary metabolites produced by plant growth-promoting bacterial endophytes. Microorganisms 10, 2008. Available from: https://doi.org/10.3390/microorganisms10102008.

Noorifar, N., Savoian, M.S., Ram, A., Lukito, Y., Hassing, B., Weikert, T.W., et al., 2021. Chitin deacetylases are required for *Epichloë festucae* endophytic cell wall remodeling during establishment of a mutualistic symbiotic interaction with *Lolium perenne*. Mol. Plant Microbe Interact. 34, 1181–1192. Available from: https://doi.org/10.1094/MPMI-12-20-0347-R.

Nuangmek, W., Aiduang, W., Kumla, J., Lumyong, S., Suwannarach, N., 2021. Evaluation of a newly identified endophytic fungus, *Trichoderma phayaoense* for plant growth promotion and biological control of gummy stem blight and wilt of muskmelon. Front. Microbiol. 12. Available from: https://doi.org/10.3389/fmicb.2021.634772.

Oljira, A.M., Hussain, T., Waghmode, T.R., Zhao, H., Sun, H., Liu, X., et al., 2020. *Trichoderma* enhances net photosynthesis, water use efficiency, and growth of wheat (*Triticum aestivum* L.) under salt stress. Microorganisms 8, 1565. Available from: https://doi.org/10.3390/microorganisms8101565.

Pandey, P., Irulappan, V., Bagavathiannan, M.V., Senthil-Kumar, M., 2017. Impact of combined abiotic and biotic stresses on plant growth and avenues for crop improvement by exploiting physio-morphological traits. Front. Plant Sci. 8. Available from: https://doi.org/10.3389/fpls.2017.00537.

Parrino, B., Schillaci, D., Carnevale, I., Giovannetti, E., Diana, P., Cirrincione, G., et al., 2019. Synthetic small molecules as anti-biofilm agents in the struggle against antibiotic resistance. Eur. J. Med. Chem. 161, 154–178. Available from: https://doi.org/10.1016/j.ejmech.2018.10.036.

Potshangbam, M., Indira Devi, S., Sahoo, D., Strobel, G.A., 2017. Functional characterization of endophytic fungal community associated with *Oryza sativa* L. and *Zea mays* L. Front. Microbiol. 8, 1–15. Available from: https://doi.org/10.3389/fmicb.2017.00325.

Qayyum, M.A., Wakil, W., Arif, M.J., Sahi, S.T., Dunlap, C.A., 2015. Infection of *Helicoverpa armigera* by endophytic *Beauveria bassiana* colonizing tomato plants. Biol. Control. 90, 200–207. Available from: https://doi.org/10.1016/j.biocontrol.2015.04.005.

Qessaoui, R., Bouharroud, R., Furze, J.N., El Aalaoui, M., Akroud, H., Amarraque, A., et al., 2019. Applications of new rhizobacteria *Pseudomonas* isolates in agroecology via fundamental processes complementing plant growth. Sci. Rep. 9, 12832. Available from: https://doi.org/10.1038/s41598-019-49216-8.

Rajesh, P.S., Ravishankar Rai, V., 2014. Quorum quenching activity in cell-free lysate of endophytic bacteria isolated from *Pterocarpus santalinus* Linn., and its effect on quorum sensing regulated biofilm in *Pseudomonas aeruginosa* PAO1. Microbiol. Res. 169, 561–569. Available from: https://doi.org/10.1016/j.micres.2013.10.005.

Ranjith, S., Kalaiselvi, T., Muthusami, M., Sivakumar, U., 2022. Maize apoplastic fluid bacteria alter feeding characteristics of herbivore (*Spodoptera frugiperda*) in maize. Microorganisms 10, 1850. Available from: https://doi.org/10.3390/microorganisms10091850.

Renuka, S., Ramanujam, B., Poornesha, B., 2016. Endophytic ability of different isolates of entomopathogenic fungi *Beauveria bassiana* (Balsamo) vuillemin in stem and leaf tissues of maize (*Zea mays* L.). Indian. J. Microbiol. 56, 126–133. Available from: https://doi.org/10.1007/s12088-016-0574-8.

Resende, C.F., Pacheco, V.S., Dornellas, F.F., Oliveira, A.M.S., Freitas, J.C.E., Peixoto, P.H.P., 2019. Responses of antioxidant enzymes, photosynthetic pigments and carbohydrates in micropropagated *Pitcairnia encholirioides* L.B. Sm. (Bromeliaceae) under *ex vitro* water deficit and after rehydration. Braz. J. Biol. 79, 53–62. Available from: https://doi.org/10.1590/1519-6984.175284.

Ripa, F.A., Cao, W., Tong, S., Sun, J., 2019. Assessment of plant growth promoting and abiotic stress tolerance properties of wheat endophytic fungi. BioMed. Res. Int. 1–12.

Ritonga, F.N., Chen, S., 2020. Physiological and molecular mechanism involved in cold stress tolerance in plants. Plants 9, 560. Available from: https://doi.org/10.3390/plants9050560.

Riva, V., Mapelli, F., Bagnasco, A., Mengoni, A., Borin, S., 2022. A meta-analysis approach to defining the culturable core of plant endophytic bacterial communities. Appl. Environ. Microbiol. 88. Available from: https://doi.org/10.1128/aem.02537-21.

Rochín-Hernández, L.S., Rochín-Hernández, L.J., Flores-Cotera, L.B., 2022. Endophytes, a potential source of bioactive compounds to curtail the formation–accumulation of advanced glycation end products: a review. Molecules 27, 4469. Available from: https://doi.org/10.3390/molecules27144469.

Rodríguez, C.E., Antonielli, L., Mitter, B., Trognitz, F., Sessitsch, A., 2020. Heritability and functional importance of the *Setaria viridis* bacterial seed microbiome. Phytobiomes J. 4, 40–52. Available from: https://doi.org/10.1094/PBIOMES-04-19-0023-R.

Rodriguez, R.J., White, J.F., Arnold, A.E., Redman, R.S., 2009. Fungal endophytes: diversity and functional roles: Tansley review. New Phytol. 182, 314–330. Available from: https://doi.org/10.1111/j.1469-8137.2009.02773.x.

Rolli, E., Vergani, L., Ghitti, E., Patania, G., Mapelli, F., Borin, S., 2021. 'Cry-for-help' in contaminated soil: a dialogue among plants and soil microbiome to survive in hostile conditions. Environ. Microbiol. 23, 5690–5703. Available from: https://doi.org/10.1111/1462-2920.15647.

Rong, Z.Y., Jiang, D.J., Cao, J.L., Hashem, A., Abd_Allah, E.F., Alsayed, M.F., et al., 2022. Endophytic fungus Serendipita indica accelerates ascorbate-glutathione cycle of white clover in response to water stress. Front. Microbiol. 13, 1–9. Available from: https://doi.org/10.3389/fmicb.2022.967851.

Sabagh, A.E.L., Mbarki, S., Hossain, A., Iqbal, M.A., Islam, M.S., Raza, A., et al., 2021. Potential role of plant growth regulators in administering crucial processes against abiotic stresses. Front. Agron. 3. Available from: https://doi.org/10.3389/fagro.2021.648694.

Samaddar, S., Chatterjee, P., Roy Choudhury, A., Ahmed, S., Sa, T., 2019. Interactions between *Pseudomonas* spp. and their role in improving the red pepper plant growth under salinity stress. Microbiol. Res. 219, 66–73. Available from: https://doi.org/10.1016/j.micres.2018.11.005.

Sánchez-Rodríguez, A.R., Raya-Díaz, S., Zamarreño, Á.M., García-Mina, J.M., del Campillo, M.C., Quesada-Moraga, E., 2018. An endophytic *Beauveria bassiana* strain increases spike production in bread and durum wheat plants and effectively controls cotton leafworm (*Spodoptera littoralis*) larvae. Biol. Control. 116, 90–102. Available from: https://doi.org/10.1016/j.biocontrol.2017.01.012.

Sandhya, V., Ali, S.Z., 2015. The production of exopolysaccharide by *Pseudomonas putida* GAP-P45 under various abiotic stress conditions and its role in soil aggregation. Microbiology 84, 512–519. Available from: https://doi.org/10.1134/S0026261715040153.

Sdiri, Y., Lopes, T., Rodrigues, N., Silva, K., Rodrigues, I., Pereira, J.A., et al., 2022. Biocontrol ability and production of volatile organic compounds as a potential mechanism of action of olive endophytes against *Colletotrichum acutatum*. Microorganisms 10, 571. Available from: https://doi.org/10.3390/microorganisms10030571.

Shaffique, S., Khan, M.A., Wani, S.H., Pande, A., Imran, M., Kang, S.-M., et al., 2022. A review on the role of endophytes and plant growth promoting rhizobacteria in mitigating heat stress in plants. Microorganisms 10, 1286. Available from: https://doi.org/10.3390/microorganisms10071286.

Shah, A.A., Aslam, S., Akbar, M., Ahmad, A., Khan, W.U., Yasin, N.A., et al., 2021. Combined effect of *Bacillus fortis* IAGS 223 and zinc oxide nanoparticles to alleviate cadmium phytotoxicity in *Cucumis melo*. Plant Physiol. Biochem. 158, 1–12. Available from: https://doi.org/10.1016/j.plaphy.2020.11.011.

Shah, A.A., Bibi, F., Hussain, I., Yasin, N.A., Akram, W., Tahir, M.S., et al., 2020. Synergistic Effect of *Bacillus thuringiensis* IAGS 199 and Putrescine on Alleviating Cadmium-Induced Phytotoxicity in *Capsicum annum*. Plants 9, 1512. Available from: https://doi.org/10.3390/plants9111512.

Shameer, S., Prasad, T.N.V.K.V., 2018. Plant growth promoting rhizobacteria for sustainable agricultural practices with special reference to biotic and abiotic stresses. Plant Growth Regul. 84. Available from: https://doi.org/10.1007/s10725-017-0365-1.

Sobia, A., Ahmad, H.M., Awan, S.I., Kang, S.A., Iqbal, M.S., Ali, M.A., 2014. Estimation of genetic variability, heritibility and correlation for some morphological traits in spring wheat. J. Biol. Agric. Healthc. 4, 10–16.

Subramaniam, G., Thakur, V., Saxena, R.K., Vadlamudi, S., Purohit, S., Kumar, V., et al., 2020. Complete genome sequence of sixteen plant growth promoting *Streptomyces* strains. Sci. Rep. 10, 10294. Available from: https://doi.org/10.1038/s41598-020-67153-9.

Sun, R.T., Zhang, Z.Z., Feng, X.C., Zhou, N., Feng, H.D., Liu, Y.M., et al., 2022a. Endophytic fungi accelerate leaf physiological activity and resveratrol accumulation in polygonum cuspidatum by up-regulating expression of associated genes. Agronomy 12, 1–12. Available from: https://doi.org/10.3390/agronomy12051220.

Sun, R.T., Zhang, Z.Z., Liu, M.Y., Feng, X.C., Zhou, N., Feng, H.D., et al., 2022b. Arbuscular mycorrhizal fungi and phosphorus supply accelerate main medicinal component production of Polygonum cuspidatum. Front. Microbiol. 13, 1–10. Available from: https://doi.org/10.3389/fmicb.2022.1006140.

Tellenbach, C., Sumarah, M.W., Grünig, C.R., Miller, J.D., 2013. Inhibition of Phytophthora species by secondary metabolites produced by the dark septate endophyte *Phialocephala europaea*. Fungal Ecol. 6, 12–18. Available from: https://doi.org/10.1016/j.funeco.2012.10.003.

Tiwari, P., Bajpai, M., Singh, L.K., Mishra, S., Yadav, A.N., 2020. Phytohormones producing fungal communities: metabolic engineering for abiotic stress tolerance in crops. Agriculturally Important Fungi for Sustainable Agriculture. Springer, Cham, 171–197. https://doi.org/10.1007/978-3-030-45971-0_8

Tufail, M.A., Touceda-González, M., Pertot, I., Ehlers, R.-U., 2021. *Gluconacetobacter diazotrophicus* Pal5 enhances plant robustness status under the combination of moderate drought and low nitrogen stress in *Zea mays* L. Microorganisms 9. Available from: https://doi.org/10.3390/microorganisms9040870.

Tyagi, J., Chaudhary, P., Mishra, A., Khatwani, M., Dey, S., Varma, A., 2022. Role of endophytes in abiotic stress tolerance: with special emphasis on *Serendipita indica*. Int. J. Environ. Res. 16, 62. Available from: https://doi.org/10.1007/s41742-022-00439-0.

Vaccaro, F., Cangioli, L., Mengoni, A., Fagorzi, C., 2022. Synthetic plant microbiota challenges in nonmodel species. Trends Microbiol. 30, 922–924. Available from: https://doi.org/10.1016/j.tim.2022.06.006.

Verma, S.K., Sahu, P.K., Kumar, K., Pal, G., Gond, S.K., Kharwar, R.N., et al., 2021. Endophyte roles in nutrient acquisition, root system architecture development and oxidative stress tolerance. J. Appl. Microbiol. 131, 2161–2177. Available from: https://doi.org/10.1111/jam.15111.

Wang, D.-C., Jiang, C.-H., Zhang, L.-N., Chen, L., Zhang, X.-Y., Guo, J.-H., 2019. Biofilms positively contribute to *Bacillus amyloliquefaciens* 54-induced drought tolerance in tomato plants. Int. J. Mol. Sci. 20, 6271. Available from: https://doi.org/10.3390/ijms20246271.

Wen, J., Okyere, S.K., Wang, S., Wang, J., Xie, L., Ran, Y., et al., 2022. Endophytic fungi: an effective alternative source of plant-derived bioactive compounds for pharmacological studies. J. Fungi 8, 205. Available from: https://doi.org/10.3390/jof8020205.

Woo, O.-G., Kim, H., Kim, J.-S., Keum, H.L., Lee, K.-C., Sul, W.J., et al., 2020. *Bacillus subtilis* strain GOT9 confers enhanced tolerance to drought and salt stresses in *Arabidopsis thaliana* and *Brassica campestris*. Plant Physiol. Biochem. 148, 359–367. Available from: https://doi.org/10.1016/j.plaphy.2020.01.032.

Wu, F.L., Qu, D.H.,, Tian, W., Wang, M.Y., Chen, F.Y., Li, K.K., et al., 2021. Transcriptome analysis for understanding the mechanism of dark septate endophyte S16 in promoting the growth and nitrate uptake of sweet cherry. J. Integr. Agric. 20, 1819–1831. Available from: https://doi.org/10.1016/S2095-3119(20)63355-X.

Yakti, W., Kovács, G.M., Vági, P., Franken, P., 2018. Impact of dark septate endophytes on tomato growth and nutrient uptake. Plant. Ecol. Divers. 11, 637–648. Available from: https://doi.org/10.1080/17550874.2019.1610912.

Yasin, N., Khan, W., Ahmad, S.R., Aamir, A., Shakil, A., Aqeel, A., 2018. Effect of *Bacillus fortis* 162 on growth, oxidative stress tolerance and phytoremediation potential of *Catharanthus roseus* under chromium stress. Int. J. Agric. Biol. 20, 1513–1522.

Yasmin, H., Naeem, S., Bakhtawar, M., Jabeen, Z., Nosheen, A., Naz, R., et al., 2020. Halotolerant rhizobacteria *Pseudomonas pseudoalcaligenes* and *Bacillus subtilis* mediate systemic tolerance in hydroponically grown soybean

(*Glycine max* L.) against salinity stress. PLoS One 15, e0231348. Available from: https://doi.org/10.1371/journal.pone.0231348.

Zahid, G., Iftikhar, S., Shimira, F., Ahmad, H.M., Aka Kaçar, Y., 2023. An overview and recent progress of plant growth regulators (PGRs) in the mitigation of abiotic stresses in fruits: A review. Sci. Hortic. 309, 111621. Available from: https://doi.org/10.1016/j.scienta.2022.111621.

Zhou, C., Ma, Z., Zhu, L., Xiao, X., Xie, Y., Zhu, J., et al., 2016. Rhizobacterial Strain *Bacillus megaterium* BOFC15 induces cellular polyamine changes that improve plant growth and drought resistance. Int. J. Mol. Sci. 17, 976. Available from: https://doi.org/10.3390/ijms17060976.

CHAPTER

19

Biostimulants in sustainable management of phytoparasitic nematodes in plants

Arvind[1], *Namita Goyat*[1], *Sukhmeet Singh*[1], *Mayur Mukut Murlidhar Sharma*[2] *and Pankaj Sharma*[3]

[1]Department of Zoology, Chaudhary Devi Lal University, Sirsa, Haryana, India [2]Department of Agriculture and Life Industry, Kangwon National University, Chuncheon, Gangwon, Republic of Korea [3]Department of Microbiology, CCS Haryana Agricultural University, Hisar, Haryana, India

19.1 Introduction

The world's expanding population needs to be fed, especially now that the agricultural sector is dealing with several challenging issues related to the effects of climate change. By upsetting regular climatic patterns like water budgets, this global phenomenon contributes to and exacerbates preexisting abiotic pressures like frequent floods, droughts, salinization, and temperature extremes (Sharma et al., 2020a,b; 2021a,b,c; Singh et al., 2019, 2021). Due to these factors' contributions to changes in the distribution of phytopathogens and weeds and a decline in the beneficial microbial population associated with plants, plant health is negatively impacted and susceptibility to biotic stress increases. Despite their detrimental effects on the environment, chemical fertilizers and pesticides are increasingly used in agriculture to ensure adequate crop yields and pest biocontrol (Hamid et al., 2021; Loyal et al., 2023; Rani et al., 2019, 2022a,b; Sangwan et al., 2023; Sharma et al., 2022a,b, 2023). Both human and soil health have suffered because of persistent exposure to agrochemicals. Biobased products, for example, biostimulants offer an ecologically effective technology or complement to their synthetic counterparts, agrochemicals, which help agricultural and horticultural crops produce consistently high yields under ideal and unfavorable environmental conditions (Rouphael and Colla, 2020). Using plant-based biostimulants is an exciting new development with positive implications for the environment.

Biostimulants in Plant Protection and Performance
DOI: https://doi.org/10.1016/B978-0-443-15884-1.00006-3

As agronomic tools, biostimulants have grown in importance for lowering fertilizer usage. Plants' natural processes can be aided by using biostimulants, which are synthetic substances or microorganisms with a biological origin. Effects on crop quality, yield, and resistance to abiotic stress, as well as the efficiency with which nutrients are absorbed, are all improved by stimulation (Franzoni et al., 2022). Additionally, biostimulants assist plants in waking up from their winter dormancy, growing larger fruits, developing deeper roots, increasing the activity of photosynthetic and other vegetative tissues, enhancing plant vigor and uniformity, controlling flowering, and hastening fruit maturation (Parađiković et al., 2019). Even though they encompass an extensive range of mineral and organic constituents that plants can use as growth regulators, metabolites, and nutrients, biostimulants cannot be considered as biofertilizers because their physiological effects depend on composition (Dipak Kumar and Aloke, 2020). Biostimulants can also be used as adjuvants in the quest to increase the efficiency of chemical inputs (Mire et al., 2016). Biostimulants can be either non-living materials like nitrobenzene, seaweed, humic acid amino acids, gibberellic acid, or live materials like compost or manure that contain beneficial rhizosphere microbiota. The value of biostimulant compounds has been brought to light. The biostimulant formulation must be chemically stable and physically homogenous under all likely storage circumstances to allow for the precise administration of the minimum effective dose to target locations. The goal is to maximize biostimulants' efficacy while ensuring their safe administration (Dipak Kumar and Aloke, 2020).

Phytoparasitic nematodes can be found in almost all types of agricultural climates and are known to infect a major variety of crops (Singh et al., 2019). They enter plant cells as parasites and pierce them with a hollow or solid spear or stylet to scavenge nutrients. In order to infect plants, phytonematodes frequently collaborate with other pathogens like fungi, bacteria, and viruses. In terms of the number of injuries it can cause, the combined pathogenic potential of several nematodes can occasionally appear to be significantly greater than that of a single pathogen. An important class of vegetable pests is represented by phytoparasitic nematodes that can cause an annual yield loss of 9%–15% across the globe (Nicol et al., 2011). These losses are mainly attributable to the root-knot nematode *Meloidogyne* spp., which reduces plant growth, decreases agricultural production and quality, and makes plants less resistant to biotic and abiotic stresses. Treatment of soil with synthetic nematicides was once used to get rid of these pests, but as concerns about crop safety have grown, their use has been phased out. As a result, there is a growing need for and interest in ecologically sound methods of pest management (D'Addabbo et al., 2019).

The experimental research study supported earlier findings from literature research by showing that biostimulants can also successfully control nematodes. The stimulation of crop defensive responses to nematode invasion was mainly linked to nematode inhibition by microbial biostimulants (Sofo et al., 2014; Vos et al., 2013). The processes underlying biostimulants' suppression of nematode growth are just theorized or only partially understood. Seaweed extracts are mostly utilized as biostimulants because of the presence of plant growth factors (Auxin, gibberellins, cytokinin, and carbohydrates) and various nutritive metallic elements such as aluminum, iron, nitrogen, potassium, and manganese. The effectiveness of seaweeds against phytoparasitic worms is commonly attributed to the presence of secondary metabolites with nematicidal activity; these include steroids, triterpenoids, alkaloids, and phenols (Dipak Kumar and Aloke, 2020). Fresh seaweeds have historically been used in agrarian practices as fertilizers, but biostimulant properties have just lately been discovered. This encourages the commercial use of seaweed extracts and purified substances like polysaccharides, alginates, carrageenans, and laminarin, in addition to by-products of their breakdown products (du Jardin, 2015).

19.2 Biostimulants and plants

The European Biostimulants Industry Council (EBIC) advocates for the biostimulants industry's ability to help farmers produce commercially viable yields of high-quality crops with minimal waste of valuable resources. To promote the role of biostimulants in achieving sustainable agricultural production, green innovation, economic growth, and other societal goals, EBIC is advocating for policies that encourage the development of a genuine European market for biostimulants. Biostimulants are "a crafted invention of biological origin that enhances plant growth and productivity as a potential outcome of the novel or emergent attributes of the complex of components, and not as an only result of the involvement of known essential micronutrients, plant growth regulators" according to the definition of Yakhin et al. (2017).

Plant productivity benefits from biostimulants because they engage with plant signaling pathways, lowering negative plant responses to stress. This concept acknowledges a wealth of recent data revealing that plant stress response is mediated by signaling molecules produced by the plant or its diverse microbial populations. Biostimulants can either actively interact with plant signaling cascades or stimulate endophytic and non-endophytic bacteria, yeast, and fungi to create plant-beneficial compounds. The biostimulant's effectiveness originates from the drop in assimilates redirected to non-productive stress response metabolism (Brown and Saa, 2015). Biostimulants have been shown to improve macronutrient intake which has been attributed to an influence on nitrogen metabolism stimulation. Biostimulants may also ameliorate the detrimental impacts of abiotic stress factors on plants, with notable effects on heat, salinity, drought chilling, oxidative, frost, and chemical and mechanical stress regulation.

19.3 Types and categories of biostimulants

Despite ongoing efforts to define biostimulant regulatory status, regulatory definition of plant biostimulants doesn't exist anywhere in the world, including the United States and the European Union. Because of this, a full categorization of the compounds and microorganisms encompassed by the notion is not possible. Despite this, some key groups of compounds and microbes are well recognized by scientists, regulators, and stakeholders (du Jardin, 2015). Useful bacteria, primarily Plant growth-promoting rhizobacterias (PGPRs), and helpful fungi are examples of microorganisms. They can exist as free-living, endosymbiotic, or rhizospheric organisms. These categories are mentioned briefly in the following section.

19.3.1 Humic substances

These are natural components of soil organic matter produced as a result of degradation of plant, animal, and microbial residues. These are heterogeneous composites that were primarily classified as humic and fulvic acids and humins. Root exudates and protons released by plant roots influence the complex association/dissociation dynamics of these compounds into supra-molecular colloids. The communication between organic matter, microbes, and roots produces humic substances. The practices employed to utilize humic substances as promoters of plant growth require the optimization of such interactions. Thereby, the random application of humic substances may result in unfavorable outcomes. For instance, HS application to plants (Rose et al., 2014) has been reported to increase the dry weight of shoots by 22.4% and roots by 21.6%, respectively.

Variability in HS effects is attributable to the source of humic substances, environmental situations, plant species, and HS dose coupled with the application method (Rose et al., 2014). Additionally, the agricultural byproducts also govern the degradation as well as oxidation by biochemical progressions, resulting in "humic-like substances" that are anticipated as a replacement for naturally occurring HS (Eyheraguibel et al., 2008). These composites are also known to improve soil fertility by acting on the soil's physical, physicochemical, and biological possessions.

19.3.2 Seaweed extract

These are extensively known as plant biostimulants used in the quest to enhance abiotic stress tolerance, improve nutrient utilization proficiency, and augment crop quality traits irrespective of their nutrient composition (du Jardin, 2015). Seaweeds, or macroalgae, are divided into three pigmentation categories: Phaeophyta (Brown), Rhodophyta (Red), and Chlorophyta (Green). Ascophyllum, Fucus, and Laminaria dominate brown seaweeds (du Jardin, 2015). Seaweed extracts are biochemically complex composites (polysaccharides, minerals, vitamins, oils, fats, acids, antioxidants, pigments, and hormones) (Craigie, 2011; Khan et al., 2009; Michalak and Chojnacka, 2014). Due to the many bioactive compounds in a single extract, unveiling their mechanistic approach is a complex task that necessitates a multidisciplinary methodology. SE can be used as a soil or foliar spray (du Jardin, 2015). They improve soil retention, remediation, soil microflora, nutrient supply, and hormonal effects (El Boukhari et al., 2020). Recently, seaweed extracts have been widely used in agriculture to boost crop productivity. Stimulating plant growth and development of physiological processes and improving product quality yields such improvement. Trivedi et al. (2018) applied *Kappaphycus alvarezii* extract and found a 15% upsurge in seed yield (g/plant) in water-optimal environments by increasing cob length and seed density. There are different and advanced strategies of seaweed extraction as compared to classical extraction.

Many approaches are used for seaweed extraction today. To preserve the biological activity of compounds, new extraction methods such as enzyme-assisted extraction (EAE), microwave-assisted extraction (MAE), ultrasound-assisted extraction (UAE), supercritical fluid extraction (SFU), and pressurized liquid extraction (PLE) are being used (Michalak and Chojnacka, 2015).

SWEs are a complex milieu of physiologically active chemicals, including natural and break-down products. More compounds may be found. A recent study has shown SWEs' diverse modes of action because to this combination of chemicals. No one collection of chemicals can explain the many growth and physiological changes SWEs cause. Understanding the biological mechanisms of SWEs will boost production. Their mechanistic part and their communications with different abiotic and biotic stresses and their response in field conditions, such as soil bacteria, and how they will interact in a mixture, such as with a fertilizer, must be well understood. SWEs have great potential as decomposable, non-toxic, non-polluting, and harmless plant biostimulants.

19.3.3 Amino acid-containing products

Applications of various protein-based products have been linked to increased plant development and resistance to both biotic and abiotic stress. It appears that the nutritional effect of an additional nitrogen source and these plant-stimulating effects are separate (Ertani et al., 2009). The notion that plants can easily absorb amino acids and peptides is underlying these investigations. Several studies, mentioned in Calvo et al. (2014), confirm that plant roots may absorb radio-labeled amino acids. In addition, it has been documented that amino acids can be taken up through the plant's leaves.

Protein-based products can be broken down into two main groups: specific amino acids and protein hydrolysates. Protein hydrolysates (pH)-based biostimulants are predominantly found to be important in the idea of plant stimulation. Enzymatic processing produces pH-derived biostimulants from numerous by-products and wastes of proteins, besides the combination of oligopeptides discharged during the cleavage is the principal active molecule (Martín et al., 2022). These pHs are synthesized by different methods involving enzymatic hydrolysis, and chemical, biological, or temperature-sensitive hydrolysis of a wide range of plant and animal remains (Calvo et al., 2014). Several of these have been developed into marketable goods, such as Isagro Sp A, Italy's Siapton R and Aminoplant, ILSA, Italy's ILSATOP products, and Bioiberica Corp., Spain's Macro-Sorb Foliar. Hydrolysates in these preparations have protein/peptide contents of 1%–85% (w/w) and free amino acid values of 2%–18% (w/w), correspondingly.

The most common amino acids are represented by glycine, alanine, arginine, glutamine, proline, valine, glutamate, and leucine (Ertani et al., 2009; Parrado et al., 2008). While glutamine and arginine are the most abundant amino acids in carob germ hydrolysate, proline and glycine are the most abundant in siapton. Bioassay results showed that alfalfa hydrolysate, which is rich in free amino acids (1.9% w/w), also contains macro and micronutrient components, and mimics the activity of plant hormones auxin and gibberellin (Schiavon et al., 2008). Endogenous IAA was found in a beef hydrolysate, and the plant growth regulator triacontanol was found in the same product (Ertani et al., 2013).

Recently, Rouphael et al. (2021) equated the outcome of PHs obtained from animal (A-PH) and plant (V-PH) sources at different conditions for assessing the growth of the Basil plant. The plant-originated pHs exhibited superior plant growth promotion attributes as compared to those originated from animals (Rouphael et al., 2021). The toxic attributes of animal-originated pHs are attributed to their high and unbalanced amino acid concentration and high salt concentration (Bartucca et al., 2022; Colla et al., 2015).

19.3.4 PGPR as biostimulants

Biofertilizers, biopesticides, and psychostimulants are all the terms used to describe biostimulants like PGPR (Yakhin et al., 2017). From seed germination to full growth and maturity, they assist plants by improving nutrient uptake, stress resistance, soil characteristics, and a healthy habitat for other microbes (Calvo et al., 2014). PGPR has the potential to alter the composition of the plant's cell wall and accumulate many solutes because they help with water retention and provide defense against osmotic stress (Kumari et al., 2018; Meena et al., 2017). PGPR has been shown to be effective against drought, salinity, extreme heat, and pH because it produces high levels of indole-3-acetic acid (IAA), which helps alleviate salt stress and maintains the production of exopolysaccharides (EPS) to sustain hydration around the roots under water scarcity (van Oosten et al., 2017). Studying the PGPR's role in both stressed and unstrained conditions has revealed that it typically stimulates growth more strongly in the face of adversity, such as drought stress (Rubin et al., 2017). When it comes to improving plant stress tolerance, ethylene is a key player for several PGPRs (Nadeem et al., 2014). The plant growth regulator PGPR secretes the enzyme 1-aminocyclopropane-1-carboxylase (ACC) deaminase, which inhibits the plant's ethylene production. Inoculating plants with PGPR that produce ACC deaminase has been shown to increase their stress tolerance. According to studies done on *Camelina sativa*, this is because PGPR prevents ethylene levels from rising to inhibitory levels (Ahemad and Kibret, 2014; Heydarian et al., 2016; Ruzzi and Aroca, 2015).

Biostimulant PGPR may indirectly affect ecosystem functions like organic matter accumulation and decomposition through interactions with the local microbiota, plants, or their individual activities (Hellequin et al., 2018). How PGPR might affect breakdown rates and nutrient release through interactions with major indigenous decomposer fungi is unknown (Kyker-Snowman et al., 2020). The buildup of organic matter may also result from interactions with plant roots. Root architecture, growth, and exudate production are all influenced by PGPR-released metabolites, VOCs, or auxins (Grover et al., 2021). In order to determine if these PGPR-root interactions can be scaled up to increase organic matter persistence, more study is required. Leakage of biosynthesized molecules by PGPR, such as extracellular enzymes, which can survive in soil for long periods of time and bind to minerals to generate mineral-associated organic matter, may also contribute to soil organic matter accumulation (Cotrufo et al., 2013). Growth and decay rates of organic matter in soil may be affected by interactions between PGPR bacteria, plants, and resident fungi (Moore et al., 2022).

19.3.5 Nanoparticles and nanomaterials as biostimulant

As a result of interactions between the surface charges of plant cells and nanoparticles (NP) and nanomaterials, plants can exhibit a spectrum of responses ranging from biostimulation to toxicity (NM). It has been shown that foliar sprays or nutrient solutions containing trace amounts of NPs and NMs improve plant composition, metabolism, and stress resistance. There is a narrow concentration window within which NPs and NMs can exert their biostimulatory effects. This limit is determined by the properties of the specific NPs or NMs in use. There seem to be two stages involved in biostimulation. The first stage, a physicochemical one, is caused by the interaction of surface charges; the next stage, a biochemical one, is caused by a cascade of biochemical stimuli initiated by the entry of NPs and NMs or by the modifications in the membranes or integral proteins that can occur in parallel with the cellular internment. The two phases are initiated by the same mechanism: the interaction of surface charges. Since the initial biostimulation process seems to be physicochemical, it follows that any NP or NM, when applied to plants at the proper concentration and with the necessary physicochemical characteristics, is capable of inducing biostimulation (Juárez-Maldonado et al., 2019).

19.4 Common uses of biostimulant

Biostimulants are by-products of biological activity that are obtained from organic raw material. When applied to plants, seeds, or growing substrates in precise formulations, plant biostimulants can modify the physiological actions of plants in such a way as to provide a competitive edge in growth, development, and/or stress response. This contrasts with nutrients and pesticides, which do not have this ability. Plant biostimulants are also known as plant growth promoters. Because of the basic materials that they consist of, a broad variety of biostimulants have been observed to have a suppressive impact on phytoparasitic nematodes.

19.4.1 Bionematicides

The majority of phytonematode biostimulants were liquid extracts and oils, granular or cake derivatives or powder seed meal. Numerous biostimulants derived from sesame seed oil (D'Addabbo et al., 2011), quillay water extract (Giannakou, 2011) or biomasses or seeds of neem and *Brassicaceae* plants (Abbasi et al., 2005; Curto et al., 2016; Rizvi et al., 2015) reduce root-knot nematode populations (RKN) on greenhouse tomato and field.

In *in-vitro* studies, *Ecklonia maxima* and *Ascophyllum nodosum* L. seaweed extracts killed root-knot nematode eggs and juveniles almost entirely (KKhan et al., 2015). Osbeck managed root-knot nematodes in tests of tomato soil (Williams et al., 2021). In addition to extract derivatives, soil amendments including *Spatoglossus schroederi* Agardh (Kützing) and *Uva lactuca* L. biomasses suppress *Meloidogyne* spp. infestations on vegetable crops or fruit due to their high phenolic and bioactive component content (El-Ansary and Hamouda, 2014; Paracer et al., 1987). In addition to Meloidogyne species, seaweed products inhibited economically significant nematode parasites of tropical and subtropical vegetable crops, such as *Helicotylenchusindicus Siddiqui*, *Radopholus similis*, and Cobb *Belonolaimus longicaudatus* Rau, (Thorne) (McDonald et al., 1988; Veronico & Melillo, 2021).

Chitosan and its derivatives inhibit root-knot nematodes (Asif et al., 2017; Escudero et al., 2017; Mota and dos Santos, 2016; Radwan et al., 2012) as well as other phytoparasitic species, such as the pinewood parasite *Bursaphelenchus xylophilus* and soybean cyst nematode *Heterodera glycines* Ichinoe (Liang et al., 2018; Mwaheb et al., 2017; Nunes da Silva et al., 2014).

Arbuscular mycorrhizal fungi made up the bulk of biostimulants for phytoparasitic nematodes (Candido et al., 2015). When applied in the field or greenhouse, these chemicals reduced the number of root-knot nematodes (Flor-Peregrn et al., 2014, 2017). *Helicotylenchus multicinctus* (Cobb) Golden and *Nacobbus aberrans* Thorneet Allen, two phytonematode parasites, were both eliminated from field bananas and greenhouse tomatoes. To improve crop tolerance to the cyst nematode *Heterodera schachtii* Schmid and, more generally, phytoparasitic soil nemato fauna, it has been reported that formulations of other fungal or bacterial biocontrol agents (*Bacillus* spp., *Trichoderma* spp., *Azobacter, Azospirillum*) or nitrogen fixers (Azobacter, Azospirillum) can control *M. incognita*.

19.4.2 Synergistic biocontrol

In addition to increasing soil fertility and crop yield, it appears that certain microbial plant biostimulants may also offer some degree of protection against plant diseases and arthropod pests. These include both the ingredient's direct pesticidal effects, which eliminate or reduce pests, and the ingredient's indirect consequences, which increase plant resistance towards pests (Tarigan et al., 2022). This is demonstrated by the fungus *Rhizophagus* (syn. Glomus) *intraradices* (Glomerales: Glomeraceae), which, on the one hand, increases crop tolerance to abiotic stresses and, consequently, yield, for instance in tomato plants; on the other hand, it is a member of the family *Glomerales* (Volpe et al., 2018). Moreover, it appears to limit the development of the larvae of *Spodoptera exigua* (Lepidoptera: Noctuidae), a species of the *Noctuidae* moth family (Shrivastava et al., 2015). On Spodoptera *litura* (Lepidoptera: Noctuidae), a species of the *Noctuidae* butterfly family, black gram was reported to have similar effects (Selvaraj et al., 2020). *Rhizophagus intraradices* are also reported to diminish oviposition of *Lissorhoptrus oryzophilus* (Coleoptera: *Curculionidae*) on rice (Mardani-Talaee et al., 2017). Both impacts were discovered through research (Cosme et al., 2011). The enhanced production of plant defense metabolites such as phenolics, lignin, superoxide dismutase, peroxidase, catalase, phenylalanine ammonia-lyase, and polyphenol oxidase has been attributed to *Rh. intraradices*. In addition to having insecticidal effects, there have been reports of fungicidal and nematocidal effects, such as those against *Meloidogyne javanica* (Tylenchida: *Heteroderidae*) in pistachios (Hafez et al., 2013).

19.4.3 Soil conditioner

Salinity-sodicity is a major cause of soil degradation (Shrivastava and Kumar, 2015). Stalinization degraded 25% of Mediterranean irrigated farmland (Zia-ur-Rehman et al., 2016). Salt stress harms soil physical and chemical characteristics, microbial activities, and plant growth (Laudicina et al., 2009; Oo et al., 2015; Tejada et al., 2006). High salts cause ionic and osmotic stress (Zia-ur-Rehman et al., 2016). High salt concentrations impair plant water transport and osmotic potential between cells and soil (Chung et al., 2005; Fageria et al., 2011; Hu et al., 2007). High Na and Cl concentrations damage plant cells and hinder root ion uptake (Machado and Serralheiro, 2017). Excess Na causes soil slaking, swelling, and dispersion (Oo et al., 2015; Zia-ur-Rehman et al., 2016).

In agricultural management, the use of biostimulants for improving soil structure and encouraging plant development has gained importance. The use of organic biostimulants to enhance soil features and nutrient accessibility for crops may permit a decline in the usage of chemical fertilizers, whose intensive use poses a significant threat to soil health. Biostimulants are substances that, in minute quantities, encourage plant development

(Schmidt et al., 2003). They have also been defined as any material or microorganism given to soils for improving plant nutritive efficacy and tolerance to abiotic strains (Du Jardin, 2012). Natural biostimulants are economic substitutes for synthetic fertilizers, crop defense agents, and plant growth supervisors. These products can enhance the chemical as well as biological qualities of the soil, promote plant growth, and maintain soil fertility (Abdel-Raouf et al., 2012). Additionally, biostimulants augment root biomass, nutrient absorption, and soil enzyme activity (Halpern et al., 2015a). There are numerous classes and varieties of biostimulants, such as enzymes, proteins, amino acids, micronutrients, and other substances (Chiaiese et al., 2018; Drobek et al., 2019; du Jardin, 2015).

Furthermore, fungi and bacteria are another important group of plant biostimulants that can modify the soil microbial ecology (Drobek et al., 2019). Prebiotics are natural items, such as sewage sludge, compost, humus, animal manure, and chitin-containing wastes that enhance the biochemical activity and microbial population of soil (Chiaiese et al., 2018; Strachel et al., 2017). Similarly, the application of prebiotics to soils can enhance crop growth by creating bioactive molecules such as hormones and enzymes, reducing soil illnesses, expediting lignin degradation, and supplying inorganic nutrients for plant absorption (Javaid, 2006). In addition, researchers assert that probiotics are regarded as favorable bacteria that, once put into the soil, should promote the development of a substantial biomass with numerous plant-beneficial features (Vassileva et al., 2020). Additionally, a fulvic acid-based biostimulant increased soil characteristics and plant development. Similarly, adding fulvic acids to soils can have significant effects on soil physical, chemical, and biological properties (Quilty and Cattle, 2011). Humic and fulvic acids contain plant-assimilable forms of nutritional cations such as potassium, calcium, and magnesium (Seyedbagheri, 2010). Regarding the biological function of humic compounds, researchers (Saviozzi et al., 2001) reported that the addition of these molecules improves soil fertility by altering the microbial population composition. Calcium lingo-sulfonate is an additional component added to biostimulants that can help improve soil quality. Calcium lingo-sulfonate might be utilized as a soil nutrient source and as a solution addition to tackle the issues of soil degradation and caking (Gezerman and Çorbacıoglu, 2014); it could also be used to improve soil structure and stabilize cohesive to non-cohesive soils (Karol, 2003). In recent years, biostimulants have become an effective method for protecting the soil and the ecosystem.

Numerous researches have investigated the impact of biostimulants on soil compositions and qualities, as well as plant growth. Biostimulants have been examined for application in soil repair (Tejada et al., 2011). The authors reported that the use of biostimulants under semiarid climatic circumstances increased soil enzymatic activity, produced changes in the microbial population, supported soil protection against erosion, and helped its regeneration. In addition, new studies have revealed that biostimulants can boost the activity of rhizosphere microorganisms and soil enzymes, as well as the production of soil growth regulators (Giannattasio, 2013; Nardi et al., n.d.).

Biostimulants can promote the decomposition and mineralization of soil organic matter, therefore enhancing the soil's nitrogen availability (Chen et al., 2002). Biostimulants assist plants in withstanding biotic and abiotic stressors and enhancing their nutrient uptake. Indicated that the application of biostimulants under semiarid climatic circumstances increased soil enzymatic activity, produced changes in the microbial population, supported soil protection against erosion, and helped its regeneration. In addition, additional studies have shown that biostimulants can boost the activity of rhizosphere microorganisms and soil enzymes, as well as the generation of soil growth regulators (Giannattasio, 2013; Nardi et al., n.d.).

Biostimulants assist plants endure biotic and abiotic challenges, boost their nutrient usage efficiency, and have positive effects on soil characteristics such as pH, electrical conductivity (EC), and nitrogen (Dick et al., 1988). Moreover, soil agronomic interventions have a greater impact on enzyme activity than other biochemical soil characteristics (Frankenberger and Dick, 1983). Due to their interaction with soil biology and the cycles of nutrients in the soil (Nardi et al., n.d.), soil enzyme activity is regarded as possible marker of soil quality (Karaca et al., 2010; Masciandaro et al., 2002). Catalase and dehydrogenase are two of the most well-known enzymes associated with the biological properties and fertility of soil.

Dehydrogenase activity is a useful indication of oxidative metabolism and microbiological activity in soils (Burns, 1982). In addition, the catalase enzyme is regarded as a significant indication of soil fertility and aerobic microorganisms (Burns, 1982). Moreover, phosphatase soil enzyme activity catalyzes the hydrolysis of organic phosphorus molecules and converts them to an assimilable inorganic form of phosphorus (Lemanowicz, 2018).

19.4.4 Induction of resistance against different kind of stresses

In the past few decades, plant nutrient management has undergone a substantial shift as a result of various techniques for inducing stress-resistant forgetter crop production. Exogenous application of biostimulants using a variety of methods is one of the methods used to enhance plant growth and yield under diverse environmental circumstances. Biostimulants stimulate plant development without containing any nutrients, herbicides, or soil conditioners (Ali et al., 2020). Biostimulants may be natural or synthetic and include microorganisms such as rhizobacteria that promote plant growth and development (PGRs) and helpful fungi (Halpern et al., 2015b; Zhang et al., 2003). The most important plant biostimulants are fulvic and humic acids, seaweed extracts, protein hydrolysates, inorganic compounds, chitosan, beneficial fungi (e.g., *arbuscular mycorrhizal* fungus and *Trichoderma* spp.), and plant growth-promoting bacteria (Canellas et al., 2015; Noman et al., 2018; Rouphael et al., 2015; Ruzzi and Aroca, 2015).

19.5 Biostimulants against nematodes

The application of biostimulants to crops and soils is a sustainable and environmentally friendly approach to managing plant parasitic nematode populations (and more specifically root-knot nematode populations) that has lately attracted attention. Biostimulants derived from sustainable sources have been utilized as fertilizers and soil conditioners for decades. It is believed that they have a biostimulant impact on plants, hence improving their tolerance to abiotic and biotic stress (Khan et al., 2009). The most commercially

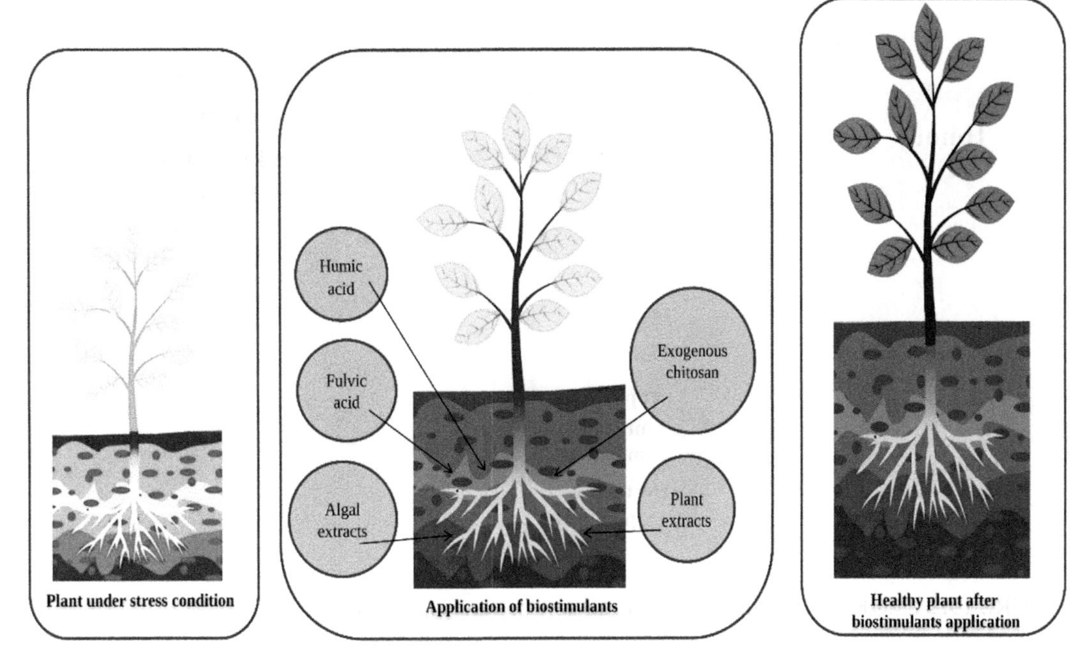

FIGURE 19.1 A portrayal depicting the plant growth promotion attributes of different types of Biostimulants.

available biostimulants are brown algae, mainly *Ascophyllum nodosum* and *Ecklonia maxima* (Craigie, 2011). It has been hypothesized that seaweed extracts may have some action against the quantity and fecundity of plant parasitic nematodes in laboratory and greenhouse conditions. While the mechanism of action for such a phenomenon is mostly unclear, the literature shows that seaweed extracts may be effective for protecting plants from biotic stress due to biostimulants components such as fucoidans and alginates that may function as inducers to prime plants for potential pathogen protection (Ali et al., 2019) (Fig. 19.1).

19.5.1 Mechanisms of biostimulant

19.5.1.1 Humic substances

Most of HS's biostimulant effects are geared toward making the roots healthier in some way. One is the enhanced uptake of macro- and micronutrients due to the elevated cation exchange capacity of soil containing polyanionic HS, and another is the enhanced availability of phosphorus due to the interference of HS with calcium phosphate precipitation (Fig. 19.2).

Another major role of HS to root nourishment is the activation of plasma membrane H^+-ATPases, that converts the free energy released by ATP hydrolysis into a transmembrane electrochemical potential that is further used for the import of nitrate and other nutrients. This Proton pumping by plasma membrane ATPases aids in nutrient absorption

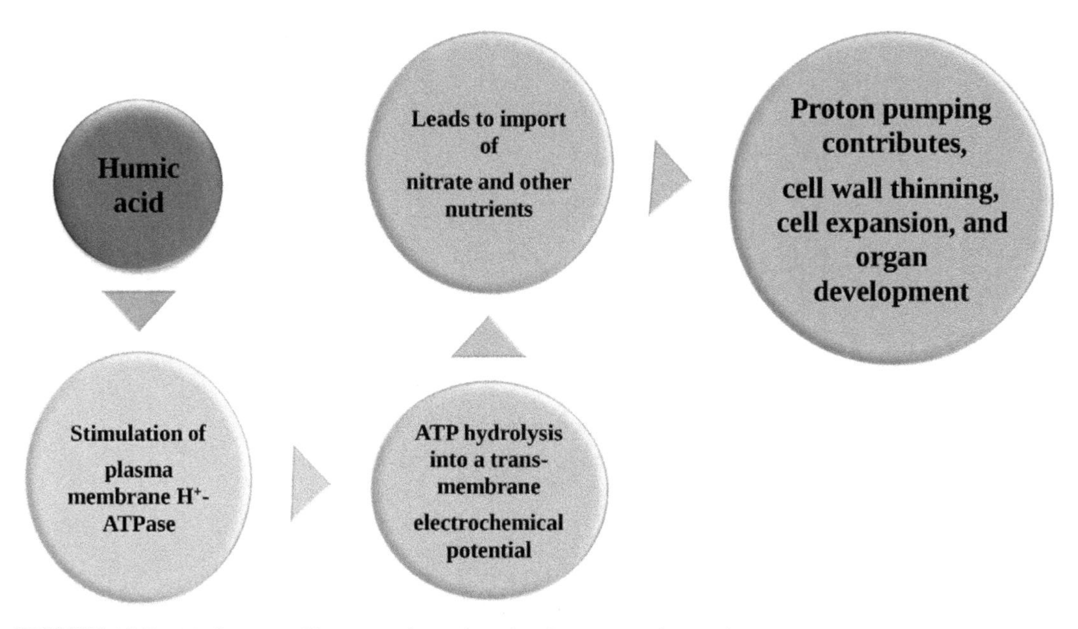

FIGURE 19.2 Mechanism of humic acid on plant development and growth.

as well as cell wall thinning, cell expansion, and organ development (Jindo et al., 2012). The availability of C substrates is likely increased by HS because of its effect on respiration and invertase activities. However, it is often not clear whether HS contain functional groups recognized by the reception/signaling complexes of plant hormonal pathways, whether they release entrapped hormonal compounds, or whether they stimulate hormone-producing microorganisms, even though their hormonal effects are also described. An additional proposed biostimulatory effect of HS is stress protection. A wide range of secondary metabolic processes and stress responses depend on the phenylpropanoid metabolism, which is also necessary to produce phenolic compounds. Research on maize seedlings grown in hydroponics shows that high-molecular-weight HS increases the activity of key enzymes of this metabolism, indicating that HS modulates the stress response (Olivares et al., 2015; Schiavon et al., 2010). Plant-parasitic nematodes that have been found to be cytotoxic or reactive to humic acid include *Meloidogyne* spp., *Rotylenchulus reniformis, Radopholus similis,* and *Helicotylenchus multicinctus*. Effective nematode control requires a concentration of humic acid between 0.04% and 2.0%. Application or drenching of the soil, dipping the roots of seedlings in a solution, and foliar spraying on the leaves are all methods of nematode control. Plant-parasitic nematodes can be controlled by humic acid through several mechanisms, such as the elimination of juveniles, the prevention of hatching, a decrease in nematode infectivity and reproduction, and the induction of systemic resistance (Nagachandrabose and Baidoo, 2021).

19.5.1.2 *Seaweed extracts*

Abiotic stresses like salinity, drought, extreme temperatures, and nutrient deficiencies reduce crop productivity. SE benefits plant abiotic stress tolerance, but the crop and climatic conditions still matter. Ionic imbalances and hyperosmotic and cause oxidative stress from antioxidant defense mechanisms and reactive oxygen species (ROS) (Debnath et al., 2011). Reduced or activated O_2 produces superoxide (O^{2-}), singlet oxygen, hydroxyl radical and hydrogen peroxide (Mittler, 2002). ROS damage DNA, lipids, and proteins (Das and Roychoudhury, 2014). Plants have several natural defenses against such damage. Enzymatic and non-enzymatic antioxidants like guaiacol peroxidase (GPX), superoxide dismutase (SOD), catalase (CAT), dehydroascorbate reductases (DHAR), ascorbate peroxidase (APX), glutathione reductase (GR), ascorbic acid (AsA), monodehydroascorbate reductase (MDHAR), phenolic compounds, tocopherol, and glutathione scavenge ROS(Das and Roychoudhury, 2014; Sreeharsha et al., 2019; Yadav et al., 2019). The application of seaweed extract to plants under abiotic stress alters the regulation of ROS scavenging, Na^+ transporter, and antiporter genes. Catalase (CAT), superoxide dismutase (SOD), and abscisic acid (ABA) are all antioxidant enzymes. Seaweed extracts may affect pathogens directly or indirectly by activating plant defense mechanisms. This is accomplished by upregulating several defense genes, thereby positively affecting specific metabolic pathways, specifically the phenylpropanoid pathway, which induces the biosynthesis of several secondary metabolites involved in disease suppression. Seaweed extracts also alter root-knot nematodes (RKN) at various stages of their life cycles. Williams et al. (2021) demonstrate that seaweed extracts can improve PPN performance in several ways, with positive impacts on the abundance of the root-knot species *M. hapla* and *M. javanica* but negative effects on *M. incognita* for some seaweeds. Reductions in the population of *Meloidogyne* spp. were observed at every stage of the life cycle that was studied. Group testing revealed that extracts of *A. nodosum* can also lessen the number of *M. incognita*. Several factors were found to influence the effect of seaweed extracts on PPN populations, including the type of seaweed used, the host crop, and the *Meloidogyne* species being targeted and elicited (Williams et al., 2021).

PGPRs promote plant development by colonizing the plant root system, but certain PGPRs have also shown nematicidal efficacy against plant-parasitic nematodes. *Pseudomonas fluorescens* CHA0 produced secondary metabolite cause the mortality of second-stage infective juveniles (J2s) and nematode eggs are a prime example (Siddiqui and Shaukat, 2003). *Bacillus subtilis, Bacillus cereus, Pseudomonas fluorescens, Pseudomonas putida*, and *Serratia proteamaculans* has shown the highest efficiency as biocontrol agents against *Meloidogyne javanica* among 860 strains of bacteria isolated from the rhizosphere (Zhao et al., 2018). In addition to promoting plant growth, PGPRs also have the capacity to directly resist pathogens such as nematodes.

The conidia suspension of *Trichoderma longibrachiatum* T6 has the potential to be employed as an effective biocontrol agent against Heterodera avenae eggs and second-stage juveniles *in vitro* and greenhouse (Zhang et al., 2017). The main mechanisms of the strain of *Trichoderma longibrachiatum* T6 against the eggs and second stage juveniles of *Heterodera avenae* were (1) the direct parasitic and lethal effect of *Trichoderma longibrachiatum* T6 on the activity of the eggs and Second stage juveniles development, and (2) the promoting effect on wheat growth and development, and the improvement of chitinase. Additionally, AO and NR can be regarded as quick vital dyes for determining the viability of *Heterodera avenae* (Zhang et al., 2017).

Recently, Vos et al. (2013) used suppression subtractive hybridization (SSH) to discover that *F. mosseae* clearly primes the tomato's defense response against *M. incognita*. Most of the genes that were found to be differentially expressed were assigned to defense, signal transduction, and protein synthesis and modification pathways. An example is the primed upregulation of chorismate synthase, the enzyme responsible for the final step in the shikimate pathway (converting the ESPS product to chorismate). Thus, the shikimate pathway appears to be involved in AMF-mediated biocontrol in various plant species against various nematodes (Hao et al., 2012; Vos et al., 2013). In addition, flavonol synthase has been reported to be primed by the shikimate pathway, which generates precursors for a wide variety of aromatic secondary metabolites generated via the phenylpropanoid pathway (Vos et al., 2013). Several products of the phenylpropanoid pathway, such as flavonols, have been shown to be detrimental to *M. incognita*, *R. similis*, and *P. penetrans* in vitro (Wuyts et al., 2006). Reducing root-knot nematode infection in mycorrhizal tomato roots was linked to SSH and ROS metabolism (Vos et al., 2013). Beneventi et al. (2013) found through pyrosequencing that an over-representation of genes containing various oxidase and peroxidase domains were upregulated in the incompatible interaction, implying a significant role for ROS generation in the resistance of soybean to *M. javanica*. The reduction of root-knot nematode infection in mycorrhizal tomato roots was similarly linked to the reactive oxygen species (ROS) metabolism via suppression subtractive hybridization (SSH) (Vos et al., 2013). Through pyrosequencing, Beneventi et al. (2013) identified an over representation of genes encoding multiple oxidase and peroxidase domains that were elevated in the incompatible interaction, which suggested a significant role for ROS generation in soybean resistance to *M. javanica*. Vos et al. (2013) found that M. incognita infection in mycorrhizal tomato plants primed genes involved in the biosynthesis of a lignin precursor, sinapoylglucose:choline sinapoyltransferase, and 1-aminocyclopropane-1-carboxylate oxidase (ACC oxidase), which catalyzes the last step in the biosynthesis of ethylene and is related.

Endophytes antagonize nematodes by a variety of mechanisms; however, they are still mostly unspecified. Endophytes can directly attack, kill, immobilize, or repel nematodes, confuse them while finding their host, interfere with nurse cell growth, compete for resources, induce plant defense responses, enhance tolerance, or a combination of things.

19.5.2 Limiting nematodes impact on plant growth

Nematode strains show increased resistance to chemical nematicides. Considering these factors, numerous approaches have been tried and evaluated in order to achieve more effective management of plant-parasitic nematodes in agricultural settings. Rhizosphere-exuded molecules from plant roots can either encourage nematode colonization of the plant's roots (attractant compounds) or deter nematode harm to the plant's roots (repellent or nematocidal chemicals) (Sikder and Vestergård, 2020). Castañeda-Ramírez et al. (2020) explored and reviewed research into the use of bioproducts derived from edible mushrooms for the management of nematodes that parasitize plants and animals. Another important biological technique for combating plant-parasitic nematodes is plant growth-promoting bacteria (PGPB) which includes rhizosphere (surrounding plant roots), rhizoplane (on root surface), and endophytic (living inside plant tissue). PGPB can boost plant growth and development, increase the nutritional value of edible seeds and fruits, and protect plants from a wide spectrum of biotic and abiotic stress (Gamalero and Glick, 2020).

19.5.3 Nematotoxic metabolites

Plants generate secondary metabolites in response to a PPN invasion. Many plants produce chlorogenic acid, a phenolic compound, as a defense mechanism against PPN infection. These plants include solanaceous, carrot, and rice species (Candido et al., 2015). Correlation exists between chlorogenic acid production and PPN resistance, but chlorogenic acid is only moderately active against the false root-knot nematode and weakly effective against M. incognita (D'Addabbo et al., 2019). This discrepancy between response and efficacy could be caused by chlorogenic acid metabolites, which have higher nematicidal activity in the target organism but may be unstable or extremely toxic to plants. Caffeic acid and chlorogenic acid can be converted into quinic acid, which can then be transformed into the PPN-toxic compound orthoquinone. Although they have been linked to PPN resistance, it is unclear exactly how orthoquinone and caffeine work. Hölscher et al. (2014) discovered that in a cultivar of banana that is resistant to the pest, phenylphenalenone anigorufone builds up at the infection sites of the burrowing nematode *Radopholus similis* (Musa sp.). Anigorufone kills nematodes very efficiently because it forms massive lipid-anigorufone complexes in the bodies of *R. similis*. An antifungal phytoalexin is another term for anigorufone.

A large class of plant chemicals known as secondary metabolites includes flavonoids. Some flavonoids function as nematicides, repellents, egg-hatching inhibitors, or nemastatic compounds to aid Phytoparasitic Nematodes (PPN) resistance (which do not kill nematodes but stop them from moving) (Chin et al., 2018). Most of the flavonoids that kill nematodes are flavonols (like myricetin, quercetin, and kaempferol), pterocarpans (e.g., glyceollin, medicarpin), and isoflavonoids. *R. similis* eggs are inhibited from hatching by Kaempferol (Wuyts et al., 2006). Quercetin, kaempferol, and myricetin repel and kill juvenile *M. incognita* (Wuyts et al., 2006), and medicarpin stops *Pratylenchus penetrans* from moving in a way that depends on the concentration. Similarly, patulitrin, patuletin, rutin, and quercetin are nematicidal properties against infective juveniles of *Heterodera zeae*, a cyst nematode CN (Faizi et al., 2011). A few flavonoids are also made when a virus infects a plant that is resistant to it. Resistant soybeans develop glyceollins, a class of prenylated pterocarpan phytoalexins that are unique to soybeans when infected with *M. incognita*. Glyceollin levels are also higher in cyst nematode-resistant soybean cultivars compared to susceptible cultivars. Glyceollin I, one of glyceollin's isomers, builds up in the tissues next to the CN head in the roots of resistant soybeans. The infection site's glyceollin accumulation is time and space-dependent.

A wide variety of indole glucosinolates and antimicrobial isothiocyanates are found in plants in the *Brassicaceae* family that have anti-PPN properties. Isothiocyanates are poisonous to RKNs and the semi-endoparasitic worm *Tylenchulus semipenetrans*, and they significantly inhibit CN and RKN hatching. Along with nematicides and nemastatic agents, PPN chemotaxis disruption may also be a productive plant response for suppressing or limiting PPN infection. Studies have shown that ethylene, which is typically produced following injury and during pathogen invasion, reduces PPN attraction to the root (Booker and DeLong, 2015; Guan et al., 2015; Holbein et al., 2019). An Arabidopsis mutant that produces too much ethylene repels PPNs less than plants treated with ethylene synthesis inhibitors or ethylene-insensitive mutants (Hu et al., 2017). These results suggest that PPN infection causes ethylene synthesis, which may prevent secondary PPN invasion by reducing attraction. The cause of the decreased attractiveness could be an increase in repellents or a decrease in attractant production. However, the molecular basis for the attraction to PPNs is still largely unknown. In order to separate RKN attractants from seed-coat mucilage and root tips, several teams have tried (Čepulytė et al., 2018; Tsai et al., 2019). By identifying chemo attractants and chemorepellents, it may be possible to gain some insight into how plants respond to nematodes in the rhizosphere during and before PPN infection.

19.6 Integrated management strategies

A key element of integrated crop management (ICM), a set of practices used to increase farming's sustainability and adaptability in the modern world, is the use of biostimulants. Abiotic stress tolerance, nutrient uptake and regulation, crop quality, and biological control are all improved by biostimulants when used on plants (Hamid et al., 2021). Plant pathogenic nematodes cause significant crop productivity losses globally. Many approaches, such as soil environment alteration and biocontrol agent treatment, have been found to be more effective in minimizing undesired environmental loads. Chemical nematicides are being employed to manage these pests with incredible success. However, due to environmental concerns, the hunt for alternative strategies for managing nematode-induced diseases has been stepped up (Ghareeb et al., 2020). In general, nematode density at-plant is an important factor in the success of retaining the nematode population below an economical threshold. As a result, combining new chemical nematicides associated with a biocontrol agent to control plant parasitic nematodes during plant vegetation may be a more effective strategy than using each treatment on its own. It was researched whether *P. lilacinum* effectivity increases when nematode populations are downregulated with at-plant nematicide therapies and whether the combination of chemical and biological nematicides increases output in a study (Dahlin et al., 2019).

19.6.1 Potential use of pesticides and minimal need for synthetic fertilizers

Synthetic fertilizers are widely used to increase the yield and productive potential of major crops in preparation for the anticipated demand for food; however, this practice causes a wide range of pollution and ecosystem damage. Increasing agricultural crop yields to meet future food concerns necessitates numerous new and ecologically friendly agriculture methods in order to maintain the sustainability of our natural ecosystem and surroundings (Majeed et al., 2018). Plant-parasitic nematodes (PPNs) are regarded as a major agricultural danger because these soil-borne diseases induce root infection in a wide range of crops, lowering yield and quality (Rani et al., 2022a,b). The anti-root-knot nematode (Meloidogyne incognita) activity of the endophytic Streptomyces enissocaesilis OM182843 was investigated, as well as its potential to produce plant growth regulators. During all experimental intervals, both caused significant mortality and prevented M. incognita eggs from hatching. *S. enissocaesilis* produced a high concentration of hydrolytic enzymes (chitinases, proteases, and gelatinase) as well as plant growth regulators (indole acetic acid (IAA), gibberellic acid(GA), salicylic acid, and proline). Furthermore, GC/Ms examination of volatile organic compounds (VOCs) produced by Strep. enissocaesilis indicated that several of them possessed nematicidal activity, with 2,4-Decadienal being the most frequent (El-Akshar et al., 2022).

Cultural practices, such as sanitation and crop rotation, and chemical methods, such as the prudent application of nematicides, are two examples of integrated approaches to nematode management. The main disadvantage of chemical management is the risk to one's health and the environment. In recent years, alternative methods with less environmental pollution and health risks have been adopted. In this context, biopesticides are critical. *A bacterium named Bacillus subtilis is frequently used to control nematodes, and Beauveria bassiana, Paecilomyces lilacinus, Trichoderma album, Bacillus megaterium, and Ascophyllum nodosum are some other commercially available biopesticides for the same purpose*. However, there is a need to expand the use and application of environmentally friendly biopesticides for nematode control in economically important crop plants (Roopa and Gadag, 2020). Plants are protected from biotic stress by biopesticides, while plants are protected from abiotic stress by biostimulants. To enhance crop production, these two categories are significantly different. Biopesticides are designed to replace chemical pesticides, and this alternative does not come with the promise of increasing yield but rather to make up crop protection safer for humans and the ecosystem. Biopesticides, some of which can be obtained locally from natural plant products and are more accessible to smallholder farmers in agricultural areas where access to chemical phytosanitary products is limited, can potentially contribute to yield increase. Plant biostimulants are expected to play a different role in yield increase than biopesticides because they do not replace existing treatments, but rather supplement existing stress alleviation and mitigation techniques (Jardin et al., 2020).

19.6.2 Improved biocontrol activity

The rising demand on vegetable farmers to optimize crop yield while limiting the use of synthetic Nitrogen fertilizer constitutes a significant issue for plant scientists working on creative techniques to ensure agricultural production in a sustainable way. *T. virens* GV41 was shown to be the top-performing microbial biostimulants in terms of crop growth and nutrient absorption on lettuce and rocket, revealing the distinct influence of diverse fungal inoculants on crop output (Fiorentino et al., 2018). The combination of *Trichoderma* spp. biostimulant treatments with fatty acid combinations increased biocontrol action, growth promotion, and yield of treated plant species. Surprisingly, these novel bioformulations were more efficient than individual component treatments (microbial or organic) not only in increasing yield but also in creating a premium-grade marketable vegetable with increased antioxidant content (Lanzuise et al., 2022). Many ways have recently been proposed to increase the actions of antagonist yeast strains, which have been used in conjunction with drugs to boost their potency, with promising results. Calcium chloride combined with chitosan and *C. laurentii*, for example, had better biocontrol effectiveness against *Penicillium expansum* in pear than chitosan or yeast treatment alone. Phytic acid and -glucan might improve *Pichia caribbica* and *C. podzolicus* effectiveness in reducing postharvest deterioration in apples, respectively. Glycine betaine improved *Cystofilobasidium infirmominiatum's* oxidative stress tolerance and population growth in apple wounds, increasing its biocontrol efficiency against *Penicillium expansum*. Furthermore, methyl jasmonate (MeJA) has been shown to boost *M. guilliermondii's* biocontrol activity against apple blue mold degradation (Sun et al., 2021).

19.7 Challenges with biostimulants

Maintaining PGPBs' effectiveness and durability when possibly used in field situations is the biggest challenge. Studies on the use of PGPBs in agriculture show that these organisms, whether in the form of direct inoculants or commercial formulations, are crucial for boosting host plant growth, immune responses, and defense against a range of biotic and abiotic stresses. Nonetheless, variability in laboratory and field experiment findings appears to be a challenge for the widespread adoption of PGPBs as an alternative to agrochemicals. The variability in the efficiency of PGPBs in lab and field circumstances is due to variances in artificial medium and the real soil environment where plants' roots communicate with those organisms in the presence of other bacteria and indigenous chemicals. Furthermore, the sensitivity of PGPBs to certain hosts remains an impediment to reaping the full benefits of such organisms. Survival and longevity of PGPBs as commercial formulations are other major difficulties that require extensive testing (Majeed et al., 2018). Thus, the development of new products by biostimulant manufacturers might be a worthwhile financial investment in such a profitable sector. However, commercialization of biostimulants for the long-term growth of phytoparasitic nematodes would necessitate fresh scientific and industrial scale-up investigations. It is uncommon to find detailed characterization of biostimulant components and their activity (Jardin et al., 2020). This suggestion would present a challenge to the scientific community as a field that has not yet been completely investigated. New studies should address the following topics: (1) identification of Biostimulant species with interesting metabolite profiles; (2) procedure selection to obtain cost-effective formulation; (3) chemical characterization of formulations; (4) modes of action; (5) efficiency studies under different environmental conditions; (6) formulation reliabilities studies; and (7) product registration and commercialization. Even though this method is lengthy and difficult, we believe that these formulations might be one of the new instruments beneficial for sustainable agriculture (Pellegrini et al., 2020). As a result, requiring such information prior to allowing biostimulants to be marketed would pose a significant hurdle to corporations, perhaps discouraging further invention and development (Jardin et al., 2020). The increased prohibition of numerous effective but artificial nematicides, all-around agricultural expansion to increase and improve food production, periodic changes in the genome of nematodes that show resistance, global warming supporting rapid Plant Parasitic Nematode reproduction and spread, and the discovery of new Plant Parasitic Nematode species are some of the most significant challenges (Abd-Elgawad, 2021).

19.8 Prospects of biostimulant in plant pest management

Global climate change and nutrient imbalance are currently having a considerable impact on agricultural productivity in many places. To achieve this aim, careful monitoring of geographic and meteorological variables, as well as mathematical modeling, will be necessary. Fewer PB source types have been investigated on an industrial scale to date, and more alternative sources are required to increase the range of products that can be made available. Because it is feasible, new PBs should be created from byproducts, as this will encourage the reuse of organic waste materials. In addition, the abundance of new PB sources ensures the discovery of novel bioactivities (Jardin et al., 2020). The widespread usage of chemical nematicides causes nematode resistance. As a result, an alternative green economic strategy to their management is required (Rani et al., 2022a,b). To fully utilize them as sustainable agricultural approaches and meet future food demands in a sustainable way, however, a number of challenges such as host specificity, commercial formulation, survival and durability, and inconsistency in a wide range of environmental conditions exist (Majeed et al., 2018). Furthermore, modern biotechnologies should be used to develop standardized formulations and conduct shelf-life studies. Future investigations of biostimulants should involve efficacy evaluations with trials in greenhouse and field operations in order to bring these formulations into agriculture (Pellegrini et al., 2020). Biostimulants derived from disease-controlling microbes might be genetically engineered for increased competitiveness and disease suppression in agriculture (Sehrawat and Sindhu, 2019). Additionally, since proteins are the most frequent targets of small molecules, future research on plant biostimulants should concentrate on creating screening techniques to find bioactive molecules that regulate protein activities. These studies, which can only be conducted in the lab, should be followed by open-air field tests evaluating potential PB compounds. A second approach starts with field observations and ends with a trip back to the lab for additional target clarification. Both methods are believed to be essential for discovering novel PBs, a task crucial for the development of the PB industry (Jardin et al., 2020).

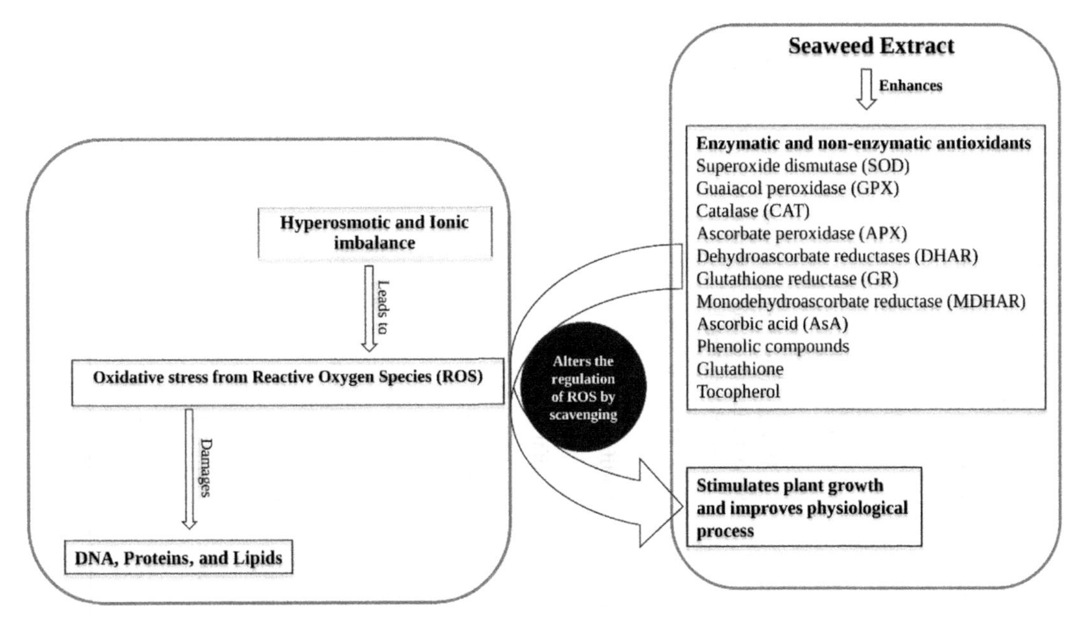

FIGURE 19.3 Mechanism of seaweed extract to combat oxidative stress.

TABLE 19.1 List of Biostimulants and their effect on respective plant species.

Plant Species	Biostimulant	Impact	References
Capsicum annuum L.	Radifarm, Megafol, Benefit, Viva	Increase in fruit production and antioxidant capacity	Paraðiković et al. (2013)
Solanum lycopersicum	Radifarm, Megafol, Viva	Enhanced growth of tomato root, stem, leaves, and increased mineral uptake	Koleška et al. (2017)
Lactuca sativa L.	Bioalgreen S-90	Effects on growth and yield parameters were beneficial, with higher levels of vitamin C and dry matter	Dudaš et al. (2016)
Lepdium sativum L.	KE Palantasalva with and without sea salt; Acker-Sahachtelhalm extract; Fermentierter Pflanzenextrakt; Biplantol Universal.	All extracts help in enhancement of germination, water uptake, and seedling development	Lisjak et al. (2015)
Fragaria x ananassa Duch	Viva, Megafol, Kendal, Porcine blood- based bistimulant.	Megafol enhances fresh and dry mass whereas Viva improved the fruit yield and increased the frost resistance	Bogunovic et al. (2016)
Allium sativum L.	Radifarm	Enhanced plant growth and development after transplantation.	Paraðiković et al. (2019)
Ocimum Basilicum L.	Radifarm	Improved root development and above-ground parts.	Paraðiković et al. (2019)
Begonia semperflores L.	Radifarm	Enhancement in all morphological parameters, nutrient uptake, and growth.	Paraðiković et al. (2017)
Rosa canina L.	Radifarm	Increased root weight and mass, shoot number, and above-ground mass of transplants	Paraðiković et al. (2019)
Tagetes patula L.	Radifarm, Bioalgreen S-90	Improved plant vitality; more flowers and fruit set.	Dudaš et al. (2016)
Tagetes erecta L.	Radifarm	Enhanced germination energy	Paraðiković et al. (2019)
Salvia splendens L.	Radifarm	Accelerated growth of both the plant's root system and its aerial parts.	Paraðiković et al. (2019)
Primula acantis	Radifarm	Enhanced growth and development of root and above-ground mass.	Paraðiković et al. (2019)

TABLE 19.2 Effects of exogenous application of different biostimulants on plant under different stress conditions.

Biostimulant Application	Type of stress	Plant	Effect	References
Humic acid exogenous application	Salt and drought	Maize	Improved biomass and increase in chlorophyll content and proline accumulation, improved antioxidant activity.	Khaled and Fawy (2011)
Foliar spray of humic acid	Salt and drought	Maize	Increased Nitrogen uptake	Kaya et al. (2018)
Humic acid exogenous application	Salt and drought	Wheat	Boosted growth and seed development	El-Bassiouny et al. (2014)
Humic acid exogenous application	Salt and drought	Lima bean	Enhanced photosynthetic activity, membrane stability, carotenoid levels.	Tadayyon et al. (2017)
Humic acid Supplementation	Any environmental	Pistachio leaves	Increased ABA accumulation and proline content	Reza et al. (2013)
Fulvic acid	Heat stress	Soyabean	Improved antioxidant activity and seedling germination	Dinler et al. (2016)
Fulvic acid as foliar spray	Drought	Rapeseed	Decrease lipid peroxidation by increasing antioxidant activity	Lotfi et al. (2015)
Exogenous chitosan	Salt	Pepper and wheat	Boosted plant growth and antioxidative defense	Guan et al. (2009)
Exogenous chitosan	Salt	Mung bean, rosemary, and maize	Increased nutrient uptake photosynthetic activity, prolineAccumulation	Helaly et al. (2018)
Exogenous chitosan	Drought	Rice, common bean, Ricinus, wheat, basil, thyme	Enhanced growth, yield and nutrient uptake	Emami Bistgani et al. (2017)
Algal extracts as foliar spray	Abiotic stress	Any plant	Plant growth and yield	Rinez (2018)
Seaweed extracts as foliar spray	Drought	Arabidopsis, soybean	Enhance stomatal conductance, water use efficacy, and relative leaf water content.	Santaniello et al. (2017), Shukla et al. (2016)
Seaweed extracts as foliar spray	Stress	Pepper, wheat, sargassum	Plant growth, nutrient uptake, and seed yield improved	Rinez (2018)
Foliar application of plant extracts	Drought and salt	Wheat	Increased nutrient uptake, proline accumulation	Farooq et al. (2017)
Algal extracts as foliar spray	Drought	Moringa, alfalfa	Crop growth and biomass synthesis	elsayed et al. (2014)
Exogenous application of plant extracts	Salt stress	Fenugreek, common bean	Plant growth and yield increased	Farooq et al. (2017), Liu et al. (2007)

The forthcoming roadmap, which must be considered as soon as possible, is the biosynthesis of nanomaterials utilizing plant extracts or microbe filtrates to reduce the population of nematodes in soil below the threshold level and to improve plant resistance against nematode infection (Ghareeb et al., 2020). More biostimulants are being found and marketed, which means safer food and better ecosystem health. It aids in the restoration of deteriorated soil caused by chemicals, as well as the prosperity of farmers (Fig. 19.3) (Tables 19.1 and 19.2).

References

Abbasi, P.A., Riga, E., Conn, K.L., Lazarovits, G., 2005. Effect of neem cake soil amendment on reduction of damping-off severity and population densities of plant-parasitic nematodes and soilborne plant pathogens. Can. J. Plant Pathol. 27, 38–45.

Abd-Elgawad, M.M.M., 2021. Optimizing safe approaches to manage plant-parasitic nematodes. Plants 10, 1911.

Abdel-Raouf, N., Al-Homaidan, A.A., Ibraheem, I.B.M., 2012. Microalgae and wastewater treatment. Saudi J. Biol. Sci. 19, 257–275.

Ahemad, M., Kibret, M., 2014. Mechanisms and applications of plant growth promoting rhizobacteria: current perspective. J. King Saud. Univ. Sci. 26, 1–20.

Ali, O., Ramsubhag, A., Jayaraman, J., 2019. Biostimulatory activities of Ascophyllum nodosum extract in tomato and sweet pepper crops in a tropical environment. PLoS One 14, e0216710.

Ali, Q., Shehzad, F., Waseem, M., Shahid, S., Hussain, A.I., Haider, M.Z., et al., 2020. Plant-based biostimulants and plant stress responses. Plant Ecophysiology and Adaptation under Climate Change: Mechanisms and Perspectives I. Springer Singapore, Singapore, pp. 625–661.

Asif, M., Ahmad, F., Tariq, M., Khan, A., Ansari, T., Khan, F., et al., 2017. Potential of chitosan alone and in combination with agricultural wastes against the root-knot nematode, Meloidogyne incognita infesting eggplant. J. Plant Prot. Res. 57, 288–295.

Bartucca, M.L., Cerri, M., del Buono, D., Forni, C., 2022. Use of biostimulants as a new approach for the improvement of phytoremediation performance—a review. Plants 11, 1946.

Beneventi, M.A., da Silva, O.B., de Sá, M.E.L., Firmino, A.A.P., de Amorim, R.M.S., Albuquerque, É.V.S., et al., 2013. Transcription profile of soybean-root-knot nematode interaction reveals a key role of phythormones in the resistance reaction. BMC Genomics 14, 322.

Bogunovic, I., Duralija, B., Gadze, J., Kisic, I., 2016. Biostimulant usage for preserving strawberries to climate damages. Horticultural Sci. 42, 132–140.

Booker, M.A., DeLong, A., 2015. Producing the ethylene signal: regulation and diversification of ethylene biosynthetic enzymes. Plant Physiol. 169, 42–50.

Brown, P., Saa, S., 2015. Biostimulants in agriculture. Front. Plant Sci. 671. Available from: https://doi.org/10.3389/fpls.2015.00671.

Burns, R.G., 1982. Enzyme activity in soil: location and a possible role in microbial ecology. Soil Biol. Biochem. 14, 423–427.

Calvo, P., Nelson, L., Kloepper, J.W., 2014. Agricultural uses of plant biostimulants. Plant Soil 3–41.

Candido, V., Campanelli, G., D'Addabbo, T., Castronuovo, D., Perniola, M., Camele, I., 2015. Growth and yield promoting effect of artificial mycorrhization on field tomato at different irrigation regimes. Sci. Hortic. 187, 35–43.

Canellas, L.P., Olivares, F.L., Aguiar, N.O., Jones, D.L., Nebbioso, A., Mazzei, P., et al., 2015. Humic and fulvic acids as biostimulants in horticulture. Sci. Hortic. 196, 15–27.

Castañeda-Ramírez, G.S., Torres-Acosta, J.F., de, J., Sánchez, J.E., Mendoza-de-Gives, P., González-Cortázar, M., et al., 2020. The possible biotechnological use of edible mushroom bioproducts for controlling plant and animal parasitic nematodes. Biomed. Res. Int. 2020, 1–12.

Čepulytė, R., Danquah, W.B., Bruening, G., Williamson, V.M., 2018. Potent attractant for root-knot nematodes in exudates from seedling root tips of two host species. Sci. Rep. 8, 10847.

Chen, S.-K., Subler, S., Edwards, C.A., 2002. Effects of agricultural biostimulants on soil microbial activity and nitrogen dynamics. Appl. Soil Ecol. 19, 249–259.

Chiaiese, P., Corrado, G., Colla, G., Kyriacou, M.C., Rouphael, Y., 2018. Renewable sources of plant biostimulation: microalgae as a sustainable means to improve crop performance. Front. Plant Sci. 9, 1782. Available from: https://doi.org/10.3389/fpls.2018.01782.

Chin, S., Behm, C., Mathesius, U., 2018. Functions of flavonoids in plant–nematode interactions. Plants 7, 85.

Chung, J., Jin, S., Cho, H., 2005. Low water potential in saline soils enhances nitrate accumulation of lettuce. Commun. Soil Sci. Plant Anal. 36, 1773–1785.

Colla, G., Nardi, S., Cardarelli, M., Ertani, A., Lucini, L., Canaguier, R., et al., 2015. Protein hydrolysates as biostimulants in horticulture. Sci. Hortic. 196, 28–38.

Cosme, M., Stout, M.J., Wurst, S., 2011. Effect of arbuscular mycorrhizal fungi (Glomus intraradices) on the oviposition of rice water weevil (Lissorhoptrus oryzophilus). Mycorrhiza 21, 651–658.

Cotrufo, M.F., Wallenstein, M.D., Boot, C.M., Denef, K., Paul, E., 2013. The Microbial Efficiency-Matrix Stabilization (MEMS) framework integrates plant litter decomposition with soil organic matter stabilization: do labile plant inputs form stable soil organic matter. Glob. Change Biol. 19, 988–995.

Craigie, J.S., 2011. Seaweed extract stimuli in plant science and agriculture. J. Appl. Phycol. 23, 371–393.

Curto, G., Dallavalle, E., Matteo, R., Lazzeri, L., 2016. Biofumigant effect of new defatted seed meals against the southern root-knot nematode, *Meloidogyne incognita*. Ann. Appl. Biol. 169, 17–26.

D'Addabbo, T., Laquale, S., Perniola, M., Candido, V., 2019. Biostimulants for plant growth promotion and sustainable management of phytoparasitic nematodes in vegetable crops. Agronomy 9.

D'Addabbo, T., Radicci, V., Lucarelli, G., Carella, A., Bernad, D., Martin, E., 2011. Effectiveness of a formulation from pedaliaceae plants (Nematon® EC) for the control of the root-knot nematode meloidogyne incognita on greenhouse tomato. Acta Hortic. 23, 233–236.

Dahlin, P., Eder, R., Consoli, E., Krauss, J., Kiewnick, S., 2019. Integrated control of Meloidogyne incognita in tomatoes using fluopyram and Purpureocillium lilacinum strain 251. Crop. Prot. 124, 104874.

Das, K., Roychoudhury, A., 2014. Reactive oxygen species (ROS) and response of antioxidants as ROS-scavengers during environmental stress in plants. Front. Environ. Sci. 2.

Debnath, M., Pandey, M., Bisen, P.S., 2011. An omics approach to understand the plant abiotic stress. OMICS 15, 739–762.

Dick, R.P., Rasmussen, P.E., Kerle, E.A., 1988. Influence of long-term residue management on soil enzyme activities in relation to soil chemical properties of a wheat-fallow system. Biol. Fertil. Soils 6, 159–164.

Dinler, B.S., Gunduzer, E., Tekinay, T., 2016. Pre-treatment of fulvic acid plays a stimulant role in protection of soybean (Glycine max L.) leaves against heat and salt stress. Acta Biol. Crac. Ser. Bot. 58, 29–41.

Dipak Kumar, H., Aloke, P., 2020. Role of biostimulant formulations in crop production: an overview. Int. J. Agric. Sci. Vet. Med. 8 (2), 1–9.

Drobek, M., Frąc, M., Cybulska, J., 2019. Plant biostimulants: importance of the quality and yield of horticultural crops and the improvement of plant tolerance to abiotic stress—a review. Agronomy 9, 335.

du Jardin, P., 2015. Plant biostimulants: definition, concept, main categories and regulation. Sci. Hortic. 196, 3–14.

Dudaš, S., Šola, I., Sladonja, B., Erhatić, R., Ban, D., Poljuha, D., 2016. The effect of biostimulant and fertilizer on "low input" lettuce production. Acta Bot. Croat. 75, 253–259.

el Boukhari, M.E.M., Barakate, M., Bouhia, Y., Lyamlouli, K., 2020. Trends in seaweed extract based biostimulants: manufacturing process and beneficial effect on soil-plant systems. Plants 9, 359.

El-Akshar, E., Elmeihy, R., Tewfike, T., Abou-Aly, H., 2022. Endophytic Streptomyces enissocaesilis as a nematicidal and biostimulant agent. Egypt. Acad. J. Biol. Sci. G Microbiol. 14, 123–133.

El-Ansary, M.S.M., Hamouda, R.A., 2014. Biocontrol of root-knot nematode infected banana plants by some marine algae. Russ. J. Mar. Biol. 40, 140–146.

El-Bassiouny, H.S.M., Bakry, B.A., El-Monem Attia, A.A., Abd Allah, M.M., 2014. Physiological role of humic acid and nicotinamide on improving plant growth, yield, and mineral nutrient of wheat (*Triticum durum*) grown under newly reclaimed sandy soil. Agric. Sci. 05, 687–700.

ElSayed, A.I., Rafudeen, M.S., Golldack, D., 2014. Physiological aspects of raffinose family oligosaccharides in plants: protection against abiotic stress. Plant Biol. 16, 1–8.

Emami Bistgani, Z., Siadat, S.A., Bakhshandeh, A., Ghasemi Pirbalouti, A., Hashemi, M., 2017. Morpho-physiological and phytochemical traits of (Thymus daenensis Celak.) in response to deficit irrigation and chitosan application. Acta Physiol. Plant 39, 231.

Ertani, A., Cavani, L., Pizzeghello, D., Brandellero, E., Altissimo, A., Ciavatta, C., et al., 2009. Biostimulant activity of two protein hydrolyzates in the growth and nitrogen metabolism of maize seedlings. J. Plant Nutr. Soil Sci. 172, 237–244.

Ertani, A., Pizzeghello, D., Altissimo, A., Nardi, S., 2013. Use of meat hydrolyzate derived from tanning residues as plant biostimulant for hydroponically grown maize. J. Plant Nutr. Soil Sci. 176, 287–295.

Escudero, N., Lopez-Moya, F., Ghahremani, Z., Zavala-Gonzalez, E.A., Alaguero-Cordovilla, A., Ros-Ibañez, C., et al., 2017. Chitosan increases tomato root colonization by Pochonia chlamydosporia and their combination reduces root-knot nematode damage. Front. Plant Sci. 8, 1415.

Eyheraguibel, B., Silvestre, J., Morard, P., 2008. Effects of humic substances derived from organic waste enhancement on the growth and mineral nutrition of maize. Bioresour. Technol. 99, 4206–4212.

Fageria, N.K., Gheyi, H.R., Moreira, A., 2011. Nutrient bioavailability in salt affected soils. J. Plant Nutr. 34, 945–962.

Faizi, S., Fayyaz, S., Bano, S., Yawar Iqbal, E., Siddiqi, H., Naz, A., 2011. Isolation of nematicidal compounds from Tagetes patula L. yellow flowers: structure–activity relationship studies against cyst nematode heterodera zeae infective stage larvae. J. Agric. Food Chem. 59, 9080–9093.

Farooq, M., Rizwan, M., Nawaz, A., Rehman, A., Ahmad, R., 2017. Application of natural plant extracts improves the tolerance against combined terminal heat and drought stresses in bread wheat. J. Agron. Crop Sci. 203, 528–538.

Fiorentino, N., Ventorino, V., Woo, S.L., Pepe, O., de Rosa, A., Gioia, L., et al., 2018. Trichoderma-based biostimulants modulate rhizosphere microbial populations and improve N uptake efficiency, yield, and nutritional quality of leafy vegetables. Front. Plant Sci. 9, 743. Available from: https://doi.org/10.3389/fpls.2018.00743.

Frankenberger, W.T., Dick, W.A., 1983. Relationships between enzyme activities and microbial growth and activity indices in soil. Soil Sci. Soc. Am. J. 47, 945–951.

Franzoni, G., Cocetta, G., Prinsi, B., Ferrante, A., Espen, L., 2022. Biostimulants on crops: their impact under abiotic stress conditions. Horticulturae 189. Available from: https://doi.org/10.3390/horticulturae8030189.

Gamalero, E., Glick, B.R., 2020. The use of plant growth-promoting bacteria to prevent nematode damage to plants. Biology (Basel) 9, 381.

Gezerman, A.O., Çorbacıoglu, B.D., 2014. Effects of calcium lignosulfonate and silicic acid on ammonium nitrate degradation. J. Chem. 2014, 1–6.

Ghareeb, R.Y., Hafez, E.E., Ibrahim, D.S.S., 2020. Current management strategies for phytoparasitic nematodes. Management of Phytonematodes: Recent Advances and Future Challenges. Springer Singapore, Singapore, pp. 339–352.

Giannakou, I.O., 2011. Efficacy of a formulated product containing Quillaja saponaria plant extracts for the control of root-knot nematodes. Eur. J. Plant Pathol. 130, 587–596.

Giannattasio, M., 2013. Microbiological features and bioactivity of a fermented manure product (preparation 500) used in biodynamic agriculture. J. Microbiol. Biotechnol. 23, 644–651.

Grover, M., Bodhankar, S., Sharma, A., Sharma, P., Singh, J., Nain, L., 2021. PGPR mediated alterations in root traits: way toward sustainable crop production. Front. Sustain. Food Syst. 4.

Guan, Y., Hu, J., Wang, X., Shao, C., 2009. Seed priming with chitosan improves maize germination and seedling growth in relation to physiological changes under low temperature stress. J. Zhejiang Univ. Sci. B 10, 427–433.

Guan, R., Su, J., Meng, X., Li, S., Liu, Y., Xu, J., et al., 2015. Multilayered regulation of ethylene induction plays a positive role in arabidopsis resistance against *Pseudomonas syringae*. Plant Physiol. 169, 299–312.

Hafez, E.E., Abdel-Fattah, G.M., El-Haddad, S.A., Rashad, Y.M., 2013. Molecular defense response of mycorrhizal bean plants infected with Rhizoctonia solani. Ann. Microbiol. 63, 1195–1203.

Halpern, M., Bar-Tal, A., Ofek, M., Minz, D., Muller, T., Yermiyahu, U., 2015a. The Use of Biostimulants for Enhancing Nutrient Uptake. Academic Press, Elsevier, Waltham, MA 02451, USA, pp. 141–174.

Halpern, M., Bar-Tal, A., Ofek, M., Minz, D., Muller, T., Yermiyahu, U., 2015b. Use Biostimulants Enhancing Nutrient Uptake 141–174.

Hamid, B., Zaman, M., Farooq, S., Fatima, S., Sayyed, R.Z., Baba, Z.A., et al., 2021. Bacterial plant biostimulants: a sustainable way towards improving growth, productivity, and health of crops. Sustainability 13, 2856.

Hao, Z., Fayolle, L., van Tuinen, D., Chatagnier, O., Li, X., Gianinazzi, S., et al., 2012. Local and systemic mycorrhiza-induced protection against the ectoparasitic nematode Xiphinema index involves priming of defence gene responses in grapevine. J. Exp. Bot. 63, 3657–3672.

Helaly, M., Farouk, S., Arafa, S., Amhimmid, N., 2018. Inducing salinity tolerance of rosemary (Rosmarinus officinalis L.) plants by chitosan or zeolite application. Asian J. Adv. Agric. Res. 5, 1–20.

Hellequin, E., Monard, C., Quaiser, A., Henriot, M., Klarzynski, O., Binet, F., 2018. Specific recruitment of soil bacteria and fungi decomposers following a biostimulant application increased crop residues mineralization. PLoS One 13, e0209089.

Heydarian, Z., Yu, M., Gruber, M., Glick, B.R., Zhou, R., Hegedus, D.D., 2016. Inoculation of soil with plant growth promoting bacteria producing 1-aminocyclopropane-1-carboxylate deaminase or expression of the corresponding acds gene in transgenic plants increases salinity tolerance in Camelina sativa. Front. Microbiol. 7.

Holbein, J., Franke, R.B., Marhavý, P., Fujita, S., Górecka, M., Sobczak, M., et al., 2019. Root endodermal barrier system contributes to defence against plant-parasitic cyst and root-knot nematodes. Plant J. 100, 221–236.

Hölscher, D., Dhakshinamoorthy, S., Alexandrov, T., Becker, M., Bretschneider, T., Buerkert, A., et al., 2014. Phenalenone-type phytoalexins mediate resistance of banana plants (*Musa* spp.) to the burrowing nematode *Radopholus similis*. Proc. Natl. Acad. Sci. USA 111, 105–110.

Hu, Y., Burucs, Z., von Tucher, S., Schmidhalter, U., 2007. Short-term effects of drought and salinity on mineral nutrient distribution along growing leaves of maize seedlings. Environ. Exp. Bot. 60, 268–275.

Hu, Y., You, J., Li, C., Williamson, V.M., Wang, C., 2017. Ethylene response pathway modulates attractiveness of plant roots to soybean cyst nematode Heterodera glycines. Sci. Rep. 7, 41282.

Jardin, P., Xu, L., Geelen, D., 2020. Agricultural functions and action mechanisms of plant biostimulants (PBs). The Chemical Biology of Plant Biostimulants. John Wiley & Sons, Inc., 111 River Street, Hoboken, NJ 07030, USA, pp. 1–30.

Javaid, A., 2006. Foliar application of effective microorganisms on pea as an alternative fertilizer. Agron. Sustain. Dev. 26, 257–262.

Jindo, K., Martim, S.A., Navarro, E.C., Pérez-Alfocea, F., Hernandez, T., Garcia, C., et al., 2012. Root growth promotion by humic acids from composted and non-composted urban organic wastes. Plant Soil 353, 209–220.

Juárez-Maldonado, A., Ortega-Ortíz, H., Morales-Díaz, A.B., González-Morales, S., Morelos-Moreno, Á., Cabrera-De la Fuente, M., et al., 2019. Nanoparticles and Nanomaterials as Plant Biostimulants. Int. J. Mol. Sci. 20, 162.

Karaca, A., Cetin, S.C., Turgay, O.C., Kizilkaya, R., 2010. Soil enzymes as indication of soil quality. In: Shukla, G., Varma, A. (Eds.), Soil Enzymology. In: Vol. 22. Springer Science & Business Media, pp. 119–148.

Karol, R.H., 2003. Chemical grouting and soil stabilization. Revised and Expanded. CRC Press, USA.

Kaya, C., Akram, N.A., Ashraf, M., Sonmez, O., 2018. Exogenous application of humic acid mitigates salinity stress in maize (*Zea mays* L.) plants by improving some key physico-biochemical attributes. Cereal Res. Commun. 46, 67–78.

Khaled, H., Fawy, H.A., 2011. Effect of different levels of humic acids on the nutrient content, plant growth, and soil properties under conditions of salinity. Soil. Water Res. 6, 21–29.

Khan, W., Rayirath, U.P., Subramanian, S., Jithesh, M.N., Rayorath, P., Hodges, D.M., et al., 2009. Seaweed extracts as biostimulants of plant growth and development. J. Plant Growth Regul. 28, 386–399.

KKhan, S.A., Abid, M., Hussain, F., 2015. Nematicidal activity of seaweeds against *Meloidogyne javanica*. Pak. J. Nematol. 33 (2), 195–203.

Koleška, I., Hasanagić, D., Todorović, V., Murtić, S., Klokić, I., Parađiković, N., et al., 2017. Biostimulant prevents yield loss and reduces oxidative damage in tomato plants grown on reduced NPK nutrition. J. Plant Interact. 12, 209–218.

Kumari, P., Meena, M., Gupta, P., Dubey, M.K., Nath, G., Upadhyay, R.S., 2018. Plant growth promoting rhizobacteria and their biopriming for growth promotion in mung bean (Vigna radiata (L.) R. Wilczek). Biocatal. Agric. Biotechnol. 16, 163–171.

Kyker-Snowman, E., Wieder, W.R., Frey, S.D., Grandy, A.S., 2020. Stoichiometrically coupled carbon and nitrogen cycling in the MIcrobial-MIneral Carbon Stabilization model version 1.0 (MIMICS-CN v1.0). Geosci. Model. Dev. 13, 4413–4434.

Lanzuise, S., Manganiello, G., Guastaferro, V.M., Vincenzo, C., Vitaglione, P., Ferracane, R., et al., 2022. Combined biostimulant applications of trichoderma spp. with fatty acid mixtures improve biocontrol activity, horticultural crop yield and nutritional quality. Agronomy 12, 275.

Laudicina, V.A., Hurtado, M.D., Badalucco, L., Delgado, A., Palazzolo, E., Panno, M., 2009. Soil chemical and biochemical properties of a salt-marsh alluvial Spanish area after long-term reclamation. Biol. Fertil. Soils 45, 691–700.

Lemanowicz, J., 2018. Dynamics of phosphorus content and the activity of phosphatase in forest soil in the sustained nitrogen compounds emissions zone. Environ. Sci. Pollut. Res. 25, 33773–33782.

Liang, W., Yu, A., Wang, G., Zheng, F., Jia, J., Xu, H., 2018. Chitosan-based nanoparticles of avermectin to control pine wood nematodes. Int. J. Biol. Macromol. 112, 258–263.

Lisjak, M., Tomić, O., Špoljarević, M., Teklić, T., Stanisavljević, A., Balas, J., 2015. Garden cress germinability and seedling vigour after treatment with plant extracts. Poljoprivreda 21, 41–46.

Liu, J., Tian, S., Meng, X., Xu, Y., 2007. Effects of chitosan on control of postharvest diseases and physiological responses of tomato fruit. Postharvest Biol. Technol. 44, 300–306.

Lotfi, R., Gharavi-Kouchebagh, P., Khoshvaghti, H., 2015. Biochemical and physiological responses of Brassica napus plants to humic acid under water stress. Russ. J. Plant Physiol. 62, 480–486.

Loyal, A., Pahuja, S.K., Sharma, P., Malik, A., Srivastava, R.K., Mehta, S., 2023. Potential environmental and human health implications of nanomaterials used in sustainable agriculture and soil improvement. Engineered Nanomaterials for Sustainable Agricultural Production, Soil Improvement and Stress Management. Academic Press, USA, pp. 387–412.

Machado, R., Serralheiro, R., 2017. Soil salinity: effect on vegetable crop growth. Management practices to prevent and mitigate soil salinization. Horticulturae 3, 30.

Majeed, A., Muhammad, Z., Ahmad, H., 2018. Plant growth promoting bacteria: role in soil improvement, abiotic and biotic stress management of crops. Plant Cell Rep. 37, 1599–1609.

Mardani-Talaee, M., Razmjou, J., Nouri-Ganbalani, G., Hassanpour, M., Naseri, B., 2017. Impact of chemical, organic and bio-fertilizers application on bell pepper, Capsicum annuum L. and biological parameters of Myzus persicae (Sulzer) (Hem.: Aphididae). Neotrop. Entomol. 46, 578–586.

Martín, M.H.J., Ángel, M.M.M., Aarón, S.L.J., Israel, B.G., 2022. Protein hydrolysates as biostimulants of plant growth and development. In: Ramawat, N., Bhardwaj, V. (Eds.), Biostimulants: Exploring Sources and Applications. Plant Life and Environment Dynamics. Springer, Singapore, pp. 141–175. Available from: https://doi.org/10.1007/978-981-16-7080-0_6.

Masciandaro, G., Ceccanti, B., Ronchi, V., Benedicto, S., Howard, L., 2002. Humic substances to reduce salt effect on plant germination and growth. Commun. Soil Sci. Plant Anal. 33, 365–378.

McDonald, A.H., de Waele, D., de Waele, E., 1988. Influence of seaweed concentrate on the reproduction of Pratylenchuszeae (Nematoda) on maize. Nematologica 34, 71–77.

Meena, M., Swapnil, P., Zehra, A., Aamir, M., Dubey, M.K., Goutam, J., et al., 2017. Beneficial microbes for disease suppression and plant growth promotion. Plant-Microbe Interactions in Agro-Ecological Perspectives. Springer Singapore, Singapore, pp. 395–432.

Michalak, I., Chojnacka, K., 2014. Algal extracts: technology and advances. Eng. Life Sci. 14, 581–591.

Michalak, I., Chojnacka, K., 2015. Algae as production systems of bioactive compounds. Eng. Life Sci. 15, 160–176.

Mire, G.le, Nguyen, M.L., Fassotte, B., Jardin, P.du, Verheggen, F., Delaplace, P., et al., 2016. Review: implementing plant biostimulants and biocontrol strategies in the agroecological management of cultivated ecosystems. BASE 299–313.

Mittler, R., 2002. Oxidative stress, antioxidants and stress tolerance. Trends Plant Sci. 7, 405–410.

Moore, J.A.M., Abraham, P.E., Michener, J.K., Muchero, W., Cregger, M.A., 2022. Ecosystem consequences of introducing plant growth promoting rhizobacteria to managed systems and potential legacy effects. N. Phytol. 234, 1914–1918.

Mota, L.C.B.M., dos Santos, M.A., 2016. Chitin and chitosan on Meloidogyne javanica management and on chitinase activity in tomato plants. Trop. Plant Pathol. 41, 84–90.

Mwaheb, M.A.M.A., Hussain, M., Tian, J., Zhang, X., Hamid, M.I., El-Kassim, N.A., et al., 2017. Synergetic suppression of soybean cyst nematodes by chitosan and Hirsutella minnesotensis via the assembly of the soybean rhizosphere microbial communities. Biol. Control 115, 85–94.

Nadeem, S.M., Ahmad, M., Zahir, Z.A., Javaid, A., Ashraf, M., 2014. The role of mycorrhizae and plant growth promoting rhizobacteria (PGPR) in improving crop productivity under stressful environments. Biotechnol. Adv. 32, 429–448.

Nagachandrabose, S., Baidoo, R., 2021. Humic acid – a potential bioresource for nematode control. Nematology 24, 1–10.

Nardi, S., Carletti, P., Pizzeghello, D., Muscolo, A., n.d. Biological activities of humic substances. In: Biophysico-Chemical Processes Involving Natural Nonliving Organic Matter in Environmental Systems. John Wiley & Sons, Inc., Hoboken, NJ, USA, pp. 305–339.

Nicol, J.M., Turner, S.J., Coyne, D.L., Nijs, L., den, Hockland, S., Maafi, Z.T., 2011. Current nematode threats to world agriculture. Genomics and Molecular Genetics of Plant-Nematode Interactions. Springer Netherlands, Dordrecht, pp. 21–43.

Noman, A., Ali, Q., Maqsood, J., Iqbal, N., Javed, M.T., Rasool, N., et al., 2018. Deciphering physio-biochemical, yield, and nutritional quality attributes of water-stressed radish (Raphanus sativus L.) plants grown from Zn-Lys primed seeds. Chemosphere 195, 175–189.

Nunes da Silva, M., Cardoso, A.R., Ferreira, D., Brito, M., Pintado, M.E., Vasconcelos, M.W., 2014. Chitosan as a biocontrol agent against the pinewood nematode (*Bursaphelenchus xylophilus*). For. Pathol. 44, 420–423.

Olivares, F.L., Aguiar, N.O., Rosa, R.C.C., Canellas, L.P., 2015. Substrate biofortification in combination with foliar sprays of plant growth promoting bacteria and humic substances boosts production of organic tomatoes. Sci. Hortic. 183, 100–108.

Oo, A.N., Iwai, C.B., Saenjan, P., 2015. Soil properties and maize growth in saline and nonsaline soils using cassava-industrial waste compost and vermicompost with or without earthworms. Land. Degrad. Dev. 26, 300–310.

Paracer, S., Tarjan, A.C., Hodgson, L.M., 1987. Effective use of marine algal products in the management of plant-parasitic nematodes. J. Nematol. 19, 194–200.

Parađiković, N., Teklić, T., Zeljković, S., Lisjak, M., Špoljarević, M., 2019. Biostimulants research in some horticultural plant species—a review. Food Energy Secur. .

Parađiković, N., Vinkovic, T., Vinković Vrček, I., Tkalec, M., 2013. Natural biostimulants reduce the incidence of BER in sweet yellow pepper plants (Capsicum annuum L.). Agric. Food Sci. 22, 307–317.

Parađiković, N., Zeljković, S., Tkalec, M., Vinković, T., Maksimović, I., Haramija, J., 2017. Influence of biostimulant application on growth, nutrient status and proline concentration of begonia transplants. Biol. Agric. Hortic. 33, 89–96.

Parrado, J., Bautista, J., Romero, E.J., García-Martínez, A.M., Friaza, V., Tejada, M., 2008. Production of a carob enzymatic extract: potential use as a biofertilizer. Bioresour. Technol. 99, 2312–2318.

Pellegrini, M., Pagnani, G., Bernardi, M., Mattedi, A., Spera, D.M., Gallo, M.del, 2020. Cell-free supernatants of plant growth-promoting bacteria: a review of their use as biostimulant and microbial biocontrol agents in sustainable agriculture. Sustainability 12, 9917.

Quilty, J.R., Cattle, S.R., 2011. Use and understanding of organic amendments in Australian agriculture: a review. Soil Res. 49, 1.

Radwan, M.A., Farrag, S.A.A., Abu-Elamayem, M.M., Ahmed, N.S., 2012. Extraction, characterization, and nematicidal activity of chitin and chitosan derived from shrimp shell wastes. Biol. Fertil. Soils 48, 463–468.

Rani, K., Sharma, P., Kumar, S., Wati, L., Kumar, R., Gurjar, D.S., et al., 2019. Legumes for sustainable soil and crop management. Sustain. Manag. soil Environ. 193–215.

Rani, K., Devi, N., Banakar, P., Kharb, P., Kaushik, P., 2022a. Nematicidal potential of green silver nanoparticles synthesized using aqueous root extract of Glycyrrhiza glabra. Nanomaterials 12, 2966.

Rani, K., Rani, A., Sharma, P., Dahiya, A., Punia, H., Kumar, S., et al., 2022b. Legumes for agroecosystem services and sustainability. Advances in Legumes for Sustainable Intensification. Academic Press, 525 B Street, Suite 1650, San Diego, CA 92101, United States, pp. 363–380.

Reza, M., Kasmani, M.B., Samavat, S., Mostafavi, M., Khalighi, A., 2013. The effect of application of humic acid foliar on biochemical parameters of pistachio under drought stress. N. Y. Sci. J. 6 (12), 26–31.

Rinez, I., 2018. Improving salt tolerance in pepper by bio-priming with Padina pavonica and Jania rubens aqueous extracts. Int. J. Agric. Biol. 20, 513–523.

Rizvi, R., Singh, G., Safiuddin, Ali Ansari, R., Ali Tiyagi, S., Mahmood, I., 2015. Sustainable management of root-knot disease of tomato by neem cake and *Glomus fasciculatum*. Cogent Food Agric. 1, 1008859.

Roopa, K.P., Gadag, A.S., 2020. Importance of biopesticides in the sustainable management of plant-parasitic nematodes. Management of Phytonematodes: Recent Advances and Future Challenges. Springer Singapore, Singapore, pp. 205–227.

Rose, M.T., Patti, A.F., Little, K.R., Brown, A.L., Jackson, W.R., Cavagnaro, T.R., 2014. A Meta-Analysis and Review of Plant-Growth Response to Humic Substances. Academic Press, USA, pp. 37–89.

Rouphael, Y., Carillo, P., Cristofano, F., Cardarelli, M., Colla, G., 2021. Effects of vegetal- versus animal-derived protein hydrolysate on sweet basil morpho-physiological and metabolic traits. Sci. Hortic. 284, 110123.

Rouphael, Y., Colla, G., 2020. Editorial: biostimulants in agriculture. Front. Plant Sci11 (40). Available from: https://doi.org/10.3389/fpls.2020.00040.

Rouphael, Y., Franken, P., Schneider, C., Schwarz, D., Giovannetti, M., Agnolucci, M., et al., 2015. Arbuscular mycorrhizal fungi act as biostimulants in horticultural crops. Sci. Hortic. 196, 91–108.

Rubin, R.L., van Groenigen, K.J., Hungate, B.A., 2017. Plant growth promoting rhizobacteria are more effective under drought: a meta-analysis. Plant Soil 416, 309–323.

Ruzzi, M., Aroca, R., 2015. Plant growth-promoting rhizobacteria act as biostimulants in horticulture. Sci. Hortic. 196, 124–134.

Sangwan, S., Sharma, P., Wati, L., Mehta, S., 2023. Effect of chitosan nanoparticles on growth and physiology of crop plants. Engineered Nanomaterials for Sustainable Agricultural Production, Soil Improvement and Stress Management. Academic Press, USA, pp. 99–123.

Santaniello, A., Scartazza, A., Gresta, F., Loreti, E., Biasone, A., di Tommaso, D., et al., 2017. Ascophyllum nodosum seaweed extract alleviates drought stress in Arabidopsis by affecting photosynthetic performance and related gene expression. Front. Plant. Sci. 8, 1362. Available from: https://doi.org/10.3389/fpls.2017.01362.

Saviozzi, A., Levi-Minzi, R., Cardelli, R., Riffaldi, R., 2001. A comparison of soil quality in adjacent cultivated, forest and native grassland soils. Plant Soil 233, 251–259.

Schiavon, M., Ertani, A., Nardi, S., 2008. Effects of an alfalfa protein hydrolysate on the gene expression and activity of enzymes of the tricarboxylic acid (TCA) cycle and nitrogen metabolism in Zea mays L. J. Agric. Food Chem. 56, 11800–11808.

Schiavon, M., Pizzeghello, D., Muscolo, A., Vaccaro, S., Francioso, O., Nardi, S., 2010. High molecular size humic substances enhance phenylpropanoid metabolism in maize (Zea mays L.). J. Chem. Ecol. 36, 662–669.

Schmidt, R.E., Ervin, E.H., Zhang, X., 2003. Questions and answers about biostimulants. Golf Course Manage. 71 (6), 91–94.

Sehrawat, A., Sindhu, S.S., 2019. Potential of biocontrol agents in plant disease control for improving food safety. Def. Life Sci. J. 4, 220–225.

Selvaraj, A., Thangavel, K., Uthandi, S., 2020. Arbuscular mycorrhizal fungi (Glomus intraradices) and diazotrophic bacterium (Rhizobium BMBS) primed defense in blackgram against herbivorous insect (Spodoptera litura) infestation. Microbiol. Res. 231, 126355.

Seyedbagheri, M.-M., 2010. Influence of humic products on soil health and potato production. Potato Res. 53, 341–349.

Sharma, P., Sharma, M.M.M., Kapoor, D., Rani, K., Singh, D., Barkodia, M., 2020a. Role of microbes for attaining enhanced food crop production. Microbial Biotechnology: Basic Research and Applications. Springer, Singapore, pp. 55–78.

Sharma, P., Sharma, M.M.M., Patra, A., Vashisth, M., Mehta, S., Singh, B., et al., 2020b. The role of key transcription factors for cold tolerance in plants. Transcription Factors for Abiotic Stress Tolerance in Plants. Academic Press, USA, pp. 123–152.

Sharma, M.M.M., Sharma, P., Kapoor, D., Beniwal, P., Mehta, S., 2021a. Phytomicrobiome community: an agrarian perspective towards resilient agriculture. Plant Performance Under Environmental Stress. Springer, Cham, pp. 493–534.

Sharma, P., Pandey, V., Sharma, M.M.M., Patra, A., Singh, B., Mehta, S., et al., 2021b. A review on biosensors and nanosensors application in agroecosystems. Nanoscale Res. Lett. 16, 1–24.

Sharma, P., Sharma, M.M.M., Malik, A., Vashisth, M., Singh, D., Kumar, R., et al., 2021c. Rhizosphere, rhizosphere biology, and rhizospheric engineering. Plant Growth-Promoting Microbes for Sustainable Biotic and Abiotic Stress Management. Springer International Publishing, Cham, pp. 577–624.

Sharma, P., Sangwan, S., Kumari, A., Singh, S., Kaur, H., 2022a. Impact of climate change on soil microorganisms regulating nutrient transformation. Plant Stress Mitigators: Action and Application. Springer Nature Singapore, Singapore, pp. 145–172.

Sharma, P., Meyyazhagan, A., Easwaran, M., Sharma, M.M.M., Mehta, S., Pandey, V., et al., 2022b. Hydrogen sulfide: a new warrior in assisting seed germination during adverse environmental conditions. Plant Growth Regul. 98 (3), 401–420.

Sharma, P., Sangwan, S., Mehta, S., 2023. Emerging role of phosphate nanoparticles in agriculture practices. Engineered Nanomaterials for Sustainable Agricultural Production, Soil Improvement and Stress Management. Academic Press, USA, pp. 71–97.

Shrivastava, P., Kumar, R., 2015. Soil salinity: a serious environmental issue and plant growth promoting bacteria as one of the tools for its alleviation. Saudi J. Biol. Sci. 22, 123–131.

Shrivastava, G., Ownley, B.H., Augé, R.M., Toler, H., Dee, M., Vu, A., et al., 2015. Colonization by arbuscular mycorrhizal and endophytic fungi enhanced terpene production in tomato plants and their defense against a herbivorous insect. Symbiosis 65, 65–74.

Shukla, P.S., Borza, T., Critchley, A.T., Prithiviraj, B., 2016. Carrageenans from red seaweeds as promoters of growth and elicitors of defense response in plants. Front. Mar. Sci. 3, 81. Available from: https://doi.org/10.3389/fmars.2016.00081.

Siddiqui, I.A., Shaukat, S.S., 2003. Plant species, host age and host genotype effects on Meloidogyne incognita biocontrol by Pseudomonas fluorescens strain CHA0 and its genetically-modified derivatives. J. Phytopathol. 151, 231–238.

Sikder, M.M., Vestergård, M., 2020. Impacts of root metabolites on soil nematodes. Front. Plant Sci. 101792. Available from: https://doi.org/10.3389/fpls.2019.01792.

Singh, A., Sharma, P., Kumari, A., Kumar, R., Pathak, D.V., 2019. Management of root-knot nematode in different crops using microorganisms. Plant Biotic Interactions. pp. 85–99.

Singh, S., Sangwan, S., Sharma, P., Devi, P., Moond, M., 2021. Nanotechnology for sustainable agriculture: an emerging perspective. J. Nanosci. Nanotechnol. 21 (6), 3453–3465.

Sofo, A., Nuzzaci, M., Vitti, A., Tataranni, G., Scopa, A., 2014. Control of biotic and abiotic stresses in cultivated plants by the use of biostimulant microorganisms. Improvement of Crops in the Era of Climatic Changes. Springer New York, New York, NY, pp. 107–117.

Sreeharsha, R.V., Mudalkar, S., Sengupta, D., Unnikrishnan, D.K., Reddy, A.R., 2019. Mitigation of drought-induced oxidative damage by enhanced carbon assimilation and an efficient antioxidative metabolism under high CO2 environment in pigeonpea (Cajanus cajan L.). Photosynth. Res. 139, 425–439.

Strachel, R., Wyszkowska, J., Baćmaga, M., 2017. The role of compost in stabilizing the microbiological and biochemical properties of zinc-stressed soil. Water Air Soil. Pollut. 228, 349.

Sun, C., Huang, Y., Lian, S., Saleem, M., Li, B., Wang, C., 2021. Improving the biocontrol efficacy of Meyerozyma guilliermondii Y-1 with melatonin against postharvest gray mold in apple fruit. Postharvest Biol. Technol. 171, 111351.

Tejada, M., Garcia, C., Gonzalez, J.L., Hernandez, M.T., 2006. Use of organic amendment as a strategy for saline soil remediation: influence on the physical, chemical and biological properties of soil. Soil Biol. Biochem. 38, 1413–1421.

Tadayyon, A., Beheshti, S., Pessarakli, M., 2017. Effects of sprayed humic acid, iron, and zinc on quantitative and qualitative characteristics of niger plant (*Guizotia abyssinica* L.). J. Plant Nutr. 40, 1644–1650.

Tejada, M., Benítez, C., Gómez, I., Parrado, J., 2011. Use of biostimulants on soil restoration: effects on soil biochemical properties and microbial community. Appl. Soil Ecol. 49, 11–17.

Tarigan, S.I., Toth, S., Szalai, M., Kiss, J., Turoczi, G., Toepfer, S., 2022. Biological control properties of microbial plant biostimulants. A review. Biocontrol Sci. Technol. 32, 1351–1371.

Trivedi, K., Vijay Anand, K.G., Vaghela, P., Ghosh, A., 2018. Differential growth, yield and biochemical responses of maize to the exogenous application of Kappaphycus alvarezii seaweed extract, at grain-filling stage under normal and drought conditions. Algal Res. 35, 236–244.

Tsai, A.Y.-L., Higaki, T., Nguyen, C.-N., Perfus-Barbeoch, L., Favery, B., Sawa, S., 2019. Regulation of root-knot nematode behavior by seed-coat mucilage-derived attractants. Mol. Plant 12, 99–112.

van Oosten, M.J., Pepe, O., de Pascale, S., Silletti, S., Maggio, A., 2017. The role of biostimulants and bioeffectors as alleviators of abiotic stress in crop plants. Chem. Biol. Technol. Agric. 4, 5.

Vassileva, M., Flor-Peregrin, E., Malusá, E., Vassilev, N., 2020. Towards better understanding of the interactions and efficient application of plant beneficial prebiotics, probiotics, postbiotics and synbiotics. Front. Plant Sci. 111068. Available from: https://doi.org/10.3389/fpls.2020.01068.

Veronico, P., Melillo, M.T., 2021. Marine organisms for the sustainable management of plant parasitic nematodes. Plants 10, 369.

Volpe, V., Chitarra, W., Cascone, P., Volpe, M.G., Bartolini, P., Moneti, G., et al., 2018. The association with two different arbuscular mycorrhizal fungi differently affects water stress tolerance in tomato. Front. Plant Sci. 91480. Available from: https://doi.org/10.3389/fpls.2018.01480.

Vos, C., Schouteden, N., van Tuinen, D., Chatagnier, O., Elsen, A., de Waele, D., et al., 2013. Mycorrhiza-induced resistance against the root–knot nematode Meloidogyne incognita involves priming of defense gene responses in tomato. Soil Biol. Biochem. 60, 45–54.

Williams, T.I., Edgington, S., Owen, A., Gange, A.C., 2021. Evaluating the use of seaweed extracts against root knot nematodes: a meta-analytic approach. Appl. Soil Ecol. 168, 104170.

Wuyts, N., Swennen, R., de Waele, D., 2006. Effects of plant phenylpropanoid pathway products and selected terpenoids and alkaloids on the behaviour of the plant-parasitic nematodes Radopholus similis, Pratylenchus penetrans and Meloidogyne incognita. Nematology 8, 89–101.

Yadav, D.S., Rai, R., Mishra, A.K., Chaudhary, N., Mukherjee, A., Agrawal, S.B., et al., 2019. ROS production and its detoxification in early and late sown cultivars of wheat under future O3 concentration. Sci. Total Environ. 659, 200–210.

Yakhin, O.I., Lubyanov, A.A., Yakhin, I.A., Brown, P.H., 2017. Biostimulants in plant science: a global perspective. Front. Plant Sci. 72049. Available from: https://doi.org/10.3389/fpls.2016.02049.

Zhang, Y., Dubé, M.A., McLean, D.D., Kates, M., 2003. Biodiesel production from waste cooking oil: 2. Economic assessment and sensitivity analysis. Bioresour. Technol. 90, 229–240.

Zhang, S., Gan, Y., Ji, W., Xu, B., Hou, B., Liu, J., 2017. Mechanisms and characterization of Trichoderma longibrachiatum T6 in suppressing nematodes (Heterodera avenae) in wheat. Front. Plant Sci. 8, 1491. Available from: https://doi.org/10.3389/fpls.2017.01491.

Zhao, Dan, Zhao, H., Zhao, Di, Zhu, X., Wang, Y., Duan, Y., et al., 2018. Isolation and identification of bacteria from rhizosphere soil and their effect on plant growth promotion and root-knot nematode disease. Biol. Control 119, 12–19.

Zia-ur-Rehman, M., Murtaza, G., Qayyum, M.F., Saifullah, Rizwan, M., Ali, S., et al., 2016. Degraded soils: origin, types and management. Soil Science: Agricultural and Environmental Prospectives. Springer Int. Publishing, Cham, pp. 23–65.

Further reading

Abdel-Baky, H.H., 2009. Enhancing antioxidant availability in grains of wheat plants grown under seawater-stress in response to microalgae extracts treatments. Afr. J. Biochem. Res. 3, 77–083.

IAS_5_3_171_172, n.d.

Ibrahim, W.M., Ali, R.M., Hemida, K.A., Sayed, M.A., 2014. Role of *Ulva lactuca* extract in alleviation of salinity stress on wheat seedlings. Sci. World J. 2014, 1–11.

Karimi, M., Avci, P., Mobasseri, R., Hamblin, M.R., Naderi-Manesh, H., 2013. The novel albumin–chitosan core–-shell nanoparticles for gene delivery: preparation, optimization and cell uptake investigation. J. Nanopart. Res. 15, 1651.

Kumar, R., Sharma, P., Gupta, R.K., Kumar, S., Sharma, M.M.M., Singh, S., et al., 2020. Earthworms for eco-friendly resource efficient agriculture. Resources Use Efficiency in Agriculture. pp. 47–84.

Malekpoor, F., Pirbalouti, A.G., Salimi, A., 2016. Effect of foliar application of chitosan on morphological and physiological characteristics of basil under reduced irrigation. Res. Crops 17, 354.

Pajares, S., Bohannan, B.J.M., 2016. Ecology of nitrogen fixing, nitrifying, and denitrifying microorganisms in tropical forest soils. Front. Microbiol. 7.

Plant_Biostimulants_final_report_bio_2012_en, n.d.

Pongprayoon, W., Roytrakul, S., Pichayangkura, R., Chadchawan, S., 2013. The role of hydrogen peroxide in chitosan-induced resistance to osmotic stress in rice (Oryza sativa L.). Plant Growth Regul. 70, 159–173.

Ray, S.R., Haque Bhuiyan, M.J., Anowar Hossain, M., Mahmud, S., Ul-Arif, T., 2016. Chitosan suppresses antioxidant enzyme activities for mitigating salt stress in mungbean varieties. IOSR J. Agric. Vet. Sci. 09, 36–41.

Shaddad, M., M. Farghl, A.A., Galal, H.R., Hassan, E.A., 2014. Influence of Seaweed Extracts (Sargassum dentifolium or Padina gymnospora) on the Growth and Physiological Activities of Faba Bean and Wheat Plants Under Salt Stress.

Tohidi Moghadam, H.R., Khalafi Khamene, M., Zahedi, H., n.d. Open Access Effect of humic acid foliar application on growth and quantity of corn in irrigation withholding at different growth stages.

Zeng, D., Luo, X., 2012. Physiological effects of chitosan coating on wheat growth and activities of protective enzyme with drought tolerance. Open J. Soil Sci. 02, 282–288.

CHAPTER

20

Applications of *Trichoderma virens* and biopolymer-based biostimulants in plant growth and productions

Divya Kapoor[1], *Mayur Mukut Murlidhar Sharma*[2], *Sheetal Yadav*[1] and *Pankaj Sharma*[1]

[1]**Department of Microbiology, CCS Haryana Agricultural University, Hisar, Haryana, India**
[2]**Department of Agriculture and Life Industry, Kangwon National University, Chuncheon, Gangwon, Republic of Korea**

20.1 Introduction

The agricultural environment is continuously facing severe pressures of constantly changing climatic conditions, increasing incidence of abiotic stresses, evolution of new cultivars of herbicide-resistant weeds, the evolution of novel plant pathogens, increased incidence of pathogenic attacks, and rapid depletion of soil nutrient status (Rani et al., 2019, 2022; Sharma et al., 2020a,b; Singh et al., 2019, 2021). The continuously increasing global population has also become a major factor that is putting continuous pressure on increasing productivity in global agricultural systems. The common agricultural practice of dealing with pathogen attacks and competition from weeds includes the application of different pesticides and herbicides. Recent decades have faced the explicit usage of agrochemicals that marks their increased persistence in the soil and the agricultural produce (Anamika et al., 2023; Kumar et al., 2020; Loyal et al., 2023; Sangwan et al., 2023; Sharma et al., 2021a,b,c; Sharma et al., 2022a,b; Sharma et al., 2023a,b,c,d). The severity of global pesticide-induced toxicity can be taken into consideration by the global number of 385 million cases that fall ill every year from pesticide poisoning (Pesticide Atlas, 2022). Furthermore, this explicit usage of chemical pesticides leading to pesticide abuse is increasing the resistance of different pathogens along with the contamination of water and

Biostimulants in Plant Protection and Performance
DOI: https://doi.org/10.1016/B978-0-443-15884-1.00008-7

soil resources. In addition, the off-target effects resulting from pesticide abuse prove to be highly detrimental to beneficial insects and soil microbiomes (Tyśkiewicz et al., 2022).

The protection of agricultural environments from pesticide-induced toxicity requires the inclusion of strategies targeting sustainable food production including integrated pest management practices. The sustainable and environmentally protective alternative for pesticides represents the inclusion of biocontrol agents (BCAs) for controlling the increased incidence of pathogen attacks. Different studies have identified a diverse array of microorganisms possessing the beneficial attribute of biocontrol (BC) by employing different mechanisms. The dominant microorganisms that have been reported for their BC potential are represented by *Gliocladium virens, Pseudomonas* spp., *Streptomyces* spp., *Bacillus* spp., *Glomus mosseae, Trichoderma* spp., *Beauveria bassiana,* and *Pythium oligandrum,* (Niu et al., 2020; Subedi et al., 2020). Along with the use of BCAs, the biostimulants of natural origin have also been reported as effective tools for increasing plant productivity by rendering it enhanced forbearance to different abiotic and biotic stresses. The biopolymer-based biostimulants are usually represented by peptides and lignosulphonates originating from some natural sources, especially plants and microorganisms (Lucini et al., 2018). The combinatorial treatment of plant systems with *Trichoderma* and biopolymer-based biostimulants can result in a synergistic interaction and a better outcome as compared to the individual applications of *Trichoderma* and biostimulants.

Among all the classes of BCAs, the genus *Trichoderma* has been documented multiple times for advantageous assets of avirulent *Trichoderma* strains that make its application much easier for sustainable agriculture as assessed in terms of biostimulation, plant security, and biofertilization (Martínez-Medina et al., 2014). The members of this genera are known to effectively dwell the plant roots, rhizoplane, and rhizosphere followed by the production of numerous metabolites owing to antimicrobial (antibiotics, cell wall damaging enzymes, and volatile composites) along with biostimulating features (Fig. 20.1).

In addition, this microorganism is highly potent for interacting with plant root and rhizosphere microbiome and has the potential to interplay with the native microbial inhabitants that puts forward the possibility of shaping the microbial dynamics in the vicinity of roots. Apart from its ability to possess plant growth promotion attributes, it has also been recognized to degrade the complex xenobiotics that highlighting its potential to rejuvenate degraded or polluted soils (Cocaign et al., 2013; Escudero-Leyva et al., 2022). The present chapter highlights the agricultural potential of *Trichoderma* and biopolymer-based biostimulants as an effective way to increase plant productivity and survival in resilient conditions. The different mechanisms opted by *Trichoderma* that render it different plant growth promotion abilities and BC potential against an extensive array of plant pathogens have also been briefly discussed.

20.2 Agricultural potential of *Trichoderma virens*

The growing population demands more food resources and with the limited agricultural lands, efficient farming practices become inevitable. Due to the ever-increasing food needs, the agricultural ecosystem is under acute pressure. Constant and expanding usage of synthetic pesticides and fertilizers is creating havoc for both the natural environment as

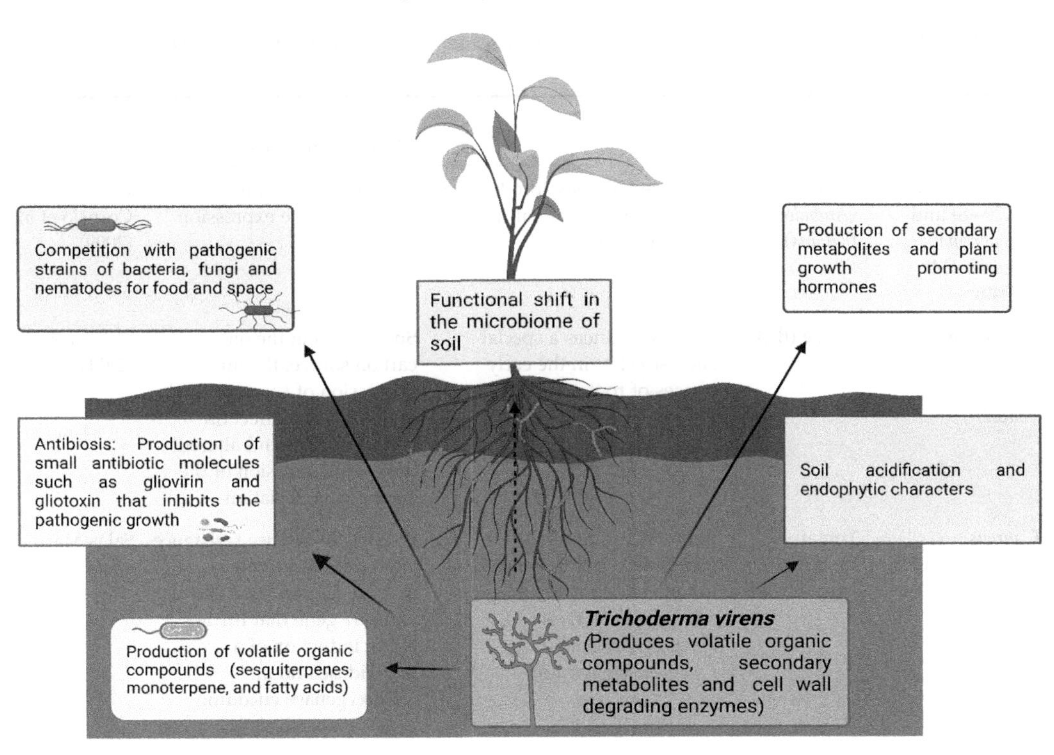

FIGURE 20.1 A portrayal depiction of different mechanisms followed by *Trichoderma* for promoting plant growth and health.

well as human health. Thus, more investigators are aiming to find ways to protect agricultural lands from the negative impact of the perpetual use of chemicals on the fields. Numerous strategies that have shown some potential for better agricultural practices include organic farming, integrated pest management (IPM), and using BCAs (BCAs) (Grasswitz 2019; Rahman et al., 2018; Thambugala et al., 2020). In the past few years, BCAs have gained wide attention and application in agriculture due to the potential of wide array of non-pathogenic fungi and bacteria to be used and commercialized as BCAs (Niu et al., 2020; Subedi et al., 2020). *Beauveria bassiana, Streptomyces* spp., *Bacillus* spp., *Glomus mosseae, Trichoderma* spp., *Gliocladium virens, Pseudomonas* spp., and *Pythium oligandrum* are the top targets to be commercialized as BCAs.

Most of the filamentous fungi that have been employed as BCAs belong to the phylum Ascomycota and moreover, most of the species belong to the genera *Trichoderma* (Gajera et al., 2013). The virulent *Trichoderma* strains acquire multiple beneficial properties that confer plant protection, bioremediation, and biofertilization (Jaroszuk-Ściseł et al., 2019; Martínez-Medina et al., 2014). Many species of *Trichoderma* produce numerous metabolites with antimicrobial and biostimulating characteristics (Table 20.1). Its evolutionary rise to inhabit the rhizosphere and rhizoplane, along with the possession of its BC attributes, indicates that *Trichoderma* is a valiant vanquisher of ecological niches, an established

TABLE 20.1 Review of literature highlighting the agricultural potential of different *Trichoderma* virens strains.

Strain	Treated Plant	Effect	Mechanism of action	References
Trichoderma virens (Gv.29–8) and *Trichoderma atroviride* (MI 206040)	Arabidopsis (*Arabidopsis thaliana*)	Increased biomass production and stimulated lateral root development	Activation of auxin-regulated gene expression	Contreras-Cornejo et al. (2009)
Trichoderma virens Gv.29–8 *(TvSut) and T. Virens (tvinv2)*	Maize (B73)	*T. virens* induces a special sucrose/H + in the early phases of root colonization	Sucrose being the only carbon source, the null expression of *tvsut* resulted in devastating effect on fungal growth and altered the gene expression of the symbiotic association	Vargas et al. (2011)
T. virens (Gv.29–8) and *T. atroviride* (IMI 206040)	Tomato (*Solanum lycopersicum*)	Both the strains could stimulate systemic protection in tomato against pathogenic strains: *Alternaria solani*, *Botrytis cinerea*, and *Pseudomonas syringae*	Increase in disease resistance was conferred to the overexpression (OE) of *epl1* and *sm1* gene that further led to induction of a peroxidase and an α-dioxygenase encoding genes, respectively	Salas-Marina et al. (2015)
Trichoderma virens (Gv29–8)	Maize or Tomato	*T. virens* colonize roots of the multiple hosts and stimulate the release of host-specific genes	Activity of CBH1 promoter was reported in a transgenic *T. virens* and depicted the reprogramming of the fungal transcriptome	Morán-Diez et al. (2015)
Trichoderma virens (Gv29–8)	Cucumber	Induction of systemic defenses against *P. syringae* in cucumber	T. virens induces 2 synthetic 18-amino-acid peptaibol isoforms (TvBI and TvBII). It also up-regulate hydroxyperoxide lyase (*hpl*), phenylalanine ammonia lyase (*pal1*) and peroxidase (*prx*) gene expression	Viterbo et al. (2007)
Trichoderma virens	Lima bean (*Phaseolus lunatus*)	Induction of plant growth promoting jasmonate and salicylate synthesis	Production of volatile compounds that are elicitors of tendril coiling	Engelberth et al. (2001)
Trichoderma virens	Cotton (*Gossypium hirsutum*) and Maize (*Zea mays*)	It induces the systemic resistance against the maize pathogen	The defense responses are conferred to the production of hydrophobin-like SSCP orthologues	Djonovic et al. (2007)

(*Continued*)

TABLE 20.1 (Continued)

Strain	Treated Plant	Effect	Mechanism of action	References
Trichoderma virens	Cotton (*Gossypium hirsutum*)	*Trichoderma* spp. enhance the activation of plant defense mechanisms	Cotton cotyledons with Sm1 (a Proteinaceous Elicitor Secreted by the Biocontrol Fungus *Trichoderma virens*) provided high levels of protection to the foliar pathogen *Colletotrichum* sp.	Djonović et al. (2006)
Trichoderma virens (TriV_JSB100)	Tomato (*Solanum lycopersicum*)	Significant enhancement in plant height, shoot, root, and fresh weight. It also led to increased nutrient uptake, and crop productivity	TriV_JSB100 BGS stimulates JA and SA signaling cascades	Jogaiah et al. (2018)
Trichoderma virens	Banana (*Musa* spp.) cv. Rasthali (AAB)	*T. virens* can control vascular wilt pathogen and thus helps in the enhancement of host plant defense	Cerato-platanin family protein 28 and a Nep-1-like protein was found to be upregulated in contact with banana roots	Muthukathan et al. (2020)
Trichoderma virens (PDR-28)	Maize	Significant increase in dry biomass of maize roots (64%) and shoots (56%). There was also an increase in chlorophyll, total soluble sugars (reducible and non-reducible), starch, and protein contents	PDR-28 led to the enhancement of PGPR traits including 1-aminocyclopropane-1-carboxylic acid (ACC) deaminase, acid phosphatase, phytase activity, siderophore production, and P solubilization	Babu et al. (2014)
Trichoderma virens (Gv 29.8)	Cotton (*Gossypium hirsutum*)	induction of phytoalexin synthesis in the roots led to development of the biocontrol activity	Induction of phytoalexin production results in biocontrol activity of *T. virens*	Howell and Puckhaber (2005)
Trichoderma virens	*Arabidopsis thaliana*	Increased antagonistic activity and the capacity to promote plant growth	*Trichoderma virens* encodes a new gene *TvCyt2* that led to synthesis of the protein homologous to the cytochrome p450 resulting in production of secondary metabolites	Ramírez-Valdespino et al. (2018)
T. harzianum (T1), *Trichoderma asperellum* (T2), *Trichoderma virens* (T3)	Lentil (Lens culinaris Medic.)	Increased dry weight and significant decline in disease severity	Antagonist effect	Akrami et al. (2011)
Trichoderma virens	*Arabidopsis thaliana*	Antagonistic activity against *R. solani*	Overexpression of the gene *tvhydii1*	Guzmán-Guzmán et al. (2017)

contender, and an insistent antagonist. Thus, it is crucial to explore and assess the potential off-target effects of *Trichoderma* strains.

Trichoderma has also been suggested as an indicator of soil health. The reports have suggested multiple uses of *Trichoderma* alone and in combination with organic compost. It helps to sustain plant growth and stimulate changes in the composition of the microbial community in terms of both structural and functional modifications. Its application has been reported to shift the whole microbial diversity to contain a maximum number of phosphorous-solubilizing bacteria resulting in stimulation of potentially beneficial microbial consortia (Qiao et al., 2019; Ros et al., 2017). *Trichoderma* can help in the preservation of microbial diversity under the influence of unfavorable growth conditions. The application of *Trichoderma* has been reported to amplify the count as well as diversity of several genera of advantageous microbes in the wheat rhizosphere that was previously damaged by explicit fertilizers application (Illescas et al., 2020) or, likewise, when *Trichoderma* was co-inoculated with endophytes that triggered an enhancement in the microbiome of plants imperiled to drought stress (He et al., 2022).

20.2.1 Trichoderma virens as direct biocontrol agent

Trichoderma has been reported as the genus with maximum BC potential among other known BC fungi. It represents 50%–60% of the known fungal-derived BCAs (Rush et al., 2021; Vishwakarma et al., 2020). At present, no less than 77 commercial *Trichoderma*- based BCA are obtainable, comprising 7 certified by the European Commission for application in the Member States of the European Union. Especially among the commercial consortiums, *Trichoderma virens* is one out of a total of 36 different species. This wide application of *Trichoderma* is thus attributed to its extensive BC and biostimulation properties (Tyśkiewicz et al., 2022). Various strains of *Trichoderma* act as BCAs against plant diseases initiated by fungi, bacteria, insects, and nematodes by either direct or indirect mechanisms. The direct mechanism is based on the specific antagonistic properties of *Trichoderma* against the phytopathogens. The direct mechanisms include the synthesis of cell wall degrading enzymes like glucanases, chitinases, and proteases, antibiosis like the production of terpenoids, polyketides, pyrones, and anthraquinones, competition for habitat and essential nutrients like iron, carbon, and nitrogen and much more. On the other hand, indirect mechanism is by mounting the plant host defense responses. *Trichoderma* can indirectly stimulate systemic plant resistance by releasing some molecules known as elicitors from the host cell walls (endoelicytors) and the infecting microbe (exoelicytors) (Woo et al., 2022). Nevertheless, it is critical to note that not all strains of *T. virens* are effective as BCAs. Conventionally, the efficiency of a BCAs or the strain depends on its biological aspects like its growth capacity, sporulation abilities, and colonizing competence in a particular ecosystem.

Trichoderma spp. is very well known to show BC activities against phytopathogenic microbes *via* multiple mechanisms. Out of which the production of volatile organic compounds (VOCs) is highly reported. Recently, a total of 43 different VOCs were discovered from seven strains of *T. virens*. They all were characterized by using gas chromatography-mass spectrometry (GC-Ms). Most of the VOCs were found to be sesquiterpenes also called antifungal chemicals (i.e., aromanderen, element, cadinene, and 2-Octanone), monoterpene, and fatty acids like oleic acid, and monopalmtin, etc. Other VOCs that were found in lesser quantities showing antifungal activity and plant growth promotion were caryophyllene and thojupsene. Furthermore, when these VOCs were tested *in vitro*, they hindered the growth of fungal pathogen *R. solani* by almost 59.4% at 5 days post-inoculation and significantly affected the morphology of fungal hypae. *T. virens* strain V3 (T.v3) and V4 (T.v4) demonstrated the capacity to reduce the growth of *R. solani* by almost 50% and had the capability to be applied as BCAs (Inayati et al., 2020).

Another major mode of BC action reported for *Trichoderma* spp. is antibiosis. It is an umbrella term that explains the production of small antagonistic substances like antibiotics, VOCs, lytic enzymes, and toxins against phytopathogens. One notable antibiotic compound reported from the strains of *T. virens* is gliovirin (Howell and Stipanovic, 1983). The processes by which gliotoxin stimulates toxicity are believed to be by coupling with thiol residues on proteins and by producing ROS due to oxidation of the decreased dithiol to the disulfide form. Gliotoxin is another antibiotic synthesized and released by *T. virens* that showed a repressive effect of *Rhizoctonia solani* and *Pytium ultimum* that initiates damping off the disease of zinnias (Lumsden and Locke, 1989).

20.2.2 Trichoderma virens as indirect biocontrol agent

Trichoderma can stimulate plant immune response against phytopathogens and play the role of a BCA indirectly. The response is mounted on stimulation by a microbe leading to a quick defense against the pathogen. In this manner, *Trichoderma* can prime the plants and ensure an enduring defense system for the plants *via* balanced phytohormone-facilitated pathways (Hermosa et al., 2012). Interestingly, the priming response of *Trichoderma* is evolutionary and has always been a part of the dawn and institution of *Trichoderma*. Its function to mount response against various biotic and abiotic stressors is secondary. Although it has always been studied widely, the reports are dubious about the molecular interactions between *Trichoderma* and plants and the nature of signal induction and transmission (Mendoza-Mendoza et al., 2018). *Trichoderma* can mount this secondary/indirect response against components of cell wall and cell membrane of the phytopathogen such as β-glucans, chitin, and sterols, also called microorganism-associated molecular patterns (MAMPs) (Mathys et al., 2012). Furthermore, on recognition of MAMPs, several hydrolytic enzymes are released by *Trichoderma* and act as damage-associated molecular patterns (DAMPs), which, once recognized by pattern-recognition receptors (PRRs), initiate MAMP-activated immunity, that seem to be sturdier as equated to pathogen-triggered immunity, thus conferring plant resistance (Hermosa et al., 2013).

During the onset of *Trichoderma* colonization in the plant roots, there is a substantial boost observed in salicylic acid levels which is a central phytohormone that regulates initial root colonization. This systemic-acquired response limits the *Trichoderma* growth in cortex and apoplastic sites of the root and has the potential to limit the colonization of other biotrophic pathogens as well (Pieterse et al., 2014). In response to this resistance mounted by the plant, *Trichoderma* ingeniously represses salicylic acid (SA) and stimulates jasmonic acid (JA) biosynthesis and jasmonic-acid receptive genes that spread systemically initiating from roots (Brotman et al., 2013). This resistance model is called jasmonic acid–ethylene-reliant induced systemic resistance and is exceptionally efficient against necrotrophic pathogens and herbivore insult (Pieterse et al., 2014). Interestingly, *Trichoderma* takes benefit of the hostility among salicylic acid and jasmonic acid to colonize the roots efficiently. Multiple reports recommend that the priming of *Trichoderma* spp. leads to induction of salicylic acid-jasmonic acid–ethylene-dependent induced systemic resistance and provides protection against plant pathogens in different plants like *Arabidopsis thaliana* grape (*Vitis vinifera*), tomato, and melon (*Cucumis melo*) (Jogaiah et al., 2018). Recently, Jogaiah et al. (2018) demonstrated that the tomato plants, on treatment with the spores of *T. virens* (TriV_JSB100) and grown on barley grains and cell-free culture filtrate selectively induce JA and SA signaling for the induction of *Fusarium oxysporum* endurance in the tomato plant.

Trichoderma spp. are reported to improve plant growth by producing various phytohormones. Furthermore, the improvement of plant health also boosts the defense ability of the host plant. Enhanced plant health, as encouraged by *Trichoderma*, also precludes nematode entry to the roots (TariqJaveed et al., 2021). In tomato roots in which *Meloidogyne* root-knot nematodes achieve their life cycle, *Trichoderma* rearrange plant protection by altering SA-JA-dependent defenses corresponding to the nematode infection stage (Medeiros et al., 2017). *T. virens* increased mung bean seedling growth assessed in terms of increment in total biomass, root weight, and length. Additionally, it also led to an improvement in the chlorophyll content and levels of IAA-synthase. *Trichoderma* virens-Tv4 has also been reported to enhance the growth promotion and increase the plant defense-related enzymes of mung bean against soil-borne pathogen *Rhizoctonia solani*. The treatment with *T. virens* either alone or in the presence of pathogen-induced mung bean defense-related enzymes was indicated by increase in the total phenolic and flavonoid content (Inayati et al., 2020).

20.3 Agriculture potential of biopolymer-based biostimulants

Use of biopolymer-based biostimulants has the potential use in the agricultural arena regarding the enhancement of growth, productivity, production, fruit set, seed set, plant development, and protection against different external adversities such as abiotic and biotic stress (Fig. 20.2) which could have the negative consequences otherwise (Colla and Rouphael, 2015). Researchers have established that biostimulants can induce an extensive array of anatomical, physiological, biochemical, and molecular responses to increase plant health and overall productivity (Rouphael et al., 2020). Among the various compounds used for their biostimulatory properties are protein hydrolysates. These chemicals are generally a blend of peptides and amino acids which are derived from plant or animal origin after a series of enzymatic, physical, and chemical processing (Shahrajabian et al., 2021).

Seaweed extracts represent a major category of biopolymer sourced from nature. Red brown and green algae are the most used macroalgae. Biopolymer extracts including, proteins, amino acids, and carbohydrates have been extracted from algal species such as *S. platensis*, *Acutodesmus dimorphus*, *Chlorella ellipsoida*, *Dunaliella salina*, *Scenedesmus quadricauda*, and *Spirulina maxima* have been used as plant biostimulant (Chiaiese et al., 2018). The stimulation of crop productivity, the enhancement of seed germination, and different parameters for plant growth and development in response to the biostimulant are

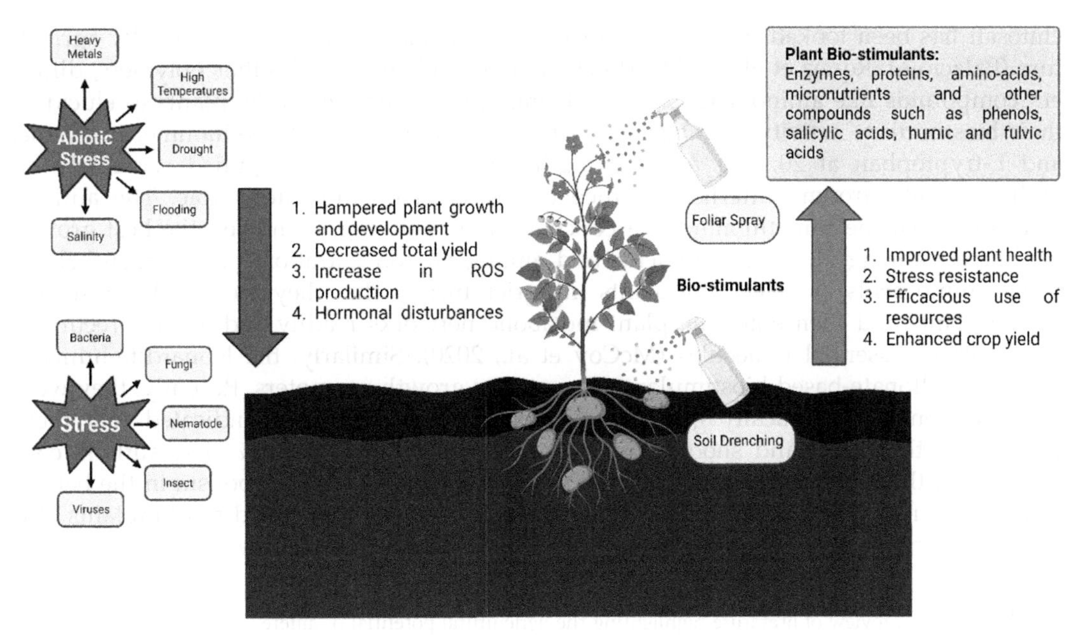

FIGURE 20.2 A schematic representation of biostimulant triggered increase in stress tolerance ability of plants.

attributed to the activity of signaling bioactive composites in the primary as well as secondary metabolism (Calvo et al., 2014). For example, Wilson et al. (2018) examined different biopolymers such as granulated gelatin, gelatin hydrolysate, and amino acid combinations simulated gelatin composition for their expected use as a biostimulant in increasing cucumber plant growth. It was concluded that gelatin hydrolysate application induced the expression of genes encoding for the transport of amino acids and nitrogen, which ultimately will be more available to the plant for better growth. Bioactive polymers have different ways to contribute as a biostimulant. For instance, some peptides also exhibit phytohormone-like properties, which help in the growth of different plant tissues. One such example is bioactive peptides isolated from soybean seeds were able to increase root length and number in a variety of crops such as clover, cabbage, and lettuce (Matsumiya and Kubo, 2011). Another such study conducted on melons demonstrated the root growth-promoting activity of a vegetal biopolymer-based biostimulant and triggered the brassinosteroids and stress-related compounds (Lucini et al., 2018). Another biopolymer that has been widely used as a biostimulant is chitosan. Chitosan is commercially derived from the shells of seafood and is being used in agriculture to protect the crop against some pathogens, owing to its stimulatory property of inducing the production of some protective molecules. However, another biostimulant activity of chitosan has also been reported, such as the augmentation of the photosynthetic activity, and protection against adverse environmental conditions like abiotic stress including heat and drought stress (Pichyangkura and Chadchawan, 2015). Due to the bulk availability of its production source and owing to its multiple biostimulatory properties with minimum side effects, chitosan has been looked upon as one of the sustainable options for eco-friendly agriculture (Palacio-MáRquez et al., 2022). Alongside in combination with other polymers, different compounds like amino acids are also being applied directly to the plants to elucidate their biostimulant activity. For instance, different amino acids (L-methionine, L-glycine, and L-tryptophan at 20, 210, and 220 mg/L, respectively) were applied separately on hydroponically grown butterhead lettuce to assess their stimulator's role. The authors concluded that the L-methionine at a concentration of 0.2 mg/L showed the best growth parameters among all the amino acids and among all concentrations. The reason behind using amino acids as direct stimulants in agriculture is that they are rapidly absorbed by the plants and then enter the plant metabolic networks readily and act as precursors of numerous essential molecules (McCoy et al., 2020). Similarly, the leonardite-humate- and lignosulfonate-based biostimulants are used as growth promoters. Recently, they were examined on hydroponically grown maize plants and the results indicated that there was a growth in root and shoot growth by 51%–140% and 5%–35%. The authors concluded that the metabolism of these compounds could be like N metabolism in the below-ground organs of the plants (Ertani et al., 2019). The biopolymer-based biostimulants also serve as ingredients to alleviate stress as has been demonstrated previously (Table 20.2).

TABLE 20.2 A review of literature highlighting the agricultural potential of different biostimulants.

Biopolymer	Treated plant	Effect	Mechanism of action	References
Quik-link (Lateral root-promoting peptides, and lignosulphonates)	Melon (*Cucumis melo* L. - cv. Giorillo)	Significant enhancement in the plant biomass	Brassinosteroids increased the root development whereas increase in phytohormone levels increased the leaf biomass	Lucini et al. (2018)
Chitosan	Banana (*Musa accuminatea* AAB)	Significant reduction in disease severity	Change in cell permeability	Jinasena et al. (2011)
Water-soluble chitosan (WSC)	Chili pepper (*Capsicum annuum* L.)	Decrease in disease prevalence and the lesion diameter	Increase in chitinase activity and phenolic content	Long et al. (2018)
Chitosan and Oligachitosan	Peach (*Prunus persica* (L.) Batsch)	Increase in antioxidants enzymes	Induction of the expression of GLU and POD genes	Ma et al. (2013)
Quik-Link	Tomato (*Solanum lycopersicum* L.)	Increase in root biomass	Increased uptake of N, P, and Fe	Colla et al. (2017)
Chitosan biopolymer	Strawberry (*Fragaria* × *annanasa*)	Increase in plant growth and yield	Increased production of flavonoids and phenolics	Rahman et al. (2018)
VBP; "Quik-link" (vegetal biopolymer-based biostimulant)	Lettuce (*Lactuca sativa* L. var. *longifolia*)	Significant increase in plant weight	Increase in antioxidant activity	Rouphael et al. (2020)

(*Continued*)

TABLE 20.2 (Continued)

Biopolymer	Treated plant	Effect	Mechanism of action	References
Bio-algeen S-90 and Megagreen	Garden cress (*Lepidium sativum*) L.	Increase in plant height and Chlorophyll content	Increase in macro and micronutrients, vitamins and amino acids	Dudaš et al. (2016)
Asahi SL, Optysil and Kelpak SL	Cucumber fruits (*Cucumis sativus* L.)	Increase in fruit yield and fruit nutritional content	Increased uptake of key elements like Fe, Mn, B, Zn and Cu	Ługowska (2019)
Radifarm + Megafol	Bell peppers (*Capsicum annuum* L.)	Increased macro- and micronutrient, especially Ca^{2+} ion concentration in leaves and fruits	–	Paradikovic et al. (2013)
Radifarm	*Salvia splendens* cv. Red Hot Sally	Increase in fresh and dry weight of roots	Increase in amino acid content	Zeljković et al. (2010)
AlgaminoPlant	*Cornus alba* "Aurea" and *C. alba* "Elegantissima"	Increase in root ball formation	Increase in levels of carbohydrate contents and free amino acids	Pacholczak et al. (2012)
Radifarm, Megafol, Benefit and Viva	Bell peppers (*Capsicum annuum* L.)	Increase in fruit yield	Increase in amino acids like arginine, asparagine, proline and tryptophan	Tkalec et al. (2010)
Viva and Megafol	Strawberry (*Fragaria* x *ananassa* Duch, cultivar Elsanta)	Increase in plant biomass	Increase in activity of antioxidant enzymes	Špoljarević et al. (2010)
Asahi SL, Bio-algreen, Tytanit	Sweet basil (*Ocimum basilicum* L.)	Increased plant growth and a lower weed infestation	Increase in the levels of micronutrients like N, P, K, Mg and Ca	Kwiatkowski and Juszczak (2011)

For example, the physiological and biochemical effects of β-(1,3)-glucan (paramylon) purified from the microalga *Euglena gracilis* on water-stress Micro-Tom were also assessed by an Italian research group. It was proved in the study that the biopolymers triggered the enhancement of photosynthesis. Furthermore, it also improved the sensitivity of the photosystem II toward the damage caused by the dehydration (Barsanti et al., 2019). In another study, melatonin was proposed as a biostimulant and plant protector. The authors summarized the response of different plant species to melatonin in different stress conditions (Arnao and Hernández-Ruiz, 2019). The biostimulants are also looked upon for improving the qualitative traits of the crops. The bio extracts of plants packed with biopolymers are exposed to different crop plants to enhance the level of antioxidants or bioactive compounds for human consumption. For example, when tomato plants were sprayed with tropical plant extract and protein hydrolysate, the plants resulted in a far greater amount of Vitamin C accumulation as compared to the control plants (Caruso et al., 2019).

20.4 *Trichoderma* for sustainable agriculture

The incessant growth of the population and their constantly increasing demands are directly/indirectly setting pressure on the agricultural environment. Thus, eco-sustainability through any methods, techniques, and/or bioproducts is inevitable to meet both the present and future needs of the growing population. To satisfy the agricultural directives and minimize the use of chemicals in the farmlands, the use of *Trichoderma* is predictable owing to its ability to help improve the shelf life, efficacy, and standards of chemical fertilizers. Also, the strategies to make stress-tolerant and agrochemical tolerant *Trichoderma*-based products will help further in achieving a sustainable agriculture system (Collinge et al., 2022; Ons et al., 2020). Application of *Tichoderma* would be a cost-effective and sustainable approach owing to strategic improvement of the yield of conidia and chlamydospores and their abiotic tolerance capabilities. It is crucial to select the strain on the basis of their prospective BC, rhizosphere proficiency, endophytic establishment attributes, and can stimulate disease endurance and/or encourage plant growth and development. *Trichoderma* is applied on the fields in the form of consortia including strains like BC bacteria (*Bacillus, Pseudomonas*), non-pathogenic fungi (*Fusarium, Rhizoctonia, Coniothyrium*), insect-killing fungi (*Beauveria, Metarhizium*), nematode insnaring fungi (*Arthrobotrys, Dactylellina, Drechslerella*), arbuscular mycorrhizal fungi, ectomycorrhizal fungi, and plant growth-promoting rhizobacteria (PGPR) (*Bacillus, Pseudomonas, Azotobacter*).

In today's world where most of the agricultural lands are overburdened and suffer from soil fatigue, it is extremely important to develop new age group biofertilizers and develop agricultural probiotic microflora to aid microflora restoration and thus improve soil health. Trichoderma in bioformulations with bioactive compounds from algae, and phyto-extracts is also an efficient alternative to improve soil health and plant growth. For example, three novel metabolites called trichorenins A-C were extracted from the *T. virens* Y13−3, an epiphyte of Gracilaria vermiculophylla (marine red algae) showed compelling inhibition against two marine phytoplankton species, *Chattonella marina* and *Karlodinium veneficum* (Shi et al., 2018). Also, Ghoneem et al. (2019) BC of cumin wilt disease (GBTV) and arbuscular mycorrhizal fungi (AM) for biological control of cumin wilt disease triggered by *Fusarium oxysporum f. sp. cumini* and also the most devastating disease, limiting the cumin production in Egypt.

Other formulations of *Trichoderma* spp. are also available at the commercial levels for applications other than agriculture directly such as industrial processes. The qualities of the *Trichoderma* can be utilized in industries by utilizing their properties of producing bioactive molecules and guarding the beneficial attributes during fermentation. Other possibilities of using Trichoderma indirectly for the betterment of agricultural health are its nanoencapsulation and nanoparticle machinery (Fraceto et al., 2018; Karuppiah et al., 2019; Vinale et al., 2017). The discharge of a thioredoxin reductase was described during the interaction of *T. virens* with maize plants (Nogueira-Lopez et al., 2018), implying that SeNPs could have an impact on the relationship between Trichoderma and plants. However, it is crucial to further investigate the outcome of this association as positive or negative.

The current research in *Trichoderma* urges thinking on the efficacious application of this fungus in agriculture and the advancement of functional uses with chosen possible strains. The applications include crop cultivation in fringe lands, upgraded crop resilience to hostile climate changes, bioremediation for the diminution of pollutants in polluted sites, and a common influence on the reduction of methane and carbon dioxide emissions in the environment.

20.5 Biopolymer-based biostimulants in sustainable agriculture

Due to the constantly increasing population and decreasing soil health, there is a constant need to find alternatives that are effective and promote sustainable agriculture. Biostimulants are suggested as a potent source of inoculants for the improvement of plant health, development, fruit yield, and soil health. Recently, numerous biostimulant materials (i.e., humic and fulvic acids, protein hydrolysates, and seaweed extracts) and microbial inoculants (i.e., mycorrhizal fungi and PGPR) have been established as an effective, secure, and ecological tool to improve root system thus enhancing crop performance, and nutrient usage effectivity along with boosting tolerance to environmental stressors (Rouphael et al., 2017a,b). Biostimulants also aid in limiting dormancy, augment the effectiveness of the root system, increase the photosynthetic rate of the responsible tissues, and enhance seed vigor, flowering, and fruit yield along with its size enhancement (Rouphael et al., 2017c). All these potential effects of biostimulants on the plants, soil, and the environment engage the interests of scientists, farmers, and policymakers to incorporate their use in the farmlands as much as possible.

The use of biostimulants is also motivating agriculturalist to decrease their dependence on chemical fertilizers that are generally overexploited in the farmlands. The seeds can be treated with these biostimulants before sowing and even the use of biostimulants in its minimum quantity of active component can accelerate germination and enhance overall plant health as compared to the non-treated seeds (Schmitt et al., 2009; Singh et al., 2018; Wilson et al., 2018). The biostimulants can also be used exogenously and will show equally decent benefits to encourage nutrition and strength with respect to environmental strains, regardless of their present nutritional composition. Their exogenous use also promotes plant growth hormone production such as cytokinins, auxins, and gibberellins (Couto et al., 2012; Qiu et al., 2020; Yaronskaya et al., 2006).

Using biostimulants to coat the seed material has also shown impressive benefits to hasten the early institution and seedling growth. These seedlings are specially called as "biostimulant seedlings" that can be sown under hostile environment conditions with poor soil health and nutrition. Finally, the understanding of agricultural qualities that encourage quality and resistance toward biotic and abiotic stressors and enhancing nutrition in the early growth cycle of the plants would permit to construct a second-generation biostimulants that will show the best attributes in plant development, soil health, synergistic and compatible bioproducts that can be implemented in the agriculture fields with minimalist environment residues and maximum beneficial products.

20.6 Conclusion and prospects

Different strategies are often employed for enhancing plant productivity like application of nitrogen and phosphorus-based fertilizers, supplementation of different macro and micronutrients along the usage of different agrochemicals. The currently prevailing agricultural practices have exhausted the soil capacity and thus have put the agricultural production to a steady state where no more increase in plant productivity can be expected. Alongside, these practices have also degraded the soil's organic content. The explicit usage of pesticides leads to their accumulation in the environment as well as the produce that has serious consequences on the environment as well as human health. The increasing environmental and human health-related concerns have directed the focus of the scientific community toward sustainable food production. The use of *Trichoderma* and biopolymer-based biostimulants represents a viable alternative that can improve plant productivity even in hostile environments. *Trichoderma* can effectively colonize the plant rhizosphere and reshape the rhizosphere microbiome that can support plant growth and health. *Trichoderma* secretes different plant hormones that promote its use as a plant growth promotion agent and secrete different molecules that render it with biocontrol potential. It also secreted different enzymes that can degrade recalcitrant xenobiotic molecules. Alongside, its application with biostimulants can improve its overall working efficacy leading to better plant productivity. However, a critical evaluation of *Trichoderma* if assessed in terms of competition with the native beneficial microbiota should be done before its final application. The different mechanisms followed by *Trichoderma* for biocontrol of plant pathogens can be non-selective for some of the beneficial native microbiota. Therefore, a prior assessment of the biocontrol attributes of *Trichoderma* against the selected classes of plant growth-promoting rhizobacteria such as nitrogen fixing and phosphate solubilizing microbes is critically essential.

References

Akrami, M., Golzary, H., Ahmadzadeh, M., 2011. Evaluation of different combinations of Trichoderma species for controlling *Fusarium* rot of lentil. Afr. J. Biotechnol. 10 (14), 2653–2658.

Anamika, Patra, A., Shehzad, S., Rani, A., Sharma, P., Mohammad, K.F., et al., 2023. Microbial elicitors for priming plant defense mechanisms. Sustainable Agriculture Reviews 60: Microbial Processes in Agriculture. Springer Nature Switzerland, Cham, pp. 175–196.

Arnao, M.B., Hernández-Ruiz, J., 2019. Melatonin as a chemical substance or as phytomelatonin rich-extracts for use as plant protector and/or biostimulant in accordance with EC legislation. Agronomy. 9 (10), 570.

Babu, A.G., Shim, J., Bang, K.S., Shea, P.J., Oh, B.T., 2014. Trichoderma virens PDR-28: a heavy metal-tolerant and plant growth-promoting fungus for remediation and bioenergy crop production on mine tailing soil. J. Environ. Manag. 132, 129–134.

Barsanti, L., Coltelli, P., Gualtieri, P., 2019. Paramylon treatment improves quality profile and drought resistance in *Solanum lycopersicum* L. cv. Micro-Tom. Agronomy 9 (7), 394. 17.

Brotman, Y., Landau, U., Cuadros-Inostroza, A., Takayuki, T., Fernie, A.R., Chet, I., et al., 2013. Trichoderma-plant root colonization: escaping early plant defense responses and activation of the antioxidant machinery for saline stress tolerance. PLoS Pathog. 9 (3), e1003221.

Calvo, P., Nelson, L., Kloepper, J.W., 2014. Agricultural uses of plant biostimulants. Plant Soil 383, 3–41.

Caruso, G., Pascale, De, Cozzolino, S., Cuciniello, E., Cenvinzo, A., Bonini, V., et al., 2019. Yield and nutritional quality of Vesuvian Piennolo tomato PDO as affected by farming system and biostimulant application. Agronomy. 9 (9), 505.

Chiaiese, P., Corrado, G., Colla, G., Kyriacou, M.C., Rouphael, Y., 2018. Renewable sources of plant biostimulation: microalgae as a sustainable means to improve crop performance. Front. Plant Sci. 9, 1782.

Cocaign, A., Bui, L.C., Silar, P., Chan Ho Tong, L., Busi, F., Lamouri, A., et al., 2013. Biotransformation of Trichoderma spp. and their tolerance to aromatic amines, a major class of pollutants. Appl. Environ. Microbiol. 79 (15), 4719–4726.

Colla, G., Cardarelli, M., Stefanoni, W., Fiorillo, A., Canaguier, R., Mariotti, R., et al., 2017. Drip application of a biopolymer-based biostimulant enhances root growth and nutrient uptake of processing tomato, VII South-Eastern Europe Symposium Veg. Potatoes, 1326. pp. 85–92.

Colla, G., Rouphael, Y., 2015. Biostimulants in horticulture. Sci. Hortic. 196, 1–134.

Collinge, D.B., Jensen, D.F., Rabiey, M., Sarrocco, S., Shaw, M.W., Shaw, R.H., 2022. Biological control of plant diseases–What has been achieved and what is the direction? Plant Pathol. 71 (5), 1024–1047.

Contreras-Cornejo, H.A., Macías-Rodríguez, L., Cortés-Penagos, C., López-Bucio, J., 2009. Trichoderma virens, a plant beneficial fungus, enhances biomass production and promotes lateral root growth through an auxin-dependent mechanism in Arabidopsis. Plant Physiol. 149 (3), 1579–1592.

Couto, C.A., Peixoto, C.P., Vieira, E.L., Carvalho, E.V., Peixoto, V.A.B., 2012. Action of cinetina, butyric acid and gibberellic acid on the emergency of sunflower under aluminum stress/Acao da cinetina, acido indolbutirico e acido giberelico na emergencia do girassol sob estresse por aluminio. Comun. Sci. 3 (3), 206–210.

Djonović, S., Pozo, M.J., Dangott, L.J., Howell, C.R., Kenerley, C.M., 2006. Sm1, a proteinaceous elicitor secreted by the biocontrol fungus Trichoderma virens induces plant defense responses and systemic resistance. Mol. Plant Microbe Interact. 19 (8), 838–853.

Djonovic, S., Vargas, W.A., Kolomiets, M.V., Horndeski, M., Wiest, A., Kenerley, C.M., 2007. A proteinaceous elicitor Sm1 from the beneficial fungus *Trichoderma virens* is required for induced systemic resistance in maize. Plant Physiol. 145 (3), 875–889.

Dudaš, S., Šola, I., Sladonja, B., Erhatić, R., Ban, D., Poljuha, D., 2016. The effect of biostimulant and fertilizer on "low input" lettuce production. Acta Bot. Croat. 75 (2), 253–259.

Engelberth, J., Koch, T., Schüler, G., Bachmann, N., Rechtenbach, J., Boland, W., 2001. Ion channel-forming alamethicin is a potent elicitor of volatile biosynthesis and tendril coiling. Cross talk between jasmonate and salicylate signaling in lima bean. Plant Physiol. 125 (1), 369–377.

Ertani, A., Nardi, S., Francioso, O., Pizzeghello, D., Tinti, A., Schiavon, M., 2019. Metabolite-targeted analysis and physiological traits of Zea mays L. in response to application of a leonardite-humate and lignosulfonate-based products for their evaluation as potential biostimulants. Agronomy. 9 (8), 445.

Escudero-Leyva, E., Alfaro-Vargas, P., Muñoz-Arrieta, R., Charpentier-Alfaro, C., Granados-Montero, M.D.M., Valverde-Madrigal, K.S., et al., 2022. Tolerance and biological removal of fungicides by Trichoderma species isolated from the endosphere of wild Rubiaceae plants. Front. Agron. 3, 117.

Fraceto, L.F., Maruyama, C.R., Guilger, M., Mishra, S., Keswani, C., Singh, H.B., et al., 2018. Trichoderma harzianum-based novel formulations: potential applications for management of Next-gen agricultural challenges. J. Chem. Technol. Biotechnol. 93 (8), 2056–2063.

Gajera, H., Domadiya, R., Patel, S., Kapopara, M., Golakiya, B., 2013. Molecular mechanism of Trichoderma as bio-control agents against phytopathogen system–a review. Curr. Res. Microbiol. Biotechnol. 1 (4), 133–142.

Ghoneem, K.M., Khalil, A.A., Rashad, E.M., Ahmed, M.I., Mahmoud, M.S., 2019. Granular bioactive formulation of Trichoderma viride and Arbuscular mycorrhizal fungi for biological control of cumin wilt disease. Egypt. J. Phytopathol. 47 (1), 175–197.

Grasswitz, T.R., 2019. Integrated pest management (IPM) for small-scale farms in developed economies: challenges and opportunities. Insects. 10 (6), 179.

Guzmán-Guzmán, P., Alemán-Duarte, M.I., Delaye, L., Herrera-Estrella, A., Monfil, V., 2017. Identification of effector-like proteins in Trichoderma spp. and role of a hydrophobin in the plant-fungus interaction and mycoparasitism. BMC Genet. 18 (1), 1–20.

He, C., Liu, C., Liu, H., Wang, W., Hou, J., Li, X., 2022. Dual inoculation of dark septate endophytes and *Trichoderma* viride drives plant performance and rhizosphere microbiome adaptations of Astragalus mongholicus to drought. Environ. Microbiol. 24 (1), 324–340.

Hermosa, R., Rubio, M.B., Cardoza, R.E., Nicolás, C., Monte, E., Gutiérrez, S., 2013. The contribution of Trichoderma to balancing the costs of plant growth and defense. Int. Microbiol. 16 (2), 69–80.

Hermosa, R., Viterbo, A., Chet, I., Monte, E., 2012. Plant-beneficial effects of Trichoderma and of its genes. Microbiology. 158 (1), 17–25.

Howell, C.R., Puckhaber, L.S., 2005. A study of the characteristics of "P" and "Q" strains of Trichoderma virens to account for differences in biological control efficacy against cotton seedling diseases. Biol. Control. 33 (2), 217–222.

Howell, C.R., Stipanovic, R.D., 1983. Gliovirin, a new antibiotic from Gliocladium virens, and its role in the biological control of Pythium ultimum. Can. J. Microbiol. 29 (3), 321–324.

Illescas, M., Rubio, M.B., Hernández-Ruiz, V., Morán-Diez, M.E., Martínez de Alba, A.E., Nicolás, C., et al., 2020. Effect of inorganic N top dressing and Trichoderma harzianum seed-inoculation on crop yield and the shaping of root microbial communities of wheat plants cultivated under high basal N fertilization. Front. Plant Sci. 11, 575861.

Inayati, A., Sulistyowati, L., Aini, L.Q., Yusnawan, E., 2020. Trichoderma virens-Tv4 enhances growth promoter and plant defense-related enzymes of mungbean (Vigna radiata) against soil-borne pathogen Rhizoctonia solani. Biodiversitas 21 (6). Available from: https://doi.org/10.13057/biodiv/d210611.

Jaroszuk-Ściseł, J., Tyśkiewicz, R., Nowak, A., Ozimek, E., Majewska, M., Hanaka, A., et al., 2019. Phytohormones (auxin, gibberellin) and ACC deaminase in vitro synthesized by the mycoparasitic Trichoderma DEMTkZ3A0 strain and changes in the level of auxin and plant resistance markers in wheat seedlings inoculated with this strain conidia. Int. J. Mol. Sci. 20 (19), 4923.

Jinasena, D., Pathirathna, P., Wickramarachchi, S., Marasinghe, E., 2011. Use of chitosan to control anthracnose on "Embul" banana. In: International Conference on Asia Agriculture and Animal, IPCBEE, vol. 13, pp. 56–60.

Jogaiah, S., Abdelrahman, M., Tran, L.S.P., Ito, S.I., 2018. Different mechanisms of Trichoderma virens-mediated resistance in tomato against Fusarium wilt involve the jasmonic and salicylic acid pathways. Mol. Plant Pathol. 19 (4), 870–882.

Karuppiah, V., Sun, J., Li, T., Vallikkannu, M., Chen, J., 2019. Co-cultivation of Trichoderma asperellum GDFS1009 and Bacillus amyloliquefaciens 1841 causes differential gene expression and improvement in the wheat growth and biocontrol activity. Front. microbiol. 10, 1068.

Kumar, R., Sharma, P., Gupta, R.K., Kumar, S., Sharma, M.M.M., Singh, S., et al., 2020. Earthworms for eco-friendly resource efficient agriculture. Resources Use Efficiency inAgriculture. Springer, Singapore, pp. 47–84.

Kwiatkowski, C.A., Juszczak, J., 2011. The response of sweet basil (Ocimum basilicum L.) to the application of growth stimulators and forecrops. Acta Agrobot. 64 (2), 69–76.

Long, L.T., Tan, L.V., Boi, V.N., Trung, T.S., 2018. Antifungal activity of water-soluble chitosan against Colletotrichum capsici in postharvest chili pepper. J. Food Process. Preserv. 42 (1), e13339.

Loyal, A., Pahuja, S.K., Sharma, P., Malik, A., Srivastava, R.K., Mehta, S., 2023. Potential environmental and human health implications of nanomaterials used in sustainable agriculture and soil improvement. Engineered Nanomaterials for Sustainable Agricultural Production, Soil Improvement and Stress Management. Academic Press, USA, pp. 387–412.

Lucini, L., Rouphael, Y., Cardarelli, M., Bonini, P., Baffi, C., Colla, G., 2018. A vegetal biopolymer-based biostimulant promoted root growth in melon while triggering brassinosteroids and stress-related compounds. Front. Plant Sci. 9, 472.

Ługowska, M., 2019. Effects of bio-stimulants on the yield of cucumber fruits and on nutrient content. Afr. J. Agric. Res. 14 (35), 2112–2118.

Lumsden, R.D., Locke, J.C., 1989. Biological control of damping-off caused by Pythium ultimum and Rhizoctonia solani with Gliocladium virens in soilless mix. Phytopathology. 79 (3), 361–366.

Ma, Z., Yang, L., Yan, H., Kennedy, J. F., Meng, X., 2013. Chitosan and oligochitosan enhance the resistance of peach fruit to brown rot. Carbohydr. Polym. 94 (1), 272–277.

Martínez-Medina, A., Del Mar Alguacil, M., Pascual, J.A., Van Wees, S.C., 2014. Phytohormone profiles induced by Trichoderma isolates correspond with their biocontrol and plant growth-promoting activity on melon plants. J. Chem. Ecol. 40, 804–815.

Mathys, J., De Cremer, K., Timmermans, P., Van Kerckhove, S., Lievens, B., Vanhaecke, M., et al., 2012. Genome-wide characterization of ISR induced in Arabidopsis thaliana by Trichoderma hamatum T382 against Botrytis cinerea infection. Front. Plant Sci. 3, 108.

Matsumiya, Y., Kubo, M., 2011. Soybean peptide: novel plant growth promoting peptide from soybean. Soybean and Nutrition. p. 1.

McCoy, R.M., Meyer, G.W., Rhodes, D., Murray, G.C., Sors, T.G., Widhalm, J.R., 2020. Exploratory study on the foliar incorporation and stability of isotopically labeled amino acids applied to turfgrass. Agronomy. 10 (3), 358.

Medeiros, H.A.D., Araújo Filho, J.V.D., Freitas, L.G.D., Castillo, P., Rubio, M.B., Hermosa, R., et al., 2017. Tomato progeny inherit resistance to the nematode Meloidogyne javanica linked to plant growth induced by the biocontrol fungus Trichoderma atroviride. Sci. Rep. 7 (1), 1–13.

Mendoza-Mendoza, A., Zaid, R., Lawry, R., Hermosa, R., Monte, E., Horwitz, B.A., et al., 2018. Molecular dialogues between Trichoderma and roots: role of the fungal secretome. Fungal Biol. Rev. 32 (2), 62–85.

Morán-Diez, M.E., Trushina, N., Lamdan, N.L., Rosenfelder, L., Mukherjee, P.K., Kenerley, C.M., et al., 2015. Host-specific transcriptomic pattern of Trichoderma virens during interaction with maize or tomato roots. BMC Genom. 16, 1–15.

Muthukathan, G., Mukherjee, P., Salaskar, D., Pachauri, S., Tak, H., Ganapathi, T.R., et al., 2020. Secretome of Trichoderma virens induced by banana roots-identification of novel fungal proteins for enhancing plant defence. Physiol Mol. Plant Pathol. 110, 101476.

Niu, B., Wang, W., Yuan, Z., Sederoff, R.R., Sederoff, H., Chiang, V.L., et al., 2020. Microbial interactions within multiple-strain biological control agents impact soil-borne plant disease. Front. Microbiol. 11, 585404.

Nogueira-Lopez, G., Greenwood, D.R., Middleditch, M., Winefield, C., Eaton, C., Steyaert, J.M., et al., 2018. The apoplastic secretome of Trichoderma virens during interaction with maize roots shows an inhibition of plant defence and scavenging oxidative stress secreted proteins. Front. Plant Sci. 9, 409.

Ons, L., Bylemans, D., Thevissen, K., Cammue, B.P., 2020. Combining biocontrol agents with chemical fungicides for integrated plant fungal disease control. Microorganisms. 8 (12), 1930.

Pacholczak, A., Szydło, W., Jacygrad, E., Federowicz, M., 2012. Effect of auxins and the biostimulator AlgaminoPlant on rhizogenesis in stem cuttings of two dogwood cultivars (Cornus alba 'Aurea'and 'Elegantissima'). Acta Sci. Pol. Hortorum Cultus 11 (2), 93–103.

Palacio-Márquez, A., Ramírez-Estrada, C.A., Sánchez, E., Ojeda-Barrios, D.L., Chávez-Mendoza, C., Sida-Arreola, J.P., et al., 2022. Use of biostimulant compounds in agriculture: chitosan as a sustainable option for plant development. Not. Sci. Biol. 14 (1), 11124.

Paradikovic, N., Vinkovic, T., Vrcek, I., Tkalec, M., 2013. Natural biostimulants reduce the incidence of BER in sweet yellow pepper plants (Capsicum annuum L.). Agric. Food Sci. 2 (21), 307–317.

Pesticide Atlas 2022. https://eu.boell.org/sites/default/files/2023-01/pesticideatlas2022_2ndedition_web.pdf.

Pichyangkura, R., Chadchawan, S., 2015. Biostimulant activity of chitosan in horticulture. Sci. Hortic. 196, 49–65.

Pieterse, C.M., Zamioudis, C., Berendsen, R.L., Weller, D.M., Van Wees, S.C., Bakker, P.A., 2014. Induced systemic resistance by beneficial microbes. Annu. Rev. Phytopathol. 52, 347–375.

Qiao, C., Penton, C.R., Xiong, W., Liu, C., Wang, R., Liu, Z., et al., 2019. Reshaping the rhizosphere microbiome by bio-organic amendment to enhance crop yield in a maize-cabbage rotation system. Appl. Soil Ecol. 142, 136–146.

Qiu, Y., Amirkhani, M., Mayton, H., Chen, Z., Taylor, A.G., 2020. Biostimulant seed coating treatments to improve cover crop germination and seedling growth. Agronomy. 10 (2), 154.

Rahman, M., Mukta, J.A., Sabir, A.A., Gupta, D.R., Mohi-Ud-Din, M., Hasanuzzaman, M., et al., 2018. Chitosan biopolymer promotes yield and stimulates accumulation of antioxidants in strawberry fruit. PLoS One 13 (9), e0203769.

Ramírez-Valdespino, C.A., Porras-Troncoso, M.D., Corrales-Escobosa, A.R., Wrobel, K., Martínez-Hernández, P., Olmedo-Monfil, V., 2018. Functional characterization of TvCyt2, a member of the p450 monooxygenases from Trichoderma virens relevant during the association with plants and mycoparasitism. Mol. Plant Microbe Interact. 31 (3), 289–298.

Rani, K., Rani, A., Sharma, P., Dahiya, A., Punia, H., Kumar, S., et al., 2022. Legumes for agroecosystem services and sustainability. Advances in Legumes for Sustainable Intensification. Academic Press, USA, pp. 363–380.

Rani, K., Sharma, P., Kumar, S., Wati, L., Kumar, R., Gurjar, D.S., et al., 2019. Legumes for sustainable soil and crop management. Sustainable Management of Soil and Environment. Springer, Singapore, pp. 193–215.

Ros, M., Raut, I., Santisima-Trinidad, A.B., Pascual, J.A., 2017. Relationship of microbial communities and suppressiveness of Trichoderma fortified composts for pepper seedlings infected by Phytophthora nicotianae. PLoS One 12 (3), e0174069.

Rouphael, Y., Cardarelli, M., Bonini, P., Colla, G., 2017a. Synergistic action of a microbial-based biostimulant and a plant derived-protein hydrolysate enhances lettuce tolerance to alkalinity and salinity. Front. Plant Sci. 8, 131.

Rouphael, Y., Colla, G., Giordano, M., El-Nakhel, C., Kyriacou, M.C., De Pascale, S., 2017b. Foliar applications of a legume-derived protein hydrolysate elicit dose-dependent increases of growth, leaf mineral composition, yield and fruit quality in two greenhouse tomato cultivars. Sci. Hortic. 226, 353–360.

Rouphael, Y., Carillo, P., Colla, G., Fiorentino, N., Sabatino, L., El-Nakhel, C., et al., 2020. Appraisal of combined applications of Trichoderma virens and a biopolymer-based biostimulant on lettuce agronomical, physiological, and qualitative properties under variable N regimes. Agronomy. 10 (2), 196.

Rouphael, Y., De Micco, V., Arena, C., Raimondi, G., Colla, G., De Pascale, S., 2017c. Effect of Ecklonia maxima seaweed extract on yield, mineral composition, gas exchange, and leaf anatomy of zucchini squash grown under saline conditions. J. Appl. Phycol. 29, 459–470.

Rush, T.A., Shrestha, H.K., Gopalakrishnan Meena, M., Spangler, M.K., Ellis, J.C., Labbé, J.L., et al., 2021. Bioprospecting Trichoderma: a systematic roadmap to screen genomes and natural products for biocontrol applications. Front. Fungal Biol. 2, 41.

Salas-Marina, M.A., Isordia-Jasso, M.I., Islas-Osuna, M.A., Delgado-Sánchez, P., Jiménez-Bremont, J.F., Rodríguez-Kessler, M., et al., 2015. The Epl1 and Sm1 proteins from Trichoderma atroviride and Trichoderma virens differentially modulate systemic disease resistance against different life style pathogens in Solanum lycopersicum. Front. Plant Sci. 6, 77.

Sangwan, S., Sharma, P., Wati, L., Mehta, S., 2023. Effect of chitosan nanoparticles on growth and physiology of crop plants. Engineered Nanomaterials for Sustainable Agricultural Production, Soil Improvement and Stress Management. Academic Press, USA, pp. 99–123.

Schmitt, A., Koch, E., Stephan, D., Kromphardt, C., Jahn, M., Krauthausen, H.J., et al., 2009. Evaluation of non-chemical seed treatment methods for the control of Phoma valerianellae on lamb's lettuce seeds. J. Plant Dis. Prot. 116 (5), 200.

Shahrajabian, M.H., Chaski, C., Polyzos, N., Petropoulos, S.A., 2021. Biostimulants application: a low input cropping management tool for sustainable farming of vegetables. Biomolecules. 11 (5), 698.

Sharma, P., Sharma, M.M.M., Kapoor, D., Rani, K., Singh, D., Barkodia, M., 2020a. Role of microbes for attaining enhanced food crop production. Microb. Biotechnol: Basic. Res. Appl. 55–78.

Sharma, P., Sharma, M.M.M., Patra, A., Vashisth, M., Mehta, S., Singh, B., et al., 2020b. The role of key transcription factors for cold tolerance in plants. Transcription factors for abiotic stress tolerance in plants. Academic Press, USA, pp. 123–152.

Sharma, M.M.M., Sharma, P., Kapoor, D., Beniwal, P., Mehta, S., 2021a. Phytomicrobiome community: an agrarian perspective towards resilient agriculture. Plant Performance Under Environmental Stress. Springer, Cham, pp. 493–534.

Sharma, P., Sharma, M.M.M., Malik, A., Vashisth, M., Singh, D., Kumar, R., et al., 2021b. Rhizosphere, rhizosphere biology, and rhizospheric engineering. Plant Growth-Promoting Microbes for Sustainable Biotic and Abiotic Stress Management. Springer International Publishing, Cham, pp. 577–624.

Sharma, P., Pandey, V., Sharma, M.M.M., Patra, A., Singh, B., Mehta, S., et al., 2021c. A review on biosensors and nanosensors application in agroecosystems. Nanoscale Res. Lett. 16, 1–24.

Sharma, P., Sangwan, S., Kumari, A., Singh, S., Kaur, H., 2022a. Impact of climate change on soil microorganisms regulating nutrient transformation. Plant Stress Mitigators: Action and Application. Springer Nature Singapore, Singapore, pp. 145–172.

Sharma, P., Meyyazhagan, A., Easwaran, M., Sharma, M.M.M., Mehta, S., Pandey, V., et al., 2022b. Hydrogen sulfide: a new warrior in assisting seed germination during adverse environmental conditions. Plant Growth Regul. 98 (3), 401–420.

Sharma, P., Sangwan, S., Mehta, S., 2023a. Emerging role of phosphate nanoparticles in agriculture practices. Engineered Nanomaterials for Sustainable Agricultural Production, Soil Improvement and Stress Management. Academic Press, USA, pp. 71–97.

Sharma, P., Patra, A., Singh, B., Mehta, S., 2023b. Microbial rejuvenation of soils for sustainable agriculture. Sustainable Agriculture Reviews 60: Microbial Processes in Agriculture. Springer Nature Switzerland, Cham, pp. 293–323.

Sharma, P., Sangwan, S., Kaur, H., Patra, A., Mehta, S., 2023c. Microbial remediation of agricultural residues. Sustainable Agriculture Reviews 60: Microbial Processes in Agriculture. Springer Nature Switzerland, Cham, pp. 325–358.

Sharma, P., Sangwan, S., Kaur, H., Patra, A., Mehta, S., 2023d. Diversity and evolution of nitrogen fixing bacteria. Sustainable Agriculture Reviews 60: Microbial Processes in Agriculture. Springer Nature Switzerland, Cham, pp. 95–120.

Shi, Z.Z., Miao, F.P., Fang, S.T., Yin, X.L., Ji, N.Y., 2018. Trichorenins A–C, algicidal tetracyclic metabolites from the marine-alga-epiphytic fungus Trichoderma virens Y13-3. J. Nat. products 81 (4), 1121–1124.

Singh, A., Sharma, P., Kumari, A., Kumar, R., Pathak, D.V., 2019. Management of root-knot nematode in different crops using microorganisms. Plant Biotic Interactions. Springer, Cham, pp. 85–99.

Singh, S., Sangwan, S., Sharma, P., Devi, P., Moond, M., 2021. Nanotechnology for sustainable agriculture: an emerging perspective. J. Nanosci. Nanotechnol. 21 (6), 3453–3465.

Singh, U.B., Malviya, D., Khan, W., Singh, S., Karthikeyan, N., Imran, M., et al., 2018. Earthworm grazed-Trichoderma harzianum biofortified spent mushroom substrates modulate accumulation of natural antioxidants and bio-fortification of mineral nutrients in tomato. Front. Plant Sci. 9, 1017.

Špoljarević, M., Štolfa, I., Lisjak, M., Stanisavljević, A., Vinković, T., Agić, D., et al., 2010. Strawberry (Fragaria x ananassa Duch) leaf antioxidative response to biostimulators and reduced fertilization with N and K. Poljoprivreda 16 (1), 50–56.

Subedi, P., Gattoni, K., Liu, W., Lawrence, K.S., Park, S.W., 2020. Current utility of plant growth-promoting rhizobacteria as biological control agents towards plant-parasitic nematodes. Plants. 9 (9), 1167.

TariqJaveed, M., Farooq, T., Al-Hazmi, A.S., Hussain, M.D., Rehman, A.U., 2021. Role of Trichoderma as a biocontrol agent (BCA) of phytoparasitic nematodes and plant growth inducer. J. Invertebrate Pathol. 183, 107626.

Thambugala, K.M., Daranagama, D.A., Phillips, A.J., Kannangara, S.D., Promputtha, I., 2020. Fungi vs. fungi in biocontrol: an overview of fungal antagonists applied against fungal plant pathogens. Front. Cell. Infect. Microbiol. 10, 604923.

Tkalec, M., Vinković, T., Baličević, R., Paraðiković, N., 2010. Influence of biostimulants on growth and development of bell pepper (Capsicum annuum L.). Acta Agric. Serb. 15 (29), 83–88.

Tyśkiewicz, R., Nowak, A., Ozimek, E., Jaroszuk-Ściseł, J., 2022. Trichoderma: the current status of its application in agriculture for the biocontrol of fungal phytopathogens and stimulation of plant growth. Int. J. Mol. Sci. 23 (4), 2329.

Vargas, W.A., Crutcher, F.K., Kenerley, C.M., 2011. Functional characterization of a plant-like sucrose transporter from the beneficial fungus Trichoderma virens. Regulation of the symbiotic association with plants by sucrose metabolism inside the fungal cells. N. Phytol. 189 (3), 777–789.

Vinale, F., Nicoletti, R., Borrelli, F., Mangoni, A., Parisi, O.A., Marra, R., et al., 2017. Co-culture of plant beneficial microbes as source of bioactive metabolites. Sci. Rep. 7 (1), 14330.

Vishwakarma, K., Kumar, N., Shandilya, C., Mohapatra, S., Bhayana, S., Varma, A., 2020. Revisiting plant–microbe interactions and microbial consortia application for enhancing sustainable agriculture: a review. Front. Microbiol. 11, 560406.

Viterbo, A.D.A., Wiest, A.R.I.C., Brotman, Y., Chet, I.L.A.N., Kenerley, C., 2007. The 18mer peptaibols from Trichoderma virens elicit plant defence responses. Mol. Plant Pathol. 8 (6), 737–746.

Wilson, H.T., Amirkhani, M., Taylor, A.G., 2018. Evaluation of gelatin as a biostimulant seed treatment to improve plant performance. Front. Plant Sci. 9, 1006.

Woo, S.L., Hermosa, R., Lorito, M., Monte, E., 2022. Trichoderma: a multipurpose, plant-beneficial microorganism for eco-sustainable agriculture. Nat. Rev. Microbiol. 1–15.

Yaronskaya, E., Vershilovskaya, I., Poers, Y., Alawady, A.E., Averina, N., Grimm, B., 2006. Cytokinin effects on tetrapyrrole biosynthesis and photosynthetic activity in barley seedlings. Planta. 224, 700–709.

Zeljković, S.B., Paraðiković, N.A., Babić, T.S., Đurić, G.D., Oljača, R.M., Vinković, T.M., et al., 2010. Influence of biostimulant and substrate volume on root growth and development of scarlet sage (Salvia splendens L.) transplants. J. Agric. Sci. 55 (1), 29–36.

CHAPTER

21

Microbial-based stimulants on plant adaptation to climate change

Wiwiek Harsonowati[1], *Dyah Manohara*[1], *Mutia Erti Dwiastuti*[1], *Sri Widawati*[2], *Suliasih*[2], *Abdul Hasyim Sodiq*[3], *Rida Oktorida Khastini*[4] and *Jati Purwani*[2]

[1]Research Center for Horticultural and Estate Crops, National Research and Innovation Agency (BRIN), Cibinong, Indonesia [2]Research Center for Applied Microbiology, National Research and Innovation Agency (BRIN), Cibinong, Indonesia [3]Department of Agrotechnology, Faculty of Agriculture, Sultan Ageng Tirtayasa University, Banten, Indonesia [4]Department of Biology Education, Faculty of Teacher Training and Education, Sultan Ageng Tirtayasa University, Banten, Indonesia

21.1 Agriculture and climate change: current status

Many people consider global warming and climate change to be synonymous, but scientists prefer to use "climate change" to describe the complex shifts currently affecting the weather and climate systems of our planet. Climatic change is defined as any major shift in climate measurements that lasts over an extended period. Climate change, in other words, encompasses significant changes in temperature, precipitation, or wind patterns, among other effects, that occur over several decades or more. In this current condition of rapid climate change, agricultural plants are increasingly threatened by abiotic and biotic stresses. Stress in plants can be defined as any external factor that negatively influences plant growth, productivity, reproductive capacity, or adaptation for survival.

The Earth is warming due to anthropogenic greenhouse gas emissions, particularly carbon dioxide (CO_2), which has reached its highest level in 800,000 years with an atmospheric concentration of over 400 ppm (Ritchie et al., 2020). Climate change includes not only rising average temperatures but also extreme weather events, shifting wildlife

Biostimulants in Plant Protection and Performance
DOI: https://doi.org/10.1016/B978-0-443-15884-1.00015-4

populations and habitats, rising sea levels, and a variety of other impacts. All of these changes are occurring as a result of humans continuing to emit heat-trapping greenhouse gases into the atmosphere (US EPA, 2022). Significant climate change has already occurred since the 1950s, and the global surface air temperature is likely to rise by 0.4°C – 2.6°C in the second half of this century. It's estimated that the demand for livestock products will grow by +70% between 2005 and 2050 (IPCC, 2021). According to the Intergovernmental Panel on Climate Change (IPCC), the current global warming rate will dramatically increase hazards to health, livelihoods, food security, water supply, human security, and economic growth (IPCC, 2021).

Agriculture both contributes to and is impacted by climate change. We need to reduce greenhouse gas emissions (GHGs) from agriculture and adapt its food-production system to cope with climate change and its related stress on the environment. Currently, agriculture is responsible for generating 19%–29% of total GHG emissions such as CO_2, methane (NH_4), and nitrous oxide (N_2O) in the atmosphere over the last 150 years (US EPA, 2022). GHGs from agriculture come from livestock such as cows, agricultural soils, including agrochemical applications, and rice production. The application of agrochemicals such as nitrogen fertilizers accounts for the majority of N_2O emissions in the United States (US EPA, 2022). Emissions can be reduced by reducing nitrogen-based fertilizer applications and applying these fertilizers more efficiently, as well as by modifying a climate-smart management practice. For example, by utilizing microbial-based technology as crop biostimulants, a healthy, eco-friendly, cost-effective, and climate-smart agricultural system can be created.

Climate-smart agriculture (CSA) is an integrated strategy for managing landscapes—cropland, livestock, forests, and fisheries—that address the interlinked challenges of food security and accelerating climate change. CSA (The World Bank, 2022) strives to achieve simultaneously four goals: (1) increased productivity in agriculture to improve nutrition security and boost incomes; (2) enhanced resilience to drought, pests, diseases, and other climate-related stress; (3) improved capacity to adapt to longer-term stresses as climate change continues and reduced emissions from agricultural activities.

21.2 Microbial technologies as biostimulants in agricultural sustainability

Global climate change-induced abiotic stresses (e.g., heat temperatures, drought, salinity, flooding, and disease outbreaks) have destabilized the fragile agroecosystems and impaired plant performance, thereby reducing crop productivity and quality. Microbial-based biostimulants, as a promising and eco-friendly technology, are widely used to address environmental concerns and fulfill the need for developing sustainable/modern agriculture. Microbial-based biostimulants have great potential to elicit plant tolerance to various climate change-related stresses and thus enhance plant growth and performance-related parameters (such as root growth/diameter, flowering, nutrient use efficiency/translocation, soil water holding capacity, and microbial activity). To successfully implement biostimulant-based agriculture in the field under changing climate conditions, an understanding of agricultural functions and biostimulant action mechanisms coping with various abiotic stresses at the physicochemical, metabolic, and molecular levels is required.

To adapt to the varied situations caused by climate change, new perspectives and methods should be explored and used in the study on CSA to establish a healthy, eco-friendly, cost-effective, and sustainable agricultural system. Considering continuous global climate change and its related stresses, a more sustainable agricultural production system is urgently required to protect soil fertility and alleviate soil biodiversity loss. According to Yakhin et al. (2017) and du Jardin (2015), biostimulants are defined as biological substances, including microbes, that, when applied to seeds, plants, or soil, stimulate and improve nutrient absorption, or provide benefits to plant development and tolerance to abiotic and biotic stress, as well as increase plant growth and nutritional quality (Kocira et al., 2018), regardless of the nutrient content of the biostimulant. Microbial biostimulants are novel and innovative microbial technologies that can provide high-nutrition crop yields while mitigating the negative effects generated by environmental changes. Prospective research areas include the effectiveness and role of microbial biostimulants as a biological tool for improving plant adaptation and resilience under climate change-related stress.

Managing ecological interactions and agro-biodiversity, as well as reducing the use of potentially harmful chemical inputs, have been primary areas of study in recent years, highlighting the importance of agroecological principles (van der Putten et al., 2016). The use of biostimulants is in line with agroecological principles. Biostimulants are products that can not only act directly on plants but also keep plants' productivity up through the selection and stimulation of beneficial soil microbes (Hellequin et al., 2020). Certain biostimulants have been shown to stimulate plant-beneficial soil microorganisms. Agricultural biostimulants have emerged as a valid alternative to chemicals to indirectly sustain plant growth and productivity.

21.3 The role of microbial stimulants in mitigating climate change-induced stresses on plants

Microbial biostimulants are an advanced and viable technology for sustaining plants subjected to abiotic challenges in the current context of rapidly evolving climate change. This chapter aimed to deliver current information about the application of microbial biostimulants in enhancing and improving plant tolerance to abiotic stresses (whether single or combined stressors) caused by climate change. Exposure to harsh and unpredictable meteorological conditions, phytosanitary dangers, and cultivation conditions leads to substantial losses in agricultural yields around the world. Climate change has both direct and indirect effects on agricultural productivity, including changing rainfall patterns, drought, flooding, and the geographical redistribution of pests and diseases, as shown in Fig. 21.1. This effect is likely to remain important as climate change continues. This section highlights climate change-induced stresses and microbial biostimulants' responses to plant adaptation and resilience, establishing smart-climate agriculture.

21.3.1 Extreme temperature/heat shock

Changes in the Heat Index driven by anthropogenic CO_2 emissions will increase global exposure to dangerous environments in the coming decades. Without CO_2 emissions

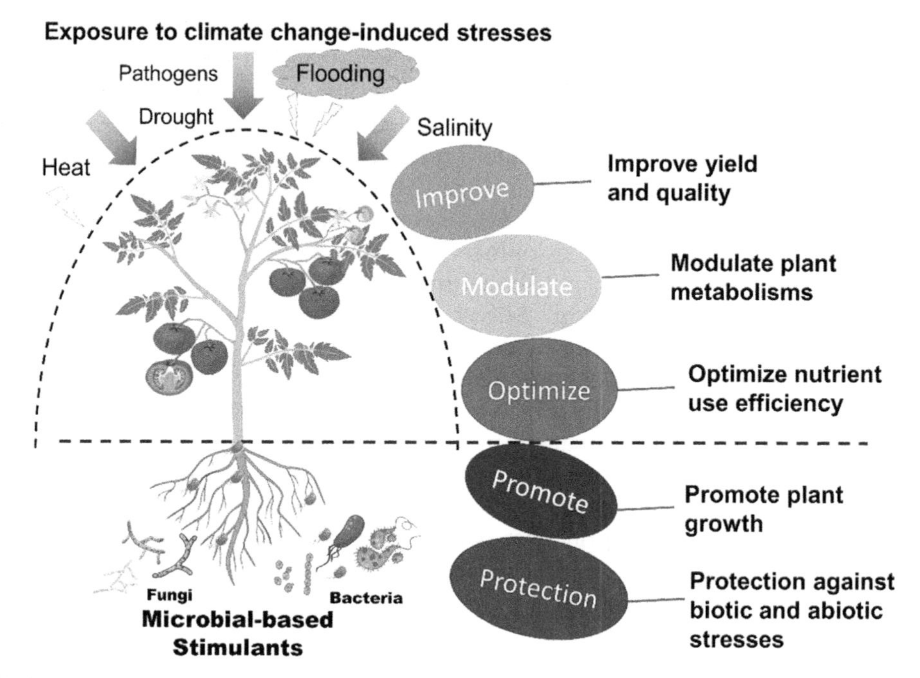

FIGURE 21.1 The role of microbial biostimulants on plant adaptation to climate change and its related stresses.

reductions, it is likely that by 2100, many people living in tropical regions will be exposed to a dangerously high Heat Index (Vargas Zeppetello et al., 2022). The dangers of extremely high temperatures generated an extreme drought and caused severe damage to the agriculture sector. These events marked a turning point in the public's understanding of climate change.

High temperatures have a significant impact on crops by causing molecular and morpho-physiological injuries such as DNA damage, oxidative stress, hormonal changes, transcriptional changes, water loss, scorching of leaves and stems, leaf abscission, leaf burning, senescence, root and shoot growth limitation, and fruit damage, which results in yield loss (El-Serafy et al., 2022). Wassie et al. (2019) reported that heat stress affected plant growth and productivity by significantly reducing the biomass, enzyme activity, antioxidant activity, relative water content, photosynthetic activity, and chlorophyll content, and increasing the electrolyte leakage and malondialdehyde (MDA) content. Furthermore, increased frequency of temperature stress can disrupt the physiological processes of plants, resulting in photosynthetic inhibition, reduced nitrogen anabolism, higher protein catabolism, and the accumulation of the end products of lipid peroxidation (Wassie et al., 2019). High temperatures negatively affect the flower initiation of seasonal plants such as strawberries (Heide and Sønsteby, 2007; Yanagi et al., 2016).

Plant thermosensing is the ability to detect even mild increases in temperature due to the presence of sensing mechanisms on their membranes (Hayes et al., 2021). Membrane sensors detect an increase in membrane fluidity in response to high-temperature stress, which triggers conformational changes and phosphorylation/dephosphorylation events (Sehgal et al., 2016). Plants' stress responses to high temperatures are regulated by hormonal signaling, such as ethylene. Ethylene is involved not only in senescence, development, plant physiology, and development but also in plant responses to multiple abiotic stressors, such as heat, salinity, and drought (Dubois et al., 2018).

Microbial biostimulants can modulate plant responses to heat stress with multiple benefits compared to agrochemical products. Duc et al. (2018) reported that the synthesis of reactive oxygen species (ROS)-degrading enzymes (e.g., superoxide, catalase, peroxidases, and dismutase) enhances heat stress tolerance and can be boosted in plants colonized by beneficial bacteria, such as *Pseudomonas* and *Bacillus*, and mycorrhizal fungi, such as *Septoglomus deserticola* and *Septoglomus constrictum*. Microbial biostimulants *P. fluorescens* and *P. aeruginosa* not only significantly improve plant tolerance to heat stress but also improve the capabilities of soil fertility and quality (Sangiorgio et al., 2020). These microbial biostimulants are further commercialized for bioremediation, phytostimulation, and soil biofertilizers. Furthermore, *Bacillus* spp. (e.g., *B. amyloliquefaciens, B. firmus, B. pumilus, B. thuringensis, B. licheniformis, B. sphaericus,* and *B. subtilis*), whether alone or in consortia, have been commercialized as a biostimulant to mitigate heat stress and another benefit as biopesticides (Fadiji et al., 2022). Endophytes-based biostimulants reported that soil augmentation was effectively increasing plant adaptation to climatic heat stress (Harsonowati et al., 2022). The root colonization of endophyte *Cladophialophora chaetospira* effectively accelerates flower initiation under the non-flower-induced condition of heat temperature, showing unpredictable traits of endophyte-based biostimulants to the host plants (Harsonowati et al., 2022).

21.3.2 Drought and salinity

The impacts of climate change on agricultural productivity are severe. Abiotic stressors, such as drought and salinity, are experienced by plants throughout their ontogenetic stages, causing crop loss ($20\% - 50\%$) and making them vulnerable in terms of survival (FAO, 2018). Drought and salinity stress are climatic stresses that significantly affect plant fitness and performance (Kaushal and Wani, 2016). Yield losses due to drought and salinity stress are increasing, mainly due to climate change and intensive agriculture that leads to soil degradation. In recent years, rising soil salinity has emerged as a major challenge for agricultural production. Salt stress causes imbalanced ion hemostasis (Na^+ and K^+) and interrupted mineral absorption in plants. Also, salinity stress leads to oxidative stress (production and accumulation of ROS) (Abid et al., 2020). An increase and accumulation of ROS are extremely dangerous for cells and eventually cause plant death (Farsaraei et al., 2020).

Halotolerant bacteria like *Bacillus amyloliquefaciens* and *Bacillus subtilis* can be used as biostimulants to reduce plant salinity stress (Abd-Allah et al., 2018; Castaldi et al., 2023). Plant growth-promoting microbes, referred to in this chapter as biostimulants, can colonize the plant root surface and alleviate salt stress by providing minerals (N, P, and K) and the production of phytohormones (auxin, cytokinin, and abscisic acid). Neshat et al. (2022) reported that bacterial biostimulants *Enterobacter* sp. and *Pseudomonas* sp. alleviate plant salinity stress by preventing the accumulation of toxic ions like Na^+ in shoots through several mechanisms, such as the exclusion of ion or preventing ions uptake and transporting them to shoots via the production of protective osmolytes like proline. Orozco-Mosqueda et al. (2020) reported the various advantages of 1-aminocyclopropane-1-carboxylate (ACC) deaminase-producing bacilli as biostimulants (plant growth-promoting) to address the problem of salinity in agricultural soils. ACC deaminase bacteria *Burkholderia phytofirmans* effectively improved the salinity tolerance of perennial ryegrass and could be incorporated into turfgrass maintenance programs in salt-affected soils (Cheng et al., 2016).

Fungal endophytes-based biostimulant *Piriformospora indica*, reported by Sabeem et al. (2022) significantly promotes growth and salinity stress tolerance by enhancing the levels of antioxidant enzymes such as catalase, superoxide dismutase, and peroxidase in plants. The fungus colonization was also associated with altered expression levels of essential Na^+ and K^+ ion channels in roots like *HKT1;5* (high-affinity potassium transporter 1) and *SOS1* (salt overly sensitive 1) genes (Sabeem et al., 2022). Several studies also reported arbuscular mycorrhizal fungi in alleviating soil salinity and drought stress (Abd-Allah et al., 2015; Ahmad et al., 2018; Baltazar-Bernal et al., 2022; Chen et al., 2017; Fattahi et al., 2021) by improving osmotic adjustment via the accumulation of compatible solutes such as proline, soluble sugars, and antioxidants and increasing the activity of multiple antioxidant enzymes, including superoxide dismutase (SOD), catalase (CAT), and ascorbate peroxidase stress. In addition, the AM alleviates the salt-induced APX and peroxidase (POD), thereby scavenging the ROS generated under salt deleterious effects by regulating the osmotic balance of the Na^+/K^+ ratio through maintaining the lesser Na^+ and K^+ under salt stress conditions (Hashem et al., 2016).

21.3.3 Flooding/waterlogged

Rainfall extremes are becoming more intense as a result of climate change, increasing the risk of flooding. Flooding has an impact on both above- and below-ground ecosystem processes, posing a significant danger to crop productivity in the face of climate change. Agriculture is extremely vulnerable to climate change due to its reliance on soil quality, irrigation, and weather patterns. As a result of climate change, floods, disasters, droughts, heat stress, unpredictable rainfall, and severe weather events have a considerable negative impact on worldwide agricultural practices. Plant stress responses to drought and flooding are comparable despite their different soil moisture levels. Plant eco-physiological parallel responses to water stress whether, drought or flooding, show an integrated framework. Drought reduces root water intake, and plants respond by closing stomata and decreasing leaf gas exchange, whereas flooding reduces root metabolism due to anoxia in the soil, which also reduces root water intake and leaf gas exchange (Chen et al., 2022). Flooding susceptibility is related to the rootstock, which is often derived from seeds (Yin et al., 2022). Drought and flooding can occur consecutively in the same system, and the resulting plant stress responses share similar mechanisms. For example, heavy rainfall can occur after prolonged periods of drought in natural ecosystems, or repetitive irrigation can be administered under persistent drought in agricultural settings.

Climate change-related stress occurs frequently, either in combination or simultaneously, causing agricultural plants to face severe damage. Francioli et al. (2022) reported that flooding reduced plant fitness and caused dramatic shifts in plant-associated fungi (mycobiota) assembly across the entire plant, i.e., phyllosphere, roots, and rhizosphere. Functional transition consists of a decline in mutualistic abundance and a richness increase in plant pathogens. Indeed, fungal pathogens associated with important cereal diseases, such as *Gibberella intricans, Mycosphaerella graminicola, Typhula incarnata,* and *Olpidium brassicae,* demonstrate the detrimental effect of flooding on the wheat mycobiota complex (Francioli et al., 2022), highlighting the urgent need to understand how climate change-associated abiotic stressors alter plant-microbe interaction in cereal crops.

21.3.4 Diseases outbreaks

Climate change is predicted to have a negative impact on global agricultural production by altering spatial and temporal disease outbreaks. The increase in heat temperature and CO_2 levels causes physiological changes in plants that increase the emergence and severity of crop diseases. Earth warming will cause shifts in agroclimatic zones, which will result in the migration of host plants into new areas and the appearance of new disease complexes. Furthermore, climate change can alter pathogen development, plant-pathogen or pathogen-pathogen interactions, and facilitate the emergence of new pathogen strains, which in turn weaken host-plant resistance. Climate change, such as rising temperatures and CO_2 levels in the atmosphere, as well as the frequency and intensity of extreme weather events such as drought and flooding, all have an impact on host-plant resistance to pathogens.

Although the outcome of host-pathogen interactions is driven by several different climate variables, temperature is the most important factor. An increase in temperature may result in greater crop vulnerability to disease. Fungi are the most common plant pathogens, with various modes of pathogenicity interaction with their host plants. The current atmospheric conditions, particularly temperature and moisture, have a significant impact on fungal pathogenesis. Any temperature change will significantly affect fungal reproduction, rate of infection, number of infection cycles, long- and short-distance dispersal, and off-season survival. For example, in cooler and subtropical rice-growing regions, rice blast (*Pyricularia oryzae*) incidence and epidemics increase with ambient temperature (Kirtphaiboon et al., 2021). Similarly, temperature sensitivity has increased the severity of Septoria leaf spot (*Septoria lycopersici*) and stem rust (*Puccinia graminis*) in oat cultivars. On the other hand, more than 40 bacterial genera have been reported to be pathogens for plants. The emergence of several bacterial genera as a serious global problem could be attributed to global warming. *Acidovorax avenae* subsp. *avenae* infects upland rice in southern Europe; *Burkholderia andropogonis* infects jojoba in eastern Australia; *B. glumae* infects rice in the southern United States; and *Dickeya zeae* infects rice in north India, particularly at high temperatures. Elevated temperature interferes with bacterial physiology, genetics, and bacterial-plant interactions (Cheng et al., 2019).

21.4 Microbial biostimulants modulation for plant adaptation

A sustainable strategy for ensuring food security with limited sources on agricultural land is to develop plant resistance and resilience to counteract climate change-induced stresses (Calvo et al., 2014). The application of microbial biostimulants could be a significant approach and innovative technology for achieving climate-smart agricultural goals, as shown in Fig. 21.2. The use of microbial-based biostimulants in agriculture has greatly increased during the past two decades as a solution to problems associated with modern agriculture, including climate change and its related stresses. Li et al. (2022) reported that the application of microbial inoculants significantly enhances crop productivity through meta-analysis studies.

Microbial biostimulants activate and regulate several defense mechanisms through different action modes and pathways. Microbial biostimulants can be applied directly to the leaves (foliar application) or the soil near the root system (soil application). The hyphae of fungal endophytes-biostimulants penetrate the plant tissues or cells and establish symbiotic associations. Biostimulants are translocated and distributed to other parts of the plant once they reach the leaves and/or roots. Different biostimulants have different mechanisms of action in plants. Several studies have shown that priming plants with microbial biostimulants results in improved plant defense responses to stresses, such as increased antioxidant enzyme activities and polyphenol and osmolyte accumulation, as shown in Table 21.1. Table 21.1 summarizes the molecular, metabolic, and physiological mechanisms underlying biostimulant-induced abiotic stress mitigation.

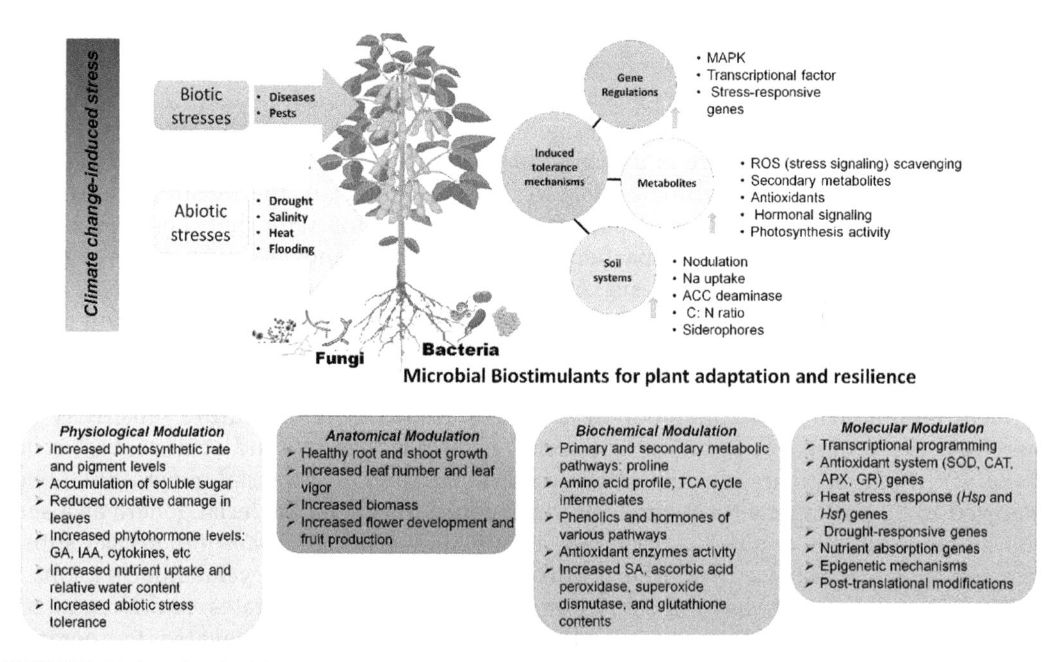

FIGURE 21.2 Microbial-based biostimulants alter plant morpho-physiological and molecular adaptation to climate change and its associated stresses.

TABLE 21.1 Mechanisms of microbial-based stimulants and plant response/adaptation to climate change-related stresses.

Biostimulant agent	Climatic stress	Mechanism and plant response/adaptation	Reference
Aspergillus aculeatus	Heat and drought	Enhancing the photosynthetic apparatus, decreased the antioxidant enzyme activities, and mitigated membrane lipid peroxidation.	Li et al. (2021)
Glomus spp.	Heat	Ensure high photosynthetic capacity by regulating photosystem (PS) II heterogeneity by converting inactive β and γ centers to active α centers, and QB nonreducing centers to reducing centers. The developed root system for absorption of water.	Mathur and Jajoo (2020)
Bacillus cereus	Heat	– Induced biosynthesis of gibberellin, IAA, and organic acids. Improved biomass, chlorophyll content and chlorophyll fluorescence. Reduced ABA and increased SA, ascorbic acid peroxidase, superoxide dismutase, and glutathione contents in soybean plants. Increased heat-shock protein (HSP) expression, and stress-responsive *GmLAX3* and *GmAKT2*. – Enhanced extracellular polymeric substances (EPS) production and reduced the adverse effects of heat on tomato growth.	Khan et al. (2020) and Mukhtar et al. (2020)
B. cereus, *Providencia rettgeri*, and *Myroides odoratimimus*	Heat	Increased plant growth, antioxidant enzyme activities, and decreased proline and MDA contents.	Bruno et al. (2020)
Aspergillus niger	Heat	Alter plant physiology in such a way that, the concentration of AAO, CAT, GR, SOD, POD, proline, and phenolics were augmented while lipid peroxidation, ROS ad ABA were reduced under thermal stress at 40°C.	Ismail et al. (2020)
Aspergillus flavus	Heat	Regulation of the concentration of ABA, proline, phenols, flavonoids, catalase, and ascorbic acid oxidase.	Ismail et al. (2019)
Themomyces lanuginosus	Heat	Maintaining maximum quantum efficiency of photosystem II, photosynthesis rate, water use efficiency, and increased root length.	Ali et al. (2018)
Bacillus amyloliquefaciens; *Azospirillum brasilense*	Heat	reduced generation of reactive oxygen species and transcript levels of several stress-related genes.	Abd El-Daim et al. (2018)

(*Continued*)

TABLE 21.1 (Continued)

Biostimulant agent	Climatic stress	Mechanism and plant response/adaptation	Reference
Glomus fasciculatum	Heat	Increase activity of antioxidative enzymes such as superoxide dismutase and ascorbate peroxidase, as well as an increase in ascorbic acid and polyphenol contents. Increase 2,2-diphenyl-1picrylhydrazyl radical scavenging activity	Maya and Matsubara (2013)
Bradyrhizobium	Heat	Thermotolerant, improved shoot dry matter, nitrogen-uptake, and seed yield of the plants.	Rahmani et al. (2009)
Phyllobacterium brassicacearum	Drought	– Induces reproductive delay and physiological changes in transpiration rate, ABA content, photosynthesis, and development for adaptation to water-use efficiency. – Increase the length of lateral roots, density and length of root hairs, and increased water flux.	Bresson et al. (2013) and Kechid et al. (2013)
Consortium of PGPR: *Bacillus cereus*, *B. subtilis*, and *Serratia* sp.	Drought	Decreased leaf monodehydroascorbate (MDA) content, enhanced: superoxide dismutase (SOD) activity; proline content, and photosynthesis activity	Wang et al. (2012)
Pseudomonas aeruginosa	Drought	Increase root and shoot dry mass, water content, ACC deaminase, and activity of SOD, POX, CAT, and regulate dehydration responsive element binding protein (*DREB2A*), catalase (*CAT1*), and dehydrin (*DHN*) drought-stress related genes	Sarma and Saikia (2014)
Funneliformis mosseae	Drought	Promote plant growth and strengthen cell wall, upregulate the expression of genes related to cellulose biosynthesis and cell growth (*XM_020312442.1*, *XM_020331230.1*, and *XM_020312442.1*).	Tarnabi et al. (2020)
Rhizophagus clarus	Drought	Improve plant photosynthesis, growth, and water status by increasing water retention and efficiency	Oliveira et al. (2022)
Glomus spp.	Drought	Promote plant growth, photosynthetic rate, stomatal conductance, gas exchange, and water use efficiency, increase the activity of SOD and catalase (CAT), and reduce the levels of MDA.	Li et al. (2019)
Enterobacter sp.	Salinity	Induce osmoregulation via inducing biosynthesis of IAA, increase plant biomass, total antioxidant capacity, ACC deaminase, upregulation of *DREB2b*, *RD29B*, *RAB18*, *P5CS1*, and *P5CS2*, *MPK3* and *MPK6*.	Kim et al. (2014)

(*Continued*)

TABLE 21.1 (Continued)

Biostimulant agent	Climatic stress	Mechanism and plant response/adaptation	Reference
Glomus sp. *Agrobacterium radiobacter*, *B. subtilis*, *Streptomyces* sp. and *Trichoderma* spp.	Salinity	Promote plant growth and biomass, leaf number, and area, and promote water and nitrogen use efficiency.	Miceli et al. (2021)
Glomus mosseae	Salinity	Increase sugars, compatible solutes (proline and glycine betaine), water use efficiency, chlorophyll, ABA levels, and nutrient uptake, enhance the activity of antioxidant enzymes, and reduce cell membrane permeability	Abdel Latef and Chaoxing (2014)

21.5 Conclusion and prospects

Globally, despite the intense agricultural production, the yield is expected to be limited by emerging infectious plant diseases and the adverse impacts of climate change. In the present era of rapidly changing climates, plants are exposed to environmental stresses more frequently. This includes exposure to extreme and unpredictably changing climatic conditions, which result in significant losses of agricultural production. The demand for sustainable agricultural practices is further augmented by the exclusion of agrochemicals such as synthetic fertilizers and pesticides. Naturally, plants coexist with microbial symbionts; some of these symbionts are important for the ecosystem and plant functions. Even in the face of stresses related to climate change, applying microbial biostimulants, which benefit from symbiotic relationships, is a long-term strategy for enhancing plant adaptation and resilience to climate change. Therefore, the application of plant biostimulants has been at the forefront as an environment-friendly approach. Due to its importance in sustainable agricultural practices, the development of a novel microbial biostimulant presents some unique challenges. Additionally, crop breeding strategies in the future should include the plant's ability to form permanent symbiotic partnerships with beneficial microbes as a very important feature, strongly connected to stress tolerance, resilience, and production.

References

Abd El-Daim, I.A., Bejai, S., Fridborg, I., Meijer, J., 2018. Identifying potential molecular factors involved in *Bacillus amyloliquefaciens* 5113 mediated abiotic stress tolerance in wheat. Plant Biol. 20, 271–279. Available from: https://doi.org/10.1111/plb.12680.

Abd-Allah, E.F., Alqarawi, A.A., Hashem, A., Radhakrishnan, R., Al-Huqail, A.A., Al-Otibi, F.O.N., et al., 2018. Endophytic bacterium *Bacillus subtilis* (BERA 71) improves salt tolerance in chickpea plants by regulating the plant defense mechanisms. J. Plant Interact. 13, 37–44. Available from: https://doi.org/10.1080/17429145.2017.1414321.

Abd-Allah, E.F., Hashem, A., Alqarawi, A.A., Bahkali, A.H., Alwhibi, M.S., 2015. Enhancing growth performance and systemic acquired resistance of medicinal plant *Sesbania sesban* (L.) Merr using arbuscular mycorrhizal fungi under salt stress. Saudi J. Biol. Sci. 22, 274–283. Available from: https://doi.org/10.1016/j.sjbs.2015.03.004.

Abdel Latef, A.A.H., Chaoxing, H., 2014. Does inoculation with *Glomus mosseae* improve salt tolerance in pepper plants? J. Plant Growth Regul. 33, 644–653. Available from: https://doi.org/10.1007/s00344-014-9414-4.

Abid, M., Zhang, Y.J., Li, Z., Bai, D.F., Zhong, Y.P., Fang, J.B., 2020. Effect of Salt stress on growth, physiological and biochemical characters of Four kiwifruit genotypes. Sci. Hortic. 271, 109473. Available from: https://doi.org/10.1016/j.scienta.2020.109473.

Ahmad, H., Hayat, S., Ali, M., Liu, T., Cheng, Z., 2018. The combination of arbuscular mycorrhizal fungi inoculation (*Glomus versiforme*) and 28-homobrassinolide spraying intervals improves growth by enhancing photosynthesis, nutrient absorption, and antioxidant system in cucumber (*Cucumis sativus*). Ecol. Evol. 8, 5724–5740. Available from: https://doi.org/10.1002/ece30.4112.

Ali, A.H., Abdelrahman, M., Radwan, U., El-Zayat, S., El-Sayed, M.A., 2018. Effect of *Thermomyces* fungal endophyte isolated from the extreme hot desert-adapted plant on heat stress tolerance of cucumber. Appl. Soil. Ecol. 124, 155–162. Available from: https://doi.org/10.1016/j.apsoil.2017.11.004.

Baltazar-Bernal, O., Spinoso-Castillo, J.L., Mancilla-álvarez, E., Bello-Bello, J.J., 2022. Arbuscular mycorrhizal fungi induce tolerance to salinity stress in taro plantlets (*Colocasia esculenta* L. Schott) during acclimatization. Plants 11. Available from: https://doi.org/10.3390/plants11131780.

Bresson, J., Varoquaux, F., Bontpart, T., Touraine, B., Vile, D., 2013. The PGPR strain *Phyllobacterium brassicacearum* STM196 induces a reproductive delay and physiological changes that result in improved drought tolerance in *Arabidopsis*. N. Phytol. 200, 558–569. Available from: https://doi.org/10.1111/nph.12383.

Bruno, L.B., Karthik, C., Ma, Y., Kadirvelu, K., Freitas, H., Rajkumar, M., 2020. Amelioration of chromium and heat stresses in *Sorghum bicolor* by Cr6+ reducing-thermotolerant plant growth promoting bacteria. Chemosphere 244, 125521. Available from: https://doi.org/10.1016/j.chemosphere.2019.125521.

Calvo, P., Nelson, L., Kloepper, J.W., 2014. Agricultural uses of plant biostimulants. Plant Soil 383, 3–41. Available from: https://doi.org/10.1007/s11104-014-2131-8.

Castaldi, S., Valkov, V.T., Ricca, E., Chiurazzi, M., Isticato, R., 2023. Use of halotolerant *Bacillus amyloliquefaciens* RHF6 as a bio-based strategy for alleviating salinity stress in *Lotus japonicus cv* Gifu. Microbiol. Res. 268, 127274. Available from: https://doi.org/10.1016/j.micres.2022.127274.

Cheng, Y.T., Zhang, L., He, S.Y., 2019. Plant-microbe interactions facing environmental challenge. Cell Host Microbe 176, 139–148. Available from: https://doi.org/10.1016/j.chom.2019.07.009.Plant-Microbe.

Cheng, L., Zhang, N., Huang, B., 2016. Effects of 1-aminocyclopropane-1-carboxylatedeaminase-producing bacteria on perennial ryegrass growth and physiological responses to salinity stress. J. Am. Soc. Hortic. Sci. 141, 233–241. Available from: https://doi.org/10.21273/jashs.141.3.233.

Chen, S., Tusscher, K., Sasidharan, R., Dekker, S., de Boer, H., 2022. Parallels between drought and flooding: an integrated framework for plant eco-physiological responses to water stress. Authorea 1–16. https://www.authorea.com/doi/full/10.22541/au.166322534.45962593/v1.

Chen, J., Zhang, H., Zhang, X., Tang, M., 2017. Arbuscular mycorrhizal symbiosis alleviates salt stress in black locust through improved photosynthesis, water status, and K+/Na+ homeostasis. Front. Plant. Sci. 8, 1–14. Available from: https://doi.org/10.3389/fpls.2017.01739.

Dubois, M., Van den Broeck, L., Inzé, D., 2018. The pivotal role of ethylene in plant growth. Trends Plant Sci. 23, 311–323. Available from: https://doi.org/10.1016/j.tplants.2018.01.003.

Duc, N.H., Csintalan, Z., Posta, K., 2018. Arbuscular mycorrhizal fungi mitigate negative effects of combined drought and heat stress on tomato plants. Plant Physiol. Biochem. 132, 297–307. Available from: https://doi.org/10.1016/j.plaphy.2018.09.011.

El-Serafy, R.S., El-Sheshtawy, A.N.A., Dahab, A.A., 2022. Fruit peel soil supplementation induces physiological and biochemical tolerance in *Schefflera arboricola* L. grown under heat conditions. J. Soil Sci. Plant Nutr. Available from: https://doi.org/10.1007/s42729-022-01102-5.

Fadiji, A.E., Babalola, O.O., Santoyo, G., Perazzolli, M., 2022. The potential role of microbial biostimulants in the amelioration of climate change-associated abiotic stresses on crops. Front. Microbiol. 12. Available from: https://doi.org/10.3389/fmicb.2021.829099.

FAO, 2018. Handbook for Saline Soil Management. Food and Agriculture Organization of the United Nations and Lomonosov Moscow State University. Food and Agriculture Organization of the United Nations, Moscow.

Farsaraei, S., Moghaddam, M., Pirbalouti, A.G., 2020. Changes in growth and essential oil composition of sweet basil in response of salinity stress and superabsorbents application. Sci. Hortic. 271, 109465. Available from: https://doi.org/10.1016/j.scienta.2020.109465.

Fattahi, M., Mohammadkhani, A., Shiran, B., Baninasab, B., Ravash, R., Gogorcena, Y., 2021. Beneficial effect of mycorrhiza on nutritional uptake and oxidative balance in pistachio (*Pistacia* spp.) rootstocks submitted to drought and salinity stress. Sci. Hortic. 281, 109937. Available from: https://doi.org/10.1016/j.scienta.2021.109937.

Francioli, D., Cid, G., Hajirezaei, M.R., Kolb, S., 2022. Response of the wheat mycobiota to flooding revealed substantial shifts towards plant pathogens. Front. Plant Sci. 13, 1–15. Available from: https://doi.org/10.3389/fpls.2022.1028153.

Harsonowati, W., Masrukhin, Narisawa, K., 2022. Prospecting the unpredicted potential traits of *Cladophialophora chaetospira* SK51 to alter photoperiodic flowering in strawberry, a perennial SD plant. Sci. Hortic. 295, 110835. Available from: https://doi.org/10.1016/j.scienta.2021.110835.

Hashem, A., Abd Allah, E.F., Alqarawi, A.A., Wirth, S., Egamberdieva, D., 2016. Arbuscular mycorrhizal fungi alleviate salt stress in lupine (*Lupinus termis* Forsik) through modulation of antioxidant defense systems and physiological traits. Legum. Res. 39, 198–207. Available from: https://doi.org/10.18805/lr.v39i20.9531.

Hayes, S., Schachtschabel, J., Mishkind, M., Munnik, T., Arisz, S.A., 2021. Hot topic: thermosensing in plants. Plant Cell Environ. 44, 2018–2033. Available from: https://doi.org/10.1111/pce.13979.

Heide, O.M., Sønsteby, A., 2007. Interactions of temperature and photoperiod in the control of flowering of latitudinal and altitudinal populations of wild strawberry (Fragaria vesca). Physiol. Plant 130, 280–289. Available from: https://doi.org/10.1111/j.1399-3054.2007.00906.x.

Hellequin, E., Monard, C., Chorin, M., Le bris, N., Daburon, V., Klarzynski, O., et al., 2020. Responses of active soil microorganisms facing to a soil biostimulant input compared to plant legacy effects. Sci. Rep. 10 (1), 15. Available from: https://doi.org/10.1038/s41598-020-70695-7.

Ismail, Hamayun, M., Hussain, A., Afzal Khan, S., Iqbal, A., Lee, I.J., 2019. *Aspergillus flavus* promoted the growth of soybean and sunflower seedlings at elevated temperature. Biomed. Res. Int. 2019. Available from: https://doi.org/10.1155/2019/1295457.

Ismail, Hamayun, M., Hussain, A., Iqbal, A., Khan, S.A., Lee, I.J., 2020. *Aspergillus niger* boosted heat stress tolerance in sunflower and soybean via regulating their metabolic and antioxidant system. J. Plant Interact. 15, 223–232. Available from: https://doi.org/10.1080/17429145.2020.1771444.

du Jardin, P., 2015. Plant biostimulants: Definition, concept, main categories and regulation. Sci. Hortic. 196, 3–14. Available from: https://doi.org/10.1016/j.scienta.2015.09.021.

Kaushal, M., Wani, S.P., 2016. Rhizobacterial-plant interactions: strategies ensuring plant growth promotion under drought and salinity stress. Agric. Ecosyst. Environ. 231, 68–78. Available from: https://doi.org/10.1016/j.agee.2016.06.031.

Kechid, M., Desbrosses, G., Rokhsi, W., Varoquaux, F., Djekoun, A., Touraine, B., 2013. The *NRT2.5* and *NRT2.6* genes are involved in growth promotion of *Arabidopsis* by the plant growth-promoting rhizobacterium (PGPR) strain *Phyllobacterium brassicacearum* STM196. N. Phytol. 198, 514–524. Available from: https://doi.org/10.1111/nph.12158.

Khan, M.A., Asaf, S., Khan, A.L., Jan, R., Kang, S.M., Kim, K.M., et al., 2020. Thermotolerance effect of plant growth-promoting *Bacillus cereus* SA1 on soybean during heat stress. BMC Microbiol. 20 (1), 14. Available from: https://doi.org/10.1186/s12866-020-01822-7.

Kim, K., Jang, Y.J., Lee, S.M., Oh, B.T., Chae, J.C., Lee, K.J., 2014. Alleviation of salt stress by *Enterobacter* sp. EJ01 in tomato and *Arabidopsis* is accompanied by up-regulation of conserved salinity responsive factors in plants. Mol. Cell 37, 109–117. Available from: https://doi.org/10.14348/molcells.20140.2239.

Kirtphaiboon, S., Humphries, U., Khan, A., Yusuf, A., 2021. Model of rice blast disease under tropical climate conditions. Chaos Solitons Fractals 143. Available from: https://doi.org/10.1016/j.chaos.2020.110530.

Kocira, S., Szparaga, A., Kocira, A., Czerwińska, E., Wójtowicz, A., Bronowicka-Mielniczuk, U., et al., 2018. Modeling biometric traits, yield and nutritional and antioxidant properties of seeds of three soybean cultivars through the application of biostimulant containing seaweed and amino acids. Front. Plant Sci. 9. Available from: https://doi.org/10.3389/fpls.2018.00388.

Li, J., Meng, B., Chai, H., Yang, X., Song, W., Li, S., et al., 2019. Arbuscular mycorrhizal fungi alleviate drought stress in C3 (*Leymus chinensis*) and C4 (*Hemarthria altissima*) grasses via altering antioxidant enzyme activities and photosynthesis. Front. Plant Sci. 10. Available from: https://doi.org/10.3389/fpls.2019.00499.

Li, J., Wang, J., Liu, H., Macdonald, C.A., Singh, B.K., 2022. Application of microbial inoculants significantly enhances crop productivity: a meta-analysis of studies from 2010 to 2020. J. Sustain. Agric. Environ. 1, 216–225. Available from: https://doi.org/10.1002/sae2.12028.

Li, X., Zhao, C., Zhang, T., Wang, G., Amombo, E., Xie, Y., et al., 2021. Exogenous *Aspergillus aculeatus* enhances drought and heat tolerance of perennial ryegrass. Front. Microbiol. 12. Available from: https://doi.org/10.3389/fmicb.2021.593722.

Mathur, S., Jajoo, A., 2020. Arbuscular mycorrhizal fungi protects maize plants from high temperature stress by regulating photosystem II heterogeneity. Ind. Crop. Prod. 143, 111934. Available from: https://doi.org/10.1016/j.indcrop.2019.111934.

Maya, M.A., Matsubara, Y.ichi, 2013. Influence of arbuscular mycorrhiza on the growth and antioxidative activity in cyclamen under heat stress. Mycorrhiza 23, 381–390. Available from: https://doi.org/10.1007/s00572-013-0477-z.

Miceli, A., Moncada, A., Vetrano, F., 2021. Use of microbial biostimulants to increase the salinity tolerance of vegetable transplants. Agronomy 11. Available from: https://doi.org/10.3390/agronomy11061143.

Mukhtar, T., Rehman, ur, Smith, S., Sultan, D., Seleiman, T., Alsadon, M.F., et al., 2020. Mitigation of heat stress in *Solanum lycopersicum* L. by ACC-deaminase and exopolysaccharide producing *Bacillus cereus*: effects on biochemical profiling. Sustainability 12. Available from: https://doi.org/10.3390/su12062159.

Neshat, M., Abbasi, A., Hosseinzadeh, A., Sarikhani, M.R., Dadashi Chavan, D., Rasoulnia, A., 2022. Plant growth promoting bacteria (PGPR) induce antioxidant tolerance against salinity stress through biochemical and physiological mechanisms. Physiol. Mol. Biol. Plants 28, 347–361. Available from: https://doi.org/10.1007/s12298-022-01128-0.

Oliveira, T.C., Cabral, J.S.R., Santana, L.R., Tavares, G.G., Santos, L.D.S., Paim, T.P., et al., 2022. The arbuscular mycorrhizal fungus *Rhizophagus clarus* improves physiological tolerance to drought stress in soybean plants. Sci. Rep. 12 (1), 15. Available from: https://doi.org/10.1038/s41598-022-13059-7.

Orozco-Mosqueda, M.del C., Glick, B.R., Santoyo, G., 2020. ACC deaminase in plant growth-promoting bacteria (PGPB): an efficient mechanism to counter salt stress in crops. Microbiol. Res. 235126439. Available from: https://doi.org/10.1016/j.micres.2020.126439.

van der Putten, W.H., Bradford, M.A., Pernilla Brinkman, E., van de Voorde, T.F.J., Veen, G.F., 2016. Where, when and how plant–soil feedback matters in a changing world. Funct. Ecol. 30, 1109–1121. Available from: https://doi.org/10.1111/1365-2435.12657.

Rahmani, H., Saleh-Rastin, N., Khavazi, K., Asgharzadeh, A., Fewer, D., Kiani, S., et al., 2009. Selection of thermotolerant bradyrhizobial strains for nodulation of soybean (*Glycine max* L.) in semi-arid regions of Iran. World J. Microbiol. Biotechnol. 25, 591–600. Available from: https://doi.org/10.1007/s11274-008-9927-8.

IPCC, 2023. Climate Change 2021: The Physical Science Basis - Summary for the Policymakers (Working Group I), Climate Change 2021: The Physical Science Basis. https://doi.org/10.1017/9781009157896.

Ritchie, H., Roser, M., Rosado, P., 2020. CO_2 and Greenhouse Gas Emissions [WWW Document]. URL https://ourworldindata.org/co2-and-other-greenhouse-gas-emissions (accessed 12.15.22).

Sabeem, M., Abdul Aziz, M., Mullath, S.K., Brini, F., Rouached, H., Masmoudi, K., 2022. Enhancing growth and salinity stress tolerance of date palm using *Piriformospora indica*. Front. Plant Sci. 13, 1–15. Available from: https://doi.org/10.3389/fpls.2022.1037273.

Sangiorgio, D., Cellini, A., Donati, I., Pastore, C., Onofrietti, C., Spinelli, F., 2020. Facing climate change: application of microbial biostimulants to mitigate stress in horticultural crops. Agronomy 10, 794.

Sarma, R.K., Saikia, R., 2014. Alleviation of drought stress in mung bean by strain *Pseudomonas aeruginosa* GGRJ21. Plant Soil 377, 111–126. Available from: https://doi.org/10.1007/s11104-013-1981-9.

Sehgal, A., Sita, K., Nayyar, H., 2016. Heat stress in plants: sensing and defense mechanisms. J. Plant Sci. Res. 32, 195–210.

Tarnabi, M.Z., Iranbakhsh, A., Mehregan, I., Ahmadvand, R., 2020. Impact of arbuscular mycorrhizal fungi (AMF) on gene expression of some cell wall and membrane elements of wheat (*Triticum aestivum* L.) under water deficit using transcriptome analysis. Physiol. Mol. Biol. Plants 26, 143–162. Available from: https://doi.org/10.1007/s12298-019-00727-8.

The World Bank, 2022. Climate-smart agriculture [WWW Document]. World Bank. URL https://www.worldbank.org/en/topic/climate-smart-agriculture#:~:text = Agriculture is a major part, is either lost or wasted. (accessed 12.30.22).

US EPA, 2022. Overview of Greenhouse Gases | US EPA [WWW Document]. United States Environ. Prot. Agency. URL https://www.epa.gov/ghgemissions/overview-greenhouse-gases (accessed 12.1.22).

Vargas Zeppetello, L.R., Raftery, A.E., Battisti, D.S., 2022. Probabilistic projections of increased heat stress driven by climate change. Commun. Earth Environ. 3 (1), 7. Available from: https://doi.org/10.1038/s43247-022-00524-4.

Wang, C.J., Yang, W., Wang, C., Gu, C., Niu, D.D., Liu, H.X., et al., 2012. Induction of drought tolerance in cucumber plants by a consortium of three plant growth-promoting rhizobacterium strains. PLoS One 7, 1–10. Available from: https://doi.org/10.1371/journal.pone.0052565.

Wassie, M., Zhang, W., Zhang, Q., Ji, K., Chen, L., 2019. Effect of heat stress on growth and physiological traits of alfalfa (*Medicago sativa* L.) and a comprehensive evaluation for heat tolerance. Agronomy 9. Available from: https://doi.org/10.3390/agronomy9100597.

Yakhin, O.I., Lubyanov, A.A., Yakhin, I.A., Brown, P.H., 2017. Biostimulants in plant science: a global perspective. Front. Plant Sci. 7. Available from: https://doi.org/10.3389/fpls.2016.02049.

Yanagi, T., Okuda, N., Okamoto, K., 2016. Effects of light quality and quantity on flower initiation of Fragaria chiloensis L. CHI-24-1 grown under 24 h day-length. Sci. Hortic. 202, 150–155. Available from: https://doi.org/10.1016/j.scienta.2016.02.035.

Yin, M.H., Gutierrez-Rodriguez, E.A., Vargas, A.I., Schaffer, B., 2022. Chemical priming with brassinosteroids to mitigate responses of avocado (*Persea americana*) trees to flooding stress. Horticulturae 8. Available from: https://doi.org/10.3390/horticulturae8121115.

Index

Note: Page numbers followed by "*f*" and "*t*" refer to figures and tables, respectively.

B

C

D

H

I

J

K

L

M

N

O

P

Q

R

S

T

9780443158841